Second Edition

HANDBOOK
FOR
PULP & PAPER
TECHNOLOGISTS

Gary A. Smook

Second Edition

HANDBOOK
FOR
PULP & PAPER
TECHNOLOGISTS

Gary A. Smook

 ANGUS WILDE PUBLICATIONS

Vancouver ❧ Bellingham

Second Printing 1994

Published by: **Angus Wilde Publications Inc.**
4543 West 11th Avenue
Vancouver, B.C. V6R 2M5

Available in the USA from:
Angus Wilde Publications Inc.
P.O. Box 1036
Bellingham, WA 98227-1036

Canadian Cataloguing in Publication Data

Smook, G.A. (Gary A.), 1934-
Handbook for Pulp & Paper Technologists
Includes index.

ISBN 0-9694628-1-6
1. Wood-pulp.   2. Paper industry I. Title.
TS1105.S66 1992   676   C92-091595-7

Printed in Canada by Friesen Printers

ABOUT THE AUTHOR

Gary A. Smook is an instructor in pulp and paper technology and unit operations at the British Columbia Institute of Technology in Burnaby, B.C., Canada.

Mr. Smook is prominent in the North American pulp and paper industry as the author of both HANDBOOK FOR PULP & PAPER TECHNOLOGISTS and HANDBOOK OF PULP AND PAPER TERMINOLOGY. He is also the author of over 30 papers on various aspects of kraft pulping, paper machine operation, technical management, and industrial training, and has contributed to three additional books. He has been awarded the Weldon Gold Medal (CPPA-TS, 1974) and TAPPI Fellowship (1987). He is a past chairman of the CPPA Professional Development Committee and is a current member of the Joint Textbook Committee of the Paper Industry.

Mr. Smook is a chemical engineering graduate from the University of California, Berkeley, and a registered professional engineer. He accepted his present teaching position in 1974 after a distinguished 18-year career in industry, and now divides his time between teaching, consulting and writing activities. His industrial consulting has focused on the areas of kraft mill process optimization, paper drying, technical information systems, and training seminars.

*To my mother -
her encouragement and
confidence in me have
been constant and
unconditional.*

CONTENTS

PREFACE TO SECOND EDITION

Compared to other pulp and paper industry technical books, the 1st Edition of Handbook for Pulp and Paper Technologists was enormously successful. It had 7 printings, sold 30,000 copies world-wide, and was translated into French and Spanish-language editions. Obviously, the book fulfilled a compelling need for a single-volume introductory treatment of the entire industry.

Rather than tamper with a successful formula, the supplemental objectives for this 2nd Edition of the "Smook Book" were to update the technology in each sector, expand the coverage in selected areas, and generally make the text as readable as possible. The basic organization and chapter headings are unchanged from the 1st Edition. However, 18 new chapter sections have been added, and the number of illustrations has increased from 650 to 680. There has been a notable expansion of the text in the area of secondary fiber recovery and utilization, reflecting a current focus of the industry. I have also given increased coverage to mechanical pulping to document the extensive process developments of the past decade. Significant innovations in kraft pulping and bleaching, and in papermaking have also been detailed. In all cases, I have been guided by the principle of covering all the material even-handedly and in sufficient depth for a basic understanding.

One criticism of the 1st Edition was the inadequacy of the glossary at the back of the book. I considered expanding the glossary, but it became apparent that even a tenfold increase in size would not provide an adequate guide to the profuse terminology of the industry. Consequently, the glossary has been omitted from the 2nd Edition of the "Smook Book". Instead, I decided to write a separate pulp and paper dictionary. The Handbook of Pulp and Paper Terminology was published in 1990; it has been favorably reviewed and I recommend it as a companion volume to users of this book.

Gary A. Smook
Vancouver, B.C. July 1992

PREFACE TO 1ST EDITION

A single volume that covers the entire technology of pulp and paper manufacture cannot be more than a detailed overview. This text does not purport to cover any aspect in depth, but is intended to provide a basic foundation on which the reader can build a more detailed knowledge.

The amount of material relating to the pulp and paper industry currently being published is enormous. Readers of this book are encouraged to become familiar with the more easily available industry publications, and to delve more deeply into those areas of current concern and interest. The specific references cited in the text and listed at the end of each chapter were carefully selected to provide a broader context. They either provide a review and consolidation of current knowledge, or they document recent breakthroughs in new technology, application of knowledge, or process control. Many published articles are concerned with highly technical studies of comparatively narrow scope and limited interest; these publications are helpful to specialists, but are not suitable references to this book.

Most modern pulp and paper textbooks are written by "committees", where each individual chapter is written by a different author or group of authors. This format frequently results in uneven coverage and emphasis, lack of continuity, and omissions of technical material. By undertaking the entire authorship myself, it is hoped that some of these problems have been resolved. At the same time, I must take full responsibility for any deficiencies perceived by reader and reviewer.

My objective was to produce a readable text with interconnecting chapters that can be read in sequence from cover to cover if desired, and also to provide a handy desktop reference. The student or technologist should be able to quickly grasp the fundamental knowledge of the process, and readily perceive the key parameters and variables in a given situation. To assist toward this goal, the key variables and criteria of performance have been tabled into a standardized form for quick recognition in each section.

A number of technical terms are introduced into the text without definition. The reader will find definitions for selected pulp and paper industry terminology in the attached glossary.

Of necessity, this textbook is concerned primarily with the pulp and paper technology of North America. Non-wood raw materials are considered only briefly, although these are important in Asian countries. Emphasis is placed on those products that are most important to the North American economy.

Gary A. Smook
Vancouver, B.C. August 1982

ACKNOWLEDGEMENTS

The author wishes to thank the members of the Joint Textbook Committee for their support and encouragement in preparing the 1st Edition, and for releasing copyright to the author for publication of the 2nd Edition. Deserving of special mention are Committee Chairman Lyle Gordon, Dr. Michael Kocurek who served as technical editor for the 1st Edition, Bert Joss of the CPPA who provided liaison assistance, and Bill Cullison of TAPPI. I would also like to thank my long-time mentor, Dr. Jasper Mardon, who encouraged an early interest in technical/industrial writing, and who strongly supported this textbook project.

On a more personal level, I would like to thank my wife, Hilda Wiebe, for her unstinting support of this project and for her assistance in proofreading, preparing graphics, and running interference on business matters.

Chapter 1

Introduction

1.1 IMPORTANCE OF PAPER

The significance of paper and paper products in modern life is obvious to everyone; no manufactured product plays a more meaningful role in every area of human activity. Paper provides the means of recording, storage and dissemination of information; virtually all writing and printing is done on paper. It is the most widely used wrapping and packaging material, and is important for structural applications.

The uses and applications for paper and paper products are virtually limitless. New specialty products are continually being developed. At the same time, the industry is aware of inroads and competition from other sectors, notably plastics and electronic media, for markets traditionally served by paper. As never before, new technology and methodology is being adopted so that the industry can remain competitive in existing markets and be receptive to new opportunities.

Aside from the output of products and services, the Pulp and Paper Industry provides employment for vast numbers of people and plays a vital role in the overall economy of both the United States and Canada.

1.2 DEFINITIONS OF PULP, PAPER AND PAPERBOARD

Paper has traditionally been defined as a felted sheet formed on a fine screen from a water suspension of fibers. Current paper products generally conform to this definition except that most products also contain non-fibrous additives. Dry forming methods are now utilized for the manufacture of a few specialty paper products.

Pulp is the fibrous raw material for papermaking. Pulp fibers are usually of vegetable origin, but animal, mineral or synthetic fibers may be used for special applications. Pulps used for chemical conversion into non-paper products are called **dissolving pulps.**

The distinction between paper and **paperboard** is based on product thickness. Nominally, all sheets above 0.3 mm thickness are classed as paperboard; but enough exceptions are applied to make the distinction somewhat hazy.

1.3 CHRONOLOGY OF TECHNOLOGICAL DEVELOPMENT

Paper derives its name from the reedy plant, papyrus. The ancient Egyptians produced the world's first writing material by beating and pressing together thin layers of the plant stem (see Figure 1-1). However, complete defibering which is characteristic of true papermaking was absent.

The first authentic papermaking originated in China as early as 100 AD, utilizing a suspension of bamboo or mulberry fibers. The Chinese subsequently developed papermaking into a highly skilled art, and many beautiful examples of ancient Chinese illustrations on paper are still in existence.

After a period of several centuries the art of papermaking extended into the Middle East and later reached Europe, where cotton and linen rags became the main raw materials. By the beginning of the 15th century a number of paper mills existed in Spain, Italy, Germany and France. The first paper mill in North America was established near Philadelphia in 1690.

Some of the significant milestones in the historical development of pulp and paper manufacture are summarized in Table 1-1. These inventions and

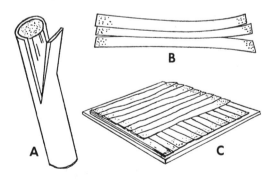

FIGURE 1-1. **Papyrus stalk is cut into 12- to 18-inch lengths**
(A) the cortex is removed and the pith split into thin strips
(B) the strips are then laid in a lattice pattern
(C) to be pounded into a single sheet.

TABLE 1-1. Milestones in pulp and paper industry development.

1798	Patent issued to Nicholas-Louis Robert for first continuous papermaking machine. (France)
1803, 1807	Patents issued to Fourdrinier brothers for improved continuous paper machine designed by Donkin. (England) [See Figure 1-2.]
1809	Patent issued to John Dickinson for the cylinder paper machine. (England)
1817	First cylinder machine in America.
1827	First Fourdrinier machine in America.
1840	Groundwood pulping method developed. (Germany)
1854	First manufacture of pulp from wood, using the soda process. (England)
1867	Patent issued to Benjamin Tilghman for the sulfite pulping process. (U.S.A.)
1870	First commercial utilization of groundwood process.
1874	First commercial utilization of sulfite process.
1884	Invention of kraft pulping process by Carl Dahl. (Germany)

pioneering prototypes provided the basis for the modern paper industry. The twentieth century has seen the rapid refinement and modification of this early and rather crude technology, along with the development of such techniques as refiner mechanical pulping, continuous cooking, continuous multistage bleaching, on-machine paper coating, twin-wire forming, and computer process control, to name just a few. Because pulp and paper operations require the continuous movement of large masses of material, the mechanization of material handling has always been an important aspect of industry development. (Refer to references 1 and 2 for further details on the early history of the industry.)

1.4 MODERN PULP AND PAPER OPERATIONS

The modern pulp and paper mill utilizes wood residuals as the basic raw material. Process operations are highly automated, and many mills now utilize computer control. Overall economics in North America usually favor large-scale units with high operator productivity. Consequently, the construction cost of a viable modern plant is extremely high. For example, a 1000-ton-per-day "greenfield" bleached kraft pulp mill is now estimated to cost in excess of 1 billion dollars. The high investment cost, typically exceeding one million dollars per worker, qualifies pulp and paper as a capital-intensive industry.

Economics also favor integrated forest products operations where logs are brought first to the wood mill for extraction of highest value lumber, plywood and particleboard. Wood residuals are then chipped and conveyed next door for conversion into pulp. Finally, the pulp is transferred to a third division of the plant complex for manufacture into paper. The high value and bulk of raw fiber dictates toward complete and optimal utilization with minimal handling.

Those segments of the industry with mechanical pulping operations are the heaviest users of energy. Other types of mills which are able to utilize wood waste materials for fuel are close to energy self-sufficiency. Large volumes of water are used by virtually all conventional pulp and paper operations, and an abundant source of water is required at the plant site. Typically, sophisticated effluent treatment is practiced before the water is returned to the receiving stream.

A breakdown of U.S. and Canadian pulp and paper production by general product category is shown in Table 1-2. Some selected industry statistics are given in Table 1-3. Canada is a world leader in the export of newsprint and market pulp. The United States is actually a net importer of paper products because of

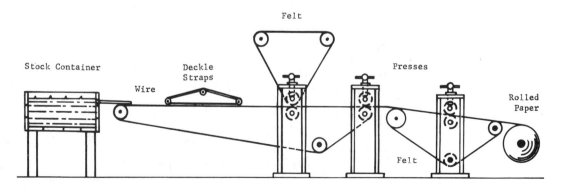

FIGURE 1-2. Improved papermaking machine of 1803.

large purchases of newsprint from Canada, but it is a leading exporter of value-added grades of paper.

By virtue of abundant timber, energy and water resources, modern technology, skilled manpower, and accessibility to markets, the North American pulp and paper industry is the world leader in terms of production and diversity of products. It can be noted that the United States and Canada together have 5% of the world's population, 15% of its paper mills, and produce 36% of its paper. Foreign producers have never been a significant factor in the domestic market, a circumstance enjoyed by few other major North American industries. In 1980, most industry observers agreed that the North American pulp and paper industry was the world's low cost producer. However, the situation has changed dramatically in a short period of time. Today, some off-shore producers can actually bring products into North America, pay shipping costs, and still be competitive in terms of product cost.

In the present era of globalization, competition for export pulp and paper markets is fierce, and the North American industry can hardly afford to rest on its laurels. Short-rotation pine and hardwood plantations are being rapidly developed in sub-temperate regions around the world that will provide vast quantities of low-cost pulpwood in future generations. Countries such as Brazil and Chile are already important producers of pulp fiber. Russia remains a question mark as a forest products competitor; its timber resource is the largest in the world, and the harvest could ultimately double the current North American cut.

Fortunately, the long-term outlook indicates continually expanding world demand for all pulp and paper products, so that outside competitive forces should not adversely affect the North American market for the foreseeable future. Per-capita consumption in other parts of the world is far below the North American level and indicates a built-in growth for many decades. Consumption data such as that shown in Figure 1-3 strongly suggest that increased utilization of paper products will be a natural consequence of economic growth among developing nations.

TABLE 1-2. 1990 production (000 short tons).

Paper	USA*	Canada**
Newsprint	6,610	9,068
Other Printing/Writing	22,371	3,599
Packaging/Wrapping	4,576	497
Tissues	5,802	495
Total	**39,359**	**13,659**
Paperboard		
Liner/Corrugating	25,097	2,045
Other Boards	14,326	761
Total	**39,423**	**2,806**
Total Paper & Paperboard	**78,748**	**16,465**
Total Pulp Production	**57,214**	**22,835**

Sources: *API; **CPPA

TABLE 1-3. Selected industry statistics*, 1989.

	USA	Canada
Number of pulp mills**	345	179
Number of paper/board mills	601	129
Number of mill employees	246,300	81,000
Product Prices, $/short ton		
Bleached softwood kraft pulp	740	
Bleached hardwood kraft pulp	690	
Newsprint (30-lb)	540	
Directory (22.5-lb)	850	
No. 1 Publication (70-lb)	1600	
Linerboard	410	
Corrugating medium	390	
Solid bleached kraft board	730	

* Various sources.
** One or more pulp mills may be at the site of a paper/board mill.

1.5 REQUIREMENTS AND SOURCES OF PAPERMAKING FIBERS

In order for fibers to be useful for papermaking, they must be conformable, i.e., capable of being matted and pressed into a uniform sheet. Strong

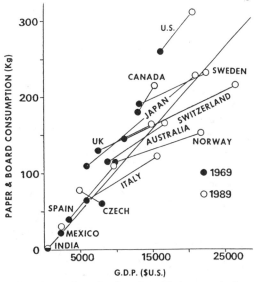

FIGURE 1-3. Graph showing high correlation between per capita gross domestic product in 1989 dollars (a measure of the average economic level of individuals in a country) and per capita paper/board consumption.

bonds must also develop at the points of contact. For some applications, the fiber structure must be stable over long periods of time. The degree of fiber conformability is characterized and measured as sheet formation, while the degree of bonding is inferred by the tensile or burst strength of the sheet.

Some valuable papermaking pulps are unusable in their raw state because the fibers are relatively non-conformable and non-bonding. These pulps must be mechanically treated to develop their papermaking properties. For example, cotton and linen rags (which are still used as pulp sources for the highest-quality durable papers) must be extensively worked to develop the desired fiber properties.

Pulp fibers can be extracted from almost any vascular plant found in nature. However, a high yield of fibers is necessary if the plant is to have economic importance. The major plant sources of pulp fibers are categorized and characterized in Table 1-4.

Wood is far and away the most abundant source of papermaking fibers, and is virtually the only source utilized in North America. As a consequence, this text will concentrate on wood as the principal raw material for pulp.

Except for seed hairs, vegetable fibers in their native state are embedded in a matrix of non-fibrous material (mostly lignin, but also containing hemicelluloses, resins and gums). Chemical and mechanical processes in the pulp mill free the fibers from the lignin matrix and leave the ultimate fiber in a relative degree of purity depending on the end use. The greater the requirement for purity, the lower will be the yield of fiber.

1.6 INTRODUCTION TO FIBER CHEMISTRY

Cellulose

In plant fibers it is the substance cellulose that determines the character of the fiber and permits its use in papermaking. Cellulose is a carbohydrate, meaning that it is composed of carbon, hydrogen and oxygen, with the latter two elements in the same proportion as in water. Cellulose is also a polysaccharide, indicating that it contains many sugar units

The chemical formula for cellulose is $(C_6H_{10}O_5)_n$, where n is the number of repeating sugar units or the degree of polymerization (DP). The value of n varies with the different sources of cellulose and the treatment received (see Table 1-5). Most paper-making fibers have a weight-averaged DP in the 600–1500 range.

The structure of cellulose is shown in Figure 1-4. The recurring unit is actually two consecutive glucose anhydride units, known as a cellobiose unit. Pure cellulose can be rather easily hydrolyzed to glucose $(C_6H_{10}O_6)$ under controlled (acidic) conditions.

The polymeric linkages during cellulose synthesis are such that the chains form in an extended manner. As a consequence, cellulose molecules fit snugly together over long segments, giving rise to powerful associative forces that are responsible for the great strength of cellulosic materials.

Cellulose in plant fibers is found in several ordered levels of orientation, as illustrated in Figure 1-5. Where the molecules fit together over long segments, regions of crystallinity develop which are difficult to penetrate by solvents or reagents. By contrast, the relatively more amorphous regions are readily penetrated and are therefore more susceptible to hydrolysis reactions. The microscopic and submicroscopic structure of cellulose is further illustrated in Figure 1-6.

TABLE 1-4. Average length, average diameter and length/diameter ratio of various pulp fibers.

	Length (mm)	Diameter (μm)	Ratio
Woods			
Coniferous (softwood)	4.0	40	100
Deciduous (hardwood)	2.0	22	90
Straws and Grasses			
Rice	0.5	9	60
Esparto	1.1	10	110
Misc. (wheat, rye, sabai)	1.5	13	120
Canes and Reeds			
Bagasse (sugar cane)	1.7	20	80
Miscellaneous	1.2	12	100
Bamboos			
Several varieties	2.8	15	180
Woody Stalks with Bast Fibers (jute, flax, kenaf, cannabis)			
Woody stems	0.25	10	25
Bast fibers*	20	20	1000
*Bast Fibers**			
Linen	55	20	2600
Ramie	130	40	3500
Leaf Fibers			
Abaca (Manila hemp)	6	24	250
Sisal	2.8	21	130
Seed Fibers			
Cotton	30	20	1500
Cotton linters	20	20	1000

* Fibers obtained from inner bark

TABLE 1-5. Degree of polymerization values (weighted averages).

Native cellulose (in situ)	3500
Purified cotton linters	1000 – 3000
Commercial wood pulps	600 – 1500
Regenerated cellulose (e.g., rayon)	200 – 600

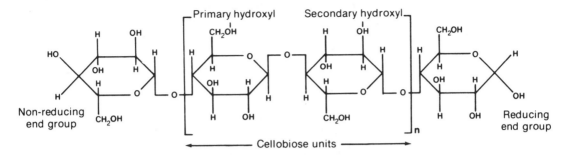

FIGURE 1-4. Cellulose structure.

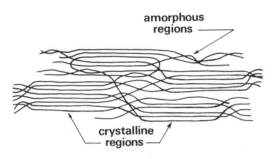

FIGURE 1-5. Schematic of molecular organization within a cellulose microfibril.

The properties of cellulosic materials are related to the DP of the constituent cellulose molecules. Decreasing the molecular weight below a certain level will cause deterioration in strength.

Long-chain cellulose is known as alpha cellulose. A number of shorter-chain polysaccharides, known collectively as hemicelluloses, also form part of the woody structure of plants. Hemicellulose (along with degraded cellulose) is further conveniently categorized (by chemical means) according to DP:
• beta cellulose - DP between 15 and 90
• gamma cellulose - DP less than 15

Hemicelluloses

By contrast to cellulose which is a polymer only of glucose, the hemicelluloses are polymers of five different sugars:
• hexoses: glucose, mannose, galactose
• pentoses: xylose, arabinose
Depending on the plant species, these sugars along with uronic acids form various polymeric structures; some are associated with the cellulosic portion of the plant, while others are more closely associated with lignin.

During chemical treatment of wood to produce pulp, the amounts, locations, and structures of the various hemicelluloses usually change dramatically.

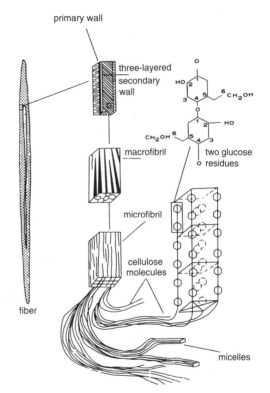

FIGURE 1-6. Microscopic and submicroscopic structure of cellulose (Bruley).

The hemicelluloses are more easily degraded and dissolved than cellulose, so their percentage is always less in the pulp than in the original wood.

Lignin

The term "holocellulose" is used to describe the total carbohydrate content of fibers. In addition to holocellulose, woody plant materials contain an amorphous, highly-polymerized substance called lignin. Its principal role is to form the middle lamella, the intercellular material which cements the fibers together. Additional lignin is also contained within the remaining cross-section of the fiber.

The chemistry of lignin is extremely complex (see Figure 1-7). The structure consists primarily of phenyl propane units linked together in three dimensions. The three linkages between the propane side chains and the benzene rings are broken during chemical pulping operations to free the cellulosic fibers.

A full treatment of cellulose and lignin chemistry is obviously beyond the scope of this book. For a comprehensive treatment, readers are referred to standard textbooks (references 3 and 4).

Extractives

In addition to holocellulose and lignin, a number of diverse substances may be present in native fibers, depending on the plant source, e.g., resin acids, fatty acids, turpenoid compounds and alcohols. Most of these substances are soluble in water or neutral organic solvents, and are collectively called extractives. Among North American wood species, many have less than 1% extractives content based in moisture-free weight. The southern pines have a notably higher content, which provides substantial

amounts of raw tall oil and turpentine as by-products from alkaline pulping operations.

The chemical composition of wood is illustrated in Figure 1-8. A schematic illustrating the chemical separation of wood components is shown in Figure 1-9.

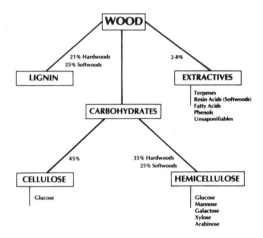

FIGURE 1-8. **Chemical components of wood.**

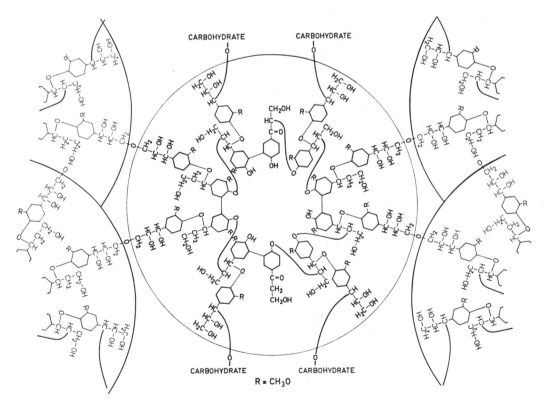

FIGURE 1-7. **Two-dimensional representation of the repeating unit of spruce lignin (by Forss, Fremor and Stenlund in 1966). Note that the actual structure is three-dimensional.**

1.7 BEHAVIOR OF CELLULOSIC FIBERS

Cellulosic fibers exhibit a number of properties which fulfill the requirements of papermaking (summarized in Table 1-6). In general, the best balance of papermaking properties occurs when most of the lignin is removed from the fibers while retaining substantial amounts of hemicellulose. Properties are also greatly optimized by a mechanical treatment (e.g., **beating** or **refining**), which causes removal of the primary fiber walls and allows the fibers to hydrate (i.e., take water into the structure) and swell, increasing their flexibility and bonding power. The typical behavior of chemical pulp handsheet strength properties during beating is illustrated in Figure 1-10.

The **hydrophillic** nature of cellulosic fibers plays an important role because the papermaking process occurs in an aqueous medium. The fibers readily absorb water and are easily dispersed in a water suspension. When wet fibers are brought together during the sheet-forming operation, bonding is promoted by the polar attraction of the water mole-cules for each other and for the hydroxyl groups covering the cellulose surface. When water is evaporated from a formed sheet, the hydroxyl groups on opposing fiber surfaces ultimately link together by means of **hydrogen bonds** as shown in Figure 1-11.

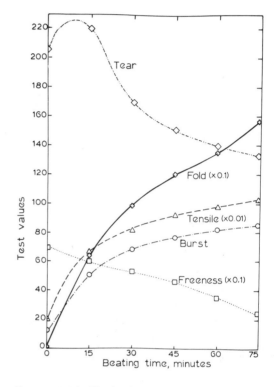

FIGURE 1-10. Typical response of softwood chemical pulp to laboratory beating.

TABLE 1-6. Properties of cellulosic fibers.

- high tensile strength
- suppleness (flexibility, conformability)
- resistance to plastic deformation
- water insoluble
- hydrophillic
- wide range of dimensions
- inherent bonding ability
- ability to absorb modifying additives
- chemically stable
- relatively colorless (white)

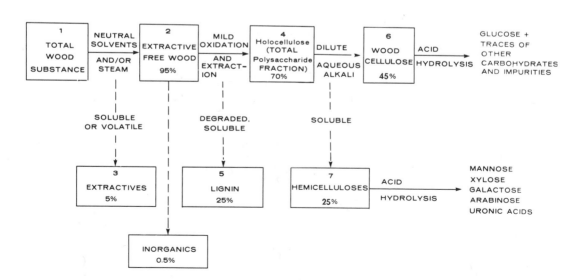

FIGURE 1-9. Schematic illustrating chemical separation of wood components (Bruley).

While individual cellulosic fibers generally have high tensile strength, the strength parameters of paper are more dependent on the bonds between fibers. Beating or refining tends to optimize bonding at the expense of individual fiber strength. Of course, the original fiber strength depends on the nature of the raw material and the method of pulping.

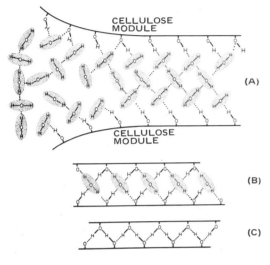

FIGURE 1-11. Illustrating different levels of hydrogen bonding:
(a) loosely through water molecules;
(b) more tightly through a monolayer of water molecules;
(c) directly.

Since **non-fibrous additives** are often used in the manufacture of paper products, the ability of the pulp fibers to retain a wide variety of modifying materials during sheet-forming operations is important. Filtration, chemical bonding, colloidal phenomena, and adsorption are involved in the retention of particles. Filtration is suitable for larger particles, but other mechanisms are more important for retaining smaller particles and colloids. Retention is enhanced by flocculation or precipitation of the insoluble material and by optimal adjustment of the electrokinetic charges within the system. The ability to absorb or adsorb soluble additives is dependent on the relative chemical "affinity" of the pulp fibers.

REFERENCES
(1) HUNTER, D. **Papermaking: The History and Technique of an Ancient Craft** (2nd Edition) Albert A. Knopf, 1957
(2) CLAPPERTON, R.H. **The Papermaking Machine. Its Invention, Evolution and Development** Pergamon Press, 1967
(3) SJOSTROM, E. **Wood Chemistry Fundamentals and Applications.** Academic Press, 1981
(4) NEVELL, T.P. and ZERONIAN, S.H. (Editors) **Cellulose Chemistry and Its Applications** John Wiley & Sons, 1985

Chapter 2

Characteristics of Wood and Wood Pulp Fibers

Wood is the principal source of cellulosic fiber for pulp and paper manufacture. At present, wood provides about 93% of the world's virgin fiber requirement, while non-wood sources, mainly bagasse, cereal straws and bamboo, provide the remainder. Approximately one-third of all paper products are recycled into secondary fiber.

2.1 TREE STRUCTURE

A tree can be considered to have three general parts:
• the crown composed of leaves and branches
• the stem
• the root system

The leaves or needles are the factories where food material is manufactured through photosynthesis to provide the tree with energy and growth. Photosynthesis is the production of carbohydrates from carbon dioxide and water in the presence of chlorophyll and light.

Although the crown is both the source of nutrients and the regulating center for wood production, wood is not produced directly by photosynthesis. Rather, wood results through cell divisions of the vascular cambium using energy derived from the products of photosynthesis. After cambial division, each successive cell undergoes enlargement, wall thickening, and lignification.

Figure 2-1 shows a cutaway sketch of a tree trunk revealing the general structure. Figure 2-2 shows a transverse cross-section. The **cambium** consists of a thin layer of tissue between the bark and the inner sapwood. In temperate climes, the rate of cambial growth varies with the seasons giving rise to the deposition of thin-wall fiber cells in the spring and more dense thick-wall fibers in the fall. The cambium is dormant during the cooler months of the year. The yearly growth cycle is reflected in the annual rings, the total number of which represent the tree's age.

The **inner bark** (phloem) is a narrow layer of tissue where the carbohydrate-containing sap moves upward and downward through sieve tubes and rays. The outer bark is a collection of dead cells which originally existed in the inner living bark; it is composed of a variety of extraneous components in addition to cellulose, hemicellulose and lignin.

The **sapwood** portion of the tree provides structural support for the crown, acts as a food storage reservoir, and provides the important

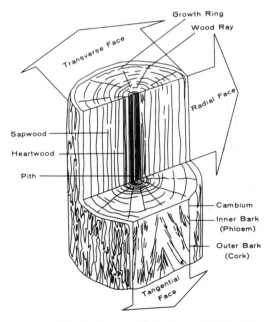

FIGURE 2-1. Cutaway block of wood illustrating wood structure and cross-sectional faces (J.H. DeGrace).

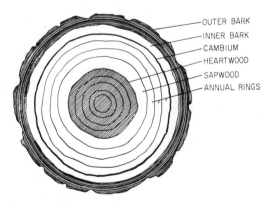

FIGURE 2-2. Cross-sectional sketch of a mature stem, showing outer bark, inner bark, sapwood and heartwood.

function of water conduction up from the roots. It is physiologically active (parenchyma cells only) and in continuous communication with the cambium and phloem through sap flow from the crown.

The inner **heartwood** is a core of dead woodcells in the center of the stem whose physiological activity has ceased. It functions only as mechanical support. Heartwood is usually much darker in color than sapwood due to deposition of resinous organic compounds in the cell walls and cavities. Such deposition makes liquor penetration during chemical pulping more difficult in heartwood than in sapwood. In a few species (most notably spruce) the color difference between heartwood and sapwood is minor. At the center of the tree is a small core of soft tissue called the pith.

2.2 CHARACTERISTICS OF WOOD

Botanically, woods are classified into two major groups. The gymnosperms are commonly called **softwoods** or conifers. The angiosperms are the **hardwoods** or broad-leafed trees, either deciduous or evergreen. The main structural features of each wood group are illustrated in Figures 2-3 and 2-4.

Softwoods

The vertical structure of conifers is composed almost entirely of long, tapering cells called **tracheids.** In some species, vertical resin canals are also present. The horizontal system is composed of narrow rays, only one cell in width but often several cells high. Ray cells are of two specialized types: ray parenchyma occur in all species, while ray tracheids are present in only certain species.

Seasonal growth is usually characterized by a denser band of tracheids at the end of the annual ring. This latewood (or summerwood) tissue has quite different properties from the earlywood (or springwood) tissue whose density may be only one-half or one-third that of the latewood. The cell wall itself has a relative density (specific gravity) of about 1.5 (oven-dry basis).

The wall of a typical tracheid or "fiber" is composed of several layers. The middle lamella, very high in lignin content, separates two contiguous tracheids. Each tracheid has a primary wall and a three-layered secondary wall with specific alignments of **microfibrils.** Microfibrils are bundles of cellulose molecules, and their orientation can

TRANSVERSE VIEW

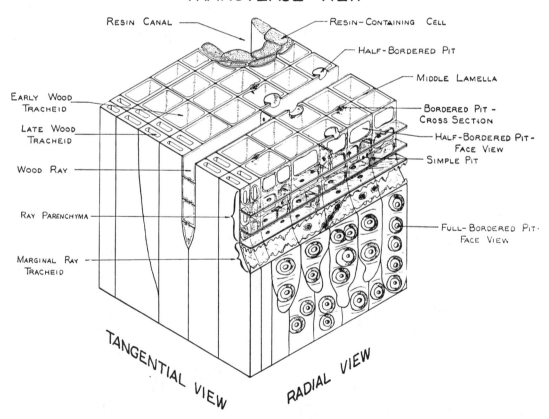

FIGURE 2-3. Composite wood block illustrating the structural features of a softwood (Hyland).

TRANSVERSE VIEW

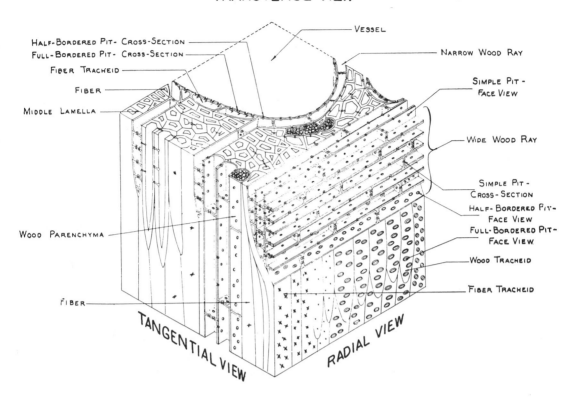

FIGURE 2-4. Composite wood block illustrating the structural features of a hardwood (Hyland).

FIGURE 2-5. (on right side) Diagram of cell wall organization.

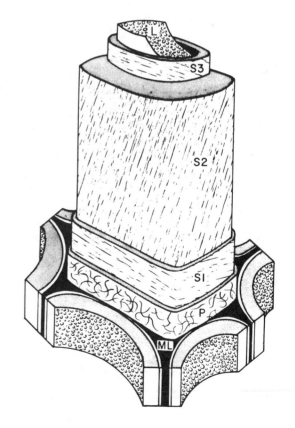

TABLE 2-1. Layers of softwood tracheid (20–40 microns diameter).

Middle Lamella (ML)	- bond between fibers, mostly lignin
Primary Wall (P)	- a thin, relatively impermeable covering about 0.05 µm thick
Secondary Wall (S)	- makes up bulk of cellwall; forms three distinct layers characterized by different fibril alignments:
	• S_1 is the outer layer of the secondary wall (0.1–0.2 µm thick)
	• S_2 forms the main body of the fiber and is from 2 to 10 µm thick
	• S_3 is the inner layer of the secondary wall (about 0.1 µm thick)
Tertiary Wall (T)	- same as S_3
Lumen (L)	- the central canal of fiber (void)

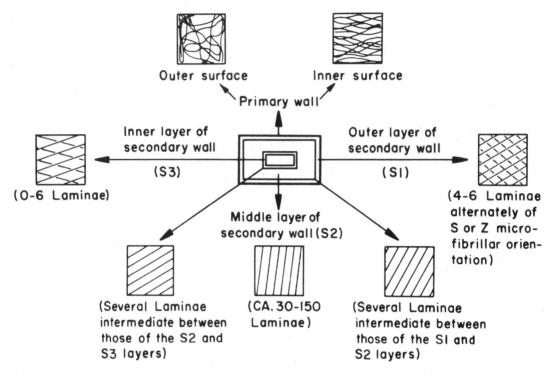

FIGURE 2-6. Cell wall organization showing microfibrillar textures (Wardrop and Harada).

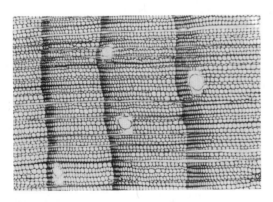

FIGURE 2-7. Cross-section of "typical pine" showing annual rings, earlywood-latewood differentiation, and resin canals (MacMillan Bloedel Research Ltd.).

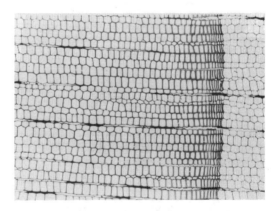

FIGURE 2-8. Cross-section of western red cedar showing annual ring and earlywood-latewood differentiation (MacMillan Bloedel Research Ltd.).

influence the characteristics of a pulp fiber. The structure of a tracheid is illustrated and explained in Table 2-1 and Figure 2-5. Note that the symbols S_3 and T are used interchangeably. Figure 2-6 shows additional detail with respect to microfibrillar layers (or laminae), alignments and textures.

Radial cross-sections of four representative North American softwoods ("typical" pine, western red cedar, Douglas fir, hemlock) are shown under increasing magnification in Figures 2-7 to 2-10. Note the distinctive features of each species and the differences in cell wall thickness between earlywood and latewood. A three-dimensional view of spruce wood is shown in Figure 2-11 and under increasing magnification in Figures 2-12 and 2-13.

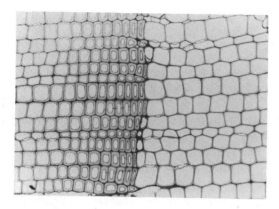

FIGURE 2-9. Cross-section of Douglas fir showing annual ring and earlywood-latewood differentiation (MacMillan Bloedel Research Ltd.).

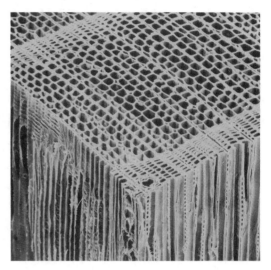

FIGURE 2-11. Sitka spruce cube (micrograph courtesy of Forintek Canada Corp., Western Laboratory).

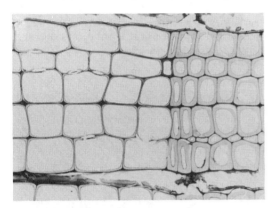

FIGURE 2-10. Cross-section of hemlock showing annual ring and earlywood-latewood differentiation (MacMillan Bloedel Research Ltd.).

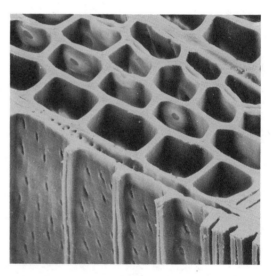

FIGURE 2-12. Sitka spruce showing cross-field pitting on the radial surface and rays on the transverse surface (micrograph courtesy of Forintek Canada Corp., Western Laboratory).

Hardwoods

The principal vertical structure of hardwoods is composed of both relatively long, narrow cells, called libriform fibers, and much shorter, wider cells, called vessels. Vessels in a typical hardwood sample are often large enough in diameter to be seen easily with the naked eye, in cross-section as "pores" or on vertical surfaces as a series of long grooves. Hardwoods also have a vertical parenchyma system and a horizontal or ray parenchyma system.

Vessel diameter varies from earlywood to latewood within an annual ring. If this difference is extreme and abrupt, the rings become easy to distinguish, and the wood is termed ring-porous (Figure 2-14). In other species where the gradation in vessel diameter is small and gradual, the annual rings are more difficult to distinguish, and the wood is termed diffuse-porous (Figure 2-15). A three-dimensional view of diffuse-porous white birch is shown in Figure 2-16.

Softwoods vs. Hardwoods

The dramatic difference between softwood and hardwood (e.g., spruce vs. birch) with respect to weight and volume percentages of the various types of fiber cells is illustrated in Table 2-2. Another major difference is the length of the fibers; a typical relationship between biological age of wood and fiber length (Figure 2-17) shows that softwood fibers are more than twice as long as hardwood fibers.

FIGURE 2-13. Transverse surface of Sitka spruce under high magnification (micrograph courtesy of Forintek Canada Corp., Western Laboratory).

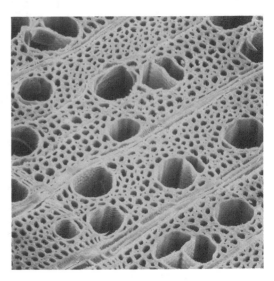

FIGURE 2-15. Transverse surface of big leaf maple illustrating diffuse-porous wood (micrograph courtesy of Forintek Canada Corp., Western Laboratory).

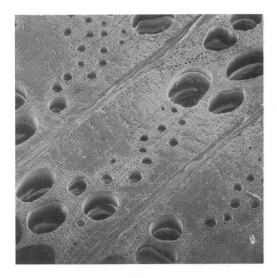

FIGURE 2-14. Transverse surface of red oak illustrating ring-porous wood (micrograph courtesy of Forintek Canada Corp., Western Laboratory).

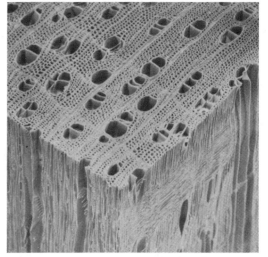

FIGURE 2-16. White birch cube (micrograph courtesy of Forintek Canada Corp., Western Laboratory).

The average **relative density** (oven-dry weight per green volume) of commonly used coniferous pulping species ranges from 0.31 for western red cedar to 0.55 in western larch. This difference in density relates to the fact that cedar has relatively little latewood tissue while larch has a large proportion. Hardwood relative density ranges from 0.30 in black cottonwood to over 0.60 for rock elm, hickory and white oak which all contain thicker-walled fibers.

A relationship exists (within the respective

TABLE 2-2. Types of cells - spruce vs. birch.

	Fibers (%)		Vessels (%)		Parenchyma (%)	
	by wt.	by vol.	by wt.	by vol.	by wt.	by vol.
spruce	99	95	–	–	1	5
birch	86	65	9	25	5	10

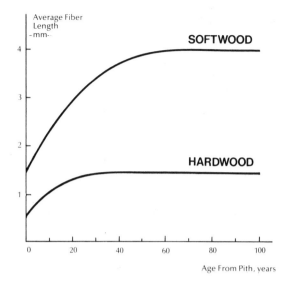

FIGURE 2-17. Effect of biological age on average fiber length for typical softwood and hardwood.

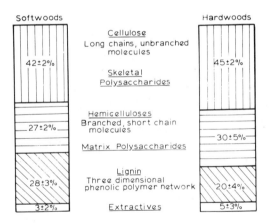

FIGURE 2-18. Average composition of softwoods and hardwoods.

TABLE 2-3. Properties of North American pulpwoods*

Species	Fiber length (mm)	Fiber diameter (microns)	Wood density (lb/cu ft)
Southern Region			
Longleaf Pine	4.9	35–45	41
Shortleaf Pine	4.6	35–45	36
Loblolly Pine	3.6	35–45	36
Slash Pine	4.6	35–45	43
Northeast Region			
Black Spruce	3.5	25–30	30
White Spruce	3.3	25–30	26
Jack Pine	3.5	28–40	30
Balsam Fir	3.5	30–40	25
Northwest Region			
Douglas Fir	3.9	35–45	34
Western Hemlock	4.2	30–40	29
Redwood	6.1	50–65	25
Red Cedar	3.5	30–40	23
Hardwoods			
Aspen	1.04	10–27	27
Birch	1.85	20–36	38
Beech	1.20	16–22	45
Oaks	1.40	14–22	46
Red Gum	1.70	20–40	34

***Isenberg, I.H. *The Pulpwoods of the United States and Canada*. Second Ed., The Institute of Paper Chemistry, Appleton, Wisconsin. 1951.**

softwoods, but a greater percentage of extractives. The average compositions and the normal range of values are shown in Figure 2-18.

Variations Within Softwood Species

Douglas fir and larch may have from 10 to 50% latewood, and the percentage range in southern pines may be even more extreme. These conifers, characterized by an abrupt earlywood/latewood transition, generally show a pattern of **specific gravity variation** with age from the pith as shown in Figure 2-19. These trees, therefore, contain a core of lower density juvenile wood having properties quite different from more mature wood. For species such as hemlock, spruce and balsam fir which do not have well-defined core wood, the variable most closely associated with density variations is the growth rate.

Specific gravity (i.e., density) is to some extent a genetically-transmitted characteristic. Greater yields of wood (and therefore of pulp on a volume input basis) can be obtained by considering specific gravity as an attribute in tree-breeding programs.

A feature of softwood growth is the formation of **compression wood** on the underside of leaning trees and branches (See Figure 2-20). Virtually every tree has some of this tissue, but the amount and severity are quite variable from one tree to another. Compression wood is characterized by high density, rounded thick-walled tracheids, large fibril angles,

softwood and hardwood groupings) between wood density and a number of pulping parameters. For example, the yield of pulp per unit volume of wood is usually directly related to density. A high wood density generally indicates a slower beating response for the pulp, lower tensile, burst and fold strengths, greater bulk, and higher tear strength. As shown in Table 2-3, hardwoods tend to have higher densities than softwoods, and the southern pine species are denser than softwoods from other growth regions.

Generally hardwoods contain a larger proportion of holocellulose and less lignin as compared to highly lignified secondary wall, shorter fiber length and spiral checks in the cell wall. All of these attributes have consequences in pulp manufacture and quality. Compression wood fibers generally do not respond well to beating or refining. However, with severe mechanical treatment, the thick layer of the secondary wall becomes unravelled into very long fibrils.

Branches give lower yields and quality of pulp because of higher percentages of both juvenile wood and compression wood.

Variations Within Hardwood Species

The **relative density** of hardwoods varies from 9 or 10% of the average value in hard maple and birch

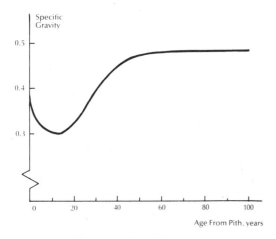

FIGURE 2-19. Variation in specific gravity with age from the pith for softwoods with abrupt earlywood/latewood transition.

FIGURE 2-20. Compression wood in the cross section of a stem.

to 23% in elm and ash. Ring-porous woods generally exhibit more variation than diffuse-porous woods, depending on the rate of growth. Slow-growth ring-porous woods typically have a large earlywood vessel volume, leading to low relative density.

A feature of hardwoods is the formation of **tension wood** on the upper side of leaning stems and branches. The slightly denser tension wood is characterized by a relative absence of vessels and by fibers having a distinct gelatinous inner layer composed of highly crystalline cellulose. Although chemical pulping yields are higher for tension wood, the fibers are difficult to beat and exhibit poor interfiber bonding. Tension wood is found to some degree in every hardwood tree, but the amount and severity depends on the growing habit of the particular tree.

Bark

Bark is the outer covering or rind of woody stems and branches. It is distinct and separable from wood. The structure is complex (as compared to wood) because bark contains three types of tissue (cortex, periderm and phloem), each of which has several types of cells. While bark generally is considered to be a contaminant in pulping operations, some types (e.g., western red cedar and aspen) contain significant quantities of fiber and can be tolerated to an extent in an alkaline pulping system.

However, certain bark constituents are resistant to typical pulping conditions, principally cork cells, dense sclereids or "stone cells" (see Figure 2-21), and cells impregnated with extractives. The extractives consume relatively large amounts of chemical, while partially pulped particles remain as dirt in the finished pulp.

Greater amounts of bark are tending to be introduced into the pulp mill with the chip furnish because of more intense tree utilization (e.g., whole-

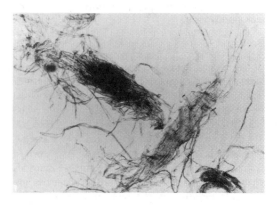

FIGURE 2-21. Schlereid fiber bundle (MacMillan Bloedel Research Ltd.)

tree chipping). Techniques are employed in the pulp mill to remove bark from the wood chips and remove bark specks from the pulp.

2.3 EFFECT OF FIBER STRUCTURE (MORPHOLOGY) ON PAPER PROPERTIES

The properties of paper are dependent on the structural characteristics of the various fibers that compose the sheet. Undoubtedly the two most important of these characteristics are fiber length and cell wall thickness. A minimum length is required for interfiber bonding, and length is virtually proportional to tear strength. A comparison of some softwood and hardwood cell types is shown in Figure 2-22.

In general, softwood tracheids with relatively thin cell walls collapse readily into ribbons during sheet formation (e.g., western red cedar in Figure 2-23). Tracheids with thicker cell walls resist collapse and do not contribute to interfiber bonding to the same extent. The thicker-walled fibers (e.g., Douglas fir in Figure 2-24) tend to produce an open, absorbent, bulky sheet with low burst/tensile strength and high tearing resistance. Within species, the thin-walled earlywood tracheids are relatively flexible, while the latewood tracheids, having as much as 60 to 90 % of their volume in cell wall material, are less conformable.

To help illustrate the basic principle, Figure 2-25 shows two types of idealized fiber structures. In the upper representation (A), the thick-walled fibers are shown as hollow cylinders; the thin-walled fibers below (B) are shown as ribbon-like elements. The numbers of fibers and contact points are the same in both structures; but the area of contact and potential bonding sites are clearly much greater in the sheet composed of thin-walled fibers.

The ratio of pulp fiber length to cell wall thickness (L/T) is sometimes used as an index of relative fiber flexibility. However, a more specific indication of a fiber's behavior is provided by its coarseness value. Fiber coarseness is defined as the weight of fiber wall material in a specified fiber length, and is usually expressed in units of milligrams per 100 meters. Some typical measurements for a number of representative pulp fibers are given in Table 2-4.

The predominant fibril angles in the three layers of the secondary wall (refer back to Figure 2-6) vary among softwood species and are related to certain paper strength parameters in a well bonded sheet. The greatest impact is on the stretch or extensibility of the paper sheets.

For hardwood pulps, the relationships between fiber measurements and sheet properties are less clearly defined because of the presence of vessel elements and other cellular components in varying

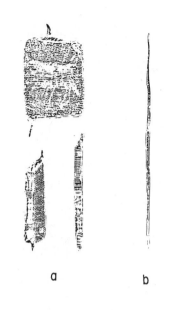

a   b   c

FIGURE 2-22. Facsimiles of some hardwood and softwood cell types. The coniferous tracheid (c) is much longer than the hardwood libriform fiber (b) and the vessel segments (a).

proportions. Refer to Figures 2-26 and 2-27 for illustrations of hardwood pulp fibers.

2.4 WOOD SPECIES IDENTIFICATION

In conjunction with the operation of a pulp and paper mill, it is often of interest to precisely identify and analyze (quantify) the parent species makeup of an unknown sample of pulp fibers or wood chips. The procedure for carrying out such determinations is based on microscopic examination of the sample. Sample preparation is described in a TAPPI Standard Procedure (1) and the overall methodology is well covered by Strelis and Kennedy (2).

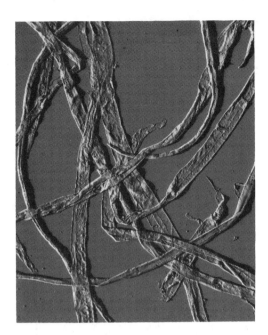

FIGURE 2-23. Western red cedar has a very thin ribbon-like fiber. Average fiber length is 3.2 mm (Weyerhaeuser Pulp Division).

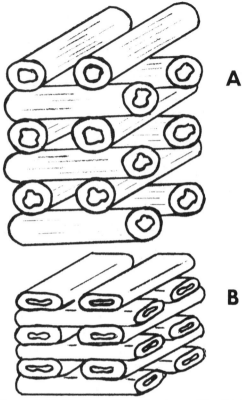

FIGURE 2-25. Idealized fiber structures. The thick-walled fibers (A) are less conformable than the thin-walled fibers (B).

FIGURE 2-24. Douglas fir fibers have relatively thick cell walls and do not collapse as readily as other softwoods. The "spiral thickenings" are an identifying characteristic.

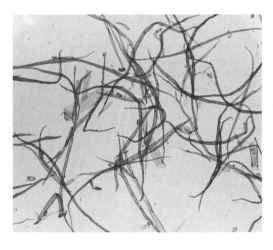

FIGURE 2-26. Mixture of aspen and maple fibers.

Wood samples are comparatively easy to identify because of the large number of diagnostic features clearly observable under the microscope. Strelis and Kennedy list 22 characteristics for softwood and 24 characteristics for hardwood. The samples must be adequately prepared and sectioned into all three

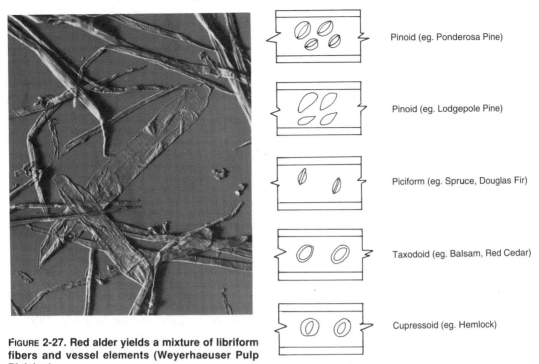

FIGURE 2-27. Red alder yields a mixture of libriform fibers and vessel elements (Weyerhaeuser Pulp Division).

Pinoid (eg. Ponderosa Pine)

Pinoid (eg. Lodgepole Pine)

Piciform (eg. Spruce, Douglas Fir)

Taxodoid (eg. Balsam, Red Cedar)

Cupressoid (eg. Hemlock)

FIGURE 2-28. Types of ray crossing pits.

TABLE 2-4. Properties of North American pulpwoods

Species	Fiber Length mm	Fiber Diameter μm	Fiber Wall Thickness, μm Earlywood Latewood	Ratio L/T	Coarseness mg/100 m
Birch	1.8	20–36	3–4	500	5–8
Red Gum	1.7	20–40	5–7	300	8–10
Black Spruce	3.5	25–30	3–4 (70%) 6–7 (30%)	700	14–19
Red Cedar	3.5	30–40	2–3	1400	15–17
Southern Pine	4.6	35–45	2–5 (50%) 8–11 (50%)	700	20–30
Douglas-Fir	3.9	35–45	2–4 (60%) 7–9 (40%)	700	25–32
Redwood	6.1	50–65	3–4	1700	25–35

planes: transverse, radial and tangential (refer back to Figure 2-1).

For pulp fibers, many of the diagnostic features of cell formation seen in wood samples are lost. However, positive identification is possible by following a logical sequence of exclusion, and finally comparing unknown fibers with authenticated fiber samples and/or photomicrographs. The technique is straightforward, especially where only a limited number of choices are possible. The principal diagnostic feature is the shape and orientation of the cell wall openings (pits) in softwood earlywood tracheids and hardwood vessels which are unique for each species. See Figure 2-28 for examples of softwood ray crossing pit types.

REFERENCES
(1) **TAPPI Standards** T263, T401
(2) STRELIS, L. and KENNEDY, R.W. **Identification of North American Commercial Pulpwoods and Pulp Fibers** University of Toronto Press (1967)

Chapter

3

Wood and Chip Handling

3.1 WOOD RESOURCE

On a world-wide basis, 35% of the forest-growing stock is in the form of long-fibered conifers (softwoods) and 65% is in shorter-fibered hardwoods. Table 3-1 indicates the location of the 323 billion m^3 resource by regions and countries. It is notable that the area formerly known as the Soviet Union has more than half of the world's coniferous resource, and about 2.5 times that contained in North America. The vast majority of the hardwood resource exists in the tropics, particularly Africa and Latin America.

Current and projected **pulpwood supply patterns** (as estimated in 1977) are shown in Table 3-2. North America accounted for about half of the world supply in 1977, and the percentage was estimated to decrease by only 5% up to 1990, while rising steeply in absolute terms. The greatest rate of increase in pulpwood supply was predicted for conifers from the (former) Soviet Union and hardwoods from Latin America and the Far East.

Greater utilization of hardwoods for pulping operations has been evident since 1950. In the years between 1950 and 1972, the proportion of broad-leafed trees

TABLE 3-2. Pulpwood supply, million m^3 (FAO, 1977).

Region	1973-1975 (Average)	1990 (Estimate)
North America	257	410
Western Europe	125	194
Japan	22	47
Oceania	8	17
Latin America	11	45
Near East, N. Africa	2	5
Africa (S. of Sahara)	5	13
Far East	6	30
Centrally Planned Economies	76	163
World Total	511	921

for pulpwood rose from 14% to 29% in the U.S. and from 3% to 16% in Nordic countries, and that general trend has continued. The increased use of hardwoods in developed countries has been abetted by a corresponding decline in fuelwood consumption.

Another definite trend in pulpwood supply has been the increased utilization of industrial wood residues. About 27% of North American pulpwood furnish is now supplied in the form of residual chips. The movement toward integrated operations has contributed to this trend.

Canada has about 13% of the world's coniferous resource from which it contributes about 11% of the world's harvest. The United States, with only 11% of the coniferous supply, is able to produce 22% of the harvest due to more accessible timber, greater industrialization, proximity to markets, and more rapid growing cycles for the Southern forests. The major forest areas in North America are identified in Figure 3-1.

3.2 WOOD HARVESTING TECHNIQUES

Supply of wood raw materials to the pulpmill is usually the responsibility of woodlands personnel or procurement specialists apart from the pulpmill staff. Nonetheless, mill technologists need to be aware of timber harvesting/handling practices and/or chip purchasing arrangements in order to appreciate the inherent problems and limitations in maintaining a consistent supply and quality of pulpwood.

TABLE 3-1. Forest-growing stock in various countries and regions (FAO, 1977).

Species Group	Region	% Growing Stock	
Conifers	Japan	0.3	
	Europe	3.0	
	N. America	8.4	
	USSR	20.7	
	Other	2.6	
	Total		35.0
Broad-Leafed	Japan	0.3	
	Europe	1.7	
	N. America	2.9	
	USSR	3.7	
	Other	3.1	
	Temperate Total		11.8
	Asia, Far East	8.4	
	Africa	18.6	
	Latin America	26.3	
	Tropical Total		53.3

There are basic differences in resource management and exploitation between the United States and Canada. In the United States about 72% of the forest land is privately owned and managed. Company operations are estimated to account for only about 14% of the pulplogs or chips delivered to the mill site. The bulk of raw material needs are supplied directly by pulplog producers and sawmill operators or by dealers who act as intermediaries.

By contrast, about 95% of the total forest area in Canada is publicly owned and administered as "crown lands". Most of the wood harvesting is carried out by forest products companies under crown license. Indeed, because much of the forest resource is remote and covers vast, unbroken areas, large-scale company operations constitute the only feasible method of harvesting.

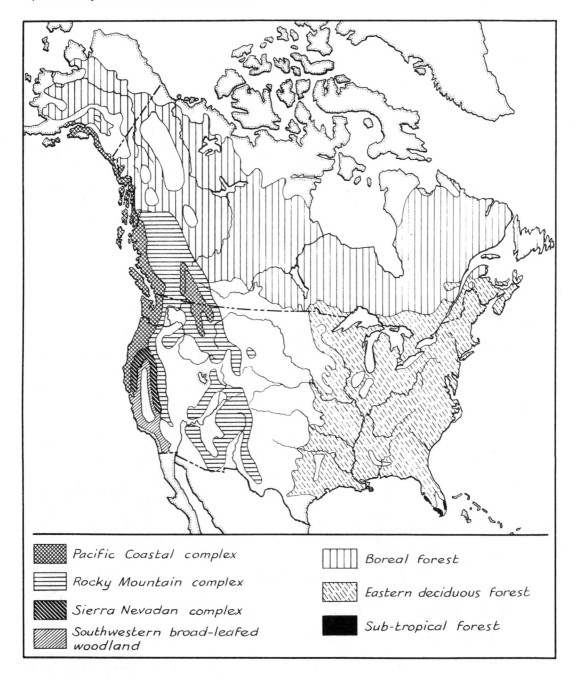

FIGURE 3-1. Major formations of original forest lands in North America (J.W. Barrett).

A limited number of forest products companies are self-sufficient with respect to pulpwood supplies. However most pulpmills find it necessary to supplement their intra-company pulpwood sources with purchased logs and/or chips from local suppliers. Some companies actually have no forest resources of their own, and must obtain all their pulpwood requirements from outside sources. A few pulpmills have no facilities for handling logs and are completely dependent on chip transfers and purchases.

Methods of **logging** and **log handling/transport** vary according to region. For example, in sections of the West or North where mills are located along the seacoast or on major rivers, logs are brought to the mill site by water. In many areas, especially where transport distances are great, rail carries a significant volume (Figure 3-2). Overall, trucks carry most of the wood (Figure 3-3).

Three principal methods of logging are used:
1) The shortwood system consists of felling, delimbing, cutting into prescribed lengths, piling, and transporting the wood to roadside. This method is traditionally labor-intensive, but some operations are now mechanized to the point where only one operator is required.

2) The full-tree system is the most basic method. The trees are felled, topped and piled; and then hauled to the roadside for further processing and/or transport to the mill. It is now common practice to delimb and debark the wood with a portable chain flail unit before transporting the logs to the mill.

FIGURE 3-4. Logs are bucked and sorted in the woods before loading onto trucks (photo by Finning).

FIGURE 3-2. Transport of short wood by rail.

FIGURE 3-3. Modern full tree logging truck (photo by Finning).

FIGURE 3-5. This older generation tree harvester cuts by shearing action, whereas more modern designs utilize a sawing mechanism. Note the high stumps with resultant waste of timber resource (photo by Finning).

FIGURE 3-6. A modern tree harvester in operation (Timberjack).

FIGURE 3-7. Detail of modern tree harvester (photo by Finning).

FIGURE 3-8. Clearing for a logging road in the Northwest (photo by Finning).

FIGURE 3-9. Skidder brings out a large cedar log in the Northwest (photo by Finning).

FIGURE 3-11. Loading logs in rough Northwestern terrain (photo by Finning).

FIGURE 3-10. Tractor skidder removes a large log from rough terrain in the Northwest (photo by Finning)

FIGURE 3-12. Log sorting in the water using "dozer boats".

3) In the tree-length system, sophisticated machines (tree combines) delimb, top, sever the tree from its stump, and pile the logs in an orderly fashion for another machine to take them to roadside. (Refer to Figures 3-4 to 3-7.)

Today larger numbers of tree-length logs are being handled, and the trend continues toward increased mechanization. However, a significant amount of timber is still cut using manual operations. Full conversion to mechanized logging involves large expenditures for the development and procurement of machines capable of efficient operation in rugged terrain (Figures 3-8 to 3-11) and/or in extreme temperatures. The major incentives are reduced labor costs and more independence from seasonal constraints. However, the logs may arrive at the mill

FIGURE 3-13. Log sorting by water.

FIGURE 3-14. Logs being sorted on land (photo by Finning).

FIGURE 3-16. Mobile chipper and chip transporters retrieve smaller logs and logging slash as chips and deposit them in roadside inventory piles.

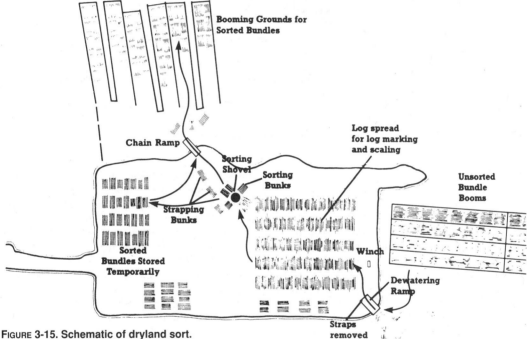

FIGURE 3-15. Schematic of dryland sort.

site in a relatively "dirty" and unsorted condition.

At some point between harvesting and wood processing operations, the logs are usually **sorted** according to species, quality, size, end-use or destination. The sorting operation may be carried out on the water (Figures 3-12 and 3-13) or on land (Figure 3-14). In areas with navigable waterways, the trend is away from water sorting to minimize generation of water-borne debris, and toward increased use of log bundling to reduce the space occupied by log booms and prevent the loss of sinker logs. Where log bundling is practiced, land sorting is a virtual necessity. A modern land-sort in the Northwest is shown schematically in Figure 3-15.

Complete-tree utilization is commonly practiced in scrub-type hardwood forests where the trees are of marginal commercial value. With this method, the entire tree (i.e., bole, top branches, leaves, etc.) is chipped at the logging site and some type of air fractionation process is utilized to separate the chipper product into pulp chips and biomass for fuel (Figure 3-16). Complete-tree utilization takes advantage of an otherwise worthless resource, but a number of potential problems must be recognized and taken into account (e.g., lower pulp strength and yield, more dirt in process). Typically, companies start with less than 10% of "whole-tree chips" and increase the percentage over a period of time.

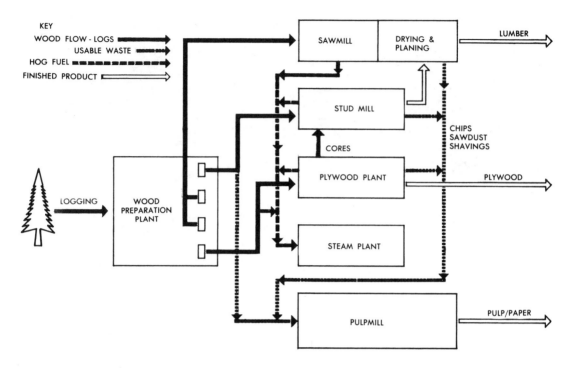

KEY
WOOD FLOW - LOGS
USABLE WASTE
HOG FUEL
FINISHED PRODUCT

SAWMILL   DRYING & PLANING   LUMBER

STUD MILL

CHIPS
SAWDUST
SHAVINGS

CORES

PLYWOOD PLANT   PLYWOOD

LOGGING   WOOD PREPARATION PLANT

STEAM PLANT

PULPMILL   PULP/PAPER

FIGURE 3-17. Wood flow for integrated forest products manufacture.

Integrated Operations

As noted earlier, the integration of forest products manufacturing operations ensures optimum economic utilization of available timber resources. It is generally wasteful and less productive to use prime sawlogs for pulp manufacture when poorer-quality pulplogs and wood residuals can be effectively utilized as pulping raw materials. Likewise, value is added to pulplogs if it is possible to extract lumber from the logs before they are chipped. Effective development of a multi-product manufacturing facility requires that log harvesting, delivery, sorting and breakdown be integrated with the subsequent conversion operations.

Obviously, the optimum economic impact is achieved when wood handling and associated plants are integrated at a single location. However, similar advantages of integration can be realized even when conversion plants are at different sites. The wood flow concept for a large integrated complex is shown in Figure 3-17. Coordination of administration and services (e.g., steam, power, effluent treatment, etc.) can provide further economies.

3.3 PULPWOOD MEASUREMENT

Aside from quality considerations (which are often not given proper weighting), the economic basis for pulpwood transfers is the weight of the dry wood. Since it is not possible to measure dry weight

directly, several methods of estimation have been devised based on measurements of either wood volume or wet weight to which a factor is applied of dry density or dry content, respectively. Unfortunately, the moisture content of logs can vary over a wide range depending on such factors as wood species, age, locality, and methods of transport and storage. It should be noted that a fluctuation between 33% and 50% moisture is equivalent to a range of water content from 0.5 to 1.0 kg per kg of dry wood. As well, the density of dry wood varies significantly, even within the same species.

At one time, the prevailing practice was to measure only the gross volume occupied by a pile of logs. The common mensuration unit was the cord, defined as the amount of wood contained within a pile of 4-foot logs stacked 8 feet across and 4 feet deep, equivalent to 128 cubic feet of total volume. The cord is an unreliable measurement of actual wood volume because the air space within the contained volume varies widely with log diameter and method of stacking. Variation in bark thickness can also be important.

Today, it is more common to utilize solid volume measurements such as the cunit (100 cubic feet of solid wood) for pulplogs. Typically, standard tables are used to estimate the solid volume of a stack of logs after the number and size of the boles are determined by scaling. Some mills utilize a water displacement method to measure wood volume directly.

Probably the most popular method of pulpwood measurement is by green weight. When pulpwood is delivered by truck, it is a simple matter to weigh the truck before and after unloading and calculate the weight by difference. This method applies equally well to pulplogs and chips; in both cases an accurate estimate of wood moisture content is required to calculate the dry weight of wood.

Chips are also bought and sold on the basis of bulk volume, the common measurement being the unit (200 cubic feet). For most wood species, a unit of chips closely approximates one short ton of oven-dry wood.

Approximate equivalent values for the various volumetric wood measurements are as follows:

1 cord	=	0.83 cunit	=	1.04 unit
1 cunit	=	1.25 unit	=	1.2 cord
1 unit	=	0.96 cord	=	0.8 cunit

3.4 WOOD PREPARATION

Wood preparation consists of a series of operations which convert pulplogs into a form suitable for the subsequent pulping operations. The ultimate product is usually wood chips. But if the groundwood process is employed for pulping, it is also necessary to reduce certain pulplogs into short blocks of wood. Some integrated woodrooms also extract lumber from the better quality logs prior to the chipping operation.

Wood is unloaded from truck or railcar by a front-end loader or by an overhead travelling crane (Figures 3-18 and 3-19). For water-driven wood, an overhead crane is usually employed. Grapples can be designed for short or long logs, as well as full-tree lengths. Intermediate storage and final sorting of logs can either be on land or in the water.

A conventional woodroom flow sequence for tree-length logs is shown schematically in Figure 3-20. The process involves transportation to the slasher deck where the logs are sorted and cut into manageable lengths, followed by debarking, chipping, chip screening and conveyance to storage. Chips are usually stored in outside piles.

The loading deck is the first element of a log-handling operation. The residence time here depends on the storage capability in the woodyard and the frequency of loading. From the loading deck, the logs are aligned and they may then be subjected to sorting and/or slashing operations (see Figure 3-21) before being conveyed to debarking. Where long logs are handled (especially entire-tree lengths), a slasher is required after the loading deck.

Some mills incorporate a deicing and/or washing stage in the log handling sequence. The logs are fed from the loading deck onto the deicing deck where they are automatically stacked up to a height of one meter or more. They are then slowly conveyed

FIGURE 3-18. Unloading long logs in the woodyard (photo by Finning).

FIGURE 3-19. Unloading shortwood logs in the woodyard.

through the deicing chamber and subjected to hot-water showers in wintertime and/or cold-water showers in summer. This operation facilitates subsequent dry debarking and also removes sand, soil and debris from the logs, thus reducing wear to the barkers, slashers and transporting systems.

Debarking

Log debarking is necessary to ensure that the woodroom chips are free of bark and dirt. Several types of mechanical debarkers are used as well as hydraulic debarkers.

The drumbarker or barking drum (Figures 3-22 and 3-23) is the most common type of mechanical debarker and is available in a number of design variations. Both short-wood tumble drums and long-wood parallel drums are used, and either may be operated wet or dry. In all designs, bark is removed from the logs by friction created from the rotating drum action as the logs rub against each other. Iron or rubber lifters inside the drum shell agitate the logs

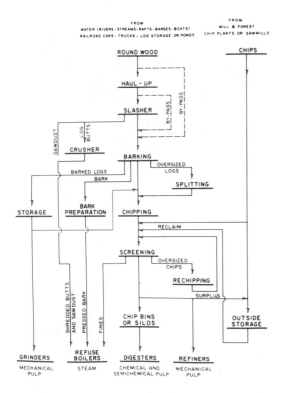

FIGURE 3-20. Flow chart of woodroom operations.

FIGURE 3-22. This 14.5 ft dia x 80 ft long barking drum handles random log lengths up to 30 ft (Fibre Making Processes, Inc.).

FIGURE 3-23. An adjustable gate at the discharge end ensures that a sizable pile of logs is maintained in the barking drum (Fibre Making Processes, Inc.).

FIGURE 3-21. Logs are cut into prescribed lengths on the slasher deck.

and help protect the drum shell from wear. In wet drumbarkers, water is added to the early solid steel portion of the drum to help loosen the bark; the remaining portion of the drum has slots to permit the removed bark to fall out while the log continues on through. In dry drumbarkers, the entire length of the drum has slots for bark removal. Dry drumbarkers are longer in length and rotate much faster than wet drumbarkers; their major advantage is that the removed bark can be fired directly into bark-burning

furnaces. Bark from a wet system must be collected in a water flume, dewatered and pressed before burning; and treatment of the resulting wastewater is difficult and costly. Drumbarkers usually create about 4 to 5% wood waste and also cause "brooming" at the ends of logs which adversely affects the quality of the resulting chips. In spite of these obvious problems and limitations, the relatively low cost and versatility of the drumbarker make it an important piece of equipment in most woodrooms or woodyards.

Ring barkers or cambium shear barkers (Figure 3-24) utilize a rotating ring carrying several arms with scraping tips that apply radial and tangential pressure against the log as it passes through the ring. This design takes advantage of the natural weakness in bonding strength between bark and wood at the cambium layer. Feed and rotor speeds can be varied to accommodate a wide range of log diameters and bark adhesion characteristics. This device does not perform well on crooked or frozen logs.

Rosser head barkers (Figure 3-25) employ a rapidly-rotating head with many cutting tools that cut and abrade the bark from the log. The log rotates during barking and moves longitudinally past the Rosser head. This device is effective for frozen logs and species with strong bark adhesion. However, wood losses are high.

Hydraulic jet barkers (Figure 3-26) are most common in the Pacific Northwest where they are used for large-diameter logs. They operate by directing high-pressure (over 1000 psig) jets of water against the log to remove the bark. Bark removal is efficient and wood losses are usually under 2%, but capital costs and energy requirements are high. Because the effluent from hydraulic barkers is difficult to clean up, many mills have switched to mechanical designs.

Chipping

After debarking (and possibly lumber extraction), the logs (or portions of logs) are reduced to chips suitable for the subsequent pulping operations. Several designs of chipper are used, the most common being the flywheel-type disc with a series of blades mounted radially along the face and projecting about 20 mm (see Figure 3-27). The logs are usually fed through a sloping spout to one side of the rotating disc so that the knives strike at an angle of 35 to 40° from the axis of the log. The logs can also be fed horizontally to a disc mounted at the proper angle. Generally, the horizontal feed provides better

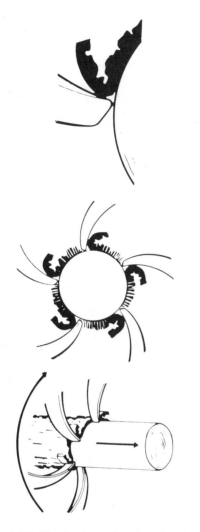

FIGURE 3-24. Illustration of tool action in ring debarker (Koch).

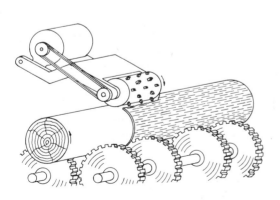

FIGURE 3-25. Rosser head barker configuration (J.R. Erickson).

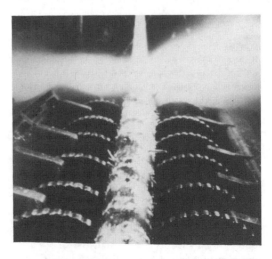

FIGURE 3-26. Hydraulic barker.

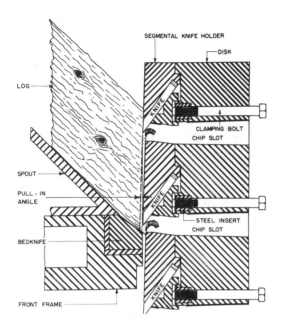

FIGURE 3-27. Section through chipper (Carthage Machine Co.).

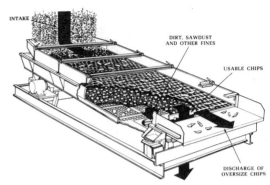

FIGURE 3-28. Slotted chip screen utilizing a rotary shaking motion (Orville Simpson Co.).

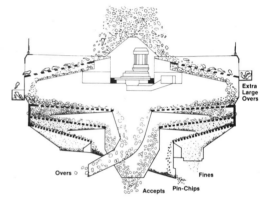

FIGURE 3-29. Four-stage chip screen.

control, but is less suitable for scrap wood pieces.

Assuming a well-designed chipper, then proper speed control and maintenance of knife sharpness will ensure satisfactory results. The ideal chip is usually considered to be about 20 mm long in the grain direction and 4 to 5 mm thick, but all chips 10 to 30 mm long and 3 to 6 mm thick are prime materials for pulping. Off-size chips adversely affect the pulping process and the quality of the resultant pulp.

Chip Screening

Chips of acceptable size must be isolated from fines and oversized pieces by passing the chips over a series of screens. The oversized chips are rejected to a conveyor which carries them to some type of comminution device for reduction into smaller fractions. The fines are usually burned along with the bark unless special pulping facilities are provided.

Older-style screens (Figures 3-28 and 3-29) segregate chips only on the basis of chip length. In recent years, chip thickness has come to be recognized as an important pulping variable, and modern disc-type or roll-type screens which segregate according to thickness are now widely accepted as the industry standard (Figure 3-30). Also, modern-design chip slicers (Figure 3-31) that concentrate on reducing chip thickness are far less damaging to wood fibers than the older-style "rechippers" which are basically hammermills.

Bark Disposal

A few years ago during the era of cheap fossil fuel, much of the bark removed from logs was considered a nuisance and a handling problem. Acceptable disposal methods included incineration (with and without steam generation), landfill and composting. Small amounts were used for horticultural mulching and as filler in building board.

Because of the high cost of fossil fuels, bark and other wood waste materials are now recognized as valuable fuel sources, and considerable attention is given to their efficient utilization in meeting the mill's energy needs. The thermal efficiency of bark-burning boilers is a direct function of the bark's moisture content. Therefore, dry debarking methods have an inherent advantage, whereas wet debarking requires mechanical pressing to remove excess water. Efforts to improve the fuel values are focused on more efficient wet pressing and on methods to economically dry the bark utilizing waste heat sources. (Also refer to Section 25.3.)

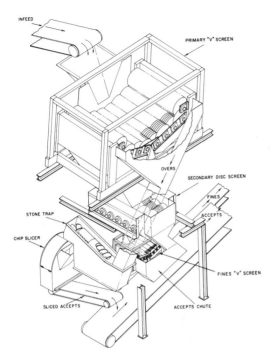

FIGURE 3-30. Chip thickness screening/ slicing system (Rader).

FIGURE 3-32. Chip truck being unloaded.

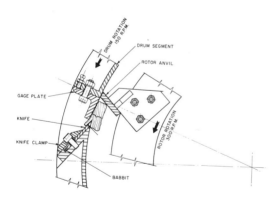

FIGURE 3-31. Chip slicer operation showing over-sized chip being split parallel to the grain (Rader).

3.5 CHIP HANDLING AND STORAGE

Within mill areas, chips are commonly transported either pneumatically within pipes or on conveyor belts. Chips are readily handled by an airveying system over distances of 300–400 meters, but power consumption is high and significant chip damage can occur. By contrast, a belt conveyor system has a much higher initial cost. Other systems such as chain and screw conveyors are also used to move chips, but usually for relatively short distances. Bucket elevators are used for vertical movement.

Residual chips from independent sawmills and other non-integrated forest products plants are commonly transported to the pulp mill by truck or rail. Barges are also important in coastal and large-river areas. Air conveying is the most efficient method of loading chips into transport carriers because of better compaction and elimination of void areas, albeit at the cost of some chip damage. Pulp mills that routinely receive chip deliveries usually are equipped with special facilities for emptying trucks and railcars into hopper bins (Figure 3-32), and also have an efficient system to transfer the chips from the bin to storage. Air vacuum systems in conjunction with tractor vehicles, as well as front-end loaders, are also utilized in some mills for chip unloading (Figure 3-33). Unloading of barges is generally performed by cranes equipped with grapples or clamshell buckets.

Chip Storage

Generally, most of the fiber raw material inventory for a pulp mill is in the form of chips because they are more economical to handle than logs (e.g., Figure 3-34). Outside chip storage has been practiced since the 1950's because large inventories can be

FIGURE 3-33. Chips being unloaded from a railcar with a front-end loader (photo by Finning).

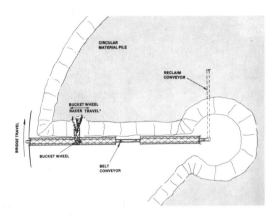

FIGURE 3-35. Ring-shaped pile assures a systematic turnover of chips (Rader).

FIGURE 3-34. Part of a woodyard showing two open-type chip storage and reclaim systems. A closed bark bin is shown in the foreground (Atlas Systems).

FIGURE 3-36. Reclaiming chips with a bucket loader (photo by Finning).

maintained without the use of bins or silos. Some disadvantages of outside storage were apparent from the beginning, for example the blowing of fines from the pile and air-borne contamination of the surface chips. However, it was not until the late 1960's that wood losses associated with outside storage were properly defined.

It is now recognized that losses of 1% wood substance per month are typical of outside storage due to a combination of respiration, chemical reactions and micro-organism activity. Considering the total world inventory of chips, the annual loss of fiber raw material is quite staggering. Considerable research has already been carried out to find a suitable chip preservative treatment, but thus far a totally effective, economical and environmentally safe method has not been identified.

In the meantime, wood losses and degradation

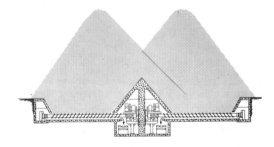

FIGURE 3-37. Rotating double-screw chip reclaim system provides a constant flow of chips, even at low storage levels (Beloit-Wennberg).

during outside chip storage can be minimized by effective chip pile management (1). (Since there are many causes of deterioration, it is perhaps not surprising to find that opinions on storage practices are sometimes contradictory.) It makes good sense to

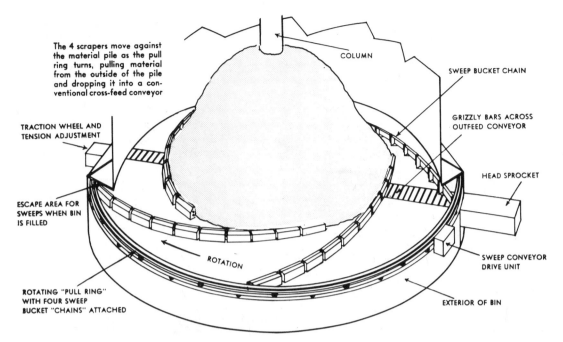

The 4 scrapers move against the material pile as the pull ring turns, pulling material from the outside of the pile and dropping it into a conventional cross-feed conveyor

COLUMN

SWEEP BUCKET CHAIN

GRIZZLY BARS ACROSS OUTFEED CONVEYOR

TRACTION WHEEL AND TENSION ADJUSTMENT

HEAD SPROCKET

ESCAPE AREA FOR SWEEPS WHEN BIN IS FILLED

ROTATION

SWEEP CONVEYOR DRIVE UNIT

ROTATING "PULL RING" WITH FOUR SWEEP BUCKET "CHAINS" ATTACHED

EXTERIOR OF BIN

FIGURE 3-38. Rotating scraper arrangement for chip reclaim is designed to provide constant feed to digester regardless of chip level (Atlas Systems).

provide a ground barrier of concrete or asphalt before building a chip pile to reduce dirt contamination and inhibit the mobility of ground organisms. Chips should be stored on a first-in/first-out basis to avoid infection of fresh chips by old chips. Storage configurations such as the ring-shaped pile (Figure 3-35) facilitate the complete separation of "old" and "new" chips. In this arrangement, the two open faces form the beginning and end of the pile, the chips being fed to one face and removed from the other face. Other arrangements where new chips are added at the top of the pile and old chips are removed from the bottom are also effective. Wind-blown concentrations of fines should be avoided because they hinder dissipation of the heat that builds up in the pile from various causes. Thermal degradation and even spontaneous combustion can result from localized heat buildup.

Optimum chip handling depends partly on pulping requirements. For example, wood extractives are a problem in sulfite pulping because of the formation of resin particles in the pulp. Since loss of extractives from chips is particularly rapid during the first two months of storage, all chips for sulfite pulping should go to storage. However, if maximum recovery of extractives-based byproducts such as tall oil and turpentine is important (as for some kraft pulping operations), then fresh chips should bypass storage wherever possible in order to maximize yield.

Chip Reclaiming

A number of methods are used to reclaim chips from storage. Older systems employ a belt or chain conveyor along the side of the pile fed by a bulldozer that pushes chips down the side of the pile. This type of arrangement is labor-intensive, usually necessitating a full-time bulldozer operator, and inevitably causes damage to the chips. Modern installations work automatically, some employing augers or chain conveyors on rotating platforms at the base of the pile (refer to Figures 3-36 to 3-38).

Chip Surge Bins

Generally, chips are not fed directly into the pulpmill process from outside storage, but are first transferred to intermediate bins or silos to facilitate metering and blending of the chip furnish. Bins of this type are essential where it is desired to blend two or three different species chips together continuously and accurately.

Modern chip silos are usually vertical cylindrical structures having conical bottom sections terminating over circular rotating table feeders. Since wood chips have notoriously poor flow characteristics, most bins are equipped with some type of "live bottom" to prevent hangups. Occasional flow problems can be anticipated even from the best-designed systems, especially if the chips have been sitting in the silo for several days.

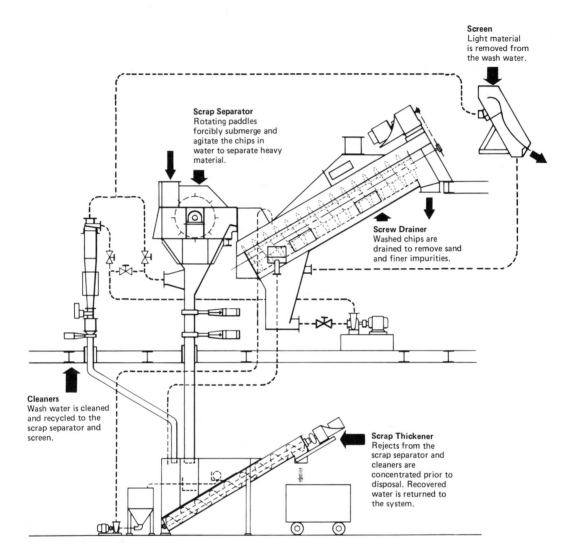

Screen
Light material
is removed from
the wash water.

Scrap Separator
Rotating paddles
forcibly submerge and
agitate the chips in
water to separate heavy
material.

Screw Drainer
Washed chips are
drained to remove sand
and finer impurities.

Cleaners
Wash water is cleaned
and recycled to the
scrap separator and
screen.

Scrap Thickener
Rejects from the
scrap separator and
cleaners are
concentrated prior to
disposal. Recovered
water is returned to
the system.

FIGURE 3-39. Defibrator chip washing system.

3.6 CHIP QUALITY CONTROL

With respect to a given wood source, the quality of chips is measured by uniformity of size (grain length and thickness) and by the relative absence of "contaminants". All chips 10 to 30 mm long and 3 to 6 mm thick are generally considered to be of good quality. Contaminants are considered to be:

• oversized chips (either length or thickness)
• pin chips (of normal length, but less than 3 mm in width and thickness)
• fines (wood particles having very short or fragmented fibers, usually taken as the fraction passing through 3 mm diameter holes)
• bark and burnt wood
• rotten wood
• dirt and foreign materials.

Oversize chips constitute a handling problem and are the main cause of screen rejects in chemical pulping. Pin chips, fines and rotten wood cause lower yields and strengths in the resultant pulps and contribute to circulation problems in chemical pulping systems. Pin chips contain good-quality fiber, but should be pulped separately if possible.

Bark and burnt wood are sources of dirt, especially in mechanical and sulfite pulp mills. The kraft process is more tolerant of bark because many bark particles are soluble in the alkaline liquor.

Utilization of whole-tree chips has stimulated development of bark removal techniques for chips, and removal efficiencies of 70% have been achieved. Methods investigated have included air separation, liquid flotation and compression debarking, along with other treatments that complement these methods. The Paprifer process is well documented (2); it entails steaming of the chips to loosen contaminants, vigorous agitation in a series of pulpers, and finally, washing to separate contaminants.

It is important that chips used in refiner mechanical pulping systems be relatively free of sand, grit and other foreign material that could accelerate refiner plate wear. As a precautionary measure, most refiner pulping systems incorporate a chip washing stage. The Defibrator system illustrated in Figure 3-39 utilizes rotating paddles to forcibly submerge and agitate the chips in water to separate heavy material; then the chips are conveyed through a screw drainer to remove sand and finer particles.

REFERENCES

(1) FULLER, W.S. **Chip Pile Storage - a Review of Practices to Avoid Deterioration and Economic Loss** *TAPPI Journal* 68:8:48–52 (August 1985)

(2) BERLYN, R.W. and SIMPSON, R.B. **Upgrading Wood Chips: the Paprifer Process** *TAPPI Journal* (March, 1988)

Chapter 4

Overview of Pulping Methodology

Pulping refers to any process by which wood (or other fibrous raw material) is reduced to a fibrous mass. Basically, it is the means by which the bonds are systematically ruptured within the wood structure. The task can be accomplished mechanically, thermally, chemically, or by combinations of these treatments. Existing commercial processes are broadly classified as mechanical, chemical or semichemical. The general characteristics of these processes are summarized in Table 4-1.

Current annual pulp production by each major process is shown in Table 4-2. Chemical pulping accounts for 70% of North American production, of which 95% is produced by the dominant kraft process. The tabulation shows that semichemical pulping accounts for only about 5% of production; however, "high-yield kraft" and "high-yield sulfite" tonnage now included in the totals for unbleached kraft and sulfite should more appropriately be included under semichemical.

A tally of North American forest-based pulp mills situated in the principal pulp-producing states and provinces (Table 4-3) helps to provide a geographical perspective. It is of interest to note that most of the sulfite mills are located in traditional pulp-producing areas (e.g., Washington, Wisconsin, Ontario, Quebec), while more-recently exploited areas (e.g., Alabama, Georgia, Louisiana, British Columbia) are relying principally on the kraft process for chemical pulping.

TABLE 4-2. North American pulp production 1990 (000 000 short tons). Sources: API & CPPA.

	USA	Canada
Dissolving	**1,293**	**221**
Chemical pulp, paper grades		
Total sulfite	**1,561**	**1,603**
Bleached hardwood kraft		1,764
Bleached softwood kraft		7,006
Total bleached kraft	27,562	8,770
Unbleached kraft	22,188	1,503
Total kraft	**49,750**	**10,273**
Semimechanical	**4,219**	**514**
Mechanical	**6,452**	**11,637**
Total	**63,275**	**25,234**

TABLE 4-1. General classification of pulping processes.

Mechanical	Hybrid	Chemical
Pulping by mechanical energy (small amount of chemicals and heat)	Pulping with combinations of chemical and mechanical treatments	Pulping with chemicals and heat (little or no mechanical energy)
High yield* (85-95%)	Intermediate yield (55-85%)	Low yield (40-55%)
Short, impure fibers • weak • unstable	"Intermediate" pulp properties (some unique properties)	Long, strong fibers • strong • stable
Good print quality		Poor print quality
Examples: • stone groundwood • refiner mechanical pulp • thermomechanical pulp	Examples: • neutral sulfite semichemical • high-yield kraft • high-yield sulfite	Examples: • kraft • sulfite • soda

$$*\text{Yield} = \frac{\text{wt. of pulp produced (o.d.)}}{\text{wt. of original wood (o.d.)}}$$

TABLE 4-3. Geographical distribution of pulp mills, 1990 (Lockwood's Directory).

	Kraft	Sulfite	Semi-chem	Mech-anical	Total
United States					
Alabama	14	-	2	4	20
Georgia	12	-	1	2	15
Louisiana	10	-	4	3	17
Maine	7	1	-	10	18
Oregon	7	-	3	11	17
Washington	7	6	3	7	23
Wisconsin	4	6	1	11	22
Canada					
Brit. Columbia	18	1	-	15	34
Ontario	9	4	2	15	30
Quebec	10	8	2	41	61

FIGURE 4-1. Spruce stone groundwood.

A complete listing of pulp mills would also include the category, "secondary fiber". Under this heading are the operations that recover usable pulp fiber from various waste paper sources. This fiber is ultimately recycled into a wide range of paper and paperboard products where strength and quality are adequate to replace more expensive "virgin pulps". The technology for secondary fiber recovery is described in Chapter 14.

4.1 INTRODUCTION TO VARIOUS PULPING METHODS

Mechanical Pulping

The oldest and still a major method of mechanical pulping is the groundwood process, where a block (or bolt) of wood is pressed lengthwise against a wetted, roughened grinding stone revolving at peripheral speeds of 1000-1200 m/min. Fibers are removed from the wood, abraded and washed away from the stone surface with water. The dilute slurry of fibers and fiber fragments (Figure 4-1) is screened to remove slivers and oversize particles, and is subsequently thickened (by removal of water) to form a pulp stock suitable for papermaking. The process is simple in principle, but the efficient production of uniform, good-quality pulp requires careful control of stone surface roughness, pressure against the stone, and shower water temperature and flow rate.

A more recent development in mechanical pulping involves shredding and defibering chips of wood between the rotating discs of a device called a refiner; the product is known as refiner mechanical pulp (RMP). RMP typically retains more long fibers than stone groundwood and yields stronger paper.

The basic RMP process has undergone extensive development in the past two decades. Most new installations now employ thermal and/or chemical presoftening of the chips to modify both the energy requirement and the resultant fiber properties. For example, when the chips are given a pressurized steam pretreatment, the resultant product, called thermomechanical pulp (TMP), is significantly stronger than RMP and contains very little screen reject material.

Mechanical pulping processes have the advantage of converting up to 95% of the dry weight of the wood into pulp, but require prodigious amounts of energy to accomplish this objective. The pulp forms a highly opaque paper with good printing properties, but the sheet is weak and discolors easily on exposure to light. To achieve adequate sheet strength, it is often necessary to add long-fibered chemical pulp to the mechanical pulp. Newsprint traditionally was made up of about 75% groundwood and 25% chemical pulp; now some newsprint is made from 100% TMP.

Mechanical pulps are commonly produced from softwood species. The smaller, thinner hardwood fibers are more severely damaged during conventional mechanical pulping and yield a finer, more flour-like pulp (Figure 4-2) that forms an exceedingly weak sheet. In spite of the obvious strength deficiencies, some exceptionally bright hardwood mechanical pulps (e.g., aspen and poplar) have sometimes been blended with softwood pulps to improve optical properties. Now, the recent development and application of the chemi-thermomechanical pulping process has finally enabled the industry to exploit certain hardwood species such as aspen and eucalypt for the production of relatively strong, short-fibered pulps that are suitable for blending into a variety of papermaking furnishes.

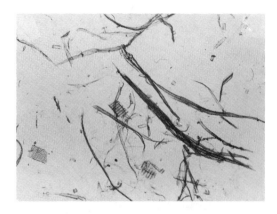

FIGURE 4-2. Poplar (cottonwood) stone groundwood.

Chemical Pulping

In chemical pulping, the wood chips are cooked with appropriate chemicals in an aqueous solution at elevated temperature and pressure. The objective is to degrade and dissolve away the lignin and leave behind most of the cellulose and hemicelluloses in the form of intact fibers (Figures 4-3 and 4-4). In practice, chemical pulping methods are successful in removing most of the lignin; they also degrade and dissolve a certain amount of the hemicelluloses and cellulose so that the yield of pulp is low relative to mechanical pulping methods, usually between 40% and 50% of the original wood substance.

The two principal chemical pulping methods are the (alkaline) kraft process and the (acidic) sulfite process. The kraft process has come to occupy the dominant position because of advantages in chemical recovery and pulp strength. The sulfite process, which was more common up to the late 1940's, appears to be in an irreversible decline. No new sulfite mills have been built in North America since the 1960's. Nonetheless, sulfite pulping still has its

adherents, and process modifications have been proposed which would appear to make the process more competitive. Table 4-4 provides a summary of chemical and semichemical pulping methodology. Each of these methods is considered briefly in this chapter. Sulfite and kraft pulping are discussed at greater length in chapters 6 and 7, respectively.

Kraft Process

The kraft process involves cooking the wood chips in a solution of sodium hydroxide (NaOH) and sodium sulfide (Na_2S). The alkaline attack causes fragmentation of the lignin molecules into smaller segments whose sodium salts are soluble in the cooking liquor. "Kraft" is the German word for strong, and kraft pulps produce strong paper products; but the unbleached pulp is characterized by a dark brown color. The kraft process is associated with malodorous gases, principally organic sulfides, which cause environmental concern.

The kraft process evolved over 100 years ago as a modification of the soda process (which utilizes only sodium hydroxide as the active chemical) when Carl S. Dahl introduced sodium sulfate into the cooking system. The subsequent conversion of sulfate to sulfide in the cooking liquor produced a dramatic improvement in reaction kinetics and pulp properties when cooking softwoods. Because sodium sulfate has been the traditional makeup chemical, the kraft process is sometimes referred to as the "sulfate process". Hardwood pulping is less affected by the presence of sodium sulfide in the alkaline cooking liquor, and some hardwood pulp is still produced by the soda process.

A number of different kraft pulp grades are produced. Unbleached grades for packaging applications are cooked to a higher yield and contain more lignin than pulps that are subsequently bleached and made into white papers.

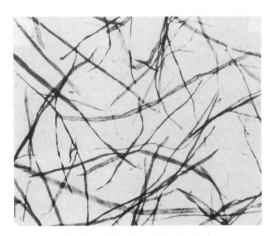

FIGURE 4-3. Softwood kraft pulp (red pine).

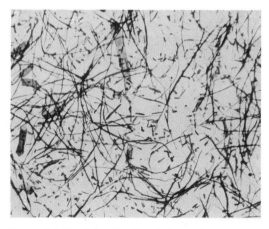

FIGURE 4-4. Hardwood kraft pulp (aspen).

TABLE 4-4. Major chemical and semichemical pulping methods.

	Kraft	Acid Sulfite	Bisulfite	NSSC
Chemicals:	NaOH Na$_2$S	H$_2$SO$_3$ M(HSO$_3$) (M = Ca, Mg, Na, NH$_4$)	M(HSO$_3$) (M = Mg, Na, NH$_4$)	Na$_2$SO$_3$ Na$_2$CO$_3$
Cooking Time:	2-4 h	4–20 h	2–4 h	1/4–1 h
Liquor pH:	13+	1–2	3–5	7–9
Cooking Temp:	170-180°C	120-135°C	140-160°C	160-180°C

Sulfite Process

In the sulfite process, a mixture of sulfurous acid (H$_2$SO$_3$) and bisulfite ion (HSO$_3^-$) is used to attack and solubilize the lignin. The sulfites combine with the lignin to form salts of lignosulfonic acid which are soluble in the cooking liquor, and the chemical structure of the lignin is left largely intact. The chemical base for the bisulfite can be ionic calcium, magnesium, sodium or ammonium. Sulfite pulping can be carried out over a wide range of pH. "Acid sulfite" denotes pulping with an excess of free sulfurous acid (pH 1–2), while "bisulfite" cooks are carried out under less acidic conditions (pH 3–5).

Sulfite pulps are lighter in color than kraft pulps and can be bleached more easily, but the paper sheets are weaker than equivalent kraft sheets. The sulfite process works well for such softwoods as spruce, fir and hemlock, and for such hardwoods as poplar and eucalyptus; but resinous softwoods and tannin-containing hardwoods are more difficult to handle. This sensitivity to wood species, along with the weaker strength and the greater difficulty in chemical recovery, are the major reasons for the decline of sulfite pulping relative to kraft. The trend toward whole-tree chipping puts sulfite at a further disadvantage because of its intolerance to bark. The relative advantages of the kraft and sulfite processes

are summarized in Table 4-5. A comparison of kraft, sulfite and bisulfite pulp strengths is shown graphically in Figure 4-5.

Several grades of sulfite pulp are produced, depending on end use. Higher-yield grades are usually prepared using a cooking liquor low in free sulfurous acid, i.e., with bisulfite ion as the predominant active chemical.

TABLE 4-5. Relative advantages of the two major chemical pulping processes (kraft vs. sulfite).

Advantages of Kraft Process
- produces highest strength pulp.
- utilizes proven technology for efficient chemical recovery.
- handles wide variety of wood species.
- tolerates bark in the pulping process.

Advantages of Sulfite Process
- produces brighter unbleached pulp.
- pulp is easier to bleach to full brightness.
- produces higher yield of bleached pulp.
- pulp is easier to refine.

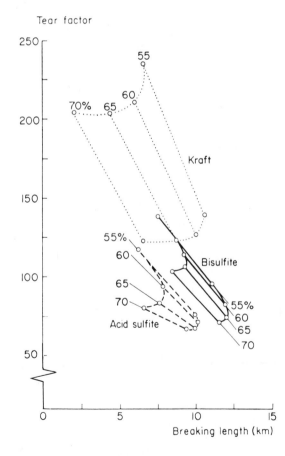

FIGURE 4-5. Comparison of tear and tensile strengths for unbeaten pulps and pulps beaten to approximately 250 CSF (data of Hartler, et al).

Semichemical Pulping

Semichemical pulping combines chemical and mechanical methods. Essentially, the wood chips are partially softened or digested with chemicals; the remainder of the pulping action is then supplied mechanically, most often in disc refiners. A typical process flow sequence is shown in Figure 4-6. Semichemical methods encompass the intermediate range of pulp yields between mechanical and chemical pulping, i.e., 55 to 85% on dry wood. The pulps have a number of end uses and some unique properties. As the prime example, pulps at about 75% yield exhibit exceptional stiffness, making them ideally suitable for the center fluted layer in corrugated container board.

Strictly speaking, any mechanical pulping process that is modified by incorporating a chemical treatment of the chips prior to or during refining should qualify as semichemical pulping. In practice, if the chemical treatment is relatively modest, the pulp is still considered to be a mechanical pulp. The general rule of thumb is that all pulps with a yield of 85% or higher are nominally considered to be mechanical pulps. As a result of this rather arbitrary classification, some categories of modern pulping processes actually overlap between mechanical and semichemical pulping. For example, so-called chemimechanical pulps and very-high-yield sulfite pulps both cover the yield range from 80% to 92%.

Included under the classification of semichemical pulping are the high-yield kraft and high-yield sulfite processes. In both instances the cooking is limited to partial delignification; the actual defibering is done mechanically. The degree of cooking controls the yield; as yield is increased, a greater amount of energy is required for defiberization.

The neutral sulfite process, applied mainly to hardwood chips, is the most widely used semichemical process. Usually abbreviated to NSSC (for Neutral Sulfite SemiChemical), this process utilizes sodium sulfite cooking liquor which is buffered with sodium carbonate (soda ash) to neutralize the organic acids liberated from the wood during cooking.

To get away from sulfur in the process, some semichemical cooking is now carried out with mixtures of sodium carbonate and sodium hydroxide (1). Other mills, especially where associated with a kraft mill, are using green liquor (sodium carbonate plus sodium sulfide) for cooking. A 1978 survey of North American corrugating medium mills (2) showed the following breakdown:

Semichemical Process	Number of Mills
NSSC	21
Green Liquor	8
"No-Sulfur"	10
Other	3

Spent liquors from semichemical pulping operations are relatively concentrated sources of chemical and organic matter, and consequently, must be subjected to some type of chemical recovery operation. Such processing can be readily accomplished in cross-recovery with an existing kraft or sulfite mill system. Where cross-recovery is not possible, fluidized bed incineration is the most common method used for combustion of organics and recovery of chemical.

A good comparison of spruce semichemical pulps produced by five different methods and covering the entire yield range is provided in Table 4-6. The approximate behavior of selected strength parameters over the full yield range for spruce kraft pulp is illustrated in Figure 4-7.

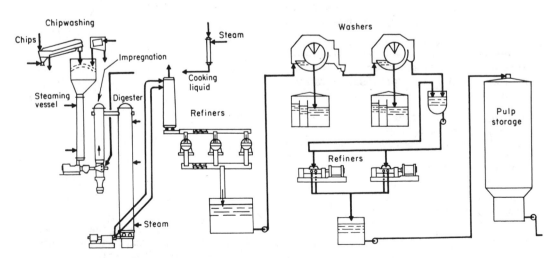

FIGURE 4-6. Flowsheet for semichemical pulping mill utilizing continuous digestion.

TABLE 4-6. Effect of pulp yield on the properties of semichemical spruce pulps produced by different methods. The comparisons were made at 500 CSF.

Type of pulp	Unbleached yield %	Lignin, % of pulp	Bright-ness %	Revolutions in PFI* (thousands)	Bulk cc/g	Burst factor	Break length km	Stretch (%)	Tear factor	Double folds
	87.3	27.6	61.2	13.1	2.02	25	5.0	1.8	52	20
	81.8	27.8	56.1	15.1	1.78	60	8.8	2.6	64	300
	76.6	26.1	55.2	11.6	1.50	75	10.4	2.9	63	580
Bisulphite	74.0	25.2	54.2	7.2	1.48	81	11.4	3.0	64	660
	65.9	23.4	55.1	7.6	1.41	93	12.4	3.1	70	1120
	57.2	19.3	61.6	5.2	1.39	91	11.5	2.8	69	1550
	55.5	12.7	64.3	2.6	1.32	94	12.3	3.2	76	1440
	89.0	28.7	60.4	6.4	2.73	17	4.0	1.8	70	0
	83.2	30.5	57.2	5.7	1.96	36	6.5	2.4	67	160
	76.5	30.0	53.6	5.1	1.78	47	7.6	2.6	74	200
Sulphite	72.8	28.9	52.5	5.7	1.62	58	8.8	2.7	67	340
	64.6	23.5	53.7	1.1	1.59	65	9.7	2.8	78	540
	59.7	—	60.3	1.6	1.50	69	10.2	3.0	76	900
	51.0	7.2	64.5	2.4	1.34	79	10.8	3.4	80	1240
	86.8	28.1	52.6	22.8	1.70	53	8.2	2.5	60	280
	79.2	—	51.8	16.6	1.61	65	9.8	2.7	61	450
Neutral sulphite	72.1	24.8	51.5	16.4	1.54	81	11.0	2.9	75	680
	69.7	23.4	50.3	17.9	1.53	92	12.5	3.0	68	880
	61.4	20.9	51.0	16.0	1.47	103	13.1	3.0	72	700
	61.7	18.7	51.2	11.9	1.42	104	14.1	3.1	74	1330
	52.3	12.8	55.5	5.40	1.47	113	14.6	3.1	76	1370
	93.4	29.2	29.7	9.2	2.43	4	1.6	0.8	23	0
	88.5	29.7	27.2	15.6	2.10	7	2.3	1.0	30	1
	76.8	30.0	18.9	19.3	1.60	26	5.3	2.2	69	45
Kraft	71.3	27.0	17.2	26.0	1.57	59	8.1	3.2	84	500
	63.8	20.9	17.8	25.0	1.43	87	10.1	3.7	96	1440
	55.1	14.4	18.6	16.1	1.40	103	12.4	3.9	100	2550
	44.7	4.3	28.7	5.4	1.30	112	12.9	3.8	117	2875
	90.2	29.6	26.6	15.0	—	—	—	—	—	—
	79.0	30.9	23.0	24.0	1.87	16	3.7	1.5	65	12
Soda	74.8	29.5	26.7	20.4	1.70	30	5.2	2.0	78	60
	60.7	22.9	17.9	17.0	1.40	66	9.65	3.2	93	680
	57.8	19.8	20.2	16.8	1.48	82	9.85	3.1	99	1240
	48.2	13.0	23.3	11.0	1.40	90	11.2	3.4	111	1840

From Liebergott & Joachimedes. Pulp & Paper Canada, 80:12:T391-395.
* Relative measurement of mechanical energy.

4.2 MARKET PULPS

Papermills can be broadly classified as either integrated (i.e., with their own pulp supply) or non-integrated. These distinctions are not always clearcut, and some qualification may be required in a given instance. For example, certain integrated mills must purchase a portion of their pulp requirement in the open market. As well, there are non-integrated mills that actually have a captive pulp supply, often from another mill within the same company or from an affiliate.

In order to appreciate the role of market pulps, it is useful to take a broad overview of the world situation as illustrated in Figure 4-8. The figures shown are order-of-magnitude; the situation is blurred because the tonnage contribution of mineral loadings and coatings in overall paper and paperboard production is not known accurately.

Also, note that a small percentage of the secondary fiber tonnage should be included in the total for market pulps.

With respect to overall paper and paperboard production, market pulps make up a relatively small portion of the world industry's fiber requirement, about 12%. Much of the market pulp is in the form of high-quality bleached softwood kraft, a vital component in many papermill furnishes. Nonetheless, because most paper production capacity is independent of outside fiber supply, demand for market pulp is especially sensitive to the cyclical ups and downs of the paper industry.

Nonpaper pulps comprise a small portion of overall virgin pulp production, under 5%; but they constitute over 20% of market pulp production. Generally, the end-uses are high value-added products, and the pulps serving these markets have special characteristics and command higher prices. Nonpaper pulps are marketed under three broad categories. The largest segment is dissolving pulp which has mature and stable markets. The fastest growing segment is fluff pulp. The balance, consisting of small and disparate uses (e.g., filters, inner shoe soles, laminates, etc.), are lumped together as specialty pulp.

Dissolving Pulp

Dissolving pulps are chemical pulps that are suitable for subsequent chemical conversion into such products as rayon, cellophane, cellulose acetate, cellulose nitrate and carboxymethyl cellulose. Production of "chemical cellulose" pulp grades represents a substantial industry in its own right.

Dissolving pulps can be manufactured by either a modified kraft or sulfite process. In each case, the objective is a relatively pure and uniform cellulose product with a controlled weight-averaged degree of polymerization. Both lignin and hemicelluloses are considered contaminants and are removed. Softwoods are the major raw material, but some hardwood is used. The highest-quality pulps (99% alpha cellulose) are manufactured from cotton linters.

The viscose-rayon process is the largest user of dissolving pulps. In the viscose process, the chemical cellulose is "steeped" in caustic solution to form alkali-cellulose, which is carefully aged and then dissolved in carbon disulfide to form a bright orange solution, called xanthate. Cellulose is regenerated by extruding the xanthate into a sulfuric acid bath. Depending on whether the xanthate is extruded through small holes or wide slots, either long filament fibers (rayon) or films (cellophane) can be produced.

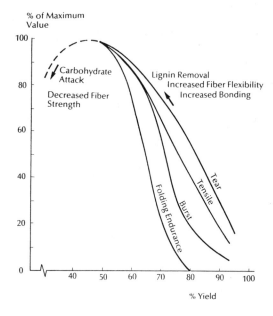

FIGURE 4-7. Effect of kraft pulp yield on selected strength parameters (at 500 CSF).

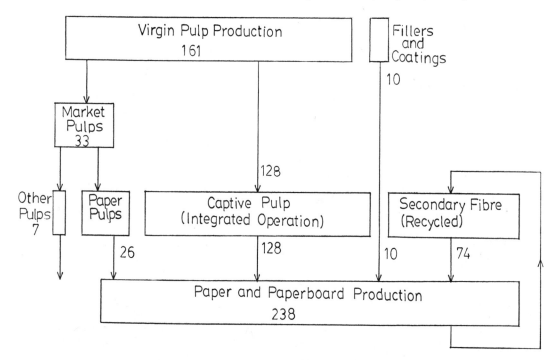

FIGURE 4-8. World pulp and paper production (000 000 metric tons per year).

Fluff Pulp

Fluff pulp is another important grade which finds a market outside of conventional papermaking applications. This product forms the absorbent medium in disposable diapers, feminine care products and hospital pads. The basic requirements for absorbency and softness have traditionally been served by long-fibered kraft or sulfite pulps, but more recently, less expensive chemimechanical grades have taken a significant share of the market.

Fluff pulp is typically supplied to the customer in roll form. The moisture level and sheet density must be strictly controlled to optimize the subsequent fluffing operation with respect to both fluff quality and energy usage. The absolute moisture content is usually controlled between 8 and 10%, a low level compared to conventional market paper pulps. At the same time, water removal by pressing must be minimized to avoid over-compaction of the sheet. These two factors impose a high evaporation load to the pulp drying equipment which causes a reduction in production rate and a high per-ton energy cost. Lower productivity and the need for additional roll winding equipment are the major reasons that most market pulp mills choose not to compete in the fluff pulp market. On the other hand, fluff pulp sells for a premium price, and the market is expected to expand more rapidly than for conventional pulp grades.

4.3 TRENDS IN PULPING

Chemical pulp mills tend to be associated with objectionable odors, which originate from the sulfur employed in the pulping operation. The odor can be greatly minimized by suitable engineering and efficient operation; emissions from modern mills are usually less than 10% the level of typical operations 25 years ago. Unfortunately, some of the organic sulfide gases are still offensive at concentrations as low as 1 part per billion. Obviously, it would be highly desirable to get completely away from sulfur as a pulping chemical. A number of non-sulfur processes are under investigation, and some appear to have commercial application. However, kraft pulping with modifications will remain the dominant chemical pulping process for both softwoods and hardwoods.

One particulary interesting development in chemical pulping is the organosolv process which uses ethyl alcohol to dissolve the lignin. Its feasibility for pulping hardwoods has been well demonstrated for a 15 ton-per-day (tpd) pilot plant, and plans are under way for a 300 tpd commercial plant. The hardwood pulp is comparable to kraft in strength and can be produced with higher yield and superior bleachability. The solvent is recycled, while lignin, furfural and wood sugars can be recovered as byproducts. This process is economically viable in plant sizes as small as 250 tpd whereas 1000 tpd may be necessary for a kraft mill; therefore, it can be considered for incremental tonnage, integrated paper mills, or as a greenfield mill where the wood resource is too small for a larger kraft mill. The process has not yet been developed to produce a kraft equivalent softwood pulp.

Overall, the fastest growing sector will be mechanical pulps, which are constantly being improved. Chemi-thermomechanical pulps (CTMP) and chemi-mechanical pulps (CMP) in particular are cleaner, brighter and have improved strength properties compared to other mechanical pulps. Consequently, they are replacing chemical pulps in a number of paper products. Again, the economy of scale favors smaller operations, so chemi-mechanical mills can be built and operated in areas that could not support a chemical pulp mill. Unfortunately, technology has not yet been developed to prevent brightness reversion. The negative impact on certain paper products can be minimized by adding suitable pigments, but brightness stability needs to be improved.

4.4 COMPARISON OF PULP PROPERTIES AND APPLICATIONS

The major pulping processes are summarized in Table 4-7. The actual papermaking properties of pulp fibers are dependent not only on the wood raw material and the pulping process employed, but also on subsequent purification, processing and conditioning treatments. More specifically, the inherent dimensional, structural and chemical properties of the liberated pulp fibers may be subsequently modified by bleaching, drying and refining.

The outstanding characteristics that determine fiber quality and suitability for papermaking applications are:
• length of fiber
• density and coarseness of fiber
• internal strength
• physical damage to fibers
• chemical damage to cellulose chains
• nature and distribution of residual lignin
• nature and distribution of hemicelluloses

The impact of each pulping process on fiber quality has been described briefly and will be elaborated upon in subsequent chapters. When properly controlled, chemical bleaching or brightening should have little effect on fiber mechanical properties, but will significantly change the optical properties of formed sheets.

Where pulp is dried prior to papermaking, significant changes in properties occur because of

TABLE 4-7. Summary of major pulping processes.

Classification	Process Name	Wood Used	Form of Wood	Yields	"Relative Strength"* Softwood	"Relative Strength"* Hardwood
Mechanical	Stone Groundwood	Softwood (mostly)	Bolts	90–95%	5	3
	RMP	Softwood (mostly)	Chips	90–95%	5–6	3
	TMP	Softwood	Chips	90	6–7	–
Chemi-mechanical	Chemigroundwood	Hardwood	Bolts	85–90%	–	5–6
	Cold Soda	Hardwood	Chips	85–90%	–	5–6
Semichemical	NSSC	Hardwood	Chips	65–80%	–	6
	High Yield Sulfite	Softwood (mostly)	Chips	55–75%	7	6
	High Yield Kraft	Softwood (mostly)	Chips	50–70%	7	6
Chemical	Kraft	both	Chips	40–50%	10	7–8
	Sulfite	both	Chips	45–55%	9	7
	Soda	Hardwood	Chips	45–55%	–	7–8

* Very "rough" ranking based on full range of pulp strengths.

irreversible internal bonds that are formed. The fiber is made stiffer and stronger internally, but is capable of less swelling and bonding to other fibers. Paper made from dried fibers is therefore bulkier and stronger in tear, but weaker in burst and tensile than paper made from never-dried fibers. Pulp prepared for shipment is usually dried to 80-90% oven dry.

All chemical pulps must be mechanically worked to develop optimum papermaking properties for various applications. As this refining takes place, the fibers collapse and become more conformable, producing denser sheets with increased burst and tensile strength. (Refer to Chapter 13 for full discussion).

Softwood kraft pulps produce the strongest papers and are preferentially utilized where strength is required. Typical applications are for wrapping, sack, and box-liner papers. Bleached kraft fibers are added to newsprint and magazine grades to provide the sheet with sufficient strength to run on high-speed printing presses. Bleached grades are also used for towelling and food boards.

Sulfite pulps find a major market in bond, writing and reproducing papers where good formation and moderate strength are required. Kraft or soda hardwood is usually added for improved formation and opacity. Sanitary and tissue papers also use large amounts of sulfite pulp to obtain the requisite softness, bulk and absorbency.

Mechanical pulp has traditionally been used primarily for newsprint and coated printing grades where it provides a well-filled and formed sheet. Because of improved quality and versatility, markets have opened for a wide range of printing grades, tissues, towelling, fluff, coating raw stocks, and food-grade boxboards.

REFERENCES

(1) PATRICK, K.L. **Stone Container Mill Brings New Pulp Mill On Line, Goes to No-Sulfur** *Pulp and Paper* (May 1979)
(2) HANSON, J.P. **No-Sulfur Pulping Pushes Out NSSC Process at Corrugating Medium Mills** *Pulp and Paper* (March 1978)

Chapter 5

Mechanical Pulping

The technology of mechanical pulping has undergone dramatic development in the past three decades. In 1960, virtually all mechanical pulp was produced by the basic stone groundwood (SGW) process. As recently as 1975, SGW still accounted for 90% of North American production. However by 1990, well over 50% of the mechanical pulp was being produced by refiner methods. The initial impetus for the rapid changeover to refiner methods was provided by three significant advantages:

(1) Utilization of sawmill residuals (chips and sawdust) in place of log bolts.

(2) Higher pulp strength.

(3) Reduced labor cost.

As refiner pulping methodology developed, further advantages became apparent, e.g., the application of larger production units, improved process control, utilization of hardwood species, and the ability to modify the process to produce a wide range of pulp properties.

In the 1960's the forest products industry was faced with the problem of how to effectively dispose of a surplus of sawmill residuals. The development of the refiner mechanical pulping process provided the opportunity to profitably utilize an available low-cost fiber source, and at the same time, to produce a stronger mechanical pulp that required less expensive chemical pulp in admixture for paper manufacture.

While advancement of refiner pulping methods has taken center stage in recent years, it must be stressed that stone grinding methods are by no means in total eclipse. Groundwood retains the advantage of lower energy consumption and higher scattering coefficient (i.e., greater sheet opacity). Also the innovative development of pressurized, high-temperature grinding has made possible the production of groundwood pulps that are more competitive with respect to strength and other properties.

Regardless of the pulping method used, a fundamental determinant of mechanical pulp quality is the amount of energy expended per unit of production, i.e., the specific energy (in units of horsepower-days per ton or megajoules per kilogram). Because specific energy is difficult to measure on a continuous basis, pulp freeness is more commonly used as the main process control parameter. (Refer to Section 22.3 for a discussion of

the freeness test). In general, the higher the per-ton energy expended, the lower will be the freeness of the pulp. Unfortunately, the relationship between the two parameters is subject to wide variation depending on the quality of the wood raw material, as illustrated by the groundwood data in Figure 5-1. Most mechanical pulps are actually manufactured to a specified range of freeness for general control of both pulp quality and stock drainability on the paper machine forming wire.

5.1 MECHANICAL PULPING NOMENCLATURE

Because of the proliferation of new and modified mechanical pulping processes, commonly accepted terminology is necessary to avoid confusion. In 1987, the TAPPI Mechanical Pulping Committee published the listing shown in Table 5-1 (1), and this nomenclature system has received wide acceptance. Each process name and abbreviation (or acronym) refers to a specific combination of temperature, chemical, pressure and refining action used to make the pulp. A family tree of mechanical pulps (which includes some high-yield sulfite semichemical pulps) is shown in Figure 5-2.

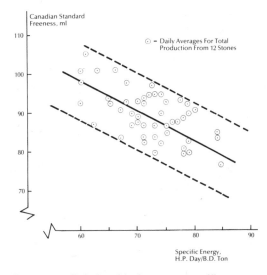

FIGURE 5-1. Relationship between specific energy and freeness for conventional groundwood.

45

TABLE 5-1. Mechanical pulp nomenclature and pulping methodology.

SGW   Stone Groundwood
       • atmospheric grinding

PGW   Pressurized Groundwood
       • grinding at temperature greater than (>)
         100°C

RMP   Refiner Mechanical Pulp
       • atmospheric refining with no
         pretreatment

TRMP   Thermo-Refiner Mechanical Pulp
       • presteaming of chips at > 100°C
       • atmospheric refining

PRMP   Pressure Refined Mechanical Pulp
       • no presteaming
       • first-stage refining at > 100°C
       • second-stage refining at > 100°C

TMP   Thermo-Mechanical Pulp
       • presteaming of chips at > 100°C
       • first-stage refining at > 100°C
       • second-stage atmospheric refining

PPTMP Pressure/Pressure Thermo-Mechanical Pulp
       • presteaming of chips at > 100°C
       • first-stage refining at > 100°C
       • second-stage refining at > 100°C

CRMP   Chemi-Refiner-Mechanical Pulp
       • atmospheric (low-temperature) chemical
         treatment
       • atmospheric refining

CTMP   Chemi-Thermo-Mechanical Pulp
       • presteaming with chemical treatment
         at > 100°C
       • first-stage refining at > 100°C
       • second-stage atmospheric refining

TCMP   Thermo-Chemi-Mechanical Pulp
       • presteaming with chemical treatment
         at > 100°C
       • atmospheric refining

TMCP   Thermo-Mechanical-Chemi Pulp
       (or OPCO Pulp)
       • first-stage refining at > 100°C
       • atmospheric chemical treatment
       • second-stage atmospheric refining

LFCMP Long Fiber Chemi-Mechanical Pulp (or)
CTLF   Chemically Treated Long Fiber
       • long fiber is separated from mechanical
         pulp
       • it is then chemically treated and refined

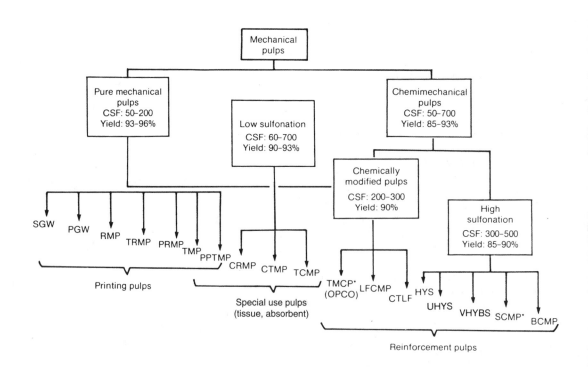

FIGURE 5-2. Family tree of mechanical pulps (W. Cooper and J.A. Kurdin).

In general, any groundwood is designated by the letters GW and any other type of mechanical pulp is designated by the letters MP. Additional letters are used to specify the combinations and sequences of processing steps. The line of demarcation between mechanical and semichemical pulps is generally considered to be 85% yield.

5.2 STONE GROUNDWOOD PROCESS

In the stone groundwood process, pulp is produced by pressing blocks of wood against an abrasive rotating stone surface. The wood blocks are oriented parallel to the axis of the stone so that the grinding action removes intact fibers.

The basic groundwood process is virtually unchanged since it was developed in the 1840's, but important developments have occurred in the design and control of grinders, in wood handling and feeding techniques, and in the manufacture of artificial pulp stones. A typical modern groundwood plant will consist of only four to six grinders to supply a large paper machine. Automatic log feeding systems have drastically reduced the labor requirement to perhaps 0.05 manhours per ton of pulp. When logs were manually handled and fed into the grinder, the labor input amounted to as much as 2 to 3 manhours per ton.

The steps in the full groundwood process are shown schematically in Figure 5-3. This section of the text will concentrate on the actual pulping operation of grinding. Screening, cleaning, thickening, and other processing steps which are

common to all pulp mill operations are discussed in Chapter 9. The purpose of screening and cleaning is to remove unseparated fiber bundles (shives), short stubby material (chop), and other undesirable components (e.g., bark and dirt particles) from the pulp stream. The objective of thickening (or deckering) is to remove water and raise the stock consistency.

The Grinder

The essential components of a grinder are illustrated in Figure 5-4. Controlled pressure is applied to the wood magazine, causing the log surfaces to be pushed against the revolving stone surface (face). The finger bars are carefully adjusted to limit the size of wood fragments that are accepted into the pit. Showers are provided to wash fibers off the stone into the pit, and also to keep the stone surface cool and clean. The stone may be further cooled by partial submergence in the pit, where an adjustable dam or weir is used to control the depth of submergence. The pulp slurry overflows the dam into a common channel leading from each grinder to a central receiving tank. From here the stock is pumped to the screen room.

The Pulpstone

The key element of the grinder is the shaft-mounted pulpstone (Figure 5-5). The quality of the produced pulp (i.e., strength and drainage properties) depends primarily on the surface characteristics of the stone. Virtually all stones used in North America are artificially manufactured using hard grit material

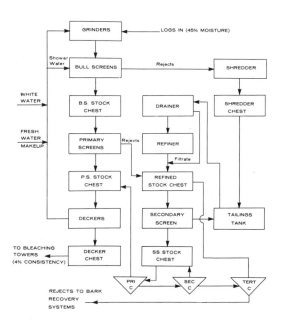

FIGURE 5-3. Typical groundwood flow sequence (Bruley).

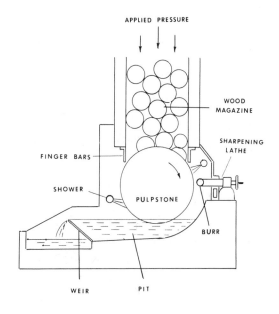

FIGURE 5-4. Arrangement of grinder showing the essential components.

FIGURE 5-5. Shaft-mounted pulpstone.

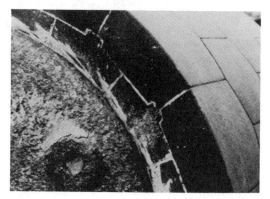

FIGURE 5-7. Closeup of grindstone showing abrasive sections.

of controlled size. However, after grinding wood for a period of time, the stone surface becomes relatively smooth and productivity falls off to the extent that it becomes necessary to roughen up the contact surface. The stone is usually "sharpened" by imprinting a pattern on the surface using a metal burr. The position of the burr is controlled by the sharpening lathe.

Abrasive segments of the pulpstone (or grindstone) are manufactured separately and bolted to a reinforced concrete core as illustrated in Figures 5-6 and 5-7. The segments are made from abrasive grits of aluminum oxide or silicon carbide held together by bonding material which has been vitrified. Typically, the abrasive layer is about 7 cm thick, and the stone is replaced when the useful layer is worn away after perhaps two years of steady use.

For manufacture of the pulpstone segments, the grit particles are fractionated by size. The average grit size, the type of vitrified bond, and the relative proportions of grits to bonding material all have an effect on the quality and properties of the pulp produced. Generally, smaller grits are used to produce low-freeness groundwood, intermediate-sized grits are used to produce newsgrade pulp, and coarse grits are used to produce pulp suitable for board grades. When a pulp mill orders a new pulpstone, the specifications must also take into account the wood species being pulped and the particular grinding conditions.

The Grinding Mechanism

At one time, the mechanism of grinding was viewed simply as the biting and tearing action of protruding grits fiberizing the wood. Although this mechanism does apply to a certain extent in the case of a freshly sharpened stone, it is now considered of minor importance with respect to typical operation by "conditioned" grits. Research has shown that the dominant action is the high-frequency compression and decompression at the interface as the grits contact the wood (Figures 5-8 and 5-9); this serves to loosen the fibers by fatigue and failure. At the same time, tremendous heat is generated by the friction of stone against wood and of wood against wood as deformation occurs. This heat softens the lignin binding the fibers together and assists in separating fibers from the wood mass.

Since virtually all the grinding energy is dissipated as heat in the grinding zone, it follows that temperature control by showering is an important aspect of the grinding operation. Too much water cooling will impede the softening action, while insufficient moisture will allow charring to occur.

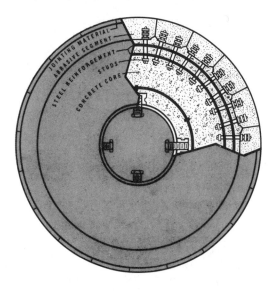

FIGURE 5-6. Construction details of a ceramic pulpstone.

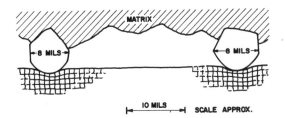

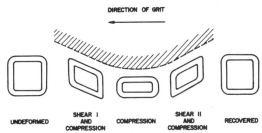

FIGURE 5-8. Fiberizing by compression of the wood utilizing "conditioned grits". The wood surface is probably deformed between 0.001 and 0.002 inches under normal grinding pressures (D. Atack).

FIGURE 5-9. As a conditioned grit traverses the wood surface, compression and shear lead to fatigue failure in the cell wall (D. Atack).

Control of Grinding

A reasonably complete listing of the variables affecting the grinding process is given in Table 5-2. The design of the grinder itself could also be listed as a variable. Generally, the major control parameter is the pulp freeness. The major groundwood grades, along with target freeness levels and comparative specific energy levels are given in Table 5-3.

Typically, the quality of the wood provided to the groundwood mill is outside the control of the operator. The pulpstone is selected to produce a suitable quality pulp from an average wood furnish. Shower water flow and temperature, as well as the degree of stone submergence, are usually optimized and maintained under close control. Stones that are sufficiently cooled and cleansed by shower action alone (i.e., without submergence) are said to employ "pitless grinding".

In practice, the only variables under the operator's direct control are magazine pressure and the stone sharpening cycle. It should be noted that magazine pressure is not the same as grinding pressure, but an obvious relationship exists between the two variables.

Periodic stone sharpening is necessary to maintain the groundwood plant production rate and stock freeness level. The freshly sharpened stone has an initial adverse effect on pulp quality as indicated by reduced specific energy and lower burst/tensile strength (refer to Figure 5-10). Low specific energy infers a freer, less-fibrillated fiber with fewer fines. Stones are usually sharpened on a staggered cycle so that overall blended pulp quality will not change abruptly; if pulp freeness or production rate falls off, the cycle can be speeded up.

TABLE 5-2. Variables affecting grinding.

Wood Condition	• species • age • moisture content • rot content • bolt diameter
Stone Specifications	• type of grit • grit size • type of bond • weight ratio, grit to bond
Showers	• water flow rate • water temperature • water pressure • number, size and position of nozzles
Stone Sharpening	• frequency • severity (depth of pattern) • pattern (type of sharpening tool)
Operational	• magazine pressure • peripheral speed of stone • submergence of stone • chemical additives

TABLE 5-3. Groundwood grades.

	Newsprint	Book (Several Grades)	Tissue (Many Grades)	Board	Specialty (Many Grades)
Specific Energy Horsepower Day/Ton	60–70	60–90	40–130	50–60	50–80
Pulp Freeness, CSF	85–95	50–90	75–140	160–175	40–150
Stone Pattern	Spiral	Spiral	Spiral Diamond Thread	Spiral	Spiral etc.
Stone Grit Size	60 Medium	60 Medium	24-coarse to 80-fine	20–46 coarse	–

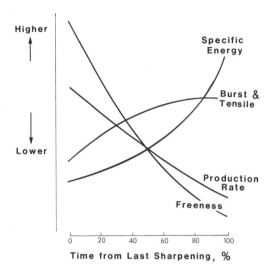

FIGURE 5-10. Effect of sharpening cycle on groundwood production and quality parameters (at constant magazine pressure).

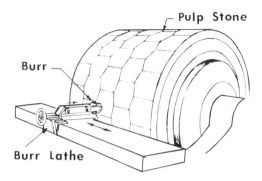

FIGURE 5-11. Stone sharpening.

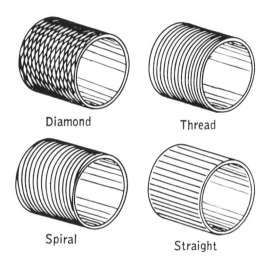

FIGURE 5-12. Various burr patterns.

Grinders are generally run at or below a specified peak motor load. The magazine pressure must be kept at a relatively low level when the stone is freshly sharpened in order to avoid overloading the motor. But, after mid-cycle, the maximum available pressure is usually not sufficient to reach the peak loading.

Stone Sharpening

Stone sharpening is achieved by moving a small patterned metal cylinder (approximately 7.5 cm diameter by 7.5 cm long), called a burr, across the surface of the rotating stone. To ensure uniform sharpening, the burr is mounted in a lathe (Figure 5-11). The burr usually traverses the stone several times in the same grooves to get the desired depth of pattern (about 1.6 mm). Various types of burr patterns are available (see Figure 5-12), of which spirals and diamonds are most widely used. An idealized cross-section of a sharpened stone is shown in Figure 5-13. Generally a new burr is used for each stone sharpening.

Effect of Wood Variables

Ideally, the wood raw material for mechanical pulping should contribute toward good strength, high brightness, absence of color, low energy consumption, and freedom from operating problems. By all these accounts the best wood species are the spruces, followed by the true firs. Among the favorable characteristics of spruces are their low wood density (due to high springwood content), light-colored heartwood, and low content of resins/extractives. Several Canadian mills are able to exploit extensive tracts of spruce and balsam fir. Hemlock and pine are used in

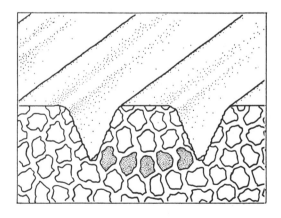

FIGURE 5-13. Cross-section of sharpened stone showing land pattern with base width equal to five grit diameters (Norton Co.).

other areas of the North and Northwest. Southern pines (e.g., loblolly and slash) constitute the principal wood furnish in the United States South. Aspen and poplar are the most commonly used hardwoods.

The advantage of spruce in relation to other species is obvious when operating data are compared (Table 5-4). Often, spruce groundwood can be used in newsprint without augmented brightening.

The southern pines are relatively dense because of their greater summerwood content. Also, the heartwood tends to be quite dark, and the entire tree has a high resin content. Southern operators have learned to compensate for these characteristics by using higher energy to achieve greater specific

surface and lower freeness, by pulping only young trees (less than 35 years old) free of heartwood, and by using alum and caustic to chemically control the pitch. However, a typical southern groundwood cannot provide the good balance of properties that are characteristic of a northern spruce groundwood.

A high wood moisture content contributes to easier grinding. 30% moisture is usually considered minimum, and 45% to 50% would be preferable. Sometimes, wood of low moisture content is soaked for a period prior to grinding.

Types of Grinders

Several designs of grinder are in use around the world, some of which are illustrated schematically in Figure 5-14. The chain and ring grinders are representative of continuous loading types. The magazine and pocket grinders are of intermittent operation with respect to the stone load which is partially removed whenever a hydraulic cylinder is retracted to refill an empty pocket.

Many of the latest mills are equipped with two-pocket grinders of the type illustrated in Figure 5-15. This design lends itself readily to fully-automatic operation and is free of wood bridging, which has been a problem with some continuous loading types. The hydraulic magazine grinder shown in Figure 5-16 is also used in some modern groundwood mills.

TABLE 5-4. Comparison of groundwood pulps made from different wood species.

	Spruce	Balsam	Hemlock	Pine
Energy required MJ/kg	5.0	5.2	5.6	6.3
Burst Index kPa m^2/g	1.4	1.3	1.1	0.9
Tear Index mN m^2/g	4.1	3.8	3.2	2.8
Brightness, %	61	59	57	56

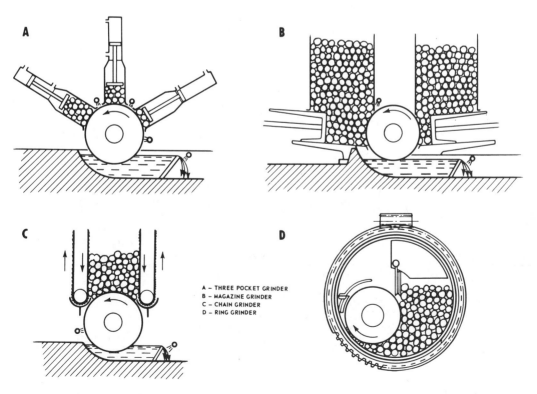

A – THREE POCKET GRINDER
B – MAGAZINE GRINDER
C – CHAIN GRINDER
D – RING GRINDER

FIGURE 5-14. Types of grinders.

Reject Refining

Until the 1950's, the objective of reject refining was simply to reduce the size of screen reject material to the point where it was no longer objectionable. There was little concern for the quality of this pulp fraction which was regarded only as a filler component. With the application of disc refining it became possible to disintegrate the fiber bundles into free fibers comparable in quality to the bulk of the groundwood. The development work on reject refining laid the foundation for the subsequent refiner mechanical pulping of wood chips.

Pressurized Grinding

In response to the success of thermomechanical pulping, a modified process to grind wood in a pressurized atmosphere was developed in Finland (by Tampella) in the late 1970's. The first commercial pressurized groundwood (PGW) system was started up in 1980. Today, more than 2.3 million tons per year of PGW capacity is installed or on order worldwide, mostly for printing and writing grades.

PGW differs from conventional SGW in two respects: grinding occurs under an overpressure of 30 psig, and the pulpstone shower water temperature is much higher, over 95°C. The overpressure, which is achieved with compressed air, makes it possible to use higher grinding temperatures. At elevated temperature, the wood lignin softens and the fibers are released from the wood matrix more readily and

in a less-damaged condition. The long fiber content of PGW is much higher than for SGW, but at the same time the fibers are well developed, i.e., fibrillated and flexible, producing superior fiber-to-fiber bonds and greater conformability. PGW tear

FIGURE 5-16. Hydraulic magazine grinder (Koehring-Waterous Ltd.).

FIGURE 5-15. Great Northern grinder (Koehring-Waterous Ltd.)

strength is typically 40 to 60% higher and tensile values are 20 to 30% higher than conventional SGW strength values.

A major advantage of PGW is that specific energy consumption is actually less than for comparable SGW and significantly less than for competitive refiner processes. Perhaps the major limitation is that PGW still requires a furnish of logs rather than chips.

The basic components of a PGW system are the pressurized grinding unit itself and the means to recycle hot water back to the grinder showers (refer to Figure 5-17). The recycle circuit from the thickener back to the showers is called the "hot loop". From the grinder, the stock flows under pressure through a shredder (where shims and slivers are reduced to matchstick size) and then through a pressure-reducing blow valve. Directly after the valve, the excess heat developed in the process is relieved from the stock as flash steam in a separating cyclone. The low-consistency (1.2 to 1.5%) stock is thickened and the hot filtrate is returned to the grinder showers.

PGW Developments

A higher-temperature modification of PGW has been dubbed "Super" PGW (or PGW-P). In the conventional PGW system, the shower temperature is limited by the thickener to a maximum under 100° C. However, the development of a new pressurized disc thickener allows the shower water to be returned at temperatures up to 140° C. Pilot trials have shown that the PGW-S process produces pulp of higher strength while allowing more significant energy recovery in comparison to the PGW process. The main drawback is a slight loss in pulp brightness due to thermal darkening.

Another development of PGW has been the introduction of various chemicals into the hot loop. The best results to date have been obtained with alkaline peroxide added to the grinder showers. In 1990, the first chemi-pressurized groundwood (CPGW) mill started up in North America utilizing alkaline peroxide during the grinding operation, followed by a second stage of peroxide bleaching. This process has been shown to be particularly effective for pulping aspen.

5.3 REFINER MECHANICAL PULPING

Commercial production of refiner mechanical pulp (RMP) was initiated in 1960. RMP is produced by the mechanical reduction of wood chips (and sometimes sawdust) in a disc refiner. The process usually involves the use of two refining stages operating in series (i.e., two-stage refining), and produces a longer-fibered pulp than conventional groundwood. As a result, it is stronger, freer, bulkier, but usually somewhat darker in color, than stone groundwood.

The heart of the RMP system is the disc refiner. The model illustrated in Figure 5-18 is a double-revolving unit, with each disc rotating in opposite directions. Other designs utilize a revolving disc opposite a stationary disc (Figure 5-19) or a revolving double-sided disc between two stationary discs (Figure 5-20). Plate clearance is of critical importance, and is accurately controlled by either an electro mechanical or hydraulic loading system. The material to be refined is introduced by a screwfeeder into the open eye of the refiner. As the material moves through the refining zone towards the periphery, the wood mass is progressively broken down into smaller particles and finally into fibers.

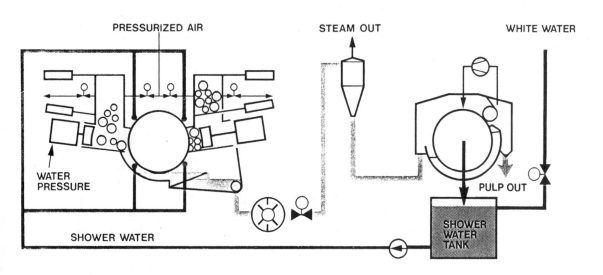

FIGURE 5-17. Schematic of pressurized groundwood process (Tampella).

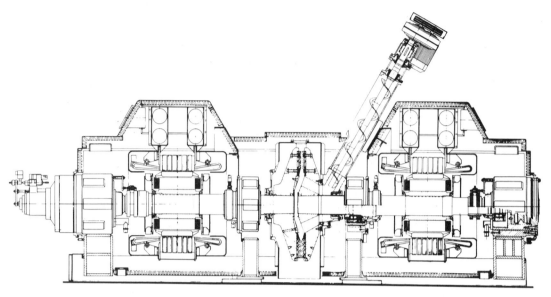

FIGURE 5-18. Double-disc refiner (Sunds Defibrator).

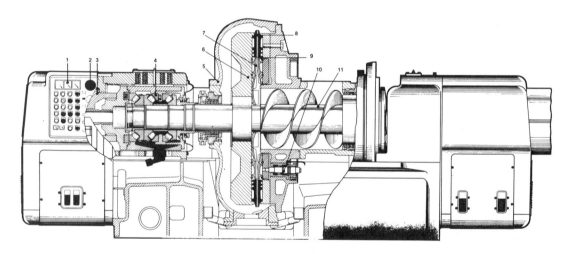

FIGURE 5-19. In this single-rotating disc design, the feedscrew (11) supplies chips to the eye of the refiner (Sunds Defibrator).

Water is supplied to the eye of the refiner to control pulp consistency; sometimes chemicals are also added. Refiners are currently available with up to 70 inch diameter and 30,000 HP (22.5 MW) of supplied power.

While the mechanism of refining is similar in some respects to grinding, there are also some basic differences. As with grinding, the lignin is softened by compression-decompression and by the friction of wood to wood and metal to wood between discs. In the refiner, the initial defiberizing process is more akin to a step-wise "unravelling" of the chips into smaller and smaller entities, and finally into fibers. This unwinding and twirling effect shows up in the freshly-produced fibers, and it is necessary to disintegrate the pulp in hot water for a period to remove this "latency" (refer to Figure 5-21). In fact, the pulp cannot be accurately characterized for freeness until latency has been removed.

Refiner Plates

Plates for the first-stage refiner are designed with widely-spaced, thick breaker bars close to the eye (see Figure 5-22). These bars shred the chips and permit the development of centrifugal forces which move and align the wood particles radially for optimum results in the refining zone. The refining

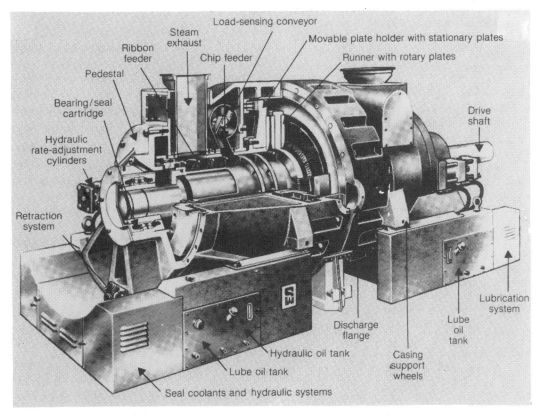

FIGURE 5-20 (above). This design utilizes both sides of the single rotating disc, thus producing two refining chambers in one unit (Sprout-Bauer).

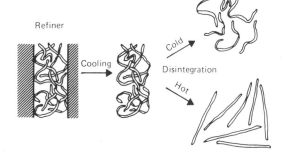

FIGURE 5-21 (to the right). The effect of hot and cold disintegration on latency.

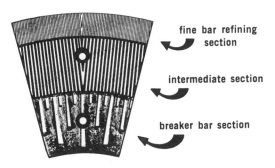

FIGURE 5-22. Representative disc segment utilized on a primary-stage refiner (Sprout Bauer).

FIGURE 5-23. Representative disc segment utilized on a secondary-stage refiner (Sunds Defibrator).

zone consists of progressively narrower bars and grooves wherein the coarse material is converted to pulp fibers. Usually the plates are tapered slightly (perhaps by 0.01 mm/mm or less) to ensure that the pulp moves evenly toward the periphery. Some plates are provided with dams to block the grooves at intervals and force the pulp to move over bars. So-called closed-periphery plates have a peripheral rim to increase the retention time of pulp between the plates.

Plates for second-stage refiners (Figure 5-23) usually have a shorter breaker bar section and a larger portion of refining surface. The breaker bars are necessary to align and impart centrifugal force to the partly refined stock. The metallurgy of refiner plates varies from low-cost nickel-hardened steel to more costly alloys and stainless steels.

Control of Chip Refining

A comprehensive listing of variables affecting chip refining is given in Table 5-5. Different conditions are used successfully, and it appears that the process is tolerant to variations within a fairly wide range. Refining is usually carried out at consistencies from 18% to 30%.

To maintain uniform operation, it is important that the steam generated within the refining zone be freely exhausted. Uneven steam exhaustion relates to "blowback", which causes interruption of chip flow and corresponding load fluctuations. Steam pressure development has been found to be affected by most operating variables, but is most easily controlled by reducing the temperature and consistency of the feed.

Chip quality is an important factor affecting mechanical pulp quality. As with grinding, "green wood" makes better pulp than dry wood. Bark, dirt and "blue stain" (i.e., blue color from fungus attack) carry through with the pulp. Odd-sized wood fragments, especially from compression wood, cause problems in feeding the refiner and with heat penetration. Chip washing is carried out in many systems for removal of stones, metal and dirt, the main objective being protection of the refiner plates, but a cleaner pulp is also produced (refer back to Section 3.6).

Some mills have utilized sawdust or pin chips as feed to the RMP system. Generally, strength and brightness are lower for pulps made from these materials, but their lower cost offsets the quality disadvantages.

The basic RMP process has now been eclipsed by thermally and/or chemically modified methods which produce better quality pulps, sometimes with special attributes. New installations virtually always incorporate the improved technology; and many older RMP mills have been retrofitted as well. However, the basic principles of RMP apply to all refiner pulping methods.

5.4 THERMOMECHANICAL PULPING

Thermomechanical pulping (TMP) was the first major modification of RMP, and is still employed on a large scale to produce high-tear pulps for newsprint and board. This process involves steaming the raw material under pressure for a short period of time prior to and during refining. The steaming serves to soften the chips, with the result that the pulp produced has a greater percentage of long fibers and fewer shives than RMP. These longer fibers produce a stronger pulp than either SGW or RMP as shown by the comparison in Table 5-6.

Most often, heating and first-stage refining are both carried out under pressure (TMP). However in a few systems, pressurized heating is followed by atmospheric first-stage refining (TRMP). The

---

TABLE 5-5. Variables affecting chip refining.

Raw Material	• species
	• general chip quality
	• moisture
	• freedom from foreign material
Plate Specifications	• material of construction (metalurgy)
	• pattern
	• taper
	• closed or open periphery
	• disc diameter
Operational	• pressure at inlet
	• temperature at inlet
	• consistency
	• applied load (gap between plates)
	• chemical additives
	• feed rate
	• change cycle of plates
	• disc rotational speed
	• back pressure at discharge of refiner

TABLE 5-6. Comparison of pulps from different mechanical pulping processes (W.B. West).

	SGW	RMP	TMP
Energy Required, MJ/kg	5.0	6.4	7.0
Burst Index, kPa m^2/g	1.4	1.9	2.3
Tear Index, mN m^2/g	4.1	7.5	9.0
Bulk, cm^3/g	2.5	2.9	2.7
R-48 (Bauer McNett*)	28	50	55
Shive Content, %	3.0	2.0	0.5
Brightness, %	61.5	59	58.5

* Relative measure of long-fibered content.

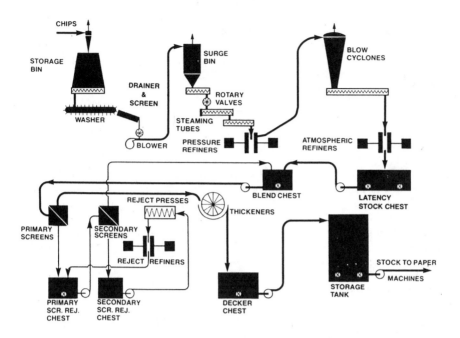

FIGURE 5-24. Basic flow diagram for TMP process (H.A. Simons).

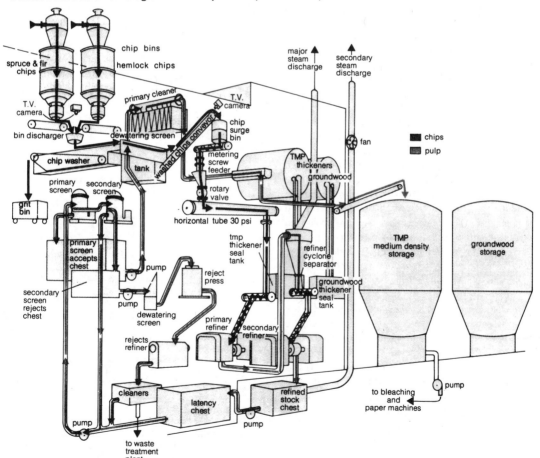

FIGURE 5-25. Layout of TMP system (Champion International, Bucksport, Maine).

secondary refining stage initially was carried out at atmospheric pressure, but is being pressurized in newer systems to facilitate heat recovery. Screening, cleaning, thickening and brightening are employed as with other mechanical pulping processes. Representative process flow schemes are illustrated in Figures 5-24 and 5-25.

It was known in the 1930's that thermal softening could be employed for making strong mechanical pulps, and the technique was applied in the manufacture of fiberboard products. However, the problem of thermal darkening discouraged application to the production of papermaking pulps. Finally, in the early 1970's a method of heating was perfected which softens the chips for improved defiberizing, but does not seriously discolor the resulting product. The major drawback is that additional power is required for optimum fiber development. On the other hand, the higher temperature of the TMP process allows substantial amounts of energy to be recovered (see Section 5.6).

Intuitively, it might be expected that TMP would save energy in comparison to RMP because less energy is required to separate the fibers. In fact, TMP requires greater energy because only a small portion is used for fiber separation; the major portion is used to develop the fibers by breaking up the primary wall and peeling the S_1 layer.

The process variables for TMP are similar to RMP, but special attention is given to controlling the temperatures during steaming and refining. Two arrangements for steaming and feeding chips to the refiner are shown in Figures 5-26 and 5-27. The chips are fed volumetrically into the steaming vessel from a presteaming bin where the chips have been heated to about 90°C using backflow steam from the refiner. Retention time in the pressure steaming vessel should be sufficient to reach the desired temperature in the center of the chip, but prolonged steaming must be avoided to minimize thermal darkening. It is important that all air carried with the wood chips be removed during presteaming/ steaming to prevent an insulating effect.

The refining itself must be carried out at a temperature below 140°C. Above 140° the fibers are easily separated at low energy consumption because the lignin has undergone a dramatic softening (Figure 5-28); but the released intact fibers are coated with soft lignin, which on cooling reverts to a glassy state that becomes an obstacle to subsequent fibrillation of the separated fibers. When chips are refined at lower temperatures, between 120 and 130°C, the lignin is sufficiently softened for good fiber separation, but fractures can now occur in the outer layers of the secondary fiber walls as

illustrated in Figures 5-29 and 5-30. The cross section of a TMP sheet is shown in Figure 5-31.

5.5 CHEMICALLY MODIFIED MECHANICAL PULPING

Chemical treatments of chips prior to refining and/or chemical additions during refining were initially investigated as means of reducing energy requirements. However, it soon became apparent that while some energy saving can result, the principal effect is to alter the qualities of the resultant pulps. Both pressure steaming and sulphonation (i.e., chemical attachment of sulfite groups onto the lignin molecules) soften the wood by reducing the interchain hydrogen bonding in the lignin. But steam-induced softening (as for TMP) is temporary; whereas the softening effect of sulfonation is

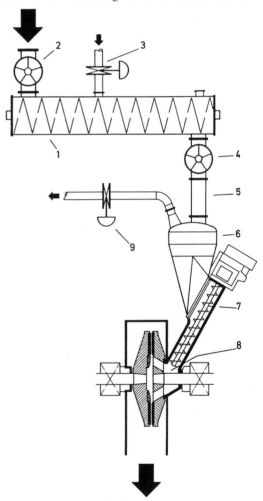

FIGURE 5-26. Steaming and feed system for Sunds TMP process. (1) steaming vessel (2 & 4) rotary valves (3) steam inlet (5) chute (6) steam vent chamber (7) screw feeder (8) double disk refiner (9) steam outlet.

permanent and has a carryover effect on the fiber properties.

There has been such a proliferation of new processes using chemical treatments, that only the major processes can be mentioned in this introductory text. For further information, the reader is referred to a general reference (2).

Chemithermomechanical Pulping (CTMP)

In recent years CTMP has evolved as a modification of the TMP process, increasing the range of products. It utilizes a modest chemical impregnation during the steaming stage, which serves to improve the bonding properties of the pulp and lower the debris content at the expense of slightly lower scattering coefficient (i.e., lower opacity). In

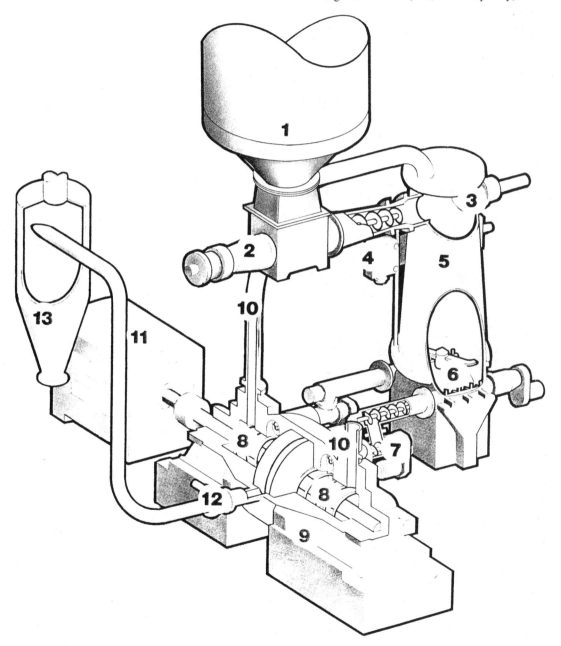

FIGURE 5-27. Steaming and feed system for Sproat Bauer process. (1) feed hopper (2) plug screw feeder (3) blow back valve (4) chip level controller (5) steaming tube (6) chip discharger (7) load sensing conveyer (8) ribbon feeder (9) refiner (10) steam equalizing line (11) drive motor (12) blow valve (13) cyclone separator.

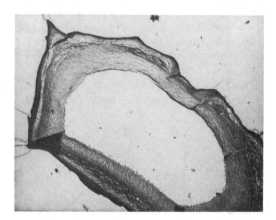

FIGURE 5-28. Electron micrograph of spruce fiber refined at 175°C, well above the glass transition for lignin (P.H. Norberg).

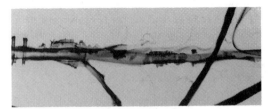

FIGURE 5-30. TMP fiber (MacMillan Bloedel Research Ltd.).

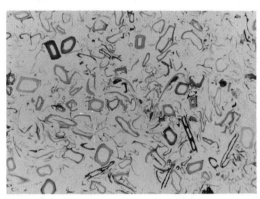

FIGURE 5-31. Cross-section of TMP sheet highly magnified (MacMillan Bloedel Research Ltd.).

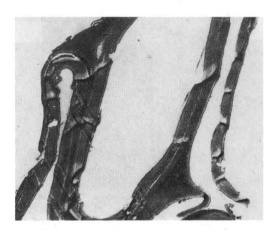

FIGURE 5-29. Electron micrograph of spruce fiber refined at 130°C shows ruptures in secondary fiber wall (P.H. Norberg).

conjunction with established peroxide bleaching technology (see Section 11.11) the CTMP process produces high-brightness pulps at reasonable cost.

A complete flow diagram for a market bleached CTMP plant utilizing two stages of refining is shown in Figure 5-32. Screen rejects are either refined separately or are combined with the stock feeding the secondary refiner. Chemical treatment of softwoods is usually with sodium sulfite solution at application levels between 1% and 5% on dry wood.

At low chemical applications, the resultant pulp could be used as the sole furnish for newsprint. The major benefit of the chemical treatment would be improved brightness and a marginal increase in strength, with negligible loss in scattering

coefficient. At somewhat higher chemical treatments, CTMP is suitable as a high-freeness pulp for the middle layer of multiply boards where it adds bulk and rigidity at lower cost than kraft pulp. Higher-sulphonated CTMP is also suitable for tissue and fluff pulp.

Hardwood CTMP grades are usually produced utilizing an alkaline sulfite impregnation and a lower-temperature thermal softening. Eucalyptus and aspen CTMP pulps are finding a market where the attributes of improved formation and greater softness are more important than strength.

Chemimechanical Pulping (CMP)

Chemimechanical pulps are produced utilizing a relatively severe chemical treatment followed by atmospheric refining. Depending on the severity of the cook (with respect to chemical application, temperature and residence time), the yield is usually in the 85 to 90% range. This type of pulp exhibits further improvement in bonding strength and fiber conformability, but losses in scattering coefficient are high. CMP grades are typically used as reinforcement pulps, most commonly in newsprint.

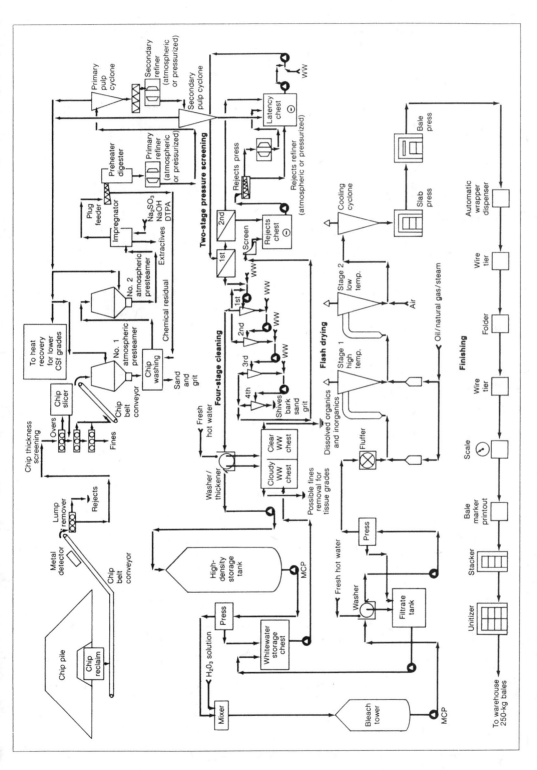

FIGURE 5-32. Flow diagram for a market bleached CTMP mill (NLK-Celpap).

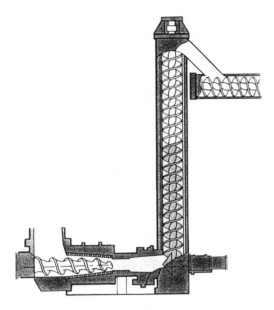

FIGURE 5-33. Prex impregnation unit (Sunds Defibrator).

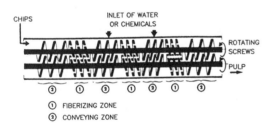

① FIBERIZING ZONE
② CONVEYING ZONE

FIGURE 5-34. Bi-viz extruder for chemical impregnation.

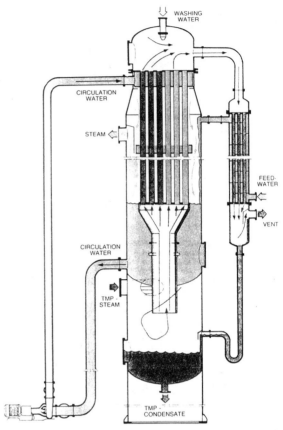

FIGURE 5-35. Heat recovery reboiler (Sproat Bauer).

considered. This is a straightforward method to improve the flexibility and bonding of the stiff long fibers from the groundwood or TMP mills. The improvements due to rejects sulfonation approach those from treatment of the whole pulp, but with the added benefit of more efficient chemical utilization and minimal loss of scattering coefficient.

Chemical Impregnation

For CTMP and CMP, efficient chemical impregnation of the wood chips is vital to the success of the pulping process. It is particularly important that the chips be free of air during this stage of the process. Air removal is best accomplished by prior atmospheric presteaming. Several types of equipment are available for chemical impregnation, some of quite recent design. Representative are the Sunds Defibrator Prex impregnation unit (Figure 5-33) and the French-developed bi-viz extruder (Figure 5-34). In both cases, the chip mass is first compressed and then allowed to expand and soak up the impregnation liquor. The objective is to achieve a uniform impregnation regardless of the initial chip moisture content.

Interstage Sulfonation

Treatment with sodium sulfite is not limited to wood chips, but can also be carried out on partially or fully fiberized pulp. Since the fibers have already been separated, some loss of tear strength is irreversible. On the other hand, the defiberization provides a larger surface area for sulfonation, resulting in a pulp with superior printing qualities. Probably the main advantage of interstage sulfonation is a significant improvement in wet web strength, which is not obtainable with chip sulfonation.

As an alternative to interstage sulfonation, the chemical treatment of all mechanical pulp mill screen rejects prior to reject refining can be

5.6 HEAT RECOVERY

The TMP and CTMP processes are the most accepted methods for producing mechanical pulps. Due to their relatively high strength and versatility, these pulps have great potential for reducing the chemical fiber content in a wide range of paper grades. On the other hand, these processes are energy-intensive, and this drawback may affect their long-term viability. Considerable development work has not been successful in reducing the energy requirements of refining. Fortunately, because of the relatively high temperature of these processes, it is quite feasible to recover a large portion of the energy (up to 75%) in a form that is useful in other sections of the mill (3).

Initially, energy recovery was limited to the production of hot water by mixing shower water with the contaminated refiner exhaust steam. But over the past decade more sophisticated systems for the generation of clean steam have evolved. To increase the amount of waste heat available, secondary refiners may also be pressurized.

The heart of a modern steam recovery system is the reboiler, a special design of heat exchanger that utilizes heat from the condensing exhaust steam to generate fresh steam. Figure 5-35 shows a representative design utilizing contaminated condensing steam on the inside surface of the tubes and a falling film of boiling water on the outside surface of the tubes. A typical overall system for heat recovery is shown in Figure 5-36.

In an integrated mill, it is often attractive to use the recovered steam for paper drying. Unfortunately, the pressure of the steam normally produced is rather low for this purpose due to the initial modest pressure from the refiner which is lowered by the necessary pressure drop across the reboiler tubes. The steam may need to be boosted to a higher pressure (e.g., with a thermocompressor or mechanical compressor) which could adversely affect the economics of utilization.

A recent innovation in refiner operation has made possible the production of higher pressure steam from the reboiler. It was discovered that the refiner can operate at a much higher pressure than the presteaming vessel without affecting pulp quality because lignin does not pass through the glass transition during its brief retention in the refiner.

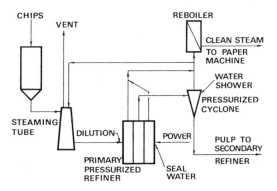

FIGURE 5-36. Schematic of pressurized refiner system showing placement of reboiler (H.A. Simons).

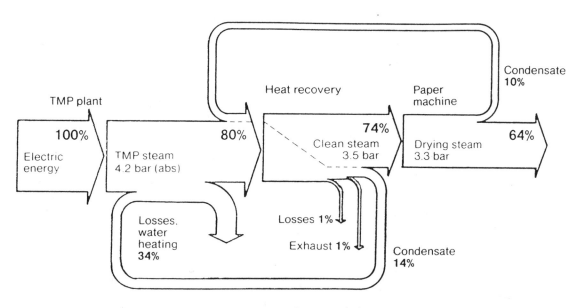

FIGURE 5-37. Energy balance of TMP steam recovery for paper drying.

Isolation of the presteaming vessel from the refiner can be accomplished using either a plug screw or rotary valve as a seal. An energy balance based on this mode of operation indicates that 64% of the original electrical energy can be recovered as clean drying steam (see Figure 5-37).

REFERENCES

(1) COOPER, W. and KURDIN, J.A. **Acronyms for Mechanical Pulp: Understanding the Alphabet Soup** *Tappi Journal* (December 1987)

(2) LEASK, R.A. and KOCUREK, M.J. (Editors) **Mechanical Pulping** (Volume 2 of Pulp and Paper Manufacture Series) Joint Textbook Committee, 1987

(3) CROPP, H.V. **Efficient Use of Recovered Energy is a Key Mechanical Pulping Goal** *Pulp & Paper* (April 1991)

Sulfite Pulping

6.1 BRIEF HISTORY OF DEVELOPMENT

The genesis of sulfite pulping dates to 1857 with the discovery by a Philadelphia chemist, Benjamin Tilghman, that wood could be softened and defibered with sulfurous acid (H_2SO_3). However, he noted that sulfurous acid acting alone produced a "burnt" or discolored pulp. Somewhat later, Tilghman undertook a systematic study and determined that the presence of a cationic base such as calcium prevented the discoloration. He was awarded the U.S. patent for the pulping process in 1867; the first commercial sulfite pulp was produced in Sweden in 1874.

The commercialization of sulfite pulping developed rapidly, and for many decades following 1890 the calcium acid sulfite process was the most important pulping method around the world. During the 1930's the kraft process became dominant because it was able to utilize various woods not suitable for sulfite pulping.

Up until about 1950, the sulfite industry was able to utilize improved equipment and better operating and control methods, but the basic chemistry and technology were virtually unchanged. Since 1950, the utilization of bases other than calcium has been a major development. These more soluble bases, namely magnesium, sodium and ammonium, are amenable to less acid cooking conditions and are readily adaptable to multi-stage processing. By appropriate choice of cooking conditions, it also became possible to handle wood species which are unsuitable for classic calcium acid sulfite pulping. These newer cooking methods produce pulps of higher yield with a wide range of properties. The relatively high cost of magnesium and sodium base chemicals has encouraged the development of efficient recovery systems, which in any case are now vital for environmental control (refer to Section 10.7).

Recovery of cooking chemicals from the calcium base system has never been practiced. Traditionally, the sulfite waste liquor was simply discharged into the nearest receiving water. Even where the liquors are now evaporated and burned for heat recovery, it is not feasible to recover usable chemicals from the predominantly calcium sulfate ash. The usual makeup chemicals, limestone and sulfur, are inexpensive and readily available.

A breakdown of U.S. and Canadian sulfite mills by chemical base in 1980 and 1990 is given in Table 6-1. It can be noted that the number of North American sulfite mills has declined markedly over the last decade.

TABLE 6-1. U.S. and Canadian sulfite mills (1990 vs. 1980).

| Pulping | U.S. | | Canada | | Total | |
Base	1980	1990	1980	1990	1980	1990
Ca	7	5	3	0	10	5
Mg	9	7	3	3	12	10
Na	4	1	25	15	29	16
NH$_3$	9	5	4	3	13	8
Total	29	18	35	21	64	39

6.2 NOMENCLATURE AND DEFINITION OF TERMS

The definitions for sulfite acid strengths and chemical applications revolve around the fact that one molecule of calcium bisulfite is chemically equivalent to one molecule each of sulfurous acid (referred to as "free SO_2") and calcium sulfite ("combined SO_2"):

$$Ca(HSO_3)_2 \longrightarrow CaSO_3 + H_2SO_3$$

The % total SO_2 in the cooking liquor is determined by straight iodometric titration. The % free SO_2 is determined by a titration with NaOH, which actually measures all the sulfurous acid plus one-half of the bisulfite according to the above reaction. The % combined SO_2 is calculated by difference. It can be seen from the foregoing that the "true free SO_2" is the free SO_2 minus the combined SO_2 (also equal to the total SO_2 minus two times the combined). An examination of the values in Table 6-2 will help to clarify the concept.

TABLE 6-2. Hypothetical sulfite cooking analysis for three samples (g/100 mL).

	Sample A	Sample B	Sample C
SO_2 Analysis:			
Free SO_2	5	7	3
Combined SO_2	5	3	7
Total SO_2	10	10	10
"True Free SO_2"	0	4	0
Liquor Composition (as SO_2):			
H_2SO_3	0	4	0
$SO_3^=$	0	0	4
HSO_3^-	10	6	6

The traditional calcium acid sulfite cook must be carried out at a low pH of about 1.5 because of the relative insolubility of calcium. A higher pH would cause scaling compounds to precipitate during cooking, leading to a condition known as "liming up" of the digester (i.e., the reaction vessel). Typically, 80% or more of the SO_2 is in the form of free SO_2.

The use of soluble bases (magnesium, sodium, ammonium) permits a greater proportion of combined SO_2 in the cooking liquor. In so-called acid bisulfite pulping, any pH within the range 1.5 to 4.0 can be achieved by controlling the ratio of free to combined SO_2. True bisulfite pulping, defined by equal amounts of free and combined SO_2, is carried out at a pH of 4.0 to 5.0. The full range of sulfite cooking also includes neutral sulfite (discussed in Chapter 4) and alkaline sulfite, which has not been used commercially but holds promise as part of a two-stage process (see Section 6.7) or with anthraquinone (see Section 7.6).

It must be noted with regret that descriptive terms for sulfite pulping are not totally standardized in spite of industry efforts (e.g, reference 1). For example, the terms acid sulfite and bisulfite are sometimes used interchangeably. The pH levels noted above are only approximate, since the actual pH during sulfite cooking is temperature-dependent.

6.3 PROCESS DESCRIPTION

A simplified flowsheet for a sulfite pulping system is shown in Figure 6-1. The cooking liquor is usually prepared by burning sulfur to produce SO_2 gas and then absorbing the SO_2 in an alkaline base solution. In the older calcium sulfite mills, limestone was used exclusively in the gas absorption tower, serving both as a packing and a chemical source of calcium to produce calcium bisulfite. In the newer mills, one of

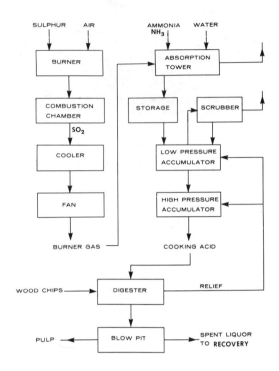

FIGURE 6-1. Sulfite pulping process with ammonia base (A.J. Bruley).

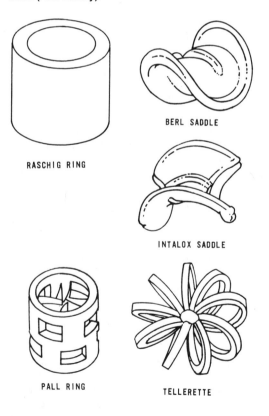

FIGURE 6-2. Common inert tower packing materials.

the soluble bases, in the form of NH_4OH, $Mg(OH)_2$ or Na_2CO_3, is used to absorb the SO_2; inert packings such as berl saddles (Figure 6-2) are used to provide contact area in the absorption tower.

The raw cooking acid after SO_2 absorption is a mixture of free SO_2 and combined SO_2 in the desired proportions. Before the raw acid is used in pulping, it is fortified with SO_2 relief gas from the digesters (see Figure 6-3). This fortification usually takes place in low- and high-pressure accumulators. Typical acid strengths (for acid sulfite) and degree of fortification are shown below (in % by weight):

	Raw Acid	Cooking Acid
Total SO_2	4.0 - 4.2	6.0 - 8.0
Free SO_2	2.8 - 2.5	5.0 - 6.8
Combined SO_2	1.2 - 1.7	1.0 - 1.2

The cooking operation is usually carried out batchwise in a pressure vessel (digester) consisting of a steel or stainless steel shell with acid-resistant lining (see Figure 6-4). The digester is first filled with chips and capped, then sufficient hot acid from the high-pressure accumulator is added to almost fill the vessel. The contents of the digester are heated according to a predetermined schedule by forced circulation of the cooking liquor through a heat exchanger. The pressure of the gas pocket at the top of the digester is allowed to rise to a predetermined level, and then is controlled by bleeding off SO_2 gas, which is absorbed in the accumulator.

As the temperature and pressure are increased, the hot acid is rapidly absorbed by the chips. The chemical reaction does not gain any momentum until the temperature has exceeded 110°C, but it is important at this stage that the wood structures are well "impregnated" with chemicals. A slow come-up time, a relatively low maximum temperature (130 - 140°C), and a long overall cook (6–8 hours) are typical for acid sulfite cooking in order to avoid undesirable lignin polycondensation reactions. More rapid heating with higher maximum temperatures and pressures can be used as the cooking liquor pH increases (i.e., for acid bisulfite cooking). Typical sulfite cooking curves of temperature and pressure are shown in Figure 6-5.

The extent of cooking is generally dictated by the amount of delignification that is desired (see Figure 6-6). Pulp for bleaching should be low in lignin (as measured by a permanganate number test), but if cooking proceeds beyond an optimum point, pulp strength, viscosity (i.e., dp) and yield will be adversely affected. The point at which to stop an individual cook is based on the operator's

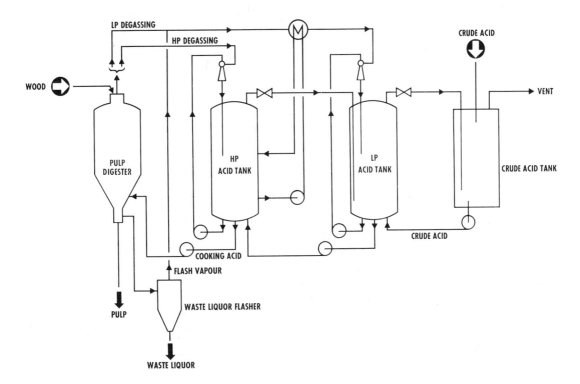

FIGURE 6-3. Representative pressurized acid system for absorbing SO_2 gases. The flash vapors are absorbed in stages, first by the high-pressure tank (accumulator); and as the digester is depressurized by the low-pressure tank.

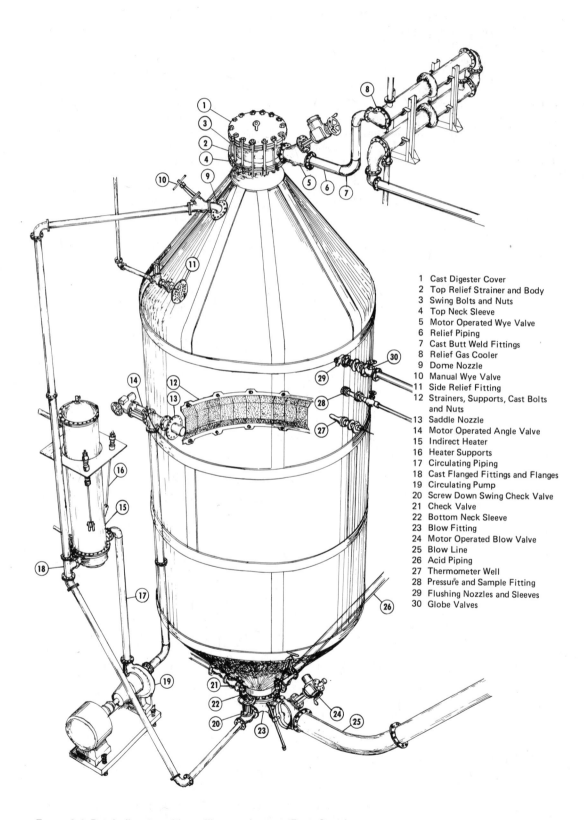

1   Cast Digester Cover
2   Top Relief Strainer and Body
3   Swing Bolts and Nuts
4   Top Neck Sleeve
5   Motor Operated Wye Valve
6   Relief Piping
7   Cast Butt Weld Fittings
8   Relief Gas Cooler
9   Dome Nozzle
10 Manual Wye Valve
11 Side Relief Fitting
12 Strainers, Supports, Cast Bolts
    and Nuts
13 Saddle Nozzle
14 Motor Operated Angle Valve
15 Indirect Heater
16 Heater Supports
17 Circulating Piping
18 Cast Flanged Fittings and Flanges
19 Circulating Pump
20 Screw Down Swing Check Valve
21 Check Valve
22 Bottom Neck Sleeve
23 Blow Fitting
24 Motor Operated Blow Valve
25 Blow Line
26 Acid Piping
27 Thermometer Well
28 Pressure and Sample Fitting
29 Flushing Nozzles and Sleeves
30 Globe Valves

FIGURE 6-4. Batch digester with ancillary equipment (Esco Corp.).

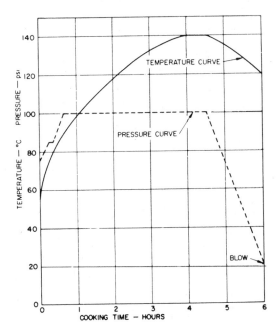

FIGURE 6-5. Typical sulfite cooking curves.

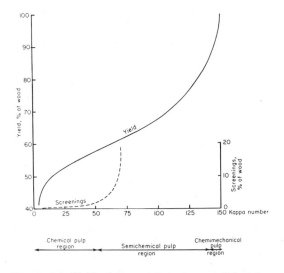

FIGURE 6-6. Yield and screenings (screen rejects) of sulfite pulps as a function of kappa number; also showing classification of pulp grades.

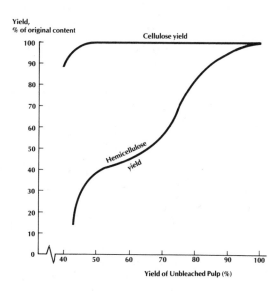

FIGURE 6-7. Yield of carbohydrate components during sulfite cooking of spruce.

judgement. Even a well-controlled mill operation will experience fluctuations in wood moisture and wood quality that cause variation in the rate of digestion. The progress of the cook is followed by observing the color of the liquor and by running periodic tests of residual SO_2.

When 1 to 1.5 hours of cooking time remain, the heating is discontinued and the pressure is gradually reduced by relieving gas and steam to the accumulator. When the pressure has been reduced to perhaps 20 or 25 psig, the contents are "blown" into a blow tank or blow pit. The gases are scrubbed to recover SO_2.

In older mills, the blow pit has a false bottom of strainer plates to allow the "red liquor" to drain off. Water is then sprayed onto the pulp and drained to achieve removal of residual liquor. This batch washing technique requires large volumes of water, and is rapidly being replaced with continuous and more efficient modern methods (see Section 9.3). Following washing, the pulp is screened and cleaned. These operations, both critical to the production of high quality sulfite pulp, are also discussed in Chapter 9.

6.4 CHEMISTRY OF SULFITE PULPING

The primary reactions taking place during sulfite cooking are generally characterized as follows:

(1) Free sulfurous acid combines with lignin to produce relatively insoluble lignosulfonic acid.

(2) In the presence of the base, lignosulfonic salts are formed which are more soluble.

(3) The sulfonated lignin undergoes cleavage into

smaller and more soluble molecular fragments due to hydrolysis reactions.

(4) Hemicellulose is hydrolyzed into soluble sugars.

All of the above reactions are desirable, except perhaps the hydrolysis of hemicellulose which results in loss of the holocellulose fraction, and hence lowers yield. Cellulose is relatively stable to chemical attack by the sulfite liquor, the major effect during cooking being a reduction in dp. However, at a yield of approximately 45%, the cellulose becomes vulnerable to hydrolysis reactions, and a portion is dissolved at lower yield levels. The comparative yield losses between cellulose and hemicellulose during sulfite cooking are illustrated in Figure 6-7.

Easier delignification (compared to the kraft process) makes the sulfite process well suited to produce pulps rich in hemicelluloses, which are ideal for making such paper grades as greaseproof and glassine. At the same time, the ease of hemicellulose removal on prolonged sulfite cooking (using somewhat higher temperature and acidity) also makes possible the production of pulp with high cellulose purity, suitable for dissolving grades and special opaque papers.

The polycondensation of lignin is a highly undesirable reaction which can occur under conditions of high acid concentration and/or high temperature. The dark-colored and insoluble compounds that are formed give rise to the apt description of "burnt" cook. Polycondensation of lignin always occurs when sulfurous acid is used alone as a pulping agent. In the presence of a base, the system is buffered and polycondensation is usually prevented. Even with a base present, problems can still occur due to more rapid penetration of SO_2 into the wood, especially with respect to the slower-diffusing divalent bases (calcium and magnesium).

Polycondensation reactions are much less likely to occur with the monovalent bases, sodium and ammonium, especially in the pH range of acid bisulfite and bisulfite cooking. Because of more rapid penetration and lower acid concentrations, faster come-up times and higher temperatures can be used, so that total cooking time can be significantly reduced in comparison to calcium acid sulfite cooks.

6.5 CHEMISTRY OF COOKING LIQUOR PREPARATION

Sulfite cooking liquor is a mixture of free sulfurous acid (H_2SO_3) and combined sulfurous acid in the form of bisulfite ion (HSO_3^-). Control of the liquor-making operation with respect to maintaining target strength and excluding contaminants is vital to the economic production of quality sulfite pulp.

SO$_2$ Preparation

Sulfur dioxide is usually obtained by burning molten sulfur in air according to the following reaction:

$$S + O_2 \longrightarrow SO_2$$

The exothermic reaction is maintained at a relatively high temperature (above 1100°C) by controlling air flow at about 10% excess oxygen in order to prevent further oxidation to SO_3. Subsequent conversion to

SO_3 is also inhibited by rapid cooling of the off-burner gas through the temperature range that favors SO_3 formation (600 - 900°C).

Absorption System Reactions

The cooled SO_2 (at about 200°C) is absorbed in water containing the base chemical. The following equations apply:

(1) Formation of sulfurous acid:

$$SO_2 + H_2O \longrightarrow H_2SO_3$$

(2) Formation of calcium bisulfite from limestone:

$$CaCO_3 + 2H_2SO_3 \longrightarrow Ca(HSO_3)_2 + CO_2 + H_2O$$

The above overall reaction can be shown as two intermediate reactions:

(2a) $CaCO_3 + H_2SO_3 \longrightarrow CaSO_3 + CO_2 + H_2O$

(2b) $CaSO_3 + H_2SO_3 \longrightarrow Ca(HSO_3)_2$

or [Combined SO_2] + [Free SO_2] $\longrightarrow$ [Bisulfite]

Intermediate formation of the monosulfite ($SO_2^=$) is important in each reaction since it explains the concept of free and combined SO_2 and can, in fact, determine actual tower operation. For example, in the formation of magnesium bisulfite liquor, the $Mg(OH)_2$ is usually added in stages to minimize buildup of $MgSO_3$.

(3) Formation of magnesium bisulfite from magnesia:

$$MgO + H_2O \longrightarrow Mg(OH)_2$$

$$Mg(OH)_2 + 2H_2SO_3 \longrightarrow Mg(HSO_3)_2 + H_2O$$

(4) Formation of sodium bisulfite from soda ash

$$Na_2CO_3 + H_2SO_3 \longrightarrow 2NaHSO_3 + CO_2 + H_2O$$

(5) Formation of ammonium bisulfite from ammonia:

$$NH_3 + H_2O \longrightarrow NH_4OH$$

$$NH_4OH + H_2SO_3 \longrightarrow NH_4HSO_3 + H_2O$$

Because of the limited solubility of SO_2 in water at atmospheric pressure (e.g., about 5.3% SO_2 at 40°C), the acid from the absorber system must be fortified under pressure. This operation takes place in the accumulators, where the cooking acid is also heated by digester relief.

6.6 OPERATION AND CONTROL OF COOKING

The basic variables for conventional sulfite cooking processes are listed in Table 6-3. The major "driving forces" are pH, free SO_2 concentration, and temperature. As the pH ranges from acid sulfite pulping (pH 1.5 -2.0) to bisulfite pulping (pH 4.0 - 5.0), the reaction rates can be maintained by using higher temperatures. 140°C is usually considered a maximum level for calcium acid sulfite cooking because of the polycondensation reactions that can occur at higher temperatures. When monovalent bases are used (usually at higher pH), the temperature can be increased to 160°C or higher.

For a given reaction rate, the cooking time must be controlled to achieve the desired degree of cooking, as measured by the yield or the lignin content (permanganate test). Where the reaction rate is low, a longer cook is required. "Slow cooks" can adversely affect pulp mill productivity if digester capacity is marginal.

In sulfite cooking, the batch digesters are totally filled with liquor and chips, except for the gas pocket at the top. A maximum weight of chips must be supplied to each digester to maintain productivity and minimize cooking variations; a full chip charge also serves to reduce the liquid volume required and to ensure the highest free SO_2 concentration. Various methods of chip packing are employed during the filling operation to help obtain a full and uniformly-distributed chip charge (refer to Section 8.1). The volume and strength of cooking acid is adjusted to ensure sufficient active chemical (in terms of % SO_2

on dry wood) for the cooking reactions and to provide a residual at the end of the cook. If the liquid-to-wood ratio is high for any reason, the concentration of active chemical will be low and the cooking rate reduced. Cooking acid is supplemented by residual cooking liquor to provide the correct liquid volume.

It is important to obtain good liquor penetration into the chips well before the maximum temperature is reached. Some mills utilize a hydraulic pressure on the digester (above the pressure exerted by the SO_2) at the beginning of the cook to reduce the penetration period, which may be as long as three hours in the case of calcium acid sulfite.

Considerable work has been done to compare the effect of different bases on sulfite cooking rates, pulping economics and pulp qualities (2). For identical conditions of temperature, pH and SO_2 concentration, the fastest cooking rates are found for the ammonium base, followed by magnesium, sodium and calcium, in descending order (see Figure 6-8). However, SO_2 consumption is highest using the ammonium base, and unbleached ammonium bisulfite pulps are distinctly lower in brightness and more difficult to bleach. The major factors influencing sulfite cooking rates are summarized in Table 6-4.

Strength properties for sulfite pulps have been shown by some investigators to be independent of

---

TABLE 6-3. Variables affecting sulfite cooking.

Wood Chips
- species
- general chip quality (size distribution, freedom from contaminants, etc.)
- moisture content
- degree of "seasoning"

Cooking Acid
- choice of base
- ratio of free/combined SO_2

Cooking Control
- chemical application (% SO_2 on o.d. wood)
- liquor-to-wood ratio (wt. liquor/wt. o.d. wood)
- pressure
- temperature cycle
- time of cooking

Control Parameters
- degree of delignification (as indicated by permanganate or kappa number test)
- color of liquor
- residual SO_2
- reduction in dp (as indicated by pulp viscosity test)

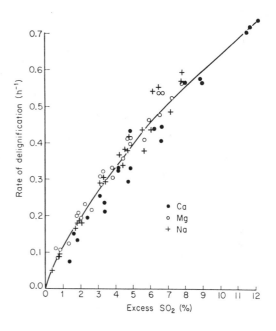

FIGURE 6-8. Rate of delignification (as inverse time to 90% delignification) vs excess SO_2 using three types of bases (data of Maass, et al.).

the base used, as for example in Figure 6-9. It must be noted that these findings apply only to identical cooking acid compositions (constant free and combined SO_2). An important point is that the soluble bases permit higher pH reactions, which do indeed produce stronger pulps. (See Figure 6-10 comparing acid sulfite and bisulfite pulp strengths.)

6.7 TWO-STAGE SULFITE PULPING

Subsequent to the introduction and application of magnesium and sodium bases, several two-stage sulfite pulping methods using these bases have been developed and utilized commercially. These methods are generally characterized by a dramatic change in cooking pH between the initial stage (for penetration and sulfonation) and the second stage (for

dissolution and removal of the lignin). In most cases, the recovery system is well integrated with the pulping process.

A listing of the most prominent two-stage processes is given in Table 6-5. The major advantages claimed for these generally longer and more complicated cooking processes are:
• more complete delignification (easier bleaching)
• better retention of hemicelluloses (higher yield)
• greater tolerance for wide variety of wood species.

TABLE 6-4. Factors influencing sulfite pulping reaction rates.

	Slow Rate		Fast Rate
pH	high	→	low
SO_2 concentration	low	→	high
Temperature	low	→	high
Base	Ca → Na → Mg → NH_4		

TABLE 6-5. Operational pH for two-stage sulfite processes.

	1st Stage	2nd Stage
Two-Stage Acid (Mg)	5.5	1.5 – 2.0
Two-Stage Magnefite (Mg)	3 – 4	6.0 – 6.5
Stora (Na)	6	3 – 4
Sivola (Na)	3 – 4	9 – 10

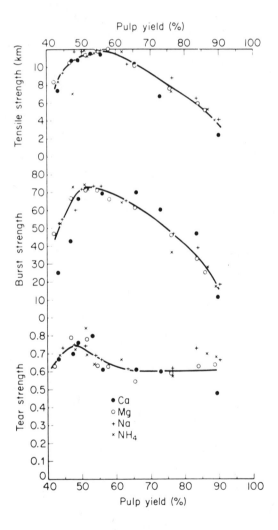

FIGURE 6-9. Pulp strength properties vs pulp yield for acid sulfite pulping of Canadian softwoods using four different bases (data of Strapp, et al.).

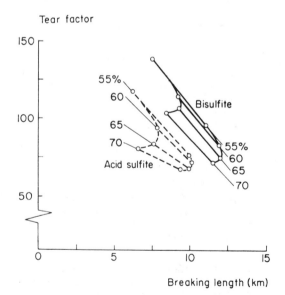

FIGURE 6-10. Comparison of tear and tensile strengths for acid sulfite and bisulfite pulps, both unbeaten and beaten to approximately 250 CSF (data of Hartler, et al.).

6.8 FUTURE PROSPECTS FOR SULFITE PULPING

During the period from 1960 to 1990, the portion of the world's annual wood pulp production attributed to traditional sulfite methods has fallen from 20% to 8%. In the face of such a statistic, there appears to be little cause for optimism about the long-term prospects for the sulfite pulping process. Nonetheless, the process retains its adherents (3), and several companies have made recent investments in new plant equipment with the obvious intention to remain competitive. Also, some researchers are predicting a brighter future based on recent process innovations.

Research efforts have concentrated on alkaline sulfite pulping. Moderately alkaline and highly alkaline sulfite processes are a comparatively recent development, and the introduction of anthraquinone (AQ) as a pulping liquor additive (to improve reaction kinetics and increase pulp yield) has made the methodology more attractive. A number of mills in Finland, Australia and Japan are now using an alkaline sulfite-AQ process. However, there are some problems to be resolved, especially with respect to liquor penetration into the chips.

One promising new sulfite pulping modification is the ASAM process (for Alkaline Sulfite process using Anthraquinone and Methanol) developed in Germany (4). The process appears to have a number of advantages (e.g., high pulp yield, high strength, insensitivity to wood species, high unbleached brightness, easy bleaching, etc.) which may make it competitive with kraft.

REFERENCES

(1) **TAPPI TIS 0607-25** (formerly **TAPPI Standard T1201**): **Definition of Terms in the Sulfite Pulping Process**

(2) ERNEST, F.M. and HARMAN, S.M. **Comparison of Several Bases in the Bisulfite Pulping of Wood** TAPPI 50:12:110-116 (December 1967)

(3) HERGERT, H.L. **Future of Sulfite Pulping is Tied to Integrated Mills, Market Pulp** *Pulp & Paper* (April 1982)

(4) BLACK, N.P. **ASAM Alkaline Sulfite Pulping Process Shows Potential for Large-Scale Operation** *Tappi Journal* (April 1991)

Chapter 7

Kraft Pulping

7.1 BRIEF HISTORY OF DEVELOPMENT

The soda process, the first recognized chemical pulping method, utilized a strongly alkaline solution of sodium hydroxide to delignify wood chips. This precursor of the kraft process was originally patented in 1854. A later patent in 1865 covered the incineration of the spent soda liquor to recover most of the alkali used in the process. The first successful soda mill went into operation in 1866. A few soda mills are still in operation around the world producing pulp from hardwoods and nonwood fibrous raw materials.

C.F. Dahl is credited with the development of the kraft (or sulfate) process. In an effort to find a substitute for expensive sodium carbonate (soda ash) as makeup for the soda process chemical cycle, he experimented by adding sodium sulfate (saltcake) to the recovery furnace. The sulfate was chemically reduced to sulfide by the action of the furnace, and sulfide was thus introduced into the liquor system. Dahl subsequently found that sulfide in the cooking liquor greatly accelerated delignification and produced a much stronger pulp; he obtained a patent for his process in 1884.

The new pulping method was first used commercially in Sweden in 1885. The superior strength properties were recognized, and the new type of papers were aptly called kraft papers. Kraft is the German word for strength. Following this development, many soda mills converted to the kraft process in order to compete with mills using the sulfite process. While sulfite pulp was stronger, cheaper and lighter in color than soda pulp, the kraft process narrowed the margin on production cost and gave a product which was stronger, though considerably darker in color. Economical recovery of pulping chemicals was a necessity for the kraft process to compete successfully against the sulfite process, which required no recovery system.

The great impetus for dominance by the kraft process came in the 1930's with the introduction of the Tomlinson recovery furnace, where final evaporation and burning of spent liquor were combined with recovery of heat and chemicals in a single process unit. Finally, the development and promotion of chlorine dioxide bleaching by Howard Rapson in the late 1940's and early 1950's paved the way to achieve bleached brightness levels on a par with sulfite pulps.

7.2 KRAFT PROCESS NOMENCLATURE AND DEFINITIONS

A simplified schematic of the kraft liquor cycle is shown in Figure 7-1. White liquor containing the active cooking chemicals, sodium hydroxide (NaOH) and sodium sulfide (Na_2S), is used for cooking the chips. The residual black liquor containing the reaction products of lignin solubilization is concentrated and burned in the recovery furnace to yield an inorganic smelt of sodium carbonate (Na_2CO_3) and sodium sulfide. The smelt is dissolved to form green liquor, which is reacted with quick lime (CaO) to convert Na_2CO_3 into NaOH and regenerate the original white liquor. The major terminology used to chemically characterize these liquor streams is defined in Table 7-1.

Most mills maintain white liquor sulfidity within the range of 25-35% (based on TTA). The critical low level for sulfidity is not well defined and may vary depending on other system parameters. However, most investigators agree that both cooking reaction rate and pulp quality are adversely affected

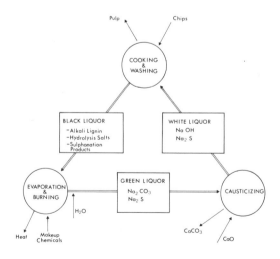

FIGURE 7-1. The kraft liquor cycle.

TABLE 7-1. Definition of kraft pulping terms.

	Term	Definition	Units
(1)	Total Alkali	Total of all "viable" sodium alkali compounds, i.e. $NaOH + Na_2S + Na_2CO_3 + Na_2SO_4 + Na_2S_2O_3 + Na_2SO_3$ (Does not include $NaCl$)	g/L as Na_2O
(2)	Total Titratable Alkali (TTA)	Total of $NaOH + Na_2S + Na_2CO_3$	g/L as Na_2O
(3)	Active Alkali (AA)	Total of $NaOH + Na_2S$	g/L as Na_2O
(4)	Effective Alkali (EA)	Total of $NaOH + 1/2 Na_2S$	g/L as Na_2O
(5)	Activity	Ratio of AA to TTA	expressed as %
(6)	Causticity	Ratio of $NaOH$ to $NaOH + Na_2CO_3$	% (on Na_2O basis)
(7)	Sulfidity	Ratio of Na_2S to AA (or to TTA) NB: The basis of sulfidity must be defined in each case.	% (on Na_2O basis)
(8)	Causticizing Efficiency (White Liquor)	Same as causticity. (However, the concentration of $NaOH$ in the green liquor should be subtracted so that the value of $NaOH$ represents only the portion produced by the causticizing reaction.)	% (on Na_2O basis)
(9)	Residual Alkali (Black Liquor)	Alkali concentration determined by acid titration.	g/L as Na_2O
(10)	Reduction Efficiency (Green Liquor)	Ratio of Na_2S to all soda sulfur compounds (sometimes simplified as ratio of Na_2S to $Na_2S + Na_2SO_4$)	% (on Na_2O basis)

at sulfidities below 15%. A higher level is maintained to provide a safety margin and allow greater use of inexpensive makeup chemicals containing sulfur (e.g., Na_2SO_4). Higher sulfidities also help to prevent loss of cellulose viscosity during cooking.

There is disagreement whether active alkali or effective alkali provides the better measurement of active chemical concentration for kraft cooking. Although both $NaOH$ and Na_2S take part in the cooking reactions, it can be shown that $NaOH$ provides the prime driving force. Since Na_2S hydrolyzes in solution:

$$Na_2S + H_2O \longrightarrow NaOH + NaSH$$

It follows that only one-half of the Na_2S is really effective in determining reaction kinetics (1).

In a mill with good sulfidity control, it makes little practical difference whether active alkali or effective alkali is used for measurement of chemical application. However, where sulfidity varies over a wide range, a constant active alkali corresponds to a variable effective alkali, and the choice of control parameter depends on consideration of their relative merit. The relationship between active alkali and effective alkali is illustrated in Figure 7-2.

7.3 DESCRIPTION OF KRAFT PROCESS

The sequential steps in the kraft pulping and recovery process are shown schematically in Figure

7-3. Following cooking (or digestion), the spent black liquor is washed from the pulp and treated in a series of steps to recover the cooking chemicals and regenerate the cooking liquor (see Chapter 10).

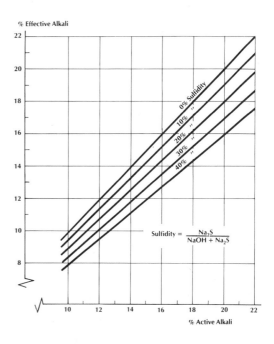

FIGURE 7-2. Conversion graph: active alkali vs effective alkali.

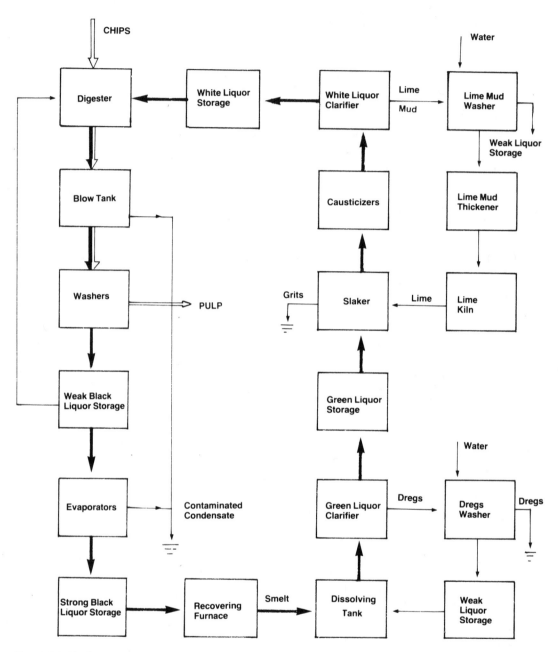

FIGURE 7-3. Kraft process.

The digestion process can be either batch or continuous. In batch cooking, the digester vessel is filled with chips and enough liquor is added to cover the chips. The contents are then heated according to a predetermined schedule, usually by forced circulation of the cooking liquor through a heat exchanger. Air and other noncondensible gases are relieved through a pressure control valve at the top of the vessel. The maximum temperature is typically reached after 1.0–1.5 hours, which allows the cooking liquor to impregnate the chips. The cook is then maintained at maximum temperature (usually about 170°C) for up to 2 hours to complete the cooking reactions. After digestion, the contents are discharged into a blow tank where the softened chips are disintegrated into fibers; the off-vapors are condensed in a heat exchanger where water is heated for pulp washing. (See Section 8.1.)

In continuous cooking, the chips are first carried through a steaming vessel where air and other noncondensibles are purged. The preheated chips and liquor then enter the continuous digester where they move through an intermediate-temperature zone (115–120°C) to allow for uniform penetration of the chemicals into the chips. As the chip mass moves through the cooking vessel, the mixture is heated to the cooking temperature, either by forced circulation of liquor through a heat exchanger or by steam injection, and maintained at this temperature for 1 - 1.5 hours. Following completion of the cook, hot spent liquor is extracted into a low-pressure tank where flash steam is generated for use in the steaming vessel. The pulp is usually quenched to below 100°C with cool liquor to prevent mechanical damage to the fibers. (See Section 8.2.)

The cooked pulp is separated from the residual liquor in a carefully controlled process known as brown stock washing. The most common method employs a series of countercurrent vacuum drum washers to provide displacement of the liquor with minimum dilution (see Section 9.3). Some continuous digesters incorporate a diffusion washing step in conjunction with spent liquor extraction and pulp cooling. Following washing, the pulp is screened and cleaned. These operations, both important to the production of high-quality kraft pulp, are also discussed in Chapter 9.

Chemical Recovery System

The recovery of chemicals and energy from the residual black liquor and reconstitution of the recovered chemicals to form white liquor are integral to the kraft mill operation. The "weak black liquor" (about 15% solids) from the brown stock washers is processed through the following steps:

(1) concentration through a series of evaporation and chemical addition steps into "heavy black liquor" at 70–75% solids.
(2) incineration of heavy black liquor in the recovery furnace to form inorganic smelt.
(3) dissolving of furnace smelt in water to form green liquor.
(4) causticizing of green liquor with reburned lime to form white liquor for the next cooking cycle.

An important function of the recovery furnace is to chemically reduce the oxidized sulfur compounds contained in the burning solids to sulfide. This aspect of furnace operation is monitored with measurements of reduction efficiency.

Control of green liquor strength (i.e., TTA concentration) is essential for smooth operation within the liquor cycle. The target level is a compromise between two factors. A higher concentration increases the inventory of soda chemicals, which will help to level out the operation and provide surge capacity against interruptions. But, a lower concentration improves causticizing efficiency, ensuring that a lower "dead load" of non-reactive Na_2CO_3 will be carried around the cycle.

The various steps in the recovery process are considered in greater detail in Chapter 10.

7.4 CHEMISTRY OF KRAFT PULPING

The present knowledge of kraft pulping chemistry has been summarized by Kleppe (2). The reactions that occur are complex and not totally understood. Essentially, the swollen lignin in the wood chips is chemically split into fragments by the hydroxyl (OH^-) and hydrosulfide (SH^-) ions present in the pulping liquor. The lignin fragments are then dissolved as phenolate or carboxylate ions. Carbohydrates, primarily hemicelluloses and some cellulose, are also chemically attacked and dissolved to some extent. During a typical cook (of bleachable-grade pulp) approximately 80% of the lignin, 50% of the hemicelluloses and 10% of the cellulose is dissolved.

Given the right conditions, the lignin fragments are able to take part in condensation reactions, either with themselves or undissolved lignin and possibly with carbohydrates. The condensed lignin is more difficult to remove from the fibers. The hydrosulfide ion is believed to reduce condensation reactions by blocking reactive groups (e.g., hydroxyl in benzyl alcohols).

The two "driving forces" for kraft pulping reactions are alkali concentration (as measured by either effective alkali or active alkali) and temperature. Within the normal cooking temperature range (155–175°C), the delignification rate more than doubles for every 10°C increase.

By arbitrarily assigning a relative reaction rate of 1 for 100°C, a method has been developed (3) for expressing the cooking time and the temperature as one single variable. When the relative reaction rate is plotted against the cooking time in hours, the area under the curve is defined as the H-factor. The concept of the H-factor has been widely applied in cooking control, but is especially useful when the temperature varies during the cooking period. A typical plot of temperature vs. time for a batch cook is shown in Figure 7-4 along with the corresponding plot of relative reaction rate vs. time.

Delignification during kraft cooking proceeds in three distinct phases, as shown in Figure 7-5. The initial very rapid lignin removal is characterized as an extraction process. Removal of the bulk of the lignin follows as a first order reaction. Kraft cooks are typically completed at a lignin content of 4-5% for softwoods and about 3% for hardwoods, well within the bulk delignification phase. If cooks were allowed to proceed further, residual delignification would occur at a much lower rate.

Superimposed on the reactions with lignin are the reactions with hemicelluloses and cellulose.

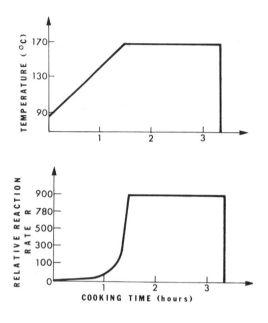

FIGURE 7-4. Temperature and relative reaction rate vs cooking time for batch cook.

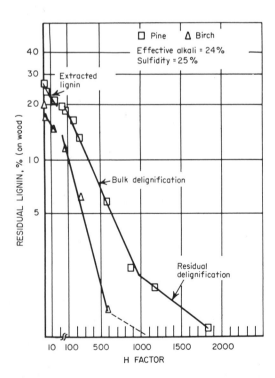

FIGURE 7-5. Lignin removal in kraft pulping of pine and birch is shown as a function of H-factor (from data of Aurell and Hartler).

Removal of each of the major wood components is shown as a function of delignification in Figure 7-6, and as a function of total pulp yield in Figure 7-7. The hemicellulose content is reduced by approximately 40% during the extraction stage of the cook, as compared to 20% of the lignin. The loss is caused by dissolution of low-molecular-weight carbohydrates, removal of acid groups, and degradation by the so-called "peeling reaction". The relatively low loss of cellulose, about 10%, is

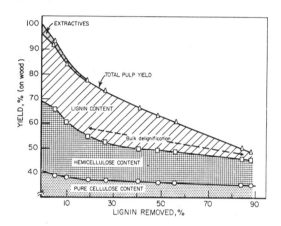

FIGURE 7-6. Yield of major wood components during kraft pulping of pine at different stages of delignification (from data of Aurell and Hartler).

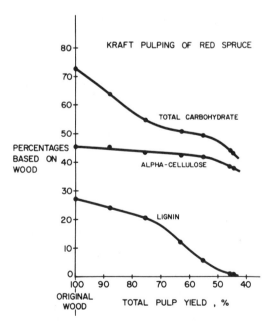

FIGURE 7-7. Removal of the major wood components during kraft pulping as a function of total pulp yield (D.W. Clayton).

explained by the low accessability of hydroxyl ions into the crystalline region of the cellulose. The acidic reaction products from the carbohydrates consume a major part of the alkali in the cooking liquor, as illustrated in Figure 7-8.

During delignification, there is also a corresponding reduction in pulp viscosity (a measure of cellulose weighted-average molecular weight). If pulp viscosity is allowed to fall below a critical level, the pulp strength drops dramatically. Maintenance of pulp viscosity is the principal reason that the kraft cook must be terminated at a point where a substantial level of residual lignin remains with the fibers. A derivation termed the G-factor (exactly analogous to the H-factor) can be applied for viscosity reduction, i.e., combinations of time and temperature that give the same G-factor can be expected to produce pulp with the same viscosity. Since the G-factor increases more rapidly with temperature than the H-factor, it follows that higher cooking temperatures have a proportionally greater effect on viscosity reduction. This is the principal reason why cooking temperatures above 180°C are avoided.

About 90% of the extractives in wood are removed during the extraction phase of the cook. Fatty and resin acids form sodium salts and are removed later from the residual liquor as tall oil soap. Volatile turpentine is recovered from the vapor relief during steaming or cooking.

Prehydrolysis Stage for Dissolving Grades

Regular (papermaking) kraft pulp contains certain carbohydrates, principally pentosans, that interfere with the chemical conversion of cellulose into rayon and acetate products. As a consequence, the practice of exposing the wood chips to acid hydrolysis prior to alkaline cooking was developed in order to reduce the pentosan content and obtain a higher proportion of alpha cellulose.

The usual method of prehydrolysis is with direct steaming. The action of the steam liberates organic acids from the wood which, at elevated temperature, hydrolyze hemicelluloses to soluble sugars. The subsequent kraft cook then produces a pulp suitable for dissolving applications. Typically, the pulp yield from a prehydrolyzed cook is 5 to 7% lower on dry wood than from a regular kraft cook.

A representative cooking schedule for batch cooking of prehydrolyzed hardwood kraft pulp is shown in Figure 7-9. Total cooking cycle time is 8 hours, including 140 minutes at maximum hydrolysis temperature and 70 minutes at maximum kraft cooking temperature.

7.5 OPERATION AND CONTROL

The basic variables affecting the kraft cooking process are listed in Table 7-2. The objective is to cook consistently to a target lignin content (as measured by the kappa number or permanganate number test) with a minimum level of screen rejects.

As with most chemical pulping methods, sufficient time must be provided at a lower temperature to achieve good liquor penetration into the chips before

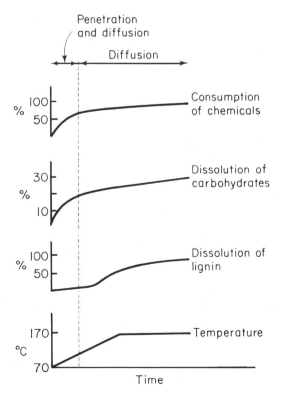

FIGURE 7-8. Consumption of chemicals, dissolution of carbohydrates, and dissolution of lignin as a function of time and temperature (Hartler).

TABLE 7-2. Variables affecting kraft cooking.

Wood Chips	• species
	• general chip quality (size distribution, freedom from contaminants, etc.)
	• moisture content
Cooking Liquor	• sulfidity
Cooking Control	• chemical application (AA or EA on o.d. wood)
	• liquor-to-wood ratio
	• temperature cycle
	• time/temperature curve (H factor)
Control Parameters	• degree of delignification (as indicated by permanganate or kappa number tests)
	• residual alkali

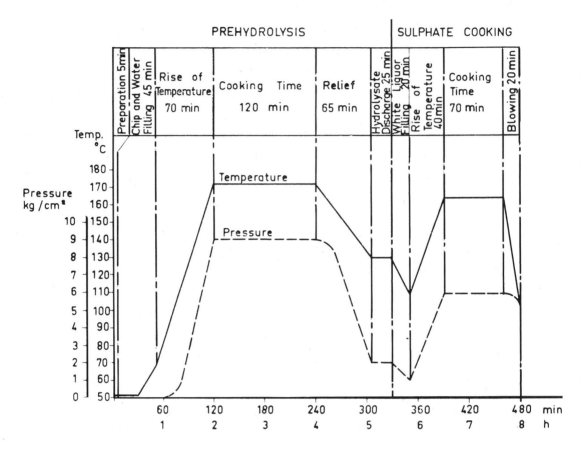

FIGURE 7-9. Prehydrolyzed kraft batch cooking schedule.

the main cooking reactions occur. The obvious mechanism of penetration is by capillary movement along the lumina (tracheids and ray cells), pits, resin ducts, and for hardwoods, the vessels; and into various longitudinal fissures. A secondary mechanism is by diffusion through the cell walls. Non-swelling or low-swelling liquids penetrate 50 to 200 times more rapidly in the longitudinal direction (by capillary movement) than in the transverse direction (by diffusion). With a swelling agent such as sodium hydroxide, the difference is much less; according to one source, only six times as great in the longitudinal direction. Therefore, chip thickness is far more important with respect to kraft liquor impregnation.

Insufficient penetration (or impregnation) lowers the degree of cooking and increases the level of screen rejects. Since air in the chips can interfere with penetration, it has been found useful in continuous systems to presteam the chips to expel the air. The increased temperature also contributes to improved liquor transport.

Effect of Chip Size

Reduction in chip thickness allows somewhat faster pulping rates and dramatically reduces the amount of screen rejects, as illustrated in Figure 7-10. A reduction in alkali usage and slightly improved retention of carbohydrates is possible when using

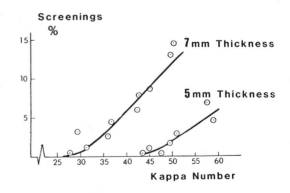

FIGURE 7-10. Effect of chip thickness and delignification on screenings content at a cooking temperature of 170°C (Hartler).

"shredded chips". However, chip shredding has been little used commercially, and it appears that the benefits do not justify the additional handling and energy requirements.

Generally, smaller-length wood fragments (i.e., pin chips, fines, sawdust) produce lower-yield, weaker pulps, and consume greater amounts of alkali. A high percentage of fine material in the chip furnish will cause poor liquor circulation in both batch and continuous digesters, and it is generally advisable to cook this material separately in specially-designed vessels (see Section 8.4).

Effect of Liquor Sulfidity

As compared to soda pulping, kraft pulping of softwoods is faster and provides a higher yield and a stronger pulp. The effects of Na_2S are quite dramatic up to about 15% sulfidity, but there appears to be only marginal additional benefit at higher sulfidities.

Effect of Alkali Charge

The normal alkali requirement for kraft pulping of softwoods is about 12 to 14% effective alkali on dry wood, while 8 to 10% is typical for hardwoods. In every situation, it is essential to provide sufficient chemical to carry the cooking reactions to completion. In practice, a slight excess of chemical is utilized to maintain a driving force and prevent redeposition of dissolved material (lignin) back onto the fibers.

The usual practice in mill operation is to utilize the minimum practicable alkali charge and vary the cooking temperature to achieve the desired reaction rate. However, the alkali application can also be used to adjust reaction rate; Figure 7-11 shows the effect of increasing effective alkali on the H-factor required for a constant kappa number. However, a higher alkali charge also causes a slight reduction in hemicellulose retention at a given kappa number (as illustrated in Figure 7-12) and changes the composition of the retained hemicelluloses.

Kraft mills producing unbleached market pulp usually employ a relatively high alkali charge. This practice may appear contradictory in view of the aforementioned effect on hemicellulose retention. However, the increased alkali provides a pulp with marginally higher brightness and lower screen rejects, allowing the mill to control the process at a significantly higher kappa number, and therefore to produce pulp at overall higher yield.

Effect of Maximum Pulping Temperature

Except for the effect on reaction rate, the choice of maximum temperature up to 180°C does not significantly affect the cooking result. Above 180°C, losses in both strength and yield become significant due to attack on the cellulose.

Effect of Liquor-to-Wood Ratio

For adequate penetration, a sufficient volume of liquor is required to ensure that all chip surfaces are wetted. In batch cooks, the digester is normally about 75% filled with liquor at the start of the cook. As the cook proceeds, chip moisture and lignin enter the liquid phase while the chip mass settles; the liquid level thus rises in relation to the chip level.

Sufficient white liquor is supplied to provide the specified alkali charge. The balance of the liquid requirement is typically made up with black liquor. Liquor-to-wood ratios range between 3 and 5. The effect of greater dilution is to decrease the

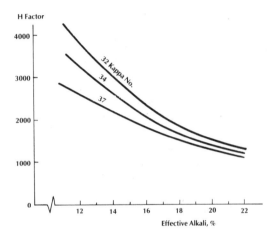

FIGURE 7-11. Required alkali and H-factor to achieve a given kappa number.

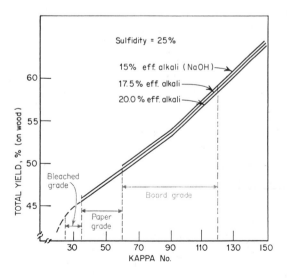

FIGURE 7-12. Total pulp yield in kraft pulping of southern pine as a function of kappa number and effective alkali application (Kleppe).

concentration of active chemical and thereby reduce the reaction rate.

Since redeposition of lignin onto the fibers can occur under certain circumstances, the wisdom of using black liquor to adjust total liquor volume might be questioned. The main reason, of course, is to avoid unnecessary dilution of the black liquor which would add to the evaporation load. There is also evidence that the significant sodium sulfide concentration remaining in the black liquor is helpful in cooking.

To ensure minimum dilution and maximum productivity from batch digesters, it is essential to employ a reliable method of chip packing. A digester with a full chip charge has less void volume to fill with liquor. (Refer to Section 8.1.)

Control Parameter

As previously noted, the control objective in kraft pulping (as well as in other chemical pulping methods) is to cook to a target kappa number. Due to the nonuniformity of a typical wood chip furnish, a certain variability in cooking result can be expected. Where a trend in the kappa number test is apparent, an offsetting change in the H-factor is usually applied to bring the test values closer to the target value. As well, at some mills, a sample of cooking liquor is extracted from the digester toward the end of a cook and analyzed for residual alkali to obtain an earlier indication of the cooking result.

A number of factors may be important in establishing the target kappa number for a pulp product. For example, the need for maximum pulp strength or a limitation on recovery furnace loading may be compelling in certain mill situations. As previously noted, the underlying objective for unbleached grades is to obtain the highest overall pulp yield. Likewise, for a bleachable grade pulp, the optimum kappa number usually corresponds to the maximum yield of bleached pulp. Figure 7-13 illustrates how the optimum range of kappa numbers is determined for a softwood kraft pulp. At lower kappa numbers (i.e., below the optimum), the unbleached screened yield is low and consequently the bleached yield is adversely affected. At higher kappa numbers, the higher screened pulp yield is more than offset by greater shrinkage (i.e., loss of yield) during bleaching.

7.6 PROCESS MODIFICATIONS

Because of significant advantages, the kraft process has become established as the world's dominant pulping method. Nonetheless, the process also has some severe shortcomings, principally the low yield from wood, the comparatively high residual lignin content of bleachable grades, and the malodors caused by reduced sulfur compounds. Various pulping modifications have been proposed over the years to overcome one or more of these shortcomings, including cooking additives, chip pretreatments, and two-stage cooks. Although progress has been slow, some of these technologies are now emerging into prominence.

Yield-Increase Modifications

Most of the earlier modifications were aimed at increased retention of carbohydrates by stabilizing the end groups of the cellulose and hemicellulose polymer chains against "peeling reactions". These reactions cause dissolution of the end groups one unit at a time and consume a significant portion of the cooking chemical. Two add-on technologies date from pre-1970, the polysulfide process (used commercially on a limited basis) and the H_2S pretreatment process (demonstrated at the pilot plant level). Both methods are successful in providing significant increases in pulp yield, up to 7% on dry wood. Unfortunately, both methods are associated with increased concentration of sulfur in the chemical cycle, which exacerbates the odor problem. Additionally, both methods require further processing steps that increase the capital and operating costs of the pulp mill.

A more recent development that holds greater promise is the use of anthraquinone (AQ) as an additive either for kraft or soda cooks (4). Because

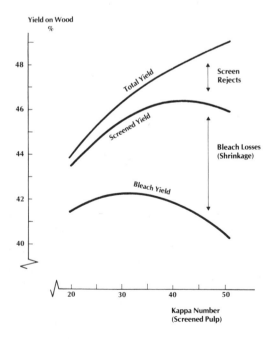

FIGURE 7-13. Yield vs kappa number relationships showing a maximum bleached yield at 28-34 kappa number.

this white, powdery organic compound is virtually insoluble in water, it is typically supplied as a dispersant. It dissolves in the reducing alkali of a kraft cook. A small addition of AQ (perhaps 1 lb/ton of pulp) accelerates the kraft pulping reactions, gives a 2 to 3% (on dry wood) increase in yield, and lowers the effective alkali requirement. When added to a soda cook, kraft-type yields, cooking rates and pulp strengths are possible.

Despite the reported advantages of anthraquinone-catalyzed pulping, the current cost of AQ (about US$ 4.50 per lb in 1991) hinders its general use, except under certain circumstances. Where wood costs are high, such as in Japan, pulping yields take on greater importance and AQ-type processes are more common. Nonetheless, in 1991, over 100 mills around the world were already using AQ or AQ-type additives. At least one mill is using anthraquinone in conjunction with polysulfide (5). More mills are certain to utilize this technology as prices for the additive are reduced. Chemical firms are pursuing a number of potential methods for producing low-cost anthraquinone catalysts, including one process based on lignin conversion (6). Exploring the full implications of AQ-type additives, both with respect to synthesis and applications, will keep investigators busy for many years to come.

Extended Delignification

To control bleach chemical cost, it has always made good sense in the production of bleachable grade pulps to delignify the wood raw material as much as possible during cooking, and thereby minimize the amount of residual delignification required during bleaching. More recently, with attention focused on the discharge of chlorinated organic compounds from bleach plants, the need to delignify pulp more completely prior to conventional bleaching has become more critical. (Refer also to Chapter 11).

The concept of extended delignification was developed in Sweden in the early 1980's. It was found that the kappa number of a softwood cook can be lowered from the conventional level of 30-32 down to about 25 without affecting pulp quality providing that the active alkali profile throughout the cook is leveled and the dissolved lignin content is low at the end of the cook. This concept has now been successfully implemented in both batch and continuous cooking systems, although extensive modifications to equipment are required. (Refer to Chapter 8 for system details.)

Oxygen delignification is another method of reducing the lignin content of pulp before conventional bleaching. This technology, which dates from the 1960's, has become a common component of bleached pulp manufacturing systems in Sweden and Japan, and has more recently been adopted in North America. Although technically independent, oxygen delignification is compatible with the kraft cooking process because its caustic effluent can be added to black liquor and processed through the recovery furnace. (Refer to Section 11.4.)

REFERENCES

(1) HINRICHS, D.D. **The Effect of Kraft Pulping Variables on Delignification** *TAPPI 50*:4:173-175 (April 1967)

(2) KLEPPE, P.J. **Kraft Pulping** (Feature Review) *TAPPI 53*:1:35-47 (January 1970)

(3) VROOM, K.E. **The H-Factor: The Means of Expressing Cooking Times and Temperatures as a Single Variable** *P & P Mag Can 58*:3:228-231 (Convention 1957)

(4) HOLTON, H. **Soda Additive Softwood Pulping: A Major New Process** *P & P Can 78*:10:19-24 (October 1977)

(5) LIGHTFOOT, W.E. **New Catalyst Improves Polysulfide Liquor Makeup, O_2 Delignification** *Pulp & Paper* (January 1990)

(6) DIMMEL, D.R. and BOZELL, J.J. **Pulping Catalysts From Lignin** *Tappi Journal* (May 1991)

Chapter

8

Cooking Equipment

This chapter is concerned primarily with alkaline pulping equipment. However, the principles discussed apply to all chemical pulping methods.

The objective for a new pulp mill design or for the expansion of an existing facility is to produce quality pulp at the lowest cost while meeting environmental standards. Quality pulp can be produced using either batch or continuous equipment, and the proper choice will depend on consideration of all relevant factors. Because of the interdependence of cooking with overall mill operations, it is not possible to generalize about the proper course of action for all installations.

When continuous digesters became operational in the late 1950's and early 1960's, a number of advantages were claimed. Among these, the claims of lower energy consumption and reduced pollution have generally proven valid; but claims of higher yield, better strength, and improved uniformity have not been substantiated by mill experience. Initially, a manpower savings was possible with continuous digester operation, but newer, fully-automated and computer-controlled batch digester installations can now be operated by just one individual per shift.

The auxiliary equipment associated with continuous digesters is more complex than that required by conventional batch digesters. One pulping line will usually depend on a single continuous digester or several batch digesters. If one batch unit is taken out of operation for any reason, the production rate is only curtailed; but if a continuous digester is down, the entire production is lost. Therefore, production availability normally favors the batch digester. Also, when frequent changes in wood furnish or degree of cooking are required, the batch digesters are more flexible. Generally, startups and shutdowns are easier for batch digesters.

Steam consumption for continuous digesters is much lower than for batch units because low-pressure flash steam from the liquor can be recycled to preheat and precondition chips. Steam demand is also more constant for continuous digesters, although fluctuation in demand from batch units can be reduced with computer control. The sizing of the power boiler is affected by both average demand and peak demand.

The relative advantages of batch and continuous digesters are summarized in Table 8-1.

TABLE 8-1. Advantages: batch vs continuous digesters.

Factors Favoring Continuous Digesters

1. Lower steam requirement (less energy).
2. More constant steam demand.
3. More compact; less space requirement.
4. Lower capacity of ancillary components because of constant loading (e.g., chip conveyor, heat recovery system).
5. Easier treatment of non-condensibles because of uniform flow.
6. Includes diffusion washing stage (Kamyr only!).
7. Adaptable for digestion of all wood subdivisions.

Factors Favoring Batch Digesters

1. More reliable operation.
2. Greater operating flexibility.
 • ability to change grades
 • ability to cook softwood and hardwood concurrently
 • ease of start-up and shut-down
3. More efficient turpentine recovery.

8.1 BATCH DIGESTERS

Several different types of pressure vessels have been used for batch cooking over the years, including stationary horizontal and spherical configurations and rotating vertical digesters. These types have been largely superseded by the stationary vertical unit (as was pictured in Figure 6-3). This discussion will be concerned only with the common vertical digester, but the principles apply to all batch units.

Modern mills are generally equipped with digesters of 7,000 to 9,400 ft^3 (200 to 265 m^3) producing up to 19 tons of pulp per batch. Larger units are more efficient up to a point; but as size is increased, the wall thickness must go up. Finally, fabrication and erection costs place a practical upper limit on unit size. Generally, kraft digesters are fabricated from carbon steel. A stainless steel liner must be used for prehydrolysis-type cooks. Sulfite digesters are usually lined with acid-resistant brick.

Screened chips are charged (usually from a transfer belt) into the top of the digester. The volumes of white liquor and black liquor required for the target alkali-to-wood and liquor-to-wood ratios, respectively, are then metered simultaneously into the digester. Some method of chip packing is usually employed to achieve maximum fill. Steam chip spreaders such as illustrated in Figures 8-1 and 8-2 are capable of packing 20% more chips in the digester than would be achieved by natural filling. Liquor filling and recirculation during chip charging is another method of packing. Direct steam injection at the bottom of the digester to shake the chip mass is also used for this purpose. These latter methods may give a 5–10% packing improvement.

When the digester is fully charged with chips and liquor, it is capped. Most batch units are now equipped with some type of automatic capping device, the most popular type being a remotely-operated ball valve (see Figure 8-3).

The modern batch digester utilizes a circulation system with an external liquor heater for bringing the contents to pulping temperature. Some older mills still use steam injection with convection mixing as illustrated in Figure 8-4. Good results can be achieved with either method, but forced circulation avoids liquor dilution, thereby maintaining higher alkali concentration during the cook and eliminating an additional load on the evaporators. A common circulation design draws liquor from a perforated

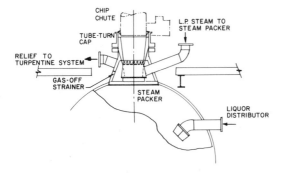

FIGURE 8-1. Rader steam packer assembly.

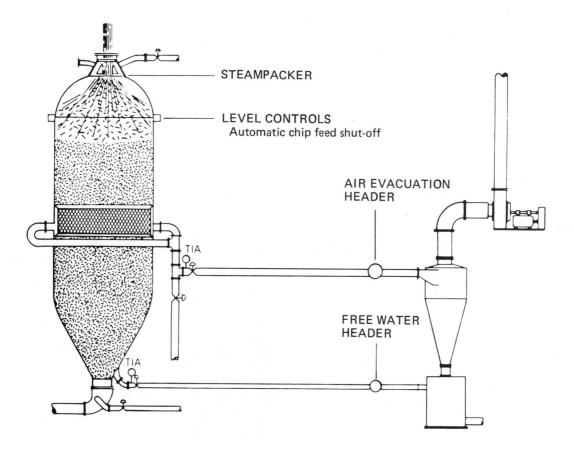

FIGURE 8-2. During steam packing, an air evacuation system draws air and other gases from the digester through the recirculation strainers (Rader).

FIGURE 8-3 Automatic capping valves can be seen in this view of the digester house operating floor.

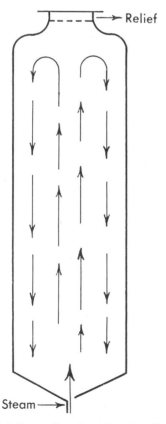

FIGURE 8-4. Convection circulation in a digester.

ring near the midpoint of the digester (refer to Figures 8-5 and 8-6), pumps it through a heat exchanger, and distributes the return flow to both the bottom and top of the vessel. The circulating pump is

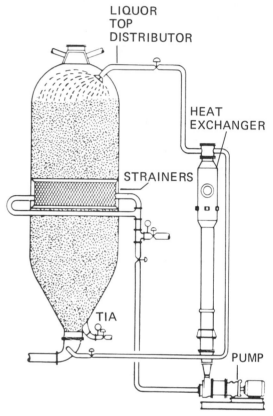

FIGURE 8-5. The indirect heating system draws liquor out of the digester through the strainer plates. The liquor is heated in the heat exchanger and returned to top and bottom inlets.

FIGURE 8-6. Strainer being installed in a brick-lined digester (Esco).

usually sized to turn over the liquid contents of the digester about once every ten minutes.

The loaded digester is brought to maximum temperature according to a preset rate (traditionally controlled by a clock-driven cam). It is then maintained at maximum temperature until the required H-factor has been reached. At that point, the remotely controlled blow valve at the bottom of the vessel is opened, and the contents are discharged under full digester pressure into the blow tank. As the cooked chip mass impacts on the tangential entry of the blow tank, the chips are disintegrated into fibers. A baffle and vapor space within the blow tank ensure good separation of stock and minimum entrainment of liquor with the flash steam. The flash steam along with organic vapors are usually carried into an accumulator heat recovery system (see Section 8.5).

Typically, the time cycle for each cook is the same, so that a staggered sequence of digester blowing and filling can be maintained for all the batch units.

8.2 MODIFICATIONS TO CONVENTIONAL BATCH KRAFT COOKING

At least 90% of batch kraft cooking is still carried out as described above. However, over the past decade some significant modifications to conventional batch cooking have been developed, and these improved methods are rapidly being incorporated into new installations and retrofitted into older mills. The new methodology involves additional equipment and processing steps, but does not extend the cooking cycle.

Displacement Heating of the Cook

Two methods of displacement heating have been developed and applied commercially to reduce the steam consumption of batch cooking. The two processes are called "Rapid Displacement Heating" (RDH) and "Super Batch" (or "Cold Blow System") by their developers, Beloit's Rader Division and Sunds Defibrator-Celleco, respectively. The methods are similar, and only the RDH system will be described in detail here.

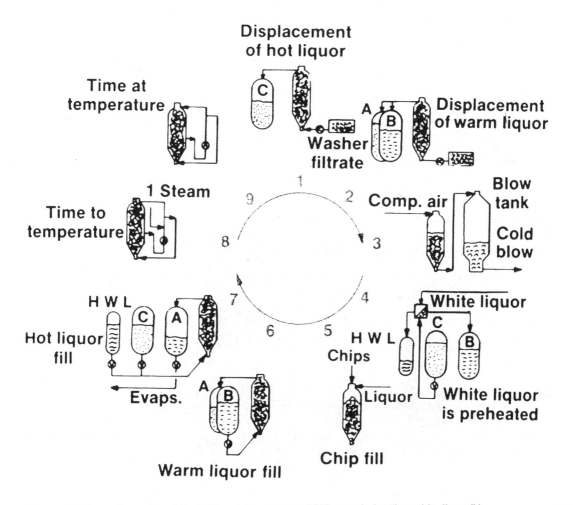

FIGURE 8-7. Operating cycle of the RDH pulping system. (HWL stands for "hot white liquor".)

The operating cycle of the RDH process is shown in Figure 8-7. It is most easily explained by starting at the end of the cook (step 1 on the diagram) where the H-factor has been reached. The spent cooking liquor is immediately displaced from the digester with brown stock washer filtrate, thereby ending the cook and cooling the chip mass. The displaced hot pulping liquor is retained under pressure in tank C. In step 2, displacement continues until the chip mass is cooled to about 80°C; the highest-temperature displaced liquors pass to pressurized tank B, and liquors below 95°C are displaced to atmospheric tank A. In step 3, the cooled chip mass is either blown with compressed air (as shown in the diagram) or pumped out as a dilute slurry (see forthcoming discussion on "new method of digester discharge").

On a continuous basis (shown as step 4), the hot black liquor in tank C is passed through a heat exchanger and gives up its heat to white liquor. The hot white liquor (at about 157°C) is stored in a pressure accumulator and the cooled black liquor passes to tank A. In step 5, the empty digester is charged with chips and packed with weak black liquor from tank A, which is introduced through special liquor nozzles. (In some systems, steam packing is used in place of liquor packing due to problems with air and soap reentrainment.) In step 6, the capping valve is closed and the digester is hydraulically filled, preheated, and pressurized with weak black liquor from tank B.

In step 7, hot white and black liquors from the accumulator and tank C, respectively, are added to the digester, displacing the warm liquor to tank A. The digester contents are thus preheated to about 150°C without adding any steam to the digester itself; however, some steam is used on a continuous basis to maintain temperature within tank C and within the white liquor accumulator. Excess liquor in tank A is pumped to the evaporators.

In step 8, the circulation pumps are started and the digester is brought up to temperature by adding steam directly to the circulation line. When the target temperature is reached (step 9), circulation is stopped, and the H-factor is accumulated. When the target H-factor has been reached, hot spent liquor is immediately displaced from the digester with washer filtrate, and the cycle is repeated.

Extended Delignification

Although originally developed as an energy saving method, it was found later (1) that the RDH process can delignify pulp to a significantly lower kappa number than the conventional kraft process without loss of strength, presumably because the chips are preimpregnated with black liquor having a high concentration of free sulfide ions. The fact that normal sulfide deficiency is offset at the start of the cook may be the reason why the RDH process is more selective in removing lignin while retaining carbohydrates. It has also been found from operating experience that cooking liquor consumption is lowered 5 to 10% due to liquor recycling. Another benefit is the partial "washing" that occurs as washer filtrate displaces spent liquor.

New Method of Digester Discharge

It was found through a series of "hanging basket" experiments in the late 1980's that pulp from cooked chips retained inside a batch digester at the end of a cook is considerably stronger than the corresponding material blown from the digester. Evidently, certain flow conditions during conventional discharge (e.g., heat, velocity, pipe friction, flashing of steam, etc.) are responsible for significant losses in pulp strength. Building on the aforementioned liquor displacement technology, a method has been devised (as illustrated in Figure 8-8) for adding dilution liquor and pumping out the contents of a cooled and depressurized digester at a low and controlled flow velocity. This new discharge method successfully delivers pulp at strength levels closely approaching those of the unblown material inside the digesters at the end of cooking (2).

8.3 CONTINUOUS DIGESTERS

The earliest design of continuous digester utilized a heated, pressurized chamber into which chips and chemicals were fed; then the mixture was carried through a series of tubes to provide retention time for the pulping reactions. Digesters of this type are still in service for producing semichemical pulps and for alkaline pulping of small wood subdivisions (fines and sawdust). A representative design is illustrated in Figure 8-9.

M & D Digester

The Messing and Durkee (M & D) design of inclined tube with conveyor flights represented a substantial improvement in operating flexibility (refer to Figure 8-10). Because the entire chamber is pressurized, the temperature at any point can be increased by steam injection or by forced circulation; thus a well-defined impregnation stage can be accomplished before maximum temperature is achieved. The retention time can be precisely controlled by the speed of the conveyor.

By using two M & D tubes in series (as shown in Figure 8-11), it is possible to maintain better control over the impregnation stage. It has been shown that, with complete impregnation, the cooking stage can be carried out very quickly in a high-temperature steam atmosphere. This technique, known as "rapid vapor phase cooking", has not, however, been carried out on a full commercial basis.

Although the M & D system can be well adapted for kraft pulping of chips, the principal commercial applications have been for sawdust kraft pulping, semichemical pulping of chips, and chip presteaming. Size is the major limitation; the digester is shop-fabricated, and the maximum diameter of an individual vessel is about 8 feet. Thus, for an economic-scale kraft pulp mill, three to five units would be required in some manner of series/parallel arrangement. It is generally more attractive in a continuous operation to carry out all pulping in one or two larger units.

Kamyr Digester

By far the most widely utilized continuous digester is the vertical downflow type developed by Kamyr. The first commercial installation was made in 1950. Approximately 350 units had been installed by 1990.

Perhaps the most common Kamyr system, dating from the late 1960's and early 1970's, is that

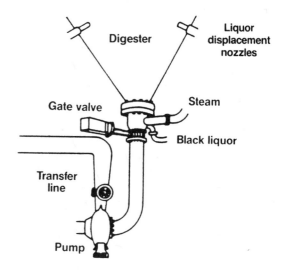

FIGURE 8-8. Modified discharge arrangement for batch digester.

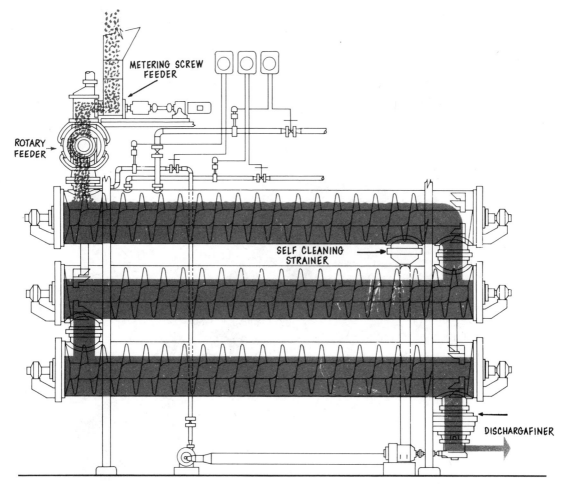

FIGURE 8-9. Horizontal tube continuous digester (Esco).

illustrated in Figure 8-12. The digester itself is a 60 - 70 m tall cylinder with increasing diameter from top to bottom. Typical capacity is about 1000 tons per day of pulp (oven-dry basis). Liquor circulation plays a dominant role in the system's operation.

The wood chips are fed from a surge bin through a volumetric chip meter and a rotary low-pressure feeder into the steaming vessel. A slowly-turning screw carries the chips through the horizontal steaming vessel where steam at 15 psig (mostly flashed from the residual liquor) preheats the chips and drives off air and noncondensibles. The chips then fall into a chute connected to the high-pressure feeder. This feeder consists of a single rotating element having a series of staggered pockets. Each pocket picks up chips while in the vertical orientation, and when the element has moved 90° to the horizontal, the chips are sluiced away with cooking liquor and carried up to the digester inlet. Here, the slurry enters a cylindrical separator; perforations in the plate allow liquor to flow to the surrounding collection ring and be returned to the feeder, while the chips are pushed downward by a rotating helical screw. When the chips enter the digester itself, they have absorbed sufficient alkali to

enable them to move downward by gravity through the liquor-filled space at the top of the vessel, until they settle atop the chip mass.

A hydraulic pressure of 165 psig is maintained on the liquid column by regulating the flows of white and black liquor to the digester. A high internal pressure is required to eliminate flashing and boiling of the liquid within the different temperature zones.

The chip mass moves downward as a uniform column without channelling (i.e., "plug flow"). The first zone at the top of the digester allows the chips to impregnate for about 45 minutes at temperatures from 105 to 130°C. In the heating zone, the temperature is raised in two steps by extracting liquor from the periphery of the vessel, pumping it through external heat exchangers, and returning the liquor to the digester through center pipes. Once the mass is up to temperature, the third zone provides sufficient retention time for the pulping reactions to be completed. Due to the exothermic character of the pulping reactions, the maximum temperature is actually about 2°C higher than that provided by external heating. A representative temperature profile for a Kamyr digester is shown in Figure 8-13.

The pulping reaction is stopped by extracting the

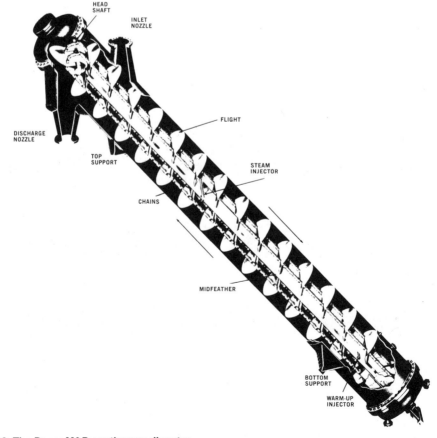

FIGURE 8-10. The Bauer M&D continuous digester.

hot residual liquor through screens at the periphery into a flash tank that supplies steam to the steaming vessel. The hot liquor is displaced by upward-flowing "wash liquor" (usually brown stock washer filtrate) which is injected near the bottom of the vessel and moves countercurrent to the chip flow to provide "diffusion washing". The digester is usually provided with sufficient height between the wash liquor inlet and extraction screens to allow at least 1.5 hours for the residual liquor to diffuse out of the chips.

At the bottom, pulp is continuously slushed from the column of cooled, softened chips by slowly-rotating paddles that are mounted on radial arms attached to the hub of the outlet device. The diluted pulp stock, which is under about 200 psig of pressure at this point, is then blown at a controlled rate, often to an atmospheric tank. The blow line typically utilizes a control valve on either side of a small pressure vessel with a variable orifice, so that there are three controlled stages of pressure drop.

The production rate for the Kamyr digester is set by the rotational speed of the volumetric chip meter. All other flow rates must be adjusted correspondingly for changes in the chip feed rate. The flow rate at the bottom of the digester is attenuated by changes in chip level at the top of the digester. The retention time in each zone of the digester is dependent on the production rate and the degree of packing. The degree of packing is defined as the ratio of chip volumes (for a specific amount of chips) in the chip meter to that in another zone of the digester. The packing is different in the impregnation, cooking and washing zones, and varies with wood species, degree of delignification and height of digester. An example of degree of packing for softwood and hardwood is also shown in Figure 8-13.

Major Modifications to Kamyr System

Although the basic operating principle of the Kamyr hydraulic digester has not changed since the initial development, a number of improvements and modifications have evolved over the last forty years. Some of the more significant developments were the following:

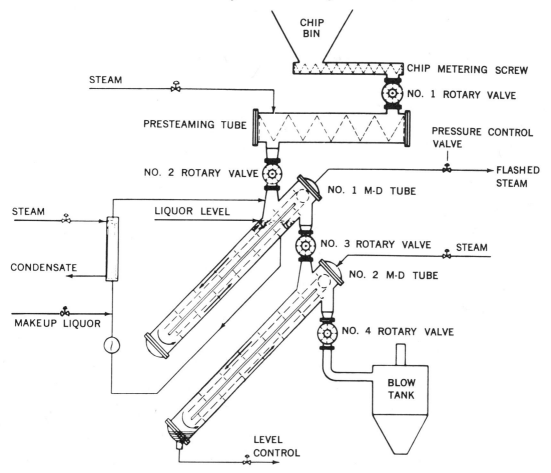

FIGURE 8-11. Double-tube arrangement of Bauer M&D continuous digesters.

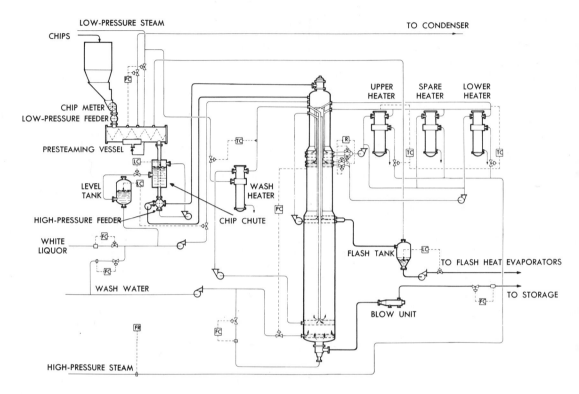

FIGURE 8-12. Kamyr continuous digester system for kraft cooking.

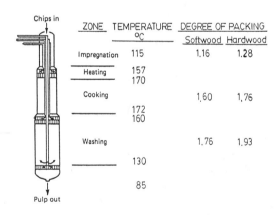

FIGURE 8-13. Temperature profile and degree of chip packing in a Kamyr digester.

1) Conversion from hot blow to cold blow to conserve pulp quality (1958).
2) Inclusion of the diffusion washing stage (1962).
3) Development of a vapor phase digester for sulfite and prehydrolyzed kraft production (1967). This unit initially utilized an external inclined separator to discharge only chips into the top of the digester. Heating was by direct addition of steam at the top of the digester.
4) Modification of the chip chute and addition of in-line drainer (1968).
5) Development of a two-vessel system employing a hydraulic impregnation vessel followed by a vapor phase digester (1972). In this system, the external inclined separator was replaced with an inverted internal separator.
6) Atmospheric presteaming of the chips in a special-design chip bin to optimize chip steaming and reduce the amount of fresh steam required in the steaming vessel (1974).
7) Development of a two-vessel system employing a hydraulic impregnation vessel followed by a hydraulic digester (1979). This system is generally considered to be an improvement over the previous two-vessel system, and most of the high-tonnage systems installed during the 1980's have been of this type.
8) Application of extended delignification to the Kamyr system (1984). The process is referred to by Kamyr as "Modified Continuous Cooking". A detailed explanation of the process is given in the next section.

Modified Continuous Cooking

Research at the Swedish Forest Products Laboratory (STFI) and the Royal Institute of

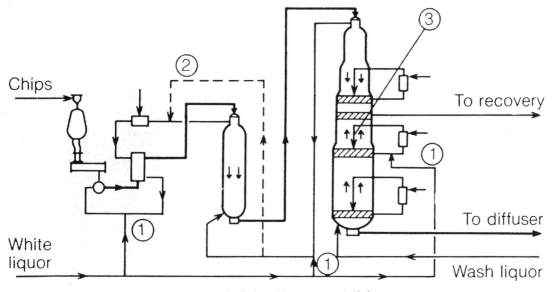

Chips

White liquor

To recovery

To diffuser

Wash liquor

1 = White liquor additions
2 = Recirculation
3 = Countercurrent zone

FIGURE 8-14. Two-vessel vapor/liquor phase digester adapted for extended delignification.

Technology, both in Stockholm, led to the development of Kamyr's Modified Continuous Cooking (MCC) system. The process is characterized by low initial concentrations of alkali and by low concentrations of lignin and sodium ions at the end of the cook. Although the main objective is to cook to lower kappa number levels, the process also produces a pulp with significantly higher strength properties (3).

Figure 8-14 shows how a two-vessel system is adapted for MCC operation. The normal process is altered in three ways:

• white liquor is added at three different points in the process.

• a liquor recirculation line is installed around the impregnation vessel.

• the final stage of the cook is carried out in a countercurrent mode.

A computer model (Figure 8-15) shows how the alkali concentration profile in the chips is levelled out by these modifications. In addition, the concentration of dissolved lignin and sodium ions decreases at the end of the cooking zone as a result of the countercurrent conditions.

Although it was originally developed for the two-vessel Kamyr systems, an adaptation of MCC has been successfully retrofitted to the single digester system (4).

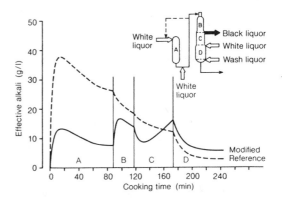

FIGURE 8-15. Concentration profile of effective alkali inside chips in a two-vessel vapor/liquid phase digester system - conventional system compared with modified system (as calculated by mathematical model).

IMPCO Digester

Another design of vertical downflow continuous digester, manufactured by IMPCO, is illustrated in Figure 8-16. This unit was first operated commercially in 1967, and some 15 of these units are now installed world wide. The IMPCO digester differs from the Kamyr design in several respects.

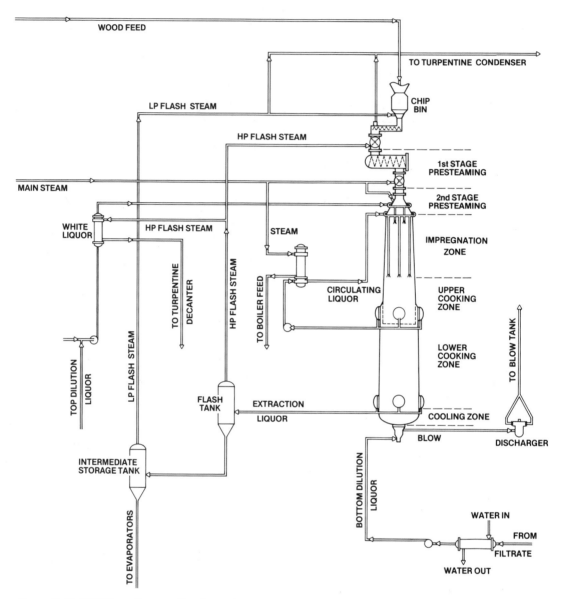

FIGURE 8-16. IMPCO continuous digester system.

After the chips are presteamed, they discharge through a high-pressure rotary feeder directly into the top of the digester, which contains high-pressure steam. The chips then enter the liquor in the impregnation zone; the chips and liquor together move downward and remain in the impregnation zone for about 20 minutes until they reach the discharge level of the downcomer pipes carrying heated circulation liquor.

The circulation liquor is extracted by the self-cleaning strainers in the upper row, and pumped through a heat exchanger to the return pipes. The flow rate is set so that the total liquor in the upper cooking zone is circulated at least twice. The chips at maximum temperature then enter the lower cooking zone for the balance of the retention time.

At the bottom, hot liquor is extracted and used to adjust entering liquor concentration, or is flashed to provide steam for the presteaming stage. The hot liquor is displaced with cold bottom dilution so that the pulp is discharged at low temperature to the blow tank.

Compared to the Kamyr digester, the IMPCO design is more tolerant of fines and sawdust in the furnish and is simpler in operation. However, it is also less flexible with respect to placement of equipment, and the totally vertical arrangement of components may be a factor in limiting size.

8.4 SAWDUST COOKING

Chemical pulping of a wood furnish composed totally of sawdust and other small wood subdivisions ("fines") is not possible in a batch digester because of difficulties in obtaining liquor circulation. A portion of sawdust, perhaps 10%, can be tolerated in a batch chip cook, but the smaller particles require less cooking time and tend to become overcooked. The disparity in cooking rate is markedly increased by differences in liquor penetration; the small particles with high specific surface (perhaps 30 times greater than for chips) absorb more than their fair share of chemical, leaving a lower concentration for the remaining chip mass.

Although sawdust chemical pulps are lower in yield and strength compared to chip pulps, the much lower cost of the sawdust raw material provides a strong incentive for its utilization. The resultant pulps can be used as partial furnish for a wide variety of paper products without any adverse effects. Therefore, some of the earliest efforts in the development of continuous digesters were directed toward sawdust pulping. The initial designs (notably by Esco, Pandia, and Defibrator) utilized the horizontal tube concept, as depicted in Figures 8-9 and 8-17. The M & D inclined tube was developed somewhat later (refer back to Figure 8-10), and was followed by vertical configurations from Kamyr (Figure 8-18) and Esco.

Some of the properties of sawdust which make it unwelcome in mixtures with chips can be readily exploited in a continuous digester system. Following or combined with the presteaming stage, the impregnation can take place very rapidly, assuming good mixing between liquor and sawdust. The chip mass is then brought up to cooking temperature and held for the requisite time period. The typical sawdust pulping system is much simpler than a comparable system handling chips, and the total retention time is about one third.

8.5 BLOW HEAT RECOVERY

A blow heat recovery system for conventional batch digesters must accomplish two objectives:
- recovery of usable heat from the flash steam
- total containment of the associated foul-smelling vapors.

The quantity of steam released during a conventional batch cook blow is substantial, amounting to about

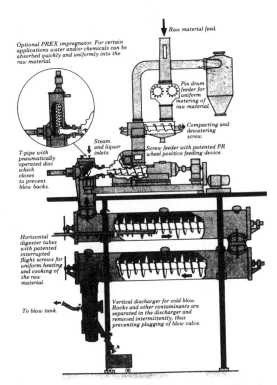

FIGURE 8-17. Horizontal digester system (Sunds Defibrator).

2000 lbs for each ton of pulp produced. The steam flashes from the liquor due to the drop in pressure from about 100 psig to atmospheric, while the cook spontaneously cools from 170°C to about 105°C.

A comparable quantity of flash steam is generated in the continuous digester system; but here it is utilized continuously to presteam the chips, thus effecting a large reduction in fresh steam usage. Direct re-use of the waste steam is not possible in the conventional batch system; but, where the heat can be effectively utilized (albeit at a lower "level"), the advantage favoring the continuous system is less extreme.

A typical kraft batch digester blow heat recovery system is depicted in Figure 8-19. The flash steam from the blow tank carries over to the top of the accumulator, where it mixes with cool condensate pumped from the bottom of the tank. The steam (along with organic vapors and some black liquor entrainment) condenses, producing hot "contaminated" condensate which flows into the top of the accumulator. This "dirty" hot water is used to indirectly heat fresh, clean water for pulp washing, after which it is returned in a relatively cool condition to the bottom of the accumulator. The flow rate of cool contaminated water to the direct-contact condenser is controlled by the outlet temperature; the flow rate cycles in response to the surges of steam from the blow tank.

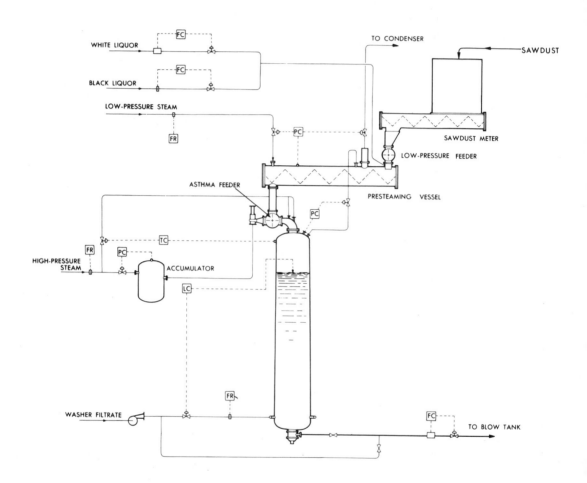

FIGURE 8-18. Kamyr continuous sawdust digester system.

The heart of the blow-heat recovery system is the accumulator tank itself which provides the necessary surge capacity. During the early stages of a blow, a large volume of hot, contaminated condensate accumulates in the top portion of the vessel. Somewhat later, as hot condensate is continuously pumped to the heat exchanger and returned as cool condensate, the interface between hot and cool layers rises, until the next blow when the interface falls. Thus, the moving interface between hot and cool sections of the tank provides for alternative accumulation and dispersion of heat, while supplying a constant flow of hot water to the heat exchanger.

The blow steam condensate is totally fouled with smelly organic and reduced sulfur compounds. The noncondensible gases which are vented from the accumulator also contain odorous components. Both these streams are concentrated sources of pollution (high BOD, strong odor) and must be given special handling. It is common practice to incinerate the noncondensibles in the lime kiln (see Section 27.3). The underflow from the accumulator is usually steam-stripped (see Section 26.2) to remove volatile organic compounds, which also are commonly burned in the lime kiln.

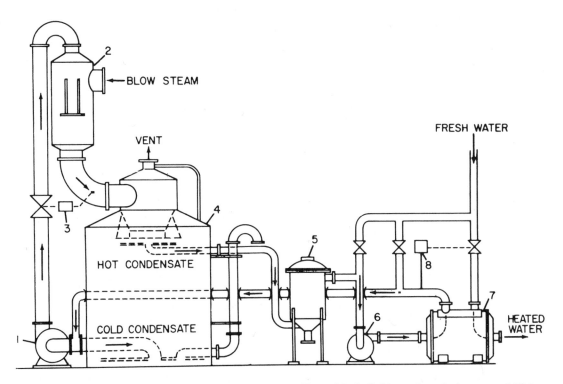

FIGURE 8-19. Representative blow heat recovery system (Rosenblad). Cold condensate is pumped (1) to the condenser (2) through a temperature-controlled flow valve (3). The accumulator (4) provides surge capacity by means of a moving interface between hot and cold sections. Hot condensate is extracted through a filter (5) and is pumped (6) to a heat exchanger (7) where the condensate is cooled before return to the accumulator. Fresh water flow to the heat exchanger is temperature controlled (8).

REFERENCES

(1) EVANS, J.C.W. **Batch Digester Heat Displacement System Reduces Steam Consumption** *Pulp & Paper* (July 1989)

(2) CYR, M.E., EMBLEY, D.F. and MACLEOD, J.M. **Stronger Kraft Softwood Pulp - Achieved** *TAPPI Journal* (October 1989)

(3) JOHANSSON, B., et al **Modified Pulping Process Shows Promising Results at Finnish Mill** *Pulp & Paper* (November 1984)

(4) WHITLEY, D.L., et al **Mill Experiences with Conversion of a Kamyr Digester to Modified Continuous Cooking** *TAPPI Journal* (January 1990)

Chapter 9

Processing of Pulps

Pulps are subjected to a wide range of processing steps, depending on their method of preparation (i.e., mechanical, semichemical, chemical virgin pulps; secondary fibers) and their end use. Screening, thickening and storage operations are necessary for virtually all pulp grades, while cleaning is usually required where appearance is important. Defibering is required for all semichemical and high-yield chemical grades. Deknotting is usually necessary in the production of clean bleachable chemical pulps. Blending operations are almost always desirable for achieving product uniformity, but are often omitted from the pulp processing scheme because of cost considerations. Pulp drying is required when pulp is stored for long periods (to prevent fungal or bacteriological activity) or shipped over considerable distances (to reduce freight costs).

A schematic flowsheet for a kraft dissolving pulp mill is shown in Figure 9-1 (next page) to illustrate a representative sequence of processing steps.

9.1 DEFIBERING

All high-yield chemical and semichemical pulps must be defiberized by mechanical means following the cooking step. At the highest yields (i.e., 80–90%), the operation is akin to chip refining (see Section 5.3), requiring high levels of applied energy to separate the fibers. In the lower-yield range (50–60%), considerably less energy is required. Defiberization of chemical pulps should be considered apart from pulp refining which is carried out in the paper mill; however, it must be recognized that the method and degree of defiberization will have an effect on later refining requirements.

Investigative work has shown that energy requirements are generally lower when the defibering can be carried out in the presence of hot residual liquor (i.e., "hot stock refining"). Since washing the chips is difficult anyway, it is almost universal practice to defiber directly from the blow tank (1). Some systems utilize a second stage of refining following washing of the pulp.

Disc refiners are commonly used for defibering. A popular configuration utilizes one double-sided rotating disc between two stationary discs. Feed is into the eye of the refiner. The stock then flows radially outward between each pair of facing discs, is collected in an annular casing, and then flows under pressure out a discharge port.

Refer to Sections 20.4 and 20.5 for specific applications of defibering.

9.2 DEKNOTTING

In a low-yield (i.e., bleachable grade) chemical pulping operation, knots are generally defined as the fraction of pulp that is retained on a 3/8" perforated plate. These rejects are most often composed of irregular shaped reaction wood pieces or overthick chips, but sometimes normal-size uncooked chips are present. Knots are removed from the pulp prior to washing, and are either discarded as waste or returned to the digester infeed.

As a control procedure, it is good practice to routinely monitor the level of knots (as % on pulp) and qualitatively characterize the reject material. A high percentage of knotter rejects, especially when showing a high proportion of uncooked chips, usually indicates poor cooking uniformity. Figure 9-2 shows some mill data on knot levels measured before and after the implementation of a process modification which was designed to improve liquor circulation.

Two types of knotters are in use. The older, vibrating screen knotters do an efficient job of knot separation, but the open-type design generates foam and liquor spatter, and needs operator attention. Foam in the stock impairs the efficiency of the subsequent washing operation and can cause other problems. The vibratory knotter is rapidly being superseded by the totally enclosed pressure screen knotter.

The pressure screen knotter (e.g., Figure 9-3) consists of a totally enclosed cylindrical, perforated screen through which accepted stock flows. A rotating foil produces a series of pressure and vacuum pulses to keep the perforations clean. Knots are retained on the entrance side of the screen and are continuously discharged, along with some good fiber. The main drawback of the pressure knotter is the requirement for secondary screening of the reject stream to return good fiber back to the system.

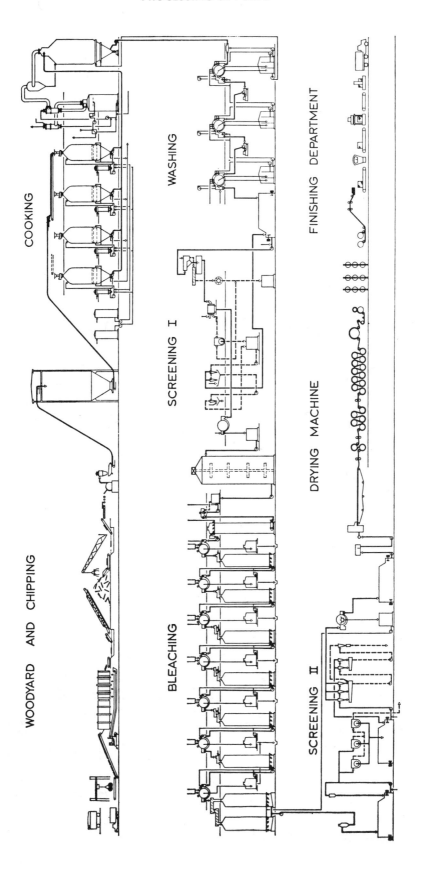

FIGURE 9-1. Schematic flowsheet for kraft dissolving mill.

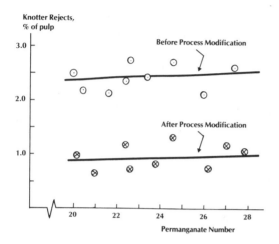

FIGURE 9-2. Knotter reject level vs permanganate number during two periods of operation.

9.3 BROWN STOCK WASHING (REFERENCES 2, 3, AND 4)

The cooked pulp from the digesters must be washed, with the objective to:

- remove residual liquor that would contaminate the pulp during subsequent processing steps.
- recover the maximum amount of spent chemicals with minimum dilution.

For many decades, the standard method of washing was to employ a series of rotary vacuum washers operating in a countercurrent flow sequence. Today, a number of alternative methods are available to challenge the predominant position of the rotary vacuum washer, the most prominent being:

- pressure and atmospheric diffusion washers
- rotary pressure washer
- horizontal belt washer
- dilution/extraction equipment.

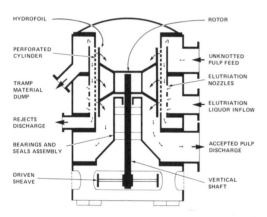

FIGURE 9-3. Cross-section schematic of pressure knotter (Impco/Ingersoll Rand).

FIGURE 9-4. Internal structure of single-stage vacuum washer (Beloit Rauma).

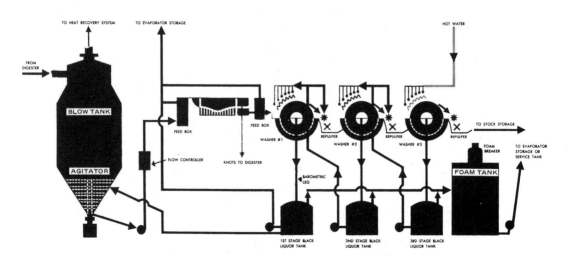

FIGURE 9-5. Typical flow sequence for countercurrent rotary brown stock washers.

Rotary Vacuum Washer

The main element of a vacuum washer is a cloth-covered cylinder that rotates in a vat containing the pulp slurry. By means of internal valving and a sealed dropleg, vacuum is applied as the rotating drum enters the stock. A thick layer of pulp builds up and adheres to the wire face as it emerges from the vat. Wash water is applied to displace the black liquor in the sheet as the drum continues to rotate. Finally, the vacuum is cut off and the washed pulp is removed from the mold. A representative design of vacuum washer is illustrated in Figure 9-4. A typical countercurrent arrangement of these washers is shown in Figures 9-5 and 9-6. (Refer also to Section 9.6.)

Although a displacement mechanism is used (as illustrated in Figure 9-7), the average efficiency of displacement for a single stage rarely exceeds 80%, due to a number of factors. Consequently, three of four stages are required to attain an overall satisfactory removal of 99% of the "washable" liquor solids. A small portion of the soda is chemically bound to the kraft pulp fibers and cannot be recovered by conventional washing techniques. A listing of the factors affecting displacement efficiency is given in Table 9-1. Reference 2 provides a good discussion of the more important variables. Perhaps the key variables are specific loading, dilution factor, and the amount of air in the stock.

Specific loading is commonly measured in oven-dry tons per day of pulp per square foot of cylinder surface (BDTPD/sq ft). Typical loading values are in

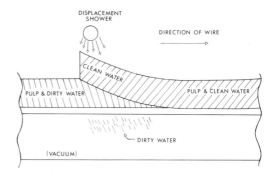

FIGURE 9-7. Illustrating displacement principle of rotary vacuum washer.

FIGURE 9-6. Brown stock vacuum rotary washers (Impco/Ingersoll Rand).

TABLE 9-1. Factors affecting brown stock washing displacement efficiency.

Fiber Characteristics
- pulping process
- stock hardness
- stock freeness
- species

Shower Characteristics
- temperature
- distribution
- method of application

Sheet Formation/Thickness
- specific loading
- vat consistency
- vat rotational speed

Operating Factors
- dilution factor
- stock temperature
- air in stock (foam)
- liquid solids level
- fabric mesh characteristics
- fabric fouling

the 0.6 to 0.8 range; overloaded conditions (common in mills with expanded production) contribute to poor washing.

Dilution factor is a measurement of the wash water applied in excess of that required for total displacement, expressed as pounds of water per pound oven-dry pulp. Generally, a higher dilution provides better washing. But the dilution factor is also a direct measure of the water being added to the liquor system. Therefore, the washing benefit achieved at higher dilution factor must be balanced against the higher evaporation loading.

In any vacuum washing operation, a certain amount of air is continuously pulled through the pulp sheet, entrained with the liquor, and carried down the dropleg into the seal tank. It is important that this air has a chance to escape before the liquor is reused to minimize foaming problems. This objective is usually accomplished by installing large seal tanks, the diameter of which may be double the height; the large surface area allows air to escape and foam bubbles to break. In some situations, mechanical foam breakers are also used.

An important design consideration that affects washing efficiency and foam generation is the method of wash liquor application. The liquid should be applied uniformly at low velocity to avoid channelling and foaming. High-pressure jets or nozzles are unsatisfactory; low-pressure nozzles usually provide uneven application and quickly become plugged with deposits. The ideal method of

application is with a weir-type shower, which usually consists of a pressurized pipe submerged in an open trough. The wash liquor is distributed into the trough through a series of submerged holes, and is then transferred without turbulence over a bent plate onto the pulp sheet. Surprisingly, weir-type showers are used on only a small portion of rotary washers.

The displacement washing effect is readily calculated from the ratio of the actual reduction in liquor solids content (compared) to the maximum possible reduction. For a single stage:

$$\text{Displacement Ratio (DR)} = \frac{C_v - C_s}{C_v - C_w}$$

where:
C_v = Vat liquor solids concentration
C_w = Wash liquor solids concentration
C_s = Solids concentration in sheet leaving washer

It can be noted that, when water is used as the wash liquor, C_w equals zero.

However, the removal of dissolved liquor solids at the brown stock washers is accomplished both by displacement and thickening. The thickening effect is calculated:

$$\text{Thickening Factor (TF)} = \frac{W_{(in)} - W_{(out)}}{W_{(in)}}$$

where:
$W_{(in)}$ = lbs of liquor per lb pulp (stock entering)
$W_{(out)}$ = lbs of liquor per lb pulp (stock leaving)

Also note:

$$W = \frac{100 - \text{Consistency}}{\text{Consistency}}$$

Thus, the overall efficiency of dissolved solids removal is calculated:

$$\% \text{ Efficiency} = [TF + (1 - TF) DR] 100$$

On a routine basis, the efficiency of washing is monitored by the amount of soda (or equivalent saltcake) remaining with the washed pulp. A washing efficiency of 99% is usually equivalent to a carryover of about 15 to 20 lbs/BDT of washable equivalent saltcake. (If total soda losses are to be calculated, the chemically-bound soda must also be considered.)

Diffusion Washing

Diffusion washing was first applied to the chip mass in the bottom of the Kamyr continuous digester (see Section 8.3). Later, the principle was adapted and applied by Kamyr to the washing of pulp; the first commercial atmospheric diffusion washers were installed in 1965; the first commercial pressure diffusion washers (independent of the digester) were installed in 1979. Diffusion washing is characterized by a relatively long period of contact between the cooked chips or pulp and the moving wash liquor, which allows time for the liquor solids to diffuse or leach from the fiber structure. These processes take place in a submerged environment, which excludes the possibility of air entrainment and foaming.

An ideal situation for diffusion washing (or "high heat washing", Kamyr's descriptive term) exists at the bottom of the Kamyr chip digester (see Figure 9-8). Here, retention times of up to four hours are possible with wash zone temperatures of 130 – 140°C and truly countercurrent flows. The level of washing efficiency that can be achieved is a function of the dilution factor and the retention time as illustrated in Figure 9-9. In practice, the downward flow of the chip mass in the digester may be impeded at higher dilution factors, especially when the digester is operating near maximum capacity. The limiting dilution factor must be determined on an individual basis.

A different situation exists in the bottom of the Kamyr sawdust digester (Figure 9-10). A true countercurrent flow cannot be used because of resistance within the closely packed pulp mass. The method used is to add wash liquor at the periphery and extract displaced liquor through a central rotating screen; the relatively fast movement of the screen prevents plugging. Typical wash zone retention is limited to about 20 minutes.

An atmospheric diffusion washer for pulp is illustrated in Figures 9-11 and 9-12. Single and

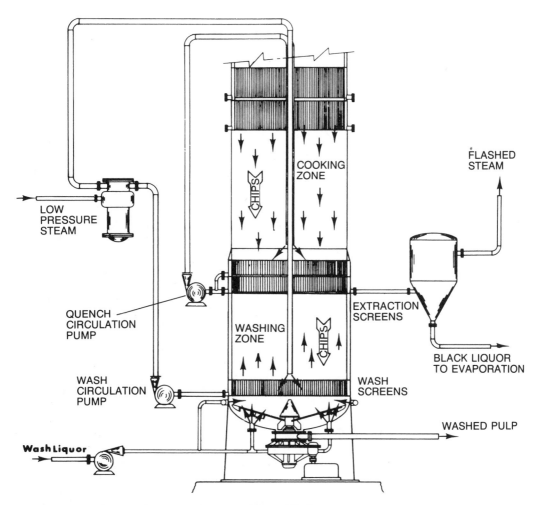

FIGURE 9-8. Internal washing zone in a typical Kamyr chip digester.

multistage units are available; Figure 9-13 depicts a total of three stages with thickener. The units consist of a series of double-sided, concentric screens with distribution nozzles between each screen. The entire screen assembly is mounted on a set of hydraulic cylinders and moves up at the same rate as the pulp. At the top of the stroke, the screen assembly is moved rapidly downward, thus providing a wiping action to keep the screen surface clean. Wash liquor

is added through the nozzles, and the displaced liquor is extracted through the screens. In a multi-stage arrangement, the pulp moves upward; the extracted liquor from an upper unit is used as wash liquor for the next stage below. Typically, the

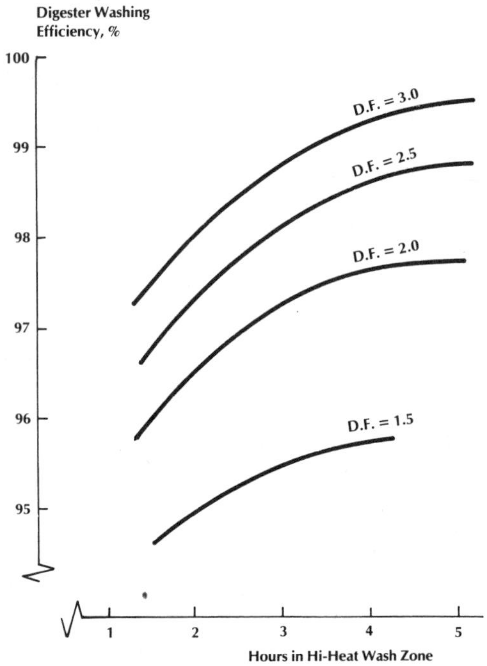

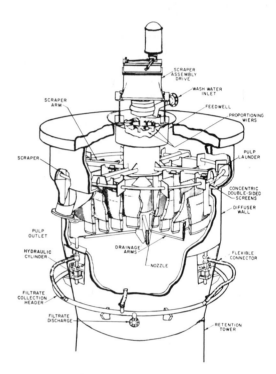

FIGURE 9-11. Cut-away drawing shows how the Kamyr diffusion washer is designed and how it works.

FIGURE 9-9. Effect of wash time and dilution factor on high-heat digester diffusion washing efficiency.

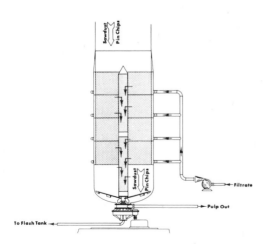

FIGURE 9-10. Internal washing zone in a Kamyr sawdust digester.

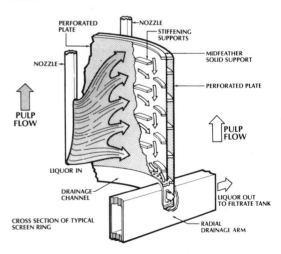

FIGURE 9-12. Schematic of Kamyr diffusion washer screen assembly. The washing medium is introduced through rotating distribution nozzles. The displaced liquor is collected through the screens and flows into the drainage arms.

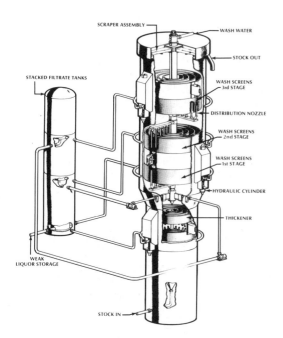

FIGURE 9-13. Three-stage diffusion washer with thickener (Kamyr).

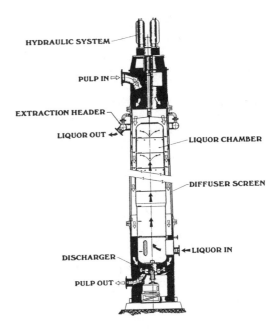

FIGURE 9-14. Kamyr pressure diffusion washer.

retention time per stage is 8 to 10 minutes.

The Kamyr pressure diffusion washer (Figure 9-14) is similar in operation to the Kamyr sawdust digester internal washer, and was designed principally for placement in series with the Kamyr digester internal washer. The pulp at blow-line consistency is introduced at the top of the vessel and is caused to flow downward through an annulus formed by a slightly tapered cylindrical screen and a central liquor chamber. The wash liquor is injected from the central chamber and is forced to flow to the screen, where it is extracted. As in other designs, the screen is moved periodically in the direction opposing the flow of pulp to wipe the pulp from the screen and prevent plugging.

Pressure Washers

Rotary pressure washers (Figures 9-15 and 9-16) are similar in operation to rotary vacuum washers, but appear to offer some significant advantages. The pulp mat is formed on the surface of the cylinder and dewatered with the aid of pressure applied inside the washer hood (i.e., outside the cylinder), as opposed to vacuum inside the cylinder. Because the driving force for mat formation, dewatering and wash liquor displacement is outside the cylinder, the interior of the washer drum can be utilized for a more sophisticated liquor collection system; therefore, a single washer can be operated with two or three displacement stages. The higher pressure also allows higher-temperature wash liquor to be used and

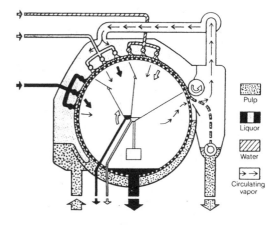

FIGURE 9-15. Basic principle and design of pressure washer (Beloit Rauma).

greatly reduces foaming. The closed vapor circulation system facilitates collection and treatment of odorous vapors.

Horizontal Belt Washer

The horizontal belt washer (Figure 9-17) resembles the fourdrinier section of a paper machine. The pulp suspension is distributed from a headbox onto a traveling filter belt (e.g., plastic screen or steel band), and formed into a mat. Wash liquor is applied to the top side of the mat while displaced filtrate is removed from the underside of the belt by suction boxes. The washer is operated in a countercurrent mode, with the filtrate from one section being returned as wash liquor to the previous section, and ultimately as dilution for the headbox furnish. In common with the vacuum washer, displacement is the principal washing mechanism, but no mixing and reforming of the pulp mat is required between stages. A total of five displacement stages are easily accommodated along the belt.

Dilution/Extraction Equipment

Dilution and extraction is the oldest method of pulp washing. It consists simply of diluting a pulp slurry with weaker liquor, and subsequent thickening. The efficiency of this type of system is dependent upon the ratio of the discharge to the incoming consistency and the amount of dilution water added to the system. Older systems were inefficient due to limitations in thickening, but modern extraction presses, some having a displacement washing capability, are capable of discharge consistencies of 30 - 40%, and multi-stage systems can be competitive with vacuum washer systems.

Figures 9-18, 9-19 and 9-20 depict three of the

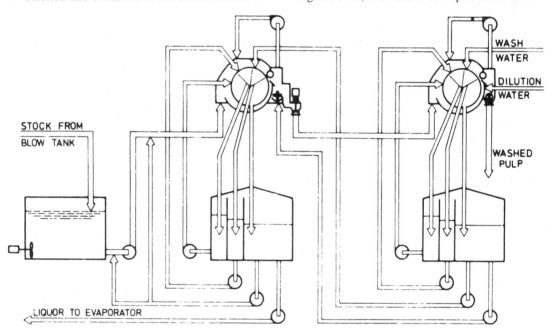

FIGURE 9-16. Flowsheet comprising two three-stage pressure washers (Beloit Rauma).

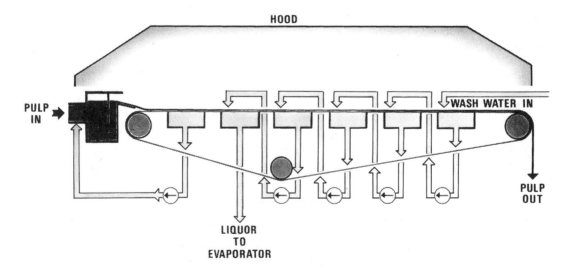

FIGURE 9-17. Schematic of horizontal belt washer system (Black Clawson Co.)

available designs for dilution/extraction equipment. This class of equipment is especially applicable for pulps that are difficult to permeate, i.e., where a displacement-type washer would not be an appropriate choice.

Norden's Method

In view of the recent proliferation of new washing methods and the possibility of connecting different types of washing equipment in series, it would be desirable to assign an efficiency number to equipment which is independent of dilution factor. Norden's method (5) has been widely applied toward this objective. It assumes that a washing stage can be likened to a number of countercurrent mixing stages connected in series. In an individual stage, the pulp and associated solids-containing liquor (from the previous stage) are mixed with lower-solids wash liquor (from the next stage); the stock is then re-thickened to the original consistency and the separated stock and liquor are passed counter-currently to their respective next stages.

The Norden efficiency factor is defined as the number of mixing stages which will give the same results as the washing equipment under consideration, when operated at the same wash liquor ratio. Table 9-2 provides a ranking of equipment utilizing a slightly modified Norden efficiency factor. A range of values is the natural consequence of different fiber types and varying operating conditions. The total Norden efficiency for a system may be found simply by adding the factors for the various components. Given the system Norden efficiency and the dilution factor, the anticipated washing efficiency of the system can be found from Figure 9-21 (6).

9.4 SCREENING

In most pulp and paper processes, some type of stock screening operation is required to remove oversized, troublesome and unwanted particles from good papermaking fibers. The major types of stock screens are vibratory, gravity centrifugal, and pressure (centrifugal or centripetal). They all depend on some form of perforated barrier to pass acceptable

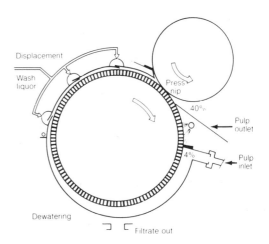

FIGURE 9-18. Operating principle of wash press (KMW).

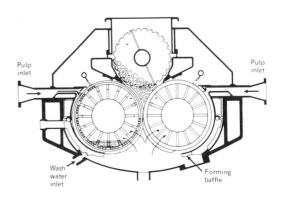

FIGURE 9-19. Extraction press unit (Impco/Ingersoll Rand).

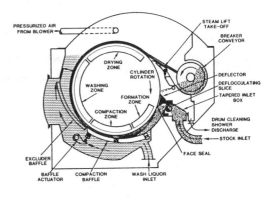

FIGURE 9-20. Schematic cross-section of compaction baffle filter (Impco/Ingersoll Rand).

TABLE 9-2. Modified Norden efficiency factors for various types of washing equipment.

	Efficiency Factor
Vacuum drum washer	2.5 – 4
Single-stage diffuser	3 – 5
Kamyr sawdust digester wash zone	5 – 9
Kamyr chip digester:	
$1\frac{1}{2}$ h in hi-heat zone	4 – 6
3 h in hi-heat zone	7 – 11

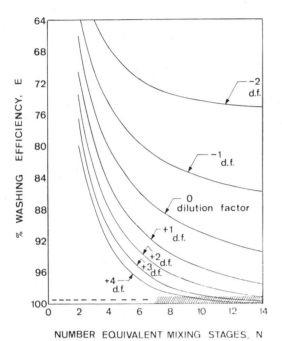

NUMBER EQUIVALENT MIXING STAGES, N

FIGURE 9-21. System washing efficiencies at different dilution factors and modified Norden numbers. Note that applicability is not valid above 99% efficiency because of diffusion/sorption effects.

fiber and reject the unwanted material. In most instances, it is the size of the perforations (usually holes or slots) that determines the minimum size of debris that will be removed.

All screens must be equipped with some type of mechanism to continuously or intermittently clean the openings in the perforated barrier. Otherwise, the screen plate would rapidly plug up. Methods of cleaning employed on current commercial screens include shaking and vibration, hydraulic sweeping action, back-flushing or, most common, pulsing the flow through the openings with various moving foils, paddles and bumps.

The typical screen is a relatively simple machine to operate. The most important consideration for stable, efficient operation is to maintain flow and consistency near optimum levels. When problems occur, they are usually due to clear overloading or underloading; but a relatively wide range of operation is possible between these extremes.

Vibratory Screens

Vibratory flat screens at one time were practically the only type used in pulp and paper mills. This screen is capable of efficient separation and concentration of reject material. However, its many disadvantages (e.g., open construction, foam problems, high maintenance, labor intensity, large floor area requirement) have rendered it obsolete for all but specialized applications. It is still the best tailings screen (for concentration of rejects). The same design concept is embodied in the laboratory flat screen, which is used for measuring debris levels on stock samples.

The rotary vibratory screen is more compact than the flat screen and requires less operator attention. But, the high maintenance cost has also rendered this design obsolete for most applications.

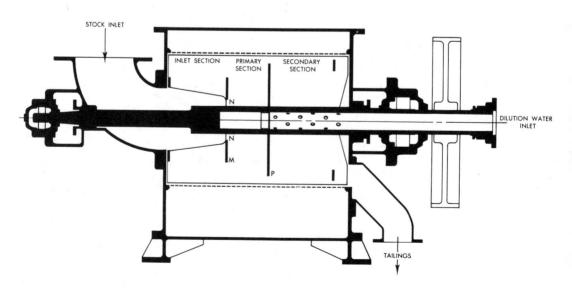

FIGURE 9-22. Example of gravity centrifugal screen (S.W. Hooper Corp.).

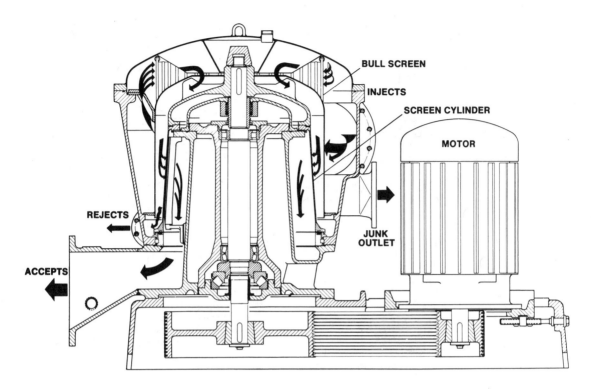

BULL SCREEN

INJECTS

SCREEN CYLINDER

MOTOR

REJECTS

JUNK OUTLET

ACCEPTS

FIGURE 9-23. Pressure screen (KMW).

Gravity Centrifugal Screens

The gravity centrifugal screen (Figure 9-22) overcame many of the problems of the vibratory screen. This design utilizes a horizontal cylindrical screen plate with round holes up to 3 mm in diameter, depending on service. A "paddle-wheel" type rotor keeps the screen plate clean. Since the unit doesn't vibrate, heavy foundations and isolation mountings are not required. Foam generation is reduced, but not eliminated.

The principle of the screen is based partly on the fact that good fiber tends to be thoroughly hydrated and has a specific gravity close to water. When the low-consistency pulp stock is rotated in the centrifugal screen, the fibers align themselves with the direction of flow, which is predominantly through the circular holes in the screen plate. The coarse materials are not fully hydrated and have a lower density; this factor limits the effect of centrifugal force and the coarse materials tend to be carried across the screen plate to discharge as rejects. The coarse material accumulates as it moves axially, and this loose mat also acts to some extent as a screening element. Gravity centrifugal screens have been applied to a wide range of stock screening applications (7).

Pressure Screens

The operating principle of pressure screens is similar to gravity centrifugal screens. The distinction is that they operate under full line pressure and the radial flow within the unit can be either centrifugal (outward), centripetal (inward) or a combination depending on design. They have the advantage of high capacity per unit, flexibility of physical location, small space requirements and economy of piping and pumping. The totally enclosed design excludes air entrainment and minimizes slime buildup.

Pressure screens were first used in paper machine approach systems where their main function was to remove gross contaminants and protect the paper machine forming fabric. More recently, with refinements and modifications in design, various configurations are being utilized in virtually all fine screening applications. Examples of current designs are illustrated in Figures 9-23 and 9-24.

All pressure screens utilize a cylindrical, perforated plate. The most common plate-cleaning mechanism is a rotating hydrofoil (Figure 9-25), but other types of rotating cleaning elements are successfully utilized, depending on screen design. Four different flow patterns are offered in commercial units (Figure 9-26); at least two designs utilize two concentric screen plates with both inward and outward flows.

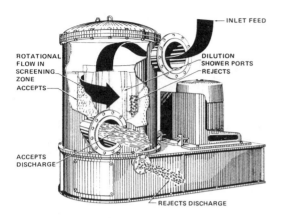

FIGURE 9-24. Pressure screen (Hooper).

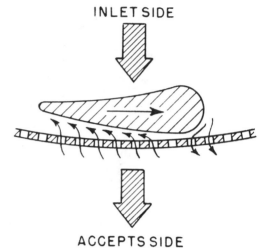

Mechanism of Debris Removal

Four possible mechanisms have been hypothesized for debris removal in a pressure screen:

1) Positive size separation - this applies only to rigid particles that are larger in all three dimensions than the screen plate openings

2) Debris orientation - long debris particles which would be accepted at a certain orientation are rejected because the "angle of approach" is steeper.

3) Fiber network - fibers and debris particles on the inlet side form a network having small openings that prevent passage of debris particles.

4) Fluid forces - elongated particles from the screen plate openings are recaptured to the inlet side due to fluid forces induced by the pulsating element.

FIGURE 9-25. The principle of a typical pressure screen is illustrated. The leading edge of the rotating foil accelerates the stock. The negative pulse under the sweeping foil momentarily reverses the flow, effectively purging the screen openings.

Most investigators focus on the first two mechanisms. The major design and operating parameters of flow configuration, screen plate openings, rotor type and speed, and stock consistency can shift the dominance from one mechanism to another.

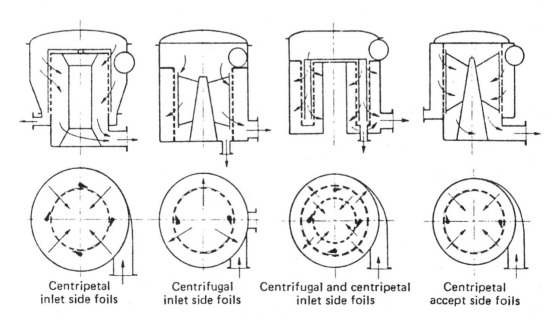

Centripetal inlet side foils   Centrifugal inlet side foils   Centrifugal and centripetal inlet side foils   Centripetal accept side foils

FIGURE 9-26. Four pressure screen flow configurations.

Design Choices

Modern pressure screen plates are precision manufactured components. They are normally fabricated from 316L stainless steel or better and are electropolished after machining to remove tiny residual burrs which might encourage "stringing". Either holes or slots may be used as screen perforations depending on requirements.

Traditionally, holes were punched in flat plates, which were then rolled into cylinders. Today, most holes are taper-drilled, i.e., cylindrical on the feed side and conical on the accept side to reduce the pressure drop across the thick plate. Hole spacing in the direction of rotation must be sufficient to avoid "hairpinning" of fibers between two holes. Because of this requirement, the open area of very fine screens may be as low as 10%.

Screen slots are normally produced using saw cutters of the desired nominal slot width. As with holes, the nominal slot width is only through a small fraction of the plate thickness on the feed side; the accept side is normally "relieved" with a much wider cutter to reduce the pressure drop. Slots are oriented perpendicular to the direction of rotor rotation so the long dimension of the debris particle is presented to the slot width. Slot widths are always less than hole diameters, so slots are more effective in removing small cubical debris. However, slotted screen open areas are generally in the 3 to 7% range, so throughput is lower than for screen cylinders with holes. Because of lower structural integrity, slotted plates are made of thicker plate.

In the late 1970's and early 1980's screen cylinders with contoured surfaces were introduced. The contoured surface in conjunction with the action of the rotor tends to introduce microturbulence which disrupts and fluidizes the mat of fibers. This action permits the opening to pass more fibers before a mat forms. Contoured surfaces generally have greater throughput, handle higher feed consistencies, and operate at lower reject rates than smooth-surfaced cylinders. Contoured screen cylinders are available in a variety of contour levels. Fluidizing and other effects are dependent on the degree of surface contour as well as on rotor design and speed.

Measuring Screen Performance

Two indices are generally accepted for measuring screening efficiency. Using nomenclature from TAPPI TIS 0605-04:

(1) Debris Reject Efficiency, (E_R)

$$E_R = \frac{S_r}{S_i} \cdot R_w$$

where:

R_w = Reject rate, decimal portion of inlet weight flow
S_r = % debris by weight in reject flow
S_i = % debris by weight in inlet (feed) flow

(2) Cleanliness Efficiency, (E_C)

$$E_C = \frac{S_i - S_a}{S_i} = 1 - \frac{S_a}{S_i}$$

where:

S_a = % debris by weight in accept flow
Note: The ratio S_a/S_i is known as the cleanliness ratio.

For relatively large debris, the efficiency can be 100%. For intermediate-size debris, the efficiency is strongly affected by the % reject rate (R_w). Both gravity and pressure screens require a significant reject rate in order to ensure a sufficiently low level of debris in the accepted stock. The typical relationship is illustrated in Figure 9-27. Numerical values of efficiency are rather meaningless unless the efficiency measurement itself is defined and unless the reject rate is reported. Obviously, the efficiencies of two screens with different reject rates can not be compared meaningfully.

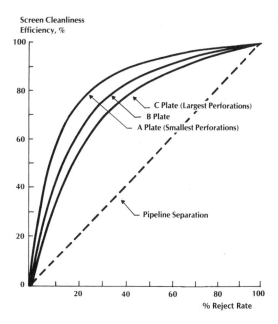

FIGURE 9-27. Effect of reject rate and plate opening on screening efficiency.

To overcome the problems that are inherent with the individual screening performance indices, Nelson (9) has proposed using the ratio E_C/E_R which he calls the "Screening Quotient" (Q). A major advantage is that Q is relatively insensitive to changes in the reject rate. It is also true that:

$$Q = \frac{S_r - S_a}{S_r}$$

where S_r equals debris in the reject flow and S_a equals debris in the accept flow, both measured as percent by weight. Therefore, only two pulp samples need to be taken and analyzed in order to calculate Q.

The traditional method of measuring debris content on a pulp sample depends on carrying out a complete screen separation in the laboratory and then drying and weighing the accepted and rejected pulp fractions. Either a Somerville or Valley laboratory flat screen is commonly used with slot widths between 0.10 and 0.25 mm (0.004 to 0.010"). Alternative methods have utilized the Pulman shive analyzer, Van Alfthan shive analyzer, or PFI mini-shive analyzer. More recently, electron image analyzers have allowed investigators to quickly determine debris contents without time-consuming "wet" laboratory tests.

When screens are rejecting debris, there is an accompanying tendency to reject long fibers. This feature of screens is sometimes exploited for stock fractionation. Rejection of long fibers can be measured quantitatively by the same types of formulas as for debris when the long-fiber fraction is defined and measurable. Since for most applications it is desirable to maximize screening efficiency (E_R) and minimize long fiber rejection (L_R), it has been proposed that the "quality of screening" can be measured by the difference between E_R and L_R at a given reject rate. Comparisons of the "quality of screening" between different systems are meaningful only when the size distributions of the stocks to be compared are similar.

Variables Affecting Screen Performance

A listing of the major design and operating variables affecting screening is given in Table 9-3. Many of these variables are strongly interrelated. For example, operating with a higher stock consistency, and/or using larger screen plate perforations will provide increased capacity, but usually at the expense of screening efficiency. In applications where high efficiency is essential, it is advisable to use the shape of perforation and the smallest perforation opening that experience has shown work best on the particular type of fiber and debris.

All gravity screens and some pressure screens require dilution on the feed side near the reject outlet because water with the inlet stock preferentially

---

TABLE 9-3. Variables affecting screening performance.

Stock Characteristics
- type of fiber
- characteristics of debris
- debris level

Design of Screen
- flow configuration
- type of plate-cleaning mechanism
- type of perforation (holes or slots)
- rotor speed (rpm)

Operating Variables
- stock flow rate (or pressure drop across screen)
- feed consistency
- reject rate
- screen plate perforation size
- stock temperature
- dilution flow to screen

---

flows through the perforations, leaving the freer pulp and debris in a concentrated form. Proper control of the amount of dilution is important to prevent rejecting excessive good fiber (too much dilution) or allowing considerable debris to pass through the screen openings (insufficient dilution).

References (9) and (10) provide a more complete discussion of the various factors affecting screen operation.

Arrangement of Screens

Because a large reject flow is required from the primary pulp screens in order to achieve adequate debris removal efficiency, additional stages of screening are performed on the reject stream to concentrate the debris and return the good fiber to the process. A simplified "cascade arrangement" of screens is shown in Figure 9-28. The final reject stream of shives or slivers is often refined into acceptable fiber, with only a small system bleed taken from the centricleaning step following refining. However, in high-quality bleach kraft mills with a history of periodic plastic contamination in the pulp stock, refining is avoided, and last-stage screen tailings are removed from the system and either burned as fuel or consigned to landfill.

Where required, mills can employ a more sophisticated flow sequence for screening and handling of refined rejects. For example, the TMP system illustrated in Figure 9-29 utilizes two primary stages in series and direct addition of cleaned, refined rejects into the accept stream. In situations where two primary stages are employed in series, it is usually advantageous to use holes for the first-stage screen plates and slots for the second stage.

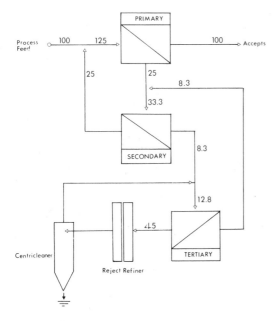

FIGURE 9-28. Three-stage cascade screening system showing a representative moisture-free stock balance. The balance is based on a feed rate of 100, 20% primary-stage reject, 25% secondary reject, 35% tertiary reject and negligible cleaner reject.

9.5 CENTRIFUGAL CLEANING

Up to the 1940's, high specific gravity contaminants such as sand and dirt solids were usually removed from pulp suspensions in a "riffler". This device was essentially a modified settling trough through which low-consistency stock was slowly channelled, allowing the heavier particles to settle out. These units were bulky, inefficient, and required frequent manual cleaning.

A better method was provided through application of the centrifugal cleaner. The classic centrifugal cleaner (illustrated in Figure 9-30) was patented in 1891, but did not come into widespread use until the 1950's. This device (also identified by such terms as liquid cyclone, hydrocyclone, vortex cleaner or, more simply, "centricleaner") consists of a conical or cylindrical-conical pressure vessel with a tangential inlet at the largest diameter of the cone or cylinder. Also centered axially at the large diameter end is the vortex finder or accepts nozzle. At the opposite end or minimum-diameter end is the underflow tip or rejects nozzle.

The centrifugal cleaner removes unwanted particles from pulp and paper stock by a combination of centrifugal force and fluid shear. Therefore, it separates both on the basis of density differences and particle shape. All centrifugal cleaners work on the principle of a free vortex generated by a pressure drop to develop centrifugal action. The power source is the pump. The stock enters the cleaner

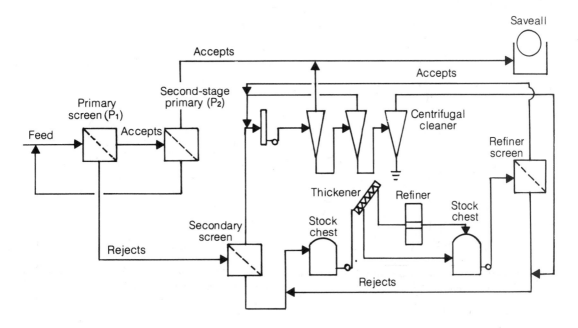

FIGURE 9-29. Two-stage primary screening system for producing high-quality TMP stock (S.W. Hooper Corp.).

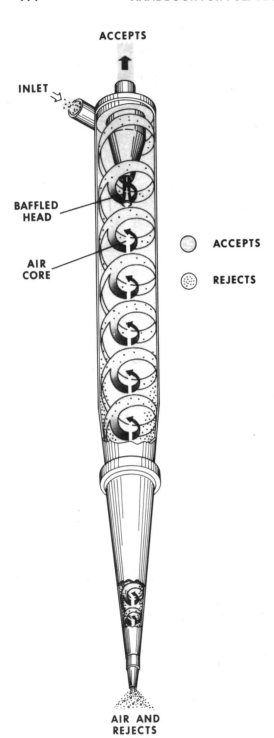

INLET

BAFFLED
HEAD

AIR
CORE

ACCEPTS

○ ACCEPTS

⦙ REJECTS

AIR AND
REJECTS

tangentially; the inlet scroll guides the flow to impart a rotating motion. As the stock flows inward, the velocity increases, resulting in high centrifugal forces near the center which carry dense particles outward and away from the accepted stock. Good fiber is carried inward and upward to the accepted stock outlet. The dirt, held in the downward current, continues toward the tip. As the diameter narrows, the flow is forced inward against increasing centrifugal force (several hundred g's), which concentrates the dirt and releases good fiber to the accepts flow.

Small-diameter cleaners develop the highest centrifugal forces, and are most effective for removing various types of small dirt particles. Where cleaners are employed principally to remove larger, less-dense particles (e.g., shives and slivers), as in paper machine approach systems, larger-diameter cleaners have proved to be more effective. The comparative efficiencies of two different-diameter cleaners with respect to particle size and shape are illustrated in Figure 9-31.

A stable air core is established in the center of the centrifugal cleaner over its entire length by the hydraulic flow patterns. The diameter of the air core is dependent on the dimensions of the cleaner, its operating conditions, and the air content of the stock. If the underflow tip is exposed directly to the

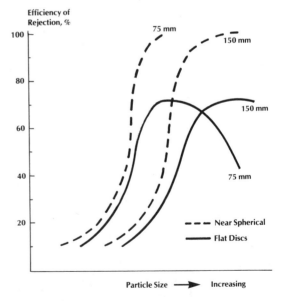

FIGURE 9-30. Schematic of a centrifugal cleaner (Bird Machine Co.).

FIGURE 9-31. Effect of particle size and shape on efficiency of rejection from 75-mm and 150-mm centrifugal cleaners.

atmosphere, the cleaner will draw additional air up the core and discharge it with the accept flow; this can be a problem in some applications, notably in the approach system to a paper machine.

In addition to the basic flow pattern within the cleaner, numerous eddies are produced which are wasteful of hydraulic energy and detrimental to cleaning efficiency. Manufacturers have modified the basic design in various ways in an effort to minimize or overcome these secondary flow patterns.

Variables Affecting Cleaner Performance

The performance of a centrifugal cleaner is usually measured in terms of its ability to remove dirt particles:

$$\% \text{ Efficiency} = \frac{\text{Dirt Count}_{(Feed)} - \text{Dirt Count}_{(Accept)}}{\text{Dirt Count}_{(Feed)}} \times 100$$

As with screens, the efficiency of dirt removal is markedly affected by how much stock is rejected. Therefore, comparisons between cleaners at different reject rates can be misleading. Typically, the reject rate is in the range of 10 to 20%. Some cleaners are equipped with a "stock saver" attachment below the cone, where some good fiber is refloated into the cleaner core using "elutriation" (dilution water).

A listing of the major design and operating variables is given in Table 9-4. Cleaning is commonly carried out at relatively low consistencies, under 1%, and efficiency in most applications is adversely affected if consistency is increased much above this level. The amount of pressure drop determines the hydraulic capacity of the cleaner and is a measure of how much centrifugal action is developed. Most units operate with a pressure drop of 30 to 35 psi. A more complete discussion of operating variables is given in reference (11).

Operating Problems

The principal operating problem with centrifugal cleaners is plugging of the reject orifice with foreign materials, fiber flocs, or high-consistency stock. The orifice is usually sized to maintain a reasonable reject level, but sometimes the orifice size must be increased to avoid plugging. The largest orifice at a given reject rate is obtained at minimum back pressure on the cleaner. Due to thickening action at the tip, the reject consistency may be three times that of the inlet; feed consistency can sometimes be lowered to reduce plugging. Some cleaner installations are "protected" with a large-diameter centrifugal cleaner to remove coarse materials likely to plug smaller units (e.g., Figure 9-32). In some cases, a deflocing screen ahead of the cleaners is helpful in eliminating fiber lumps that might contribute to plugging.

TABLE 9-4. Variables affecting centrifugal cleaner performance.

Stock Characteristics
- type of fiber
- characterization of contaminants (size, shape, density)
- dirt level

Design of Cleaner
- body diameter
- feed inlet configuration
- diameter of accept nozzle/vortex finder
- height of cylindrical section
- angle of cone
- application of spiral grooves
- method of reject rate control (fixed orifice and back pressure)

Operating Variables
- stock flow rate
- pressure drop across cleaner
- feed consistency
- reject rate
- stock temperature
- air in stock
- elutriation (when used)
- back pressure
- discharge chamber configuration

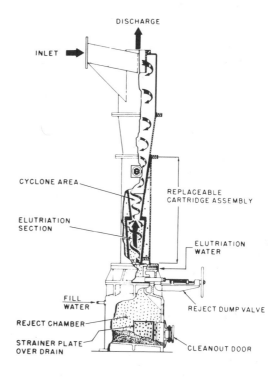

FIGURE 9-32. Centrifugal cleaner designed for heavy trash removal (Black Clawson).

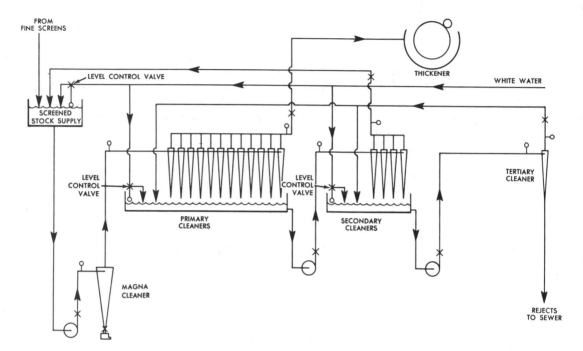

FIGURE 9-33. Flow chart for three-stage centrifugal cleaning system (Sproat Bauer).

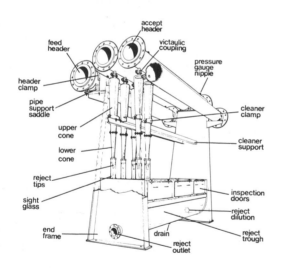

FIGURE 9-34. Module of vertical centrifugal cleaners (Hymac Ltd.).

Arrangement Of Cleaners

Cleaners are typically employed in a cascade sequence similar to that used for screens. Because a large reject flow is necessary for good operating efficiency, additional stages are required to concentrate the dirt in the reject stream and return good fiber to the process. A three-stage arrangement is illustrated in Figure 9-33. The amount of rejected fiber is usually less than 1%.

The early cleaner installations consisted of large numbers of individual units in an open vertical arrangement with hose attachments to the feed and accept headers and open discharge to the reject trough. A more updated open vertical grouping is depicted in Figure 9-34. Most of the newer installations are made up of containerized groupings of cleaners in housings and cassettes of varying design. The stacked, horizontal arrangement of cleaners provides significant space savings, while the enclosed construction promotes good housekeeping. Examples of a modern housing and an individual cleaner unit are shown in Figures 9-35 and 9-36.

Design Modifications

Three types of centrifugal cleaners are used in the pulp and paper industry. The original application, the removal of heavy debris (specific gravity greater than 1.0), employs the classic design of centrifugal cleaner, which is now more precisely designated as "forward cleaning". Forward cleaners still account for the great majority of installations.

Reverse-flow and through-flow cleaners are used principally to remove light debris (specific gravity less than 1.0) in secondary fiber pulping operations. Reverse-flow cleaners came into common use in the late 1970's, but since the mid 1980's have been displaced by through-flow cleaners because of lower pressure drop requirements and low hydraulic reject

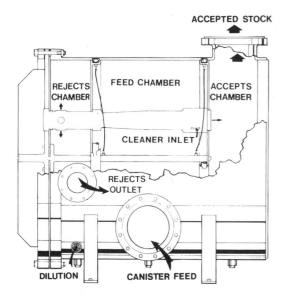

FIGURE 9-35. Modern canister housing for centrifugal cleaners (Black Clawson Co.).

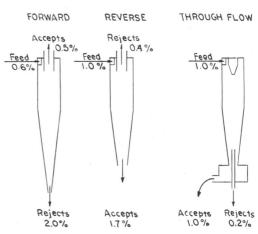

FIGURE 9-37. Flow configurations for forward, reverse-flow and through-flow cleaners, also showing typical operating consistencies.

rates (12). Figure 9-37 illustrates the general orientation and typical consistencies for each type of cleaner.

9.6 THICKENING

Following low-consistency operations such as cleaning and screening, it is necessary to thicken the stock (with or without washing) prior to the next process operation. A variety of equipment is available for this purpose (see Table 9-5) depending on requirements.

In many instances, a simple increase in stock consistency to the 4 – 8% range is required. For this service, a gravity thickener (usually called a decker) is commonly used, as illustrated in Figure 9-38. Water flows into the cylinder by virtue of the difference in liquid level between the vat and cylinder; pulp is retained on the rotating cylinder and is couched off by a rubber roll.

Less commonly, an increase in consistency to the 3.5 – 4.0% range is sufficient and can be achieved using a slusher. This type of thickener is similar to the decker except no couch roll is used. Rather, the stock moves from the inlet side of the vat through the dewatering zone to the other side of the vat, where the thickened stock is discharged, usually with the help of a continuous shower. The principle of operation is illustrated in comparison with a decker in Figure 9-39.

TABLE 9-5. Thickening equipment applications.

Equipment	Discharge Consistency (%)	Washing Capability
Slusher	3.5 – 4	none
Gravity Thickener	4 – 8	none
Valveless Filter (Internal Dropleg)	9 – 12	some
Vacuum Filter	12 – 15	great
Multidisc Filter	10 – 12	none
Screw Extractor	>20	none
Various Press Designs	>20	none → great

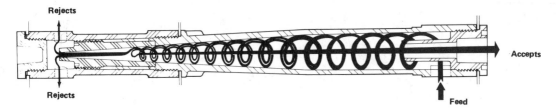

FIGURE 9-36. Individual cleaner unit mounted in canister housing (Black Clawson Co.).

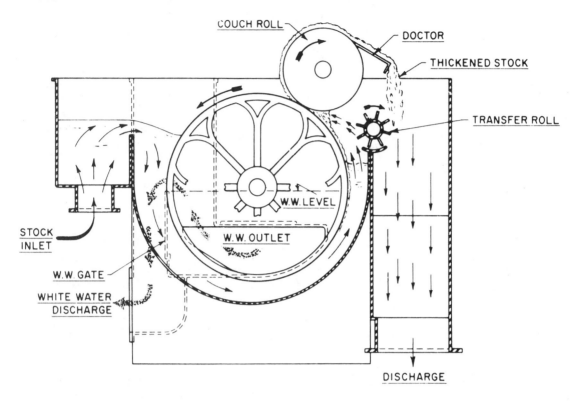

FIGURE 9-38. Gravity couch roll-type thickener for pulp stock (Black Clawson Co.).

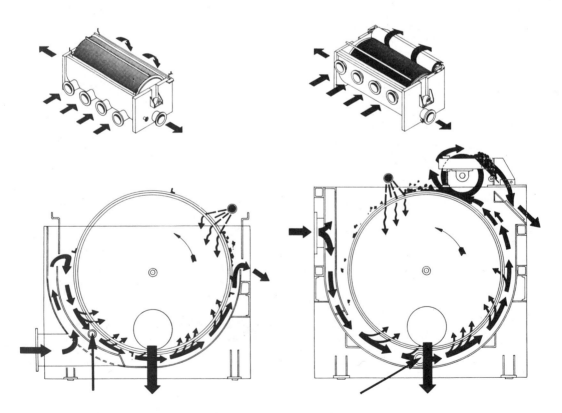

FIGURE 9-39. Comparison between slusher and gravity thickener (Hymac Ltd.).

For intermediate levels of thickening (10–12%), the so-called valveless washer can be used (see Figures 9-40 and 9-41). When thickening from a low consistency below 0.7% to a level of 10 - 12%, a two-stage operation may be required as illustrated in Figure 9-42.

To achieve consistencies of 12 – 16% and also wash the stock, it is necessary to use a vacuum washer. All equipment of this type functions by applying vacuum over the forming and washing zones, then cutting off the vacuum by means of an internal valving arrangement to allow the thickened sheet to be discharged. The suction effect is typically produced by the filtrate dropleg which discharges into a seal tank (refer to Figure 9-43). The pulp mat is discharged from the washer mould by takeoff rolls or doctors. The mesh face is often shower-cleaned prior to re-submergence in the vat.

Vacuum washers differ in design with respect to the valving arrangement. In Figure 9-43, the valve is anchored to one end of the vat. In Figure 9-44, the independent box-shaped valve is riding on its seat at

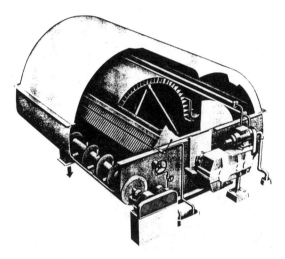

FIGURE 9-40. Valveless filter with internal dropleg (Beloit Rauma).

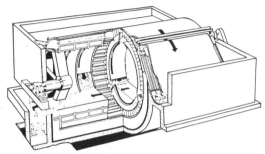

FIGURE 9-41. Valveless vacuum filter (Impco/ Ingersoll Rand).

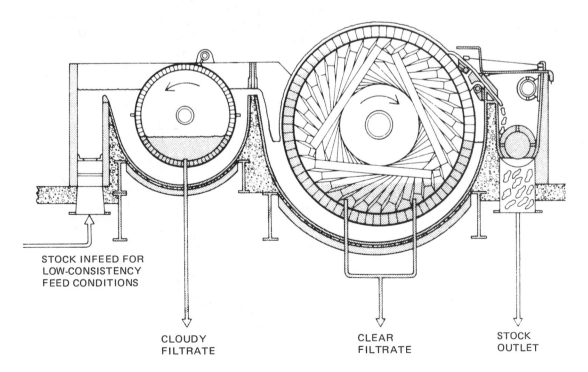

STOCK INFEED FOR
LOW-CONSISTENCY
FEED CONDITIONS

CLOUDY
FILTRATE

CLEAR
FILTRATE

STOCK
OUTLET

FIGURE 9-42. Two-stage thickening system consisting of a slusher followed by a valveless filter (Beloit Corp.).

the outer center of the cylinder. In Figure 9-45, the valve is held stationary by a fixture passing through the rear trunnion. Another design shown previously (refer back to Figure 9-4) utilizes a valve located directly on the trunnion.

For simple thickening of very dilute stocks up to 12% consistency, multidisc thickeners of the type illustrated in Figure 9-46 can be used. A fiber mat forms on the face of each sector as it submerges in the vaᵗ of stock. After complete submergence, vacuum is applied and more pulp is deposited while filtrate is drawn through the mat. The initial filtrate is relatively cloudy and can be segregated (by the action of the end valve) for dilution uses. The later

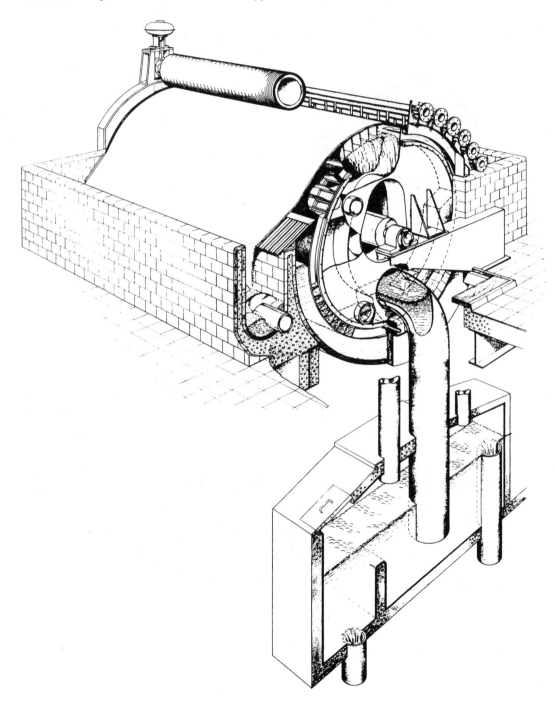

FIGURE 9-43. Vacuum washer illustrating configuration of dropleg and seal tank (Sandy Hill Corp.).

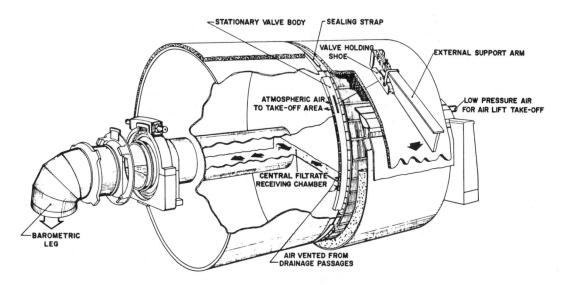

FIGURE 9-44. Circumferential valve vacuum filter (Impco/Ingersoll Rand).

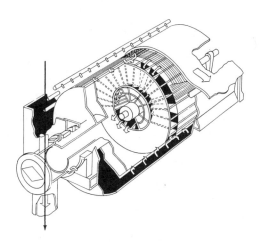

FIGURE 9-45. Center valve vacuum filter (Dorr-Oliver Inc.).

(clear) filtrate is sufficiently free of fiber to be used for shower water. Following emergence from the vat, the pulp mat is further dewatered and finally removed from the filter media by a knockoff shower. Multidisc thickeners are most commonly used as saveall devices, i.e., to recover fine fibers from white water and reuse the water.

To achieve stock consistencies above 15%, some type of screw extractor or press arrangement is usually employed. Representative examples of screw presses are shown in Figures 9-47 and 9-48.

9.7 STOCK PUMPING AND HANDLING

Efficient movement of stock through the various processing steps is at the heart of a pulp and paper mill operation, and no mill can operate successfully without reliable pumping units. Because centrifugal pumps have only one moving part and are usually driven directly from a synchronous motor, they are used wherever possible for stocks up to 6 - 7% consistency.

Centrifugal pumps for pulp and paper stocks are modelled on conventional water pump designs (Figure 9-49), but usually have modified impeller shapes (Figure 9-50) and wide clearances to assist the flow of stock through the pump passages. Several important factors must be considered when selecting a pump design, including consistency, fiber length, stock freeness, and the presence of additives.

For higher-consistency stock pumping, a number of special designs have traditionally been used. Figure 9-51 shows sections through a pump with double-meshing rotors. Figure 9-52 depicts a cutaway view of a double-rotor screw pump. Figure 9-53 illustrates a single-drive unit designed to pump over a wide range of consistencies up to 30%. These pumps are positive displacement and produce some pulsing; they all require the use of a feed chute or standpipe. A typical application is when pulp falls into a chute from a drum or diffusion washer and is pumped to the next stage of the system. Unfortunately, these devices are incapable of pumping high-consistency stock directly from a tank or bleach tower.

FIGURE 9-46. Multidisc filter utilized as a paper machine saveall (Impco/Ingersoll Rand).

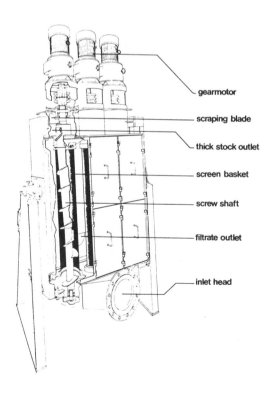

FIGURE 9-47. Three-screw extractor (Hymac Ltd.).

gearmotor

scraping blade

thick stock outlet

screen basket

screw shaft

filtrate outlet

inlet head

Medium Consistency Technology

Traditionally, the discharging/transporting of pulp from thick-stock storage or bleach towers involves bottom dilution, effective mixing, and pumping of a high-volume low-consistency stock. Often, the pulp must be re-thickened prior to the next processing step. These transport systems with auxiliary filtrate tanks, dilution pumps, controls, etc. are both capital- and energy-intensive. It has long been a goal within the industry to develop a single machine with which stock could be discharged from storage and transported without prior dilution and subsequent thickening.

The break-through occurred in the early 1980's with the introduction of a new line of equipment from Kamyr based on development work by Johan Gullichsen of Finland (13). Very quickly, a number of competitive medium consistency pumps and mixers were introduced, and the number of mill installations has soared. All units work on the same general principle: they generate shear forces high enough to fluidize pulp suspensions up to 12% consistency so that they behave like Newtonian fluids. Power consumption is minimized by keeping the active volume low. The principle of operation for the Kamyr unit is illustrated in Figure 9-54. Note that air separation and discharge are essential features of the system, since air would otherwise accumulate in the eye of the pump. A competitive pumping unit is depicted in Figure 9-55.

The same principle has now been successfully applied toward the development of a device for screening pulp suspensions at medium consistency (8 - 15%). A prerequisite for a screening separation is the ability of fibers, knots, shives and other debris to move freely within the suspension. This is easily achieved at low consistency, but can also be accomplished by applying shear forces of an intensity sufficient for complete fluidization of the suspension at the face of the separating screen. The medium-consistency screen has a particulary great impact in mechanical pulping processes. Since latency can be removed by fluidization of medium-consistency fiber suspensions, there is no need to go below 10% consistency in any part of a RMP, TMP, CTMP or CMP line. This translates into a considerable reduction in the requirement for dewatering equipment.

9.8 PULP STORAGE AND BLENDING

In pulp and paper mills, storage of fiber stocks is necessary at appropriate points in the process to provide surge capacity and allow for interruptions in either supply or demand. A basic mechanical pulp mill may need substantial storage only at the end of the process. But, in a bleached kraft mill where significant interruptions can occur at several points in the process, storage capacity is a requirement at least after cooking and bleaching. Generally the blow tank provides several hours of storage, which is especially important with batch digesters to ensure a constant flow to the brown stock washers. (Refer also to Section 12.2.)

A traditional "high-density" storage chest (i.e., at 12 - 15% consistency) and pump-out arrangement is illustrated in Figure 9-56. The stock, usually conveyed from a washer/thickener, is dropped into the top of the chest. The pulp then moves downward as a plug with essentially no intermixing. At the bottom of the chest, a side-entry agitator provides mixing energy and a number of nozzles around the periphery supply dilution water. This arrangement, referred to as "zone agitation", reduces the pulp consistency to a fairly uniform level below 4%, suitable for conventional pumping. Additional dilution water is added at the pump suction for more precise consistency control.

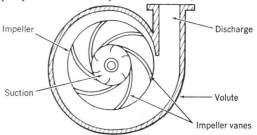

FIGURE 9-49. Typical centrifugal pump.

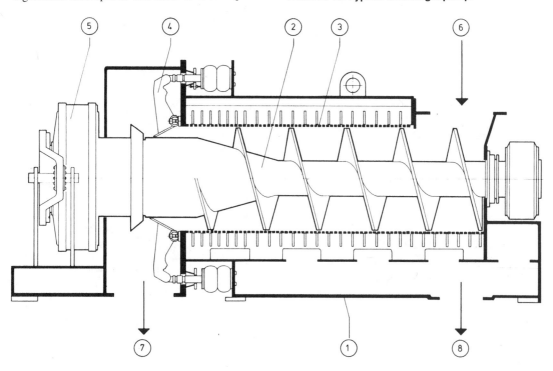

FIGURE 9-48. Screw press for dewatering pulp suspensions. (1) housing (2) screw (3) screen (4) discharging device (5) drive (6) inlet (7) outlet (8) filtrate (KMW).

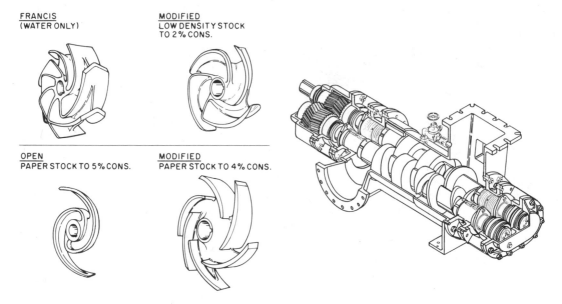

FIGURE 9-50. Centrifugal pump impellers for water and pulp stocks (Black Clawson Co.).

FIGURE 9-52. High density pump (Warren Pumps Inc.).

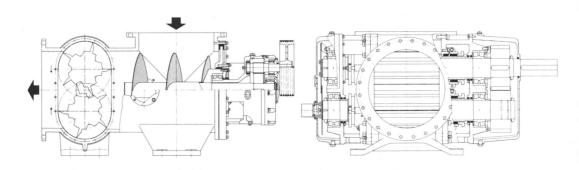

FIGURE 9-51. Thick stock pump utilizing double-meshing rotors (Beloit Rauma).

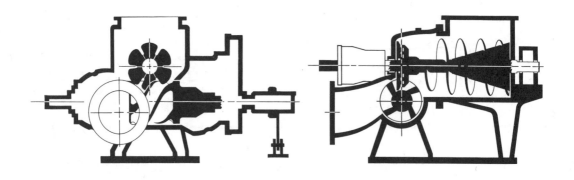

FIGURE 9-53. High density pump (Impco/Ingersoll Rand).

In newer installations that utilize the latest medium-consistency technology (refer to previous section), the arrangement can be greatly simplified. The medium-consistency pump can be coupled directly to the storage chest without the necessity of a dilution zone.

Blending

Because of a non-uniform wood furnish and variations in the pulping process, there is always corresponding variation in the quality and "processability" of mechanical and chemical pulps. Papermakers are well aware that thorough blending of the stock furnish ahead of the paper machine is essential toward avoiding upsets and producing a uniform product. It is surprising, therefore, that blending is often ignored as a process option within the pulp mill.

In the bleached kraft mill, a logical place for blending is following screening and prior to bleaching. Blending at this point will significantly reduce kappa number variation and make possible

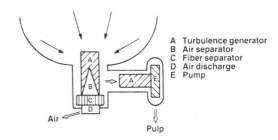

A Turbulence generator
B Air separator
C Fiber separator
D Air discharge
E Pump

FIGURE 9-54. Schematic arrangement for Kamyr medium consistency pump.

better bleaching control. The ideal arrangement is a large chest providing 3 to 4 hours of retention at a consistency below 4%, with sufficient agitation to ensure a high turnover rate. One method of blending within such a tank is illustrated in Figures 9-57 and 9-58. Another method of low-consistency stock blending utilizing a novel chest design is illustrated in Figure 9-59.

In mills that have more than one high-density

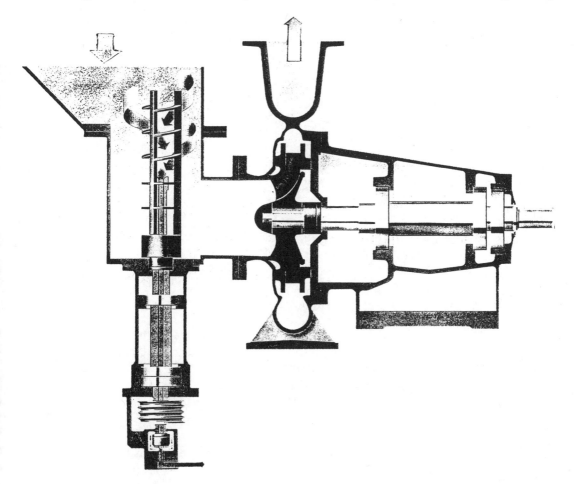

FIGURE 9-55. Medium consistency pump (Andritz).

storage tank at a given stage of the process, a degree of blending can be achieved by filling the tanks in series and pumping out in a parallel configuration. In a mill utilizing medium-consistency technology, some blending can be achieved by employing an over-size pump for the storage tank discharge and returning a portion of the discharge flow back to the top of the same storage tank, thus intermixing pulp from two production periods.

9.9 PREPARING PULP FOR SHIPMENT

In integrated mills, pulp is usually stored at 10 – 14% consistency before use on the paper machine. For non-integrated operations, the pulp must be further dewatered to decrease transportation costs. If the transport distance is small, the pulp may be handled as a high-consistency "crumb", or most commonly in "wetlap" form. Occasionally, the web from a vacuum filter is simply carried on a felt through a series of roll presses to raise the consistency to 40 – 45%, and is then either wound into rolls or cut into sheets for shipment. A complete wetlap system utilizing a double-wire press for initial dewatering and pressing is shown in Figure 9-60.

In most instances, it is necessary to dry the web prior to shipment. Pulp deliveries are commonly made at 90 – 95% air dry (equivalent to 81 - 86% oven dry) in the form of baled sheets. The dominant method, developed from the paper machine, utilizes a sheet-forming wet end (usually a fourdrinier, but sometimes a cylinder former), a press section, a drying section, and usually an on-machine slitting and sheet-cutting operation. The operations of forming, pressing, and steam-cylinder drying will be discussed under "Paper Manufacture" in Chapters 16 and 17. Two other methods of drying are more often used for pulp, and these will be discussed here.

Air Float Dryer

An exterior view of an air float dryer showing the position of the fans is provided in Figure 9-61. The principle of operation is illustrated in Figures 9-62 through 9-64. The pulp web is carried in a number of passes through a chamber in which heated air is used both for drying and for supporting the web. Capital and operating costs are similar to the steam cylinder system, but most pulp producers favor the air float dryer because it is easier to operate and maintain.

In the earlier models of the air dryer, the pulp web is carried on mechanical conveyors and hot air is

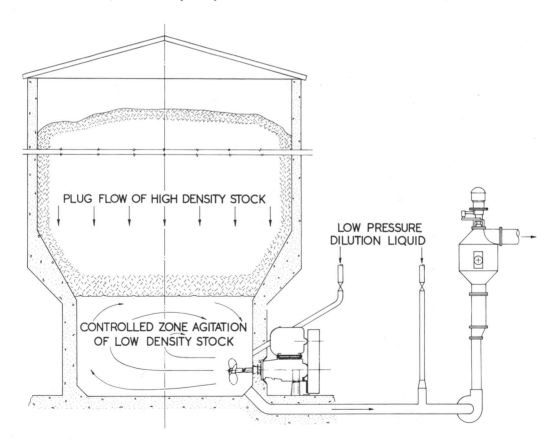

PLUG FLOW OF HIGH DENSITY STOCK

LOW PRESSURE DILUTION LIQUID

CONTROLLED ZONE AGITATION OF LOW DENSITY STOCK

FIGURE 9-56. High density storage system illustrating zone agitation in the pump out section (Greey Mixing Equipment Ltd.)

blown between the passes. The later designs incorporate true air impingement above and below the web; this principle provides much higher evaporation rates and eliminates the mechanical conveyor. Typically, the pulp web enters at the top, makes a number of horizontal passes turning around rollers at each end, and leaves at the bottom of the dryer at the opposite end. The fresh makeup air,

which has been preheated in the economizer section (utilizing waste heat from the exhaust air) enters the bottom of the dryer. The air is circulated several times by means of numerous fans on both sides of the dryer and eventually exits at the top through the air-to-air heat exchanger of the economizer unit. On each pass, the air is heated by a steam coil and pumped through the blow boxes where the air jets issue from "eyelid openings" in the top and bottom to impinge on the pulp web.

The earlier models utilize a threading table at the entrance to the dryer for attaching the tail (a 6- to 10-inch wide continuous strip of pulp) to the "kite" (a large piece of cotton with pockets to pick up air) during startup of the drying operation. The newer units utilize an automatic system for carrying the tail through the dryer consisting of twin bands or a folded tape running beside the normal position of the pulp web at the operating side of the dryer.

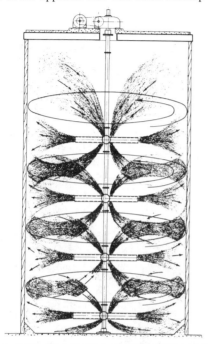

FIGURE 9-57. The Cowan stock agitator takes material from the center of the tank and distributes it to the periphery (S.W. Hooper Corp.).

FIGURE 9-58. Illustrating the principle of the Cowan stock agitator (S.W. Hooper Corp.).

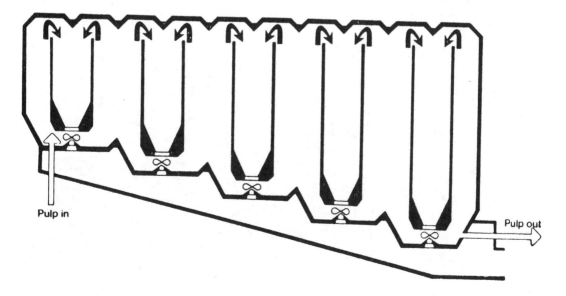

FIGURE 9-59. Low-consistency multichannel stock blending chest (Varkaus Paper Mill).

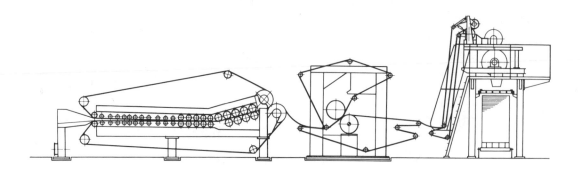

FIGURE 9-60. This wet-lap machine utilizes a double-wire press for initial dewatering of the stock suspension to achieve a consistency of up to 40% bone dry. The dryness is raised to 48 - 50% bone-dry by means of a heavy-duty press. The pulp web is then slit and cut into sheets by a rotating knife drum. The sheets drop onto a pallet. (Andritz)

FIGURE 9-61. Airborne pulp dryer (ABB Flakt Canada Ltd.).

Sheet tension in the drying unit is controlled by nip and dance rolls at the entrance and exit. The turning rolls at the end of each pass are driven during threading, but turn free when the full-width web is in the dryer.

Criteria of Performance and Operating Variables

Two factors are of paramount importance when considering the relative performance of any pulp dryer: evaporation rate (a measure of capacity) and steam economy (a measure of energy efficiency). Typical units of measurement for these criteria applied to an air dryer are:

Area Evaporation Rate   lb water evap/sq ft/hr
Steam Economy   lb steam/lb water evap

Typical values of area evaporation rate are 0.9 to 1.0 for the older, mechanically-conveyed pulp dryer and 1.7 to 1.8 for the air impingement dryer. Steam economy can vary from as low as 1.2 to as high as 1.7. The main variables that affect these criteria are summarized in Table 9-6.

Generally, higher evaporation rates can be achieved when water is near the surface of the sheet, as with low-weight sheets at high moisture content. Of course, in terms of overall water removal, it is economically attractive to remove as much water as possible mechanically (by pressing) before applying evaporative drying.

Cutter and Layboy

The sheet from the dryer is typically conveyed to a series of slitters which cut the sheet to width, then a rotating fly knife cuts the sheets to length. The cut sheets are conveyed on tapes to the layboy boxes where they are stacked on the layboy table. A hydraulic system lowers the table as the stacks get higher. When the stacks are the correct weight, layboy "fingers" come out to support the sheets while the table lowers and discharges to a conveyor. The stacks of sheets are then conveyed to the finishing area where the stacks are weighed, compressed into bales, wrapped, wired, and labelled. Refer to Figure 9-65.

Flash Drying

Flash drying refers to the process where pressed pulp is first "fluffed" (i.e., the compressed mat is separated into fibers, thus greatly increasing the exposed surface area) and then injected into a stream of hot gases. The high-temperature heat of the gas stream causes the moisture to flash into vapor. A representative system is illustrated in Figure 9-66. The dried pulp product may be compressed directly into a bale, or into smaller pieces which are then formed into a bale (Figure 9-67).

Flash drying is attractive because of lower capital cost and less space requirement, even though operating costs may be comparatively higher. Some quality problems were experienced with flash drying systems of the late 1960's and early 1970's, primarily loss of brightness and formation of hard "pills" or "nodules" that were difficult to separate. Efficient operation of the fluffer was found to be the critical factor, since good separation and large specific surface allow the evaporation to take place more rapidly and at lower temperature.

A variation of flash drying is embodied in the so-called "steam dryer", where fluffed pulp is dried in a medium of slightly superheated low-pressure steam rather than in hot air (15). Heat is transferred to the pulp and transport steam by indirectly condensing higher-pressure steam. The water evaporated from the pulp is removed from the system as low-pressure steam suitable for process use in the pulp or paper mill. It is claimed that net heat consumption can be as low as one-fourth of that used in a conventional drying system.

REFERENCES

(1) HARTLER, N., DANIELSSON, O., RURBERG, G. **Mechanical Fiber Separation in Kraft Pulping Systems** *TAPPI* 59:9:105-108 (September 1976)
(2) KORHONEN, O. **Brownstock Washing: A Review of Current Technology** *Pulp & Paper* (September 1979)

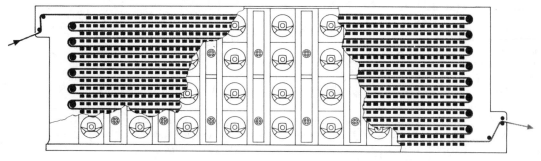

FIGURE 9-62. The pulp web moves over the dryer decks via turning rolls at the end sections (ABB Flakt Inc.).

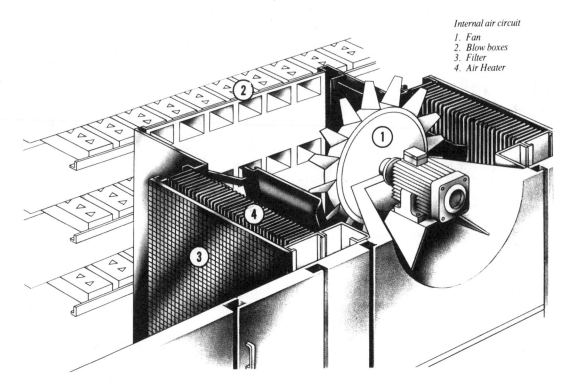

Internal air circuit
1. *Fan*
2. *Blow boxes*
3. *Filter*
4. *Air Heater*

FIGURE 9-63. Longitudinally, the airborne dryer is composed of intermediate sections with steam coils for heating and fans for distributing the air to the blow boxes (ABB Flakt Inc.).

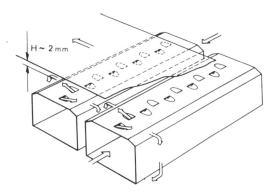

H ~ 2 mm

FIGURE 9-64. The air impingement system lifts the web clear of the deck and locates it at a fixed height.

TABLE 9-6. Variables that affect evaporation rate and steam economy.

Factors which favor high evaporation rate:
- low weight sheet (greater surface area)
- high ingoing sheet moisture
- high outgoing sheet moisture
- high air temperatures (high steam pressure in heaters)
- high rate of makeup air (low humidity in exhaust air)
- low stock pH

Factors which favor low steam economy:
- low rate of makeup air (high humidity in exhaust air)
- properly designed and maintained economizer
- proper balance between supply and exhaust air flows (i.e., minimize infiltration air and out-leakage of hot air)

(3) CROTOGINO, R.H., POIRIER, N.A. and TRINH, D.T. **The Principles of Pulp Washing** *Tappi Journal* (June 1987)

(4) SILANDER, R. **New Technologies in Brownstock Washing Replacing Vacuum Filters** *Pulp & Paper* (May 1987)

(5) NORDEN, H.V., et al **Statistical Analysis of Pulp Washing on an Industrial Rotary Drum** *P&P Canada 74*:10:T329 (October 1973)

(6) PHILLIPS, J.R. and NELSON, J. **Diffusion Washing System Performance** *P&P Canada 81*:1:T24-27 (January 1980)

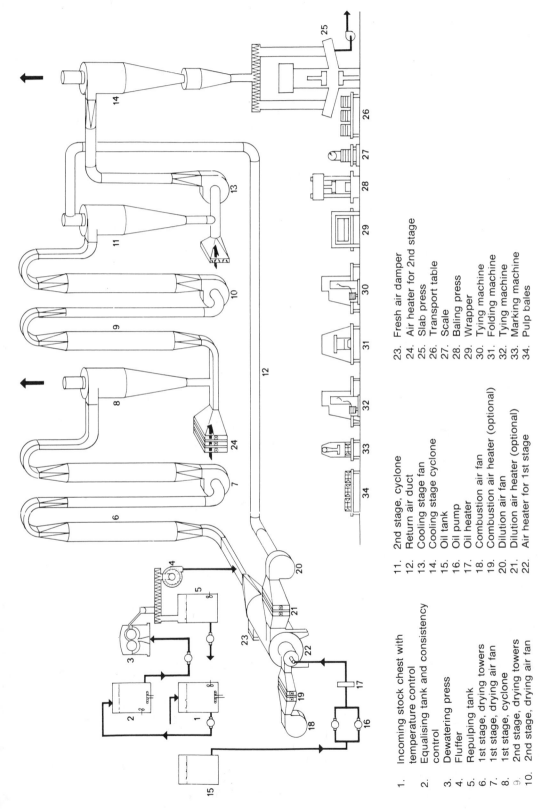

1.   Incoming stock chest with
     temperature control
2.   Equalising tank and consistency
     control
3.   Dewatering press
4.   Fluffer
5.   Repulping tank
6.   1st stage, drying towers
7.   1st stage, drying air fan
8.   1st stage, cyclone
9.   2nd stage, drying towers
10. 2nd stage, drying air fan

11. 2nd stage, cyclone
12. Return air duct
13. Cooling stage fan
14. Cooling stage cyclone
15. Oil tank
16. Oil pump
17. Oil heater
18. Combustion air fan
19. Combustion air heater (optional)
20. Dilution air fan
21. Dilution air heater (optional)
22. Air heater for 1st stage

23. Fresh air damper
24. Air heater for 2nd stage
25. Slab press
26. Transport table
27. Scale
28. Baling press
29. Wrapper
30. Tying machine
31. Folding machine
32. Tying machine
33. Marking machine
34. Pulp bales

FIGURE 9-66. Flow chart for the Flakt flash dryer.

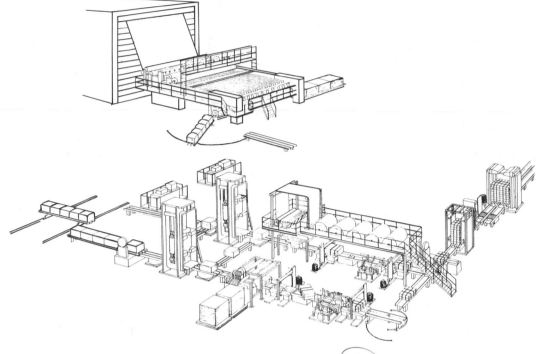

FIGURE 9-65. Pulp finishing line (Lamb).

(7) ATKEISON, C.A. **How to Achieve the Best Operation from Free-Discharge-Type Screens** *Pulp & Paper* (March 1979)

(8) NELSON, G.L. **The Screening Index: A Better Index for Screening Performance** *TAPPI* 64:5:133 (May 1981)

(9) MCCARTHY, C. **Various Factors Affect Pressure Screen Operation and Capacity** *Pulp & Paper* (December 1988)

(10) FORBES, D.R. **Pulp Mill Screening is Key to Pulp Quality, Maximum Fiber Utilization** *Pulp & Paper* (November 1987)

(11) BLISS, T.L. **Stock Preparation, Chapter 6 - Cleaning (Volume 6, Pulp & Paper Manufacture)** Joint Textbook Committee, 1992

(12) BLISS, T.L. **Through-Flow Cleaners Offer Good Efficiency With Low Pressure Drop** *Pulp & Paper* (March 1985)

(13) GULLICHSEN, J., et al **Medium Consistency Technology: Storage Dischargers and Centrifugal Pumps** *TAPPI* 64:9:113-116 (September 1981)

(14) GULLICHSEN, J., et al **Medium-Consistency Technology: the MC Screen** *TAPPI Journal* (November 1985)

(15) SVENSON, C. **Novel Pulp Dryer Promises Energy Saving** *P & P International* (June 1980)

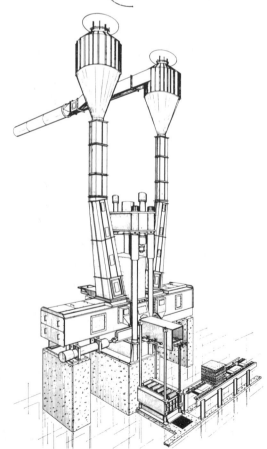

FIGURE 9-67. Vertical sheet former used in conjunction with D&S flash drying system (Vickers Canada).

Chapter 10

Chemical Recovery

The recovery of chemicals from the spent cooking liquor, the reconstitution of these chemicals to form fresh cooking liquor, the realization of energy from the incineration of organic residuals, and minimization of air and water pollution are vital parts of any chemical pulp mill recovery process. These objectives are shared by both kraft and sulfite mills, but there are some marked differences in the respective methodology. This chapter will concentrate on the more common kraft mill recovery process, with only brief coverage of sulfite recovery techniques in Section 10.7.

The kraft recovery process has not changed in any fundamental way since it was patented in 1884, but has evolved and been refined, especially since the 1930's, into a series of well-defined unit operations. A general flowsheet of the overall kraft mill liquor cycle is shown in Figure 10-1. Starting with "weak black liquor" from the brown stock washers, the steps involved in chemical recovery are as follows:

(1) Concentration of the residual liquor in multiple-effect evaporators to form "strong black liquor".
(2) Black liquor oxidation (if required).
(3) Further concentration of the residual liquor to form "heavy black liquor". (Saltcake can be added at this point to make up soda loss.)
(4) Incineration of liquor in the recovery furnace.
(5) Dissolving smelt from the furnace to form green liquor. (Soda ash can be added at this point to make up soda loss.)
(6) Causticizing green liquor with lime to form white liquor. (Caustic soda can be added at this point to make up soda loss.)
(7) Burning of lime mud to recover lime.

These steps are delineated in the following discussion.

10.1 BLACK LIQUOR OXIDATION

Black liquor oxidation converts sulfide to thiosulfate according to either of the following reactions:

$$2Na_2S + 2O_2 + H_2O \longrightarrow Na_2S_2O_3 + 2NaOH$$

$$2NaHS + 2O_2 \longrightarrow Na_2S_2O_3 + H_2O$$

Oxidation was initiated in the 1950's as an odor-reduction step to "stabilize" the sulfur in the liquor as thiosulfate. Otherwise, odorous hydrogen sulfide is stripped from the liquor by the hot combustion gases in the direct-contact evaporator according to the following reaction:

$$2NaHS + CO_2 + H_2O \longrightarrow Na_2CO_3 + H_2S$$

Black liquor oxidation systems are expensive to install and operate. The oxidation step also robs the liquor of fuel value. Most recovery furnaces installed since 1975 no longer incorporate a direct contact evaporator in their design, thus eliminating the requirement for oxidation.

Oxidation is usually accomplished by contacting the liquor with air. A variety of equipment has been used for this purpose including spray towers, bubble plate columns, and sparge rings in tanks. The objective is to provide intimate contact between liquor and air while minimizing foam generation. Nonetheless, foam is always a problem, especially when treating pulping liquors from resinous woods. Auxiliary equipment to handle the foam is usually part of the overall oxidation system design.

Oxidation was initially carried out on weak black liquor (e.g., Figure 10-2) in order to reduce odorous emissions at both the multiple-effect evaporators and the direct-contact evaporators. Later, when better methods of handling the noncondensibles from the multiple-effect evaporators were developed, it became more attractive to oxidize the lesser volume of strong black liquor (e.g., Figure 10-3). Also, it was discovered that some sulfide typically reforms following weak black liquor oxidation due to reversion of elemental sulfur, which is itself a partial product of the oxidation process. Reformation of sulfide following strong black liquor oxidation is negligible.

10.2 EVAPORATION

Weak black liquor usually leaves the brown stock washers with a solids content between 13 and 17%, while the heavy black liquor is burned at between 60 and 80% solids. This difference in solids level represents a large amount of water (between 5 and 7

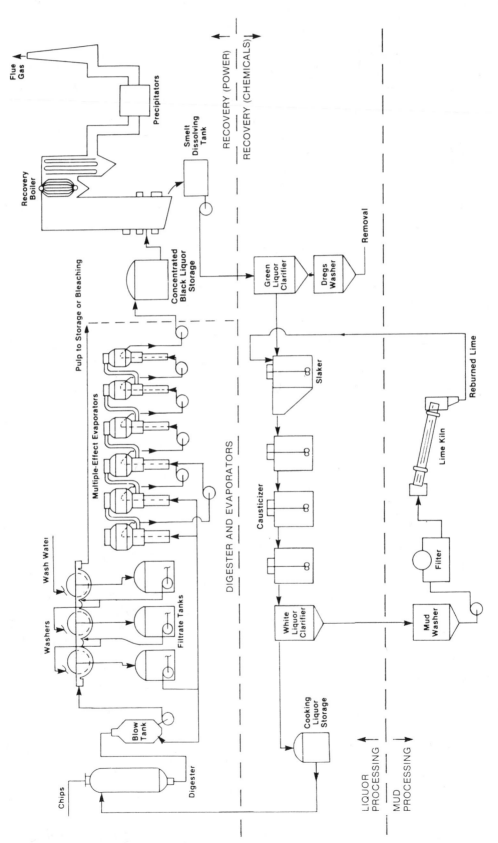

FIGURE 10-1. Kraft process.

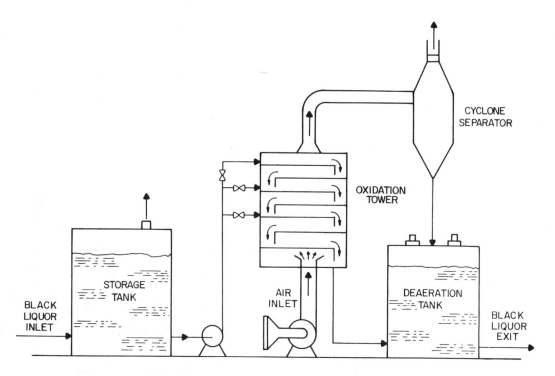

FIGURE 10-2. Trobeck-Ahlen multiple sieve tray weak black liquor oxidation system.

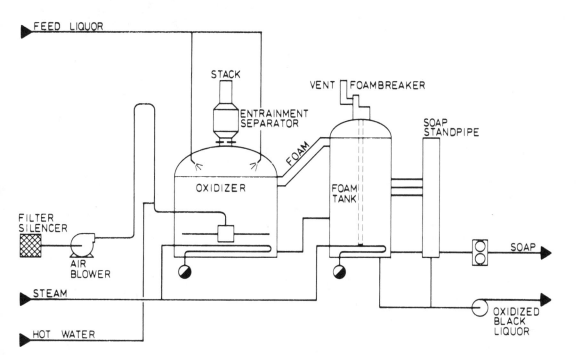

FIGURE 10-3. Strong black liquor oxidation system (MoDo-Chemetics).

lb water per lb dry solids) which must be evaporated economically in order to realize maximum net fuel value from the burning.

Multiple-Effect Evaporators

The bulk of the water removal is carried out in the multiple-effect evaporators, a series of reboilers operated at different pressures so that the vapor from one evaporator body becomes the steam supply to the next unit. The prime advantage of the multiple-effect design is high steam economy, which can be as high as 5.5 lb water evaporated per lb feed steam for a seven-effect system (see Figure 10-4).

A representative set of six-effect evaporators (of traditional long-tube vertical design) is shown in Figure 10-5 and schematically in Figure 10-6. The effects are numbered with respect to the steam and vapor flow sequence. The steam feed goes to the shell side of the pressurized first effect which usually contains two liquor effects (1A and 1B) on the tube side. The weak black liquor feed is usually split between the first two effects at the other end (6 and

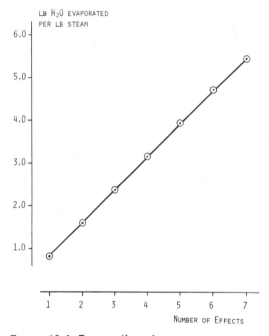

FIGURE 10-4. Evaporation steam economy as a function of number of effects.

FIGURE 10-5. Typical installation of sextuple-effect evaporators.

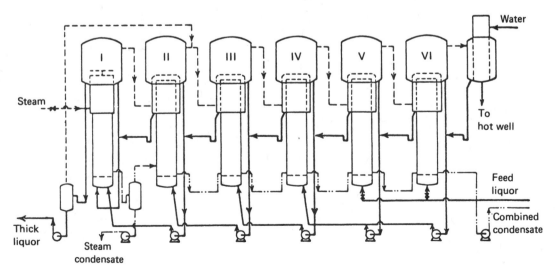

FIGURE 10-6. Flowsheet for sextuple-effect evaporators.

5) where the liquor boils at lower temperature under vacuum conditions. As the liquor moves from one effect to the next, the vapor pressure increases, the boiling temperature increases, the % solids increases, and the volume of liquor decreases. The pressures in the intermediate effects are determined by the initial steam pressure and by the vacuum applied to the vapor space of the last effect.

Representative operating data for a six-effect set of evaporators are given in Table 10-1. As the solids content increases, so does the difference between the boiling liquor temperature and the condensing temperature of its associated vapor. This temperature difference (called the "boiling point rise") significantly reduces the available driving force for heat transfer in the higher-solids effects and is a major factor limiting the number of effects which are feasible in a set of evaporators.

A typical long-tube vertical evaporator body is illustrated in Figure 10-7. The entering liquid boils and thickens as it travels rapidly up the tubes (i.e., as it absorbs heat from the vapor condensing on the outside surfaces of the tubes). The two-phase mixture overflows the tubesheet and impinges on the deflector, which breaks up the liquor and vapor. The vapor goes through a centrifugal separator (sometimes in series with a wire mesh pad) to remove entrained liquid droplets and is carried forward to the next evaporator shell. The liquor flows over the edge into a downcomer and is circulated by a pump through the closed afterheater section, where the liquor picks up sensible heat before going into the next effect.

The liquor that discharges from the last effect is superheated with respect to atmospheric pressure. Additional vapor boils off in a flash tank before the strong black liquor at 50 - 60% solids is taken to storage. This liquor must be further concentrated to

TABLE 10-1. Representative operating data for six-effect evaporators.

	Flash Tank	IA	IB	II	III	IV	V	VI
Steam feed, lb/h		20 500	30 700					
Steam pressure psi		31.5	31.5	9.4	1.2			
vacuum, in Hg.						8.3	16.2	22.0
Steam temp., °F		276	276	238	216	196	175	152
Working-temp. drop, °F		24	26	13	13	15	17	24
Liquor temp., °F	232	252	250	225	203	181	158	128
Boiling point rise, °F	14	13	11	7	5	4	3	3
Saturated vapor temp., °F	218	239	239	218	198	177	155	125
Vapor pressure psi		9.8	9.8	1.8				
vacuum, in Hg					7.4	15.6	21.4	26.0
Latent heat of vapor, Btu/lb		953	953	967	979	992	1005	1023
Pressure drop psi		0.4	0.4	0.6				
vacuum, in Hg					0.9	0.6	0.6	
Vapor pressure Next effect, psi		9.4	9.4	1.2				
Vacuum, in Hg					8.3	16.2	22.0	
Feed, lb/h	90 605	109 775	135 825	178 875	219 585	252 995	166 520	166 520
Discharge, lb/h	89 200	90 605	109 775	135 825	178 875	219 585	130 475	122 520
Total solids, %	51.2	42.3	34.2	25.9	21.1	18.3	13.9	13.9
Total solids out, %	52.0	51.2	42.3	34.2	25.9	21.1	17.8	18.9
Heating surface, ft²		4 400	4 400	8 800	8 800	8 800	8 800	8 800
Heat transfer coefficient U		174	242	392	386	316	240	190

65 - 70% solids (or higher) for burning, either by means of direct-contact evaporation or in a "concentrator".

Many evaporator body designs are used (1). Most existing equipment is of the "rising-film" type. However, "falling-film" designs (such as illustrated in Figure 10-8) are becoming increasingly popular because these systems are generally more energy-efficient and have greater turndown capability. At least one manufacturer utilizes plate elements rather than tubes for the heat exchange surface.

The traditional flow sequence for multiple effects is a straight countercurrent arrangement, as was illustrated in Figure 10-6. Alternative flow sequences are now being used in an effort to optimize heat transfer and reduce fouling (i.e., the buildup of deposits on the tube surfaces). A switching arrangement between effects is sometimes used so that lower-solids liquor can "wash away" the deposits left by high-solids liquor and extend overall service before a general washing with water (i.e., a "boilout") is required.

A variety of evaporation equipment is available to augment or supplement the multiple-effect evaporators. Among the choices are blow heat evaporators (2), mechanical vapor recompression evaporators and thermal vapor recompression evaporators. Selection should be based on careful consideration of the relative merits of each alternative and a thorough review of the overall economics, especially with respect to energy conservation.

Blow heat evaporators typically take the form of a two-effect system in which flash steam (or hot condensate from the accumulator) is used as the heating medium for the first effect and vapors from the first effect are used as the heating medium in the second effect. Mechanical vapor recompression (MVR) raises vapor temperature through mechanical compression so that it can be re-used as the heating medium for evaporation. A typical MVR flow scheme is shown in Figure 10-9. Thermal recompression methods utilize high-pressure steam to entrain low-pressure vapor so that the resultant mixture (at intermediate pressure) can be used as the heat source for evaporation.

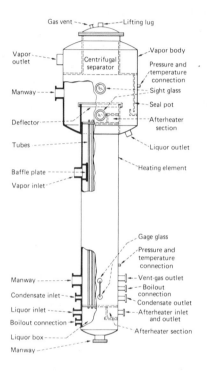

FIGURE 10-7. Internals of typical long-tube evaporator effect (HPD Inc.).

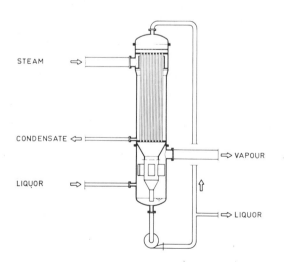

FIGURE 10-8. Falling-film type evaporator.

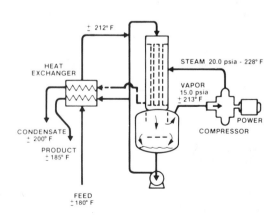

FIGURE 10-9. Evaporator utilizing mechanical vapor recompression (Unitech, Div. of Ecodyne Corp.).

Direct-Contact Evaporator

Direct-contact evaporators were incorporated into the design of virtually all recovery boilers up until 1968. They are positioned in the gas ducts just downstream of the boiler economizer tube section, and serve to further concentrate the strong black liquor utilizing residual heat from the combustion gases to evaporate water from the liquor. The intimate contact between liquor and hot gases enables the liquor to capture some gas-borne particulates, and also allows the furnace gases to strip reduced sulfur from the liquor (refer back to Section 10.1). Elimination of the direct-contact evaporator on newer recovery boiler designs has resulted in alternative approaches toward utilizing the residual heat in the combustion gases (see Section 10.3).

The cascade evaporator (Figure 10-10) consists of a rotating assembly of tubes that are sequentially submerged in liquor and exposed to the hot gases. Units are arranged in the gas ducts in parallel or series configuration, depending on the condition of the flue gas and quantity of evaporation required. This type of evaporator was commonly incorporated into the design of Combustion Engineering recovery boilers.

The cyclone evaporator (Figure 10-11) is a vertical cylindrical vessel with a conical bottom. Flue gas enters tangentially and contacts liquor which has been sprayed across the entrance opening. The gas leaves through a top duct while the concentrated liquor drains from the bottom. This type of evaporator was commonly incorporated into the design of Babcock and Wilcox recovery boilers.

Concentrator

Most mills without a direct-contact evaporator use a specially designed single- or multiple-stage, indirectly-heated evaporator, called a "concentrator", for increasing the solids level to 65 - 70%. Because the solubility limit of inorganic components is usually exceeded at higher solids concentrations, a relatively large volume of liquor must be pumped around to reduce deposits on tubes. Even so, a single-stage concentrator is usually cleaned frequently, and some mills require a spare unit for alternate service and cleaning.

Different designs are used, but in all cases a prime objective is to minimize fouling and scaling. Probably the most common type is the falling-film concentrator with recirculation as illustrated in Figure 10-12. In such a design it is crucial that the liquor be uniformly distributed over the heating surface. Evaporation takes place at the surface of the liquor and not on the heating surface.

The forced-circulation submerged-tube concentrator (not illustrated) operates by pumping large volumes of liquor through a tubular heater. The liquor is prevented from flashing in the heater by maintaining a positive head on the heater and minimizing the temperature rise across the heater. Therefore, evaporation in this design takes place in the vapor space and not in the tubes.

Some modern multiple-effect evaporation systems have integrated the concentrator into the flow

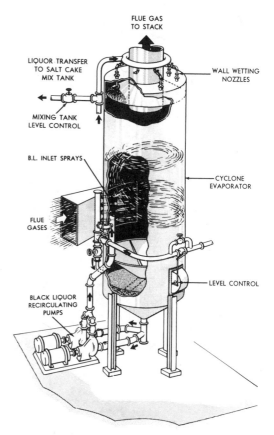

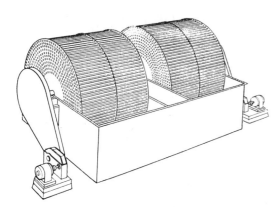

FIGURE 10-10. Cascade evaporator.   FIGURE 10-11. Cyclone evaporator.

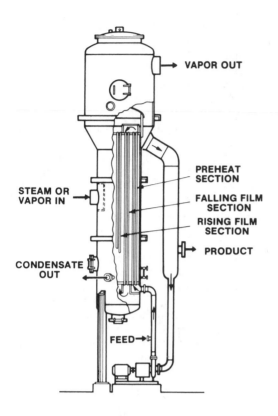

VAPOR OUT

PREHEAT
SECTION

STEAM OR
VAPOR IN

FALLING FILM
SECTION

RISING FILM
SECTION

PRODUCT

CONDENSATE
OUT

FEED→

FIGURE 10-12. This concentrator utilizes a preheat section with falling-film section and partial recirculation (Unitech, Div. of Ecodyne Corp.).

sequence with the low-solids effects so that each concentrator section is "washed" periodically with lower-solids liquor. An example of an integrated system is shown in Figure 10-13.

10.3 RECOVERY BOILER

The recovery furnace/boiler is at the heart of the kraft recovery process and fulfills the following essential functions:
1) Evaporates residual moisture from the liquor solids.
2) Burns the organic constituents.
3) Supplies heat for steam generation.
4) Reduces oxidized sulfur compounds to sulfide.
5) Recovers inorganic chemicals in molten form.
6) Conditions the products of combustion to minimize chemical carryover.

The heavy black liquor from either the direct-contact evaporator or the concentrator is sprayed into the furnace. The liquor droplets dry and partially pyrolyze before falling onto the char bed. Incomplete combustion in the porous char bed causes carbon and carbon monoxide to act as reducing agents, thus converting sulfate and thiosulfate to sulfide. The heat is sufficient to melt the sodium salts, which filter through the char bed to the floor of the furnace. The smelt then flows by gravity through water-cooled spouts to the dissolving tanks.

Evolution of Recovery Boiler Design

The development of the modern recovery boiler is generally credited to G.H. Tomlinson of Howard

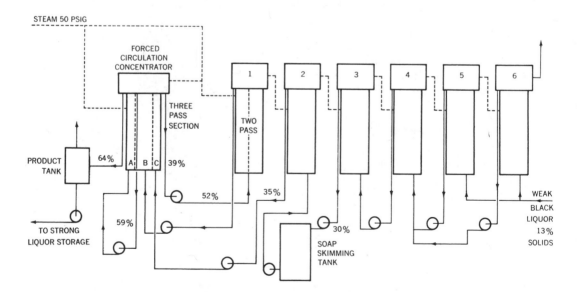

FIGURE 10-13. Multiple-effect evaporator system with three-section concentrator. One concentrator section is continuously washed with liquor from the second effect, while the other two sections concentrate the liquor to 64% solids.

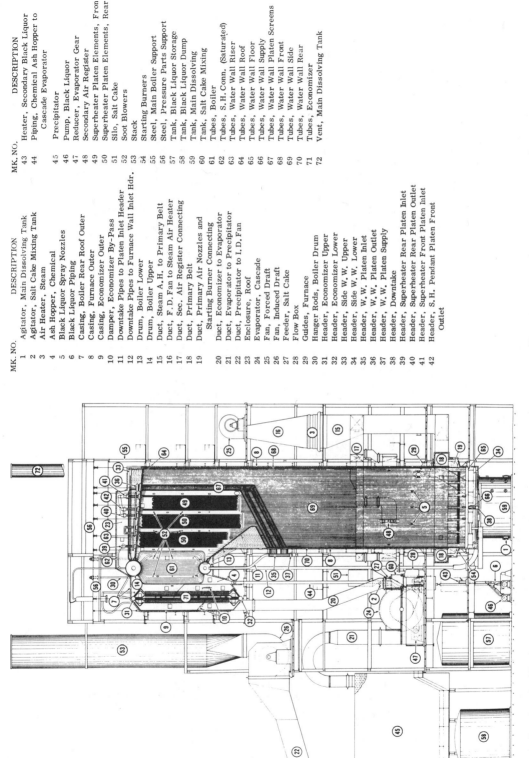

MK. NO.	DESCRIPTION
1	Agitator, Main Dissolving Tank
2	Agitator, Salt Cake Mixing Tank
3	Air Heater, Steam
4	Ash Hopper, Chemical
5	Black Liquor Spray Nozzles
6	Black Liquor Piping
7	Casing, Boiler Rear Roof Outer
8	Casing, Furnace Outer
9	Casing, Economizer Outer
10	Damper, Economizer By-Pass
11	Downtake Pipes to Platen Inlet Header
12	Downtake Pipes to Furnace Wall Inlet Hdr.
13	Drum, Boiler Lower
14	Drum, Boiler Upper
15	Duct, Steam A.H. to Primary Belt
16	Duct, F.D. Fan to Steam Air Heater
17	Duct, Sec. Air Register Connecting
18	Duct, Primary Belt
19	Duct, Primary Air Nozzles and Starting Burner Connecting
20	Duct, Economizer to Evaporator
21	Duct, Evaporator to Precipitator
22	Duct, Precipitator to I.D. Fan
23	Enclosure, Roof
24	Evaporator, Cascade
25	Fan, Forced Draft
26	Fan, Induced Draft
27	Feeder, Salt Cake
28	Flow Box
29	Guides, Furnace
30	Hanger Rods, Boiler Drum
31	Header, Economizer Upper
32	Header, Economizer Lower
33	Header, Side W.W. Upper
34	Header, Side W.W. Lower
35	Header, W.W. Platen Inlet
36	Header, W.W. Platen Outlet
37	Header, W.W. Platen Supply
38	Header, Downtake
39	Header, Superheater Rear Platen Inlet
40	Header, Superheater Rear Platen Outlet
41	Header, Superheater Front Platen Inlet
42	Header, S.H. Pendant Platen Front Outlet

MK. NO.	DESCRIPTION
43	Heater, Secondary Black Liquor
44	Piping, Chemical Ash Hopper to Cascade Evaporator
45	Precipitator
46	Pump, Black Liquor
47	Reducer, Evaporator Gear
48	Secondary Air Register
49	Superheater Platen Elements, Fron
50	Superheater Platen Elements, Rear
51	Silo, Salt Cake
52	Soot Blowers
53	Stack
54	Starting Burners
55	Steel, Main Boiler Support
56	Steel, Pressure Parts Support
57	Tank, Black Liquor Storage
58	Tank, Black Liquor Dump
59	Tank, Main Dissolving
60	Tank, Salt Cake Mixing
61	Tubes, Boiler
62	Tubes, S.H. Conn. (Saturated)
63	Tubes, Water Wall Riser
64	Tubes, Water Wall Roof
65	Tubes, Water Wall Floor
66	Tubes, Water Wall Supply
67	Tubes, Water Wall Platen Screens
68	Tubes, Water Wall Front
69	Tubes, Water Wall Side
70	Tubes, Water Wall Rear
71	Tubes, Economizer
72	Vent, Main Dissolving Tank

FIGURE 10-14. Combustion Engineering recovery boiler (circa 1968) with marks to identify various components.

Smith Paper in Ontario. Babcock and Wilcox (B&W) collaborated with Tomlinson in the construction of the first prototype in 1934. B&W further refined the Tomlinson concept in subsequent designs. Combustion Engineering (C-E) entered the field in the period 1936 - 1938 based on earlier work with rotary smelters. Over the years, B&W and C-E have remained the two dominant manufacturers of recovery boilers in North America.

Examples of C-E and B&W recovery boilers with direct-contact evaporators, circa 1968, are shown in Figures 10-14 and 10-15 (on next page) respectively. The major design differences are summarized in Table 10-2. The basic recovery furnace technology was also licensed to Scandinavian boiler manufacturers (Gotaverken of Sweden and Tampella of Finland) who, over the years, have incorporated their own modifications. Virtually all recovery boiler systems include an electrostatic precipitator for removing entrained soda fume from the exhaust gases (shown in Figure 10-14, but omitted from Figure 10-15). The operating principles of the precipitator are explained in Chapter 27.

The comparatively high cost of fuel oil in Scandinavia in the early 1960's led to demands for higher thermal efficiency. Subsequent development of higher-solids evaporation systems rendered obsolete the thermally-inefficient direct-contact evaporator. Since about 1968, most new recovery furnaces in North America have been installed without direct-contact evaporators. Most of these so-called "low-odor" designs incorporate a larger economizer section to cool the outlet gases (Figures 10-16 and 10-17).

The kraft chemical recovery unit has continued to

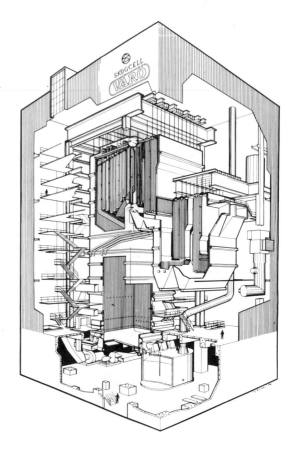

FIGURE 10-16. Gotaverken recovery boiler with large economizer section.

TABLE 10-2. Comparison of kraft recovery furnace traditional designs.

Item	Babcock & Wilcox	Combustion Engineering
Black Liquor Sprays		
• Number of nozzles	1, 2, or 4	multiple
• Pressure	20–40 psig	12–20 psig
• Oscillation		
- Vertical	15° down to 5° above	horizontal to down
- Horizontal	20° left to 20° right	(none)
• Spraying Method	onto walls	into center
Method of Drying	wall drying (solids fall to hearth from wall)	spray drying (solids in free fall to hearth)
Hearth Design	sloping to smelt spouts	flat, with smelt spouts slightly above hearth (decantation)
Air Distribution		
• Primary	~ 50% (up to 3 feet from floor)	~ 65% (up to 4 feet from floor)
• Secondary	~ 30% (below spray zone)	~ 35% above spray zone
• Tertiary	~ 20% (above spray zone)	(none)

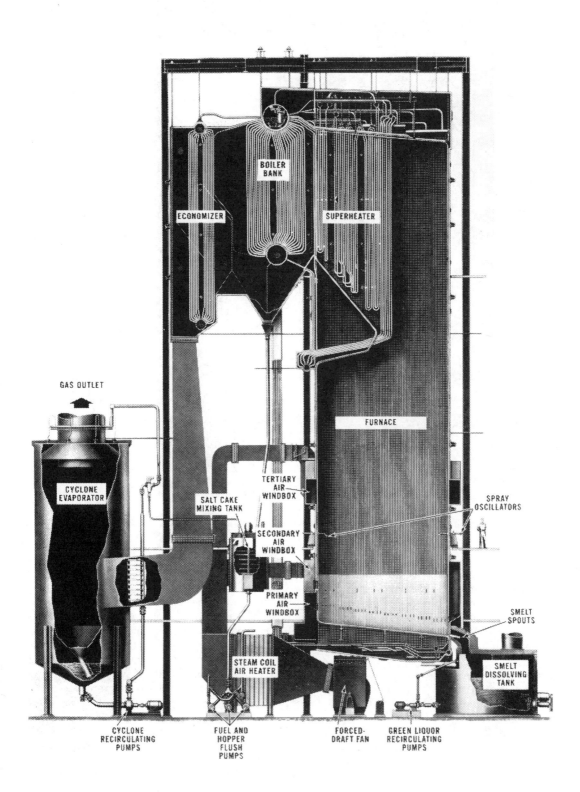

FIGURE 10-15. Babcock and Wilcox recovery furnace (circa 1968).

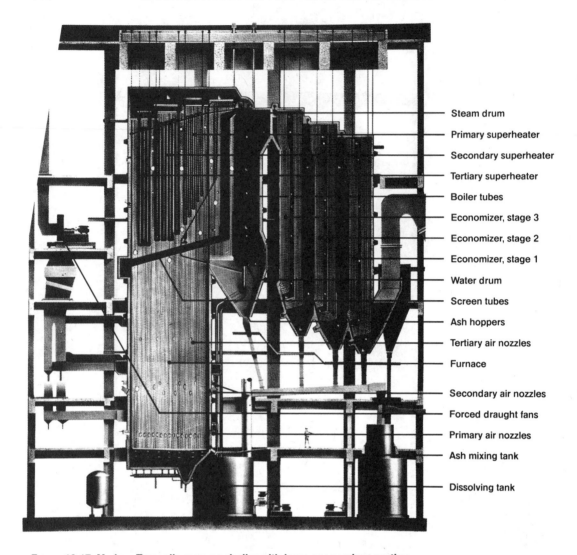

Steam drum
Primary superheater
Secondary superheater
Tertiary superheater
Boiler tubes
Economizer, stage 3
Economizer, stage 2
Economizer, stage 1
Water drum
Screen tubes
Ash hoppers
Tertiary air nozzles
Furnace
Secondary air nozzles
Forced draught fans
Primary air nozzles
Ash mixing tank
Dissolving tank

FIGURE 10-17. Modern Tampella recovery boiler with large economizer section.

evolve over the years. For example, the latest modifications offered by C-E include such features as stationary firing, three-level air system, single-drum design, upgraded furnace framing, improved protection against furnace corrosion, and higher liquor solids firing capability.

Combustion Air Flow

Air is introduced into the furnace through two or three sets of ports designated from the bottom upward as primary, secondary and tertiary air (Figures 10-18 and 10-19). The primary air ports are located a few feet above the hearth and extend around the four walls of the furnace to provide as low a velocity as practical, while supplying 50 - 65% of the air requirement. Secondary and tertiary air is usually introduced at higher velocity to ensure

complete mixing and combustion of the unburned gases.

In recovery boilers installed up to 1984, secondary and tertiary air is commonly introduced through jets located so that their center lines are tangential to an imaginary circle (Figure 10-20). While appearing to promote turbulent mixing with the unburned gases, recent measurements have shown that this type of entry actually promotes the formation of a cone-shaped body of combustibles at the core of the furnace and enhances entrainment of liquor particles. A recent paper (3) reviewed the problems and concluded that most of today's air port configurations require redesign.

The furnace ahead of the main heat-absorbing sections can be considered as consisting of three distinct zones: a drying zone where the liquor is

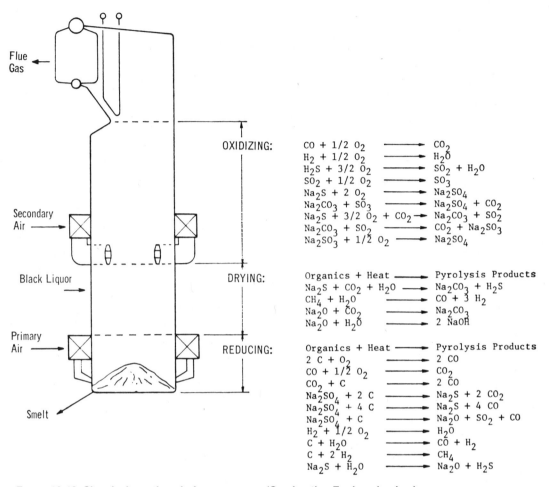

OXIDIZING:

$$CO + 1/2\ O_2 \longrightarrow CO_2$$
$$H_2 + 1/2\ O_2 \longrightarrow H_2O$$
$$H_2S + 3/2\ O_2 \longrightarrow SO_2 + H_2O$$
$$SO_2 + 1/2\ O_2 \longrightarrow SO_3$$
$$Na_2S + 2\ O_2 \longrightarrow Na_2SO_4$$
$$Na_2CO_3 + SO_3 \longrightarrow Na_2SO_4 + CO_2$$
$$Na_2S + 3/2\ O_2 + CO_2 \longrightarrow Na_2CO_3 + SO_2$$
$$Na_2CO_3 + SO_2 \longrightarrow CO_2 + Na_2SO_3$$
$$Na_2SO_3 + 1/2\ O_2 \longrightarrow Na_2SO_4$$

DRYING:

$$Organics + Heat \longrightarrow Pyrolysis\ Products$$
$$Na_2S + CO_2 + H_2O \longrightarrow Na_2CO_3 + H_2S$$
$$CH_4 + H_2O \longrightarrow CO + 3\ H_2$$
$$Na_2O + CO_2 \longrightarrow Na_2CO_3$$
$$Na_2O + H_2O \longrightarrow 2\ NaOH$$

REDUCING:

$$Organics + Heat \longrightarrow Pyrolysis\ Products$$
$$2\ C + O_2 \longrightarrow 2\ CO$$
$$CO + 1/2\ O_2 \longrightarrow CO_2$$
$$CO_2 + C \longrightarrow 2\ CO$$
$$Na_2SO_4 + 2\ C \longrightarrow Na_2S + 2\ CO_2$$
$$Na_2SO_4 + 4\ C \longrightarrow Na_2S + 4\ CO$$
$$Na_2SO_4 + C \longrightarrow Na_2O + SO_2 + CO$$
$$H_2 + 1/2\ O_2 \longrightarrow H_2O$$
$$C + H_2O \longrightarrow CO + H_2$$
$$C + 2\ H_2 \longrightarrow CH_4$$
$$Na_2S + H_2O \longrightarrow Na_2O + H_2S$$

FIGURE 10-18. Chemical reactions in furnace zones (Combustion Engineering Inc.).

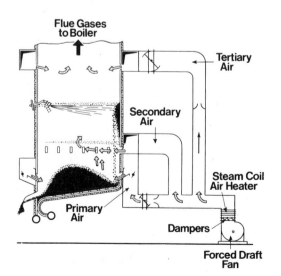

FIGURE 10-19. Char bed and air distribution for Babcock and Wilcox recovery furnace.

fired, a reduction zone at the bottom, and the oxidation zone in the turbulent upper section. The dominant chemical reactions in each zone are given in Figure 10-18. The appearance of the char bed is illustrated in both Figures 10-18 and 10-19.

Air is supplied to the furnace by one or two forced-draft fans. The flue gases are drawn through the unit by one or two large induced-draft fans at the exhaust of the precipitator. In order to maintain a safe environment for workers in the vicinity of the liquor nozzles and smelt spouts, it is necessary that the furnace operate under a draft (i.e., negative pressure). Consequently, a small amount of infiltration air enters through these openings in the tubewalls.

Water and Steam Flow

The floor and walls of a modern furnace consist of adjoining water tubes. C-E uses tangent wall construction where 3-inch tubes are seal-welded together along the line of contact; B&W employs membrane wall construction (Figure 10-21) also using 3-inch tubes, but with 1-inch filler bars between the tubes. The boiler feedwater initially enters the economizer tubes to pick up low-level heat, then flows to the steam drum (or upper drum) of the generating unit, and down one bank of closely-spaced vertical tubes to the water drum (or lower drum). From the water drum, the flow is through large-bore downcomers to a system of distribution headers below the furnace, then through supply tubes into the boiler wall headers. The water flows up the walls of the boiler and back into the steam drum. If so-called "screen tubes" are used as the first heat absorbing surface above the furnace, they utilize a separate water circuit (or sometimes a steam circuit). Steam is separated from water by the internals of the steam drum and travels through the superheater tubes, where the final steam temperature may reach 480°C. (Refer to Section 25.3 for general information on boilers.)

Gas Flow and Heat Transfer

The boiler unit can be considered as a gas and solids cooler. It is important that the solids entrained with the combustion gases are cooled below their fusion temperature prior to contact with the superheater tubes, so that ash and fume will not adhere strongly to the tubes and form an insulating layer. The dust deposits are periodically removed from the tubes by traversing high-pressure soot blowers. The critical temperature for sodium salts is usually about 800°C, but is lower in areas where NaCl gets into the chemical cycle. The typical recovery furnace is relatively tall to allow for sufficient cooling of gas and entrained solids by the water wall tubes.

After the combustion gases have travelled to the top of the furnace, they pass around a "nose baffle" and through the widely-spaced tube "screen". The flue gas then passes through the superheater and boiler generating sections. Finally, the gas is cooled in the economizer section. After the economizer, the remaining sensible heat may be used to evaporate liquor (in boilers with direct-contact evaporators) or, less commonly, to heat the incoming air. Otherwise, the gas goes directly to the precipitator, where the dust particles are ionized, collected on electrodes, and discharged either into a pool of strong black liquor (in a wet-bottom unit) or via a conveyor into a mix tank (in a dry-bottom precipitator). The flow sequences for feedwater/steam and combustion gases are illustrated in Figures 10-22 and 10-23.

Rated Capacity of a Recovery Unit

As the liquor solids loading to the furnace increases, so does the temperature of the gas and entrained solids entering the superheater section. The optimum capacity of the furnace has been exceeded when the gas temperature in the superheater is sufficiently high that the ash particles in suspension are sticky and tacky, and fouling of screen and superheater cannot be controlled with mechanical soot blowers.

Capacity ratings are usually in terms of weight of solids burned in a 24-hour period. Newer furnaces

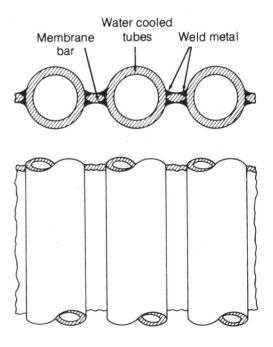

FIGURE 10-21. Tube wall construction (Babcock and Wilcox).

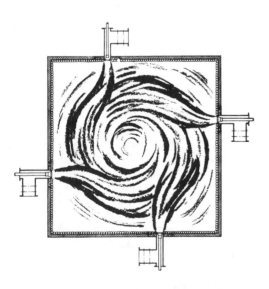

FIGURE 10-20. Plan view of older-generation Combustion Engineering furnace showing tangential entry and mixing patterns for secondary air.

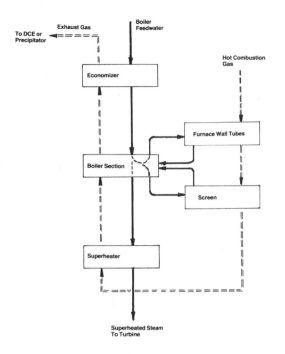

FIGURE 10-22. Simplified water/steam and combustion gas flow sequence for recovery boiler.

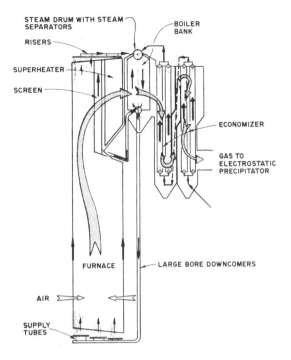

FIGURE 10-23. Gas flow and circulation schematic (courtesy Sandwell and Co. Ltd.).

may be rated in terms of the thermal loading, i.e., the heat released in the furnace. Generally, the older units were designed conservatively; and it is not uncommon to find some of these operating at 140% of rated capacity. The newer recovery boilers usually are provided with less of a capacity "cushion".

Variables of Operation

Aside from furnace loading (which may, in fact, be outside the control of the operator except to avoid overloading), the main operating variable is the amount and distribution of combustion air. A certain level of excess oxygen (usually 1 to 2%, equivalent to 5 - 10% excess air) should be carried to ensure complete combustion and minimize odor emission. The reduction efficiency is usually controlled by the relative amount of primary air; care must be exercised to avoid "starving" the char bed to the extent that large quantities of elemental carbon are contained in the smelt leaving the furnace.

Other variables that can affect the operation are the composition, solids level and temperature of the liquor, and nozzle pressure.

Thermal Efficiency

Thermal efficiency for a recovery boiler is usually defined as the percentage of black liquor fuel value which is actually realized in the generated steam. A heat balance for an older-generation boiler without black liquor oxidation (as illustrated with a Sankey diagram in Figure 10-24) shows a thermal efficiency of 61%. A modern low-odor design may achieve a thermal efficiency as high as 68%. Among the major factors that affect thermal efficiency are exit gas temperature, liquor solids concentration, and saltcake addition.

Generally, a minimum temperature of about 130°C is required for the boiler outlet gases to prevent the formation of condensation products (e.g., sulfuric acid) which are very corrosive. But, a temperature much above this level represents a loss of potentially useful heat. As a rule of thumb, a 15°C increase in gas temperature is equivalent to a 1% drop in thermal efficiency.

A lower solids concentration means that more heat is expended in evaporating water. A decrease of 2% in liquor concentration (within the normal range) is roughly equivalent to a 1% drop in thermal efficiency.

The reduction of oxidized sulfur compounds at the furnace hearth is an endothermic reaction that removes useful thermal energy from the system, and thus has a negative impact on thermal efficiency. Saltcake addition to the heavy black liquor has the same effect, causing a 1% drop in thermal efficiency for an addition rate of 75 lb per ton of pulp.

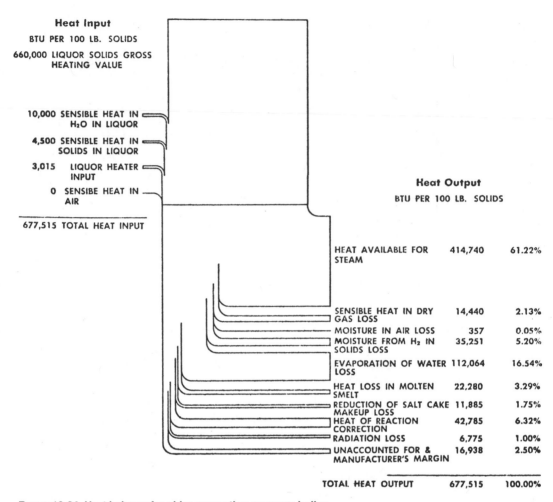

Heat Input

BTU PER 100 LB. SOLIDS

660,000 LIQUOR SOLIDS GROSS HEATING VALUE

10,000 SENSIBLE HEAT IN H_2O IN LIQUOR

4,500 SENSIBLE HEAT IN SOLIDS IN LIQUOR

3,015 LIQUOR HEATER INPUT

0 SENSIBE HEAT IN AIR

677,515 TOTAL HEAT INPUT

Heat Output

BTU PER 100 LB. SOLIDS

HEAT AVAILABLE FOR STEAM	414,740	61.22%
SENSIBLE HEAT IN DRY GAS LOSS	14,440	2.13%
MOISTURE IN AIR LOSS	357	0.05%
MOISTURE FROM H_2 IN SOLIDS LOSS	35,251	5.20%
EVAPORATION OF WATER LOSS	112,064	16.54%
HEAT LOSS IN MOLTEN SMELT	22,280	3.29%
REDUCTION OF SALT CAKE MAKEUP LOSS	11,885	1.75%
HEAT OF REACTION CORRECTION	42,785	6.32%
RADIATION LOSS	6,775	1.00%
UNACCOUNTED FOR & MANUFACTURER'S MARGIN	16,938	2.50%
TOTAL HEAT OUTPUT	677,515	100.00%

FIGURE 10-24. Heat balance for older-generation recovery boiler.

Dust Reclaim

A simplified sketch of a typical black liquor flow sequence on a recovery unit with direct-contact evaporator and wet-bottom precipitator is shown in Figure 10-25. The strong black liquor from storage first goes to the bottom of the precipitator where it "picks up" the chemical particulates (mainly $Na_2CO_3 + Na_2SO_4$) dropped from the electrodes. (If the precipitator is a "dry bottom" type, a series of screw conveyors or scraper-chain conveyors moves the material into the chemical ash tank.) Next, the liquor is pumped to the direct-contact evaporator where the solids concentration increases from both evaporation and further pickup of dust from the contacted gases. Saltcake makeup may be added to the liquor at the mix tank, along with chemical dust reclaimed from the boiler hoppers (from settling and soot blowing).

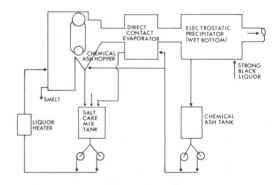

FIGURE 10-25. Black liquor flow sequence in recovery furnace system.

Chemical Recovery Efficiency

Most established kraft mills require soda makeup rates equivalent to 20 – 30 lb Na_2O per oven-dry ton of pulp. These rates are indicative of soda recovery cycle efficiencies of 97 – 98%. Recovery efficiency may be considered as the single-pass retention of soda through the pulping and recovery cycle.

Saltcake (Na_2SO_4) was the traditional chemical used in the kraft mill to make up for losses of both soda and sulfur, but it has now been largely replaced by caustic soda or soda ash (Na_2CO_3) in order to reduce the sulfur input to the liquor cycle. In a typical 1960's kraft mill, the rate of sulfur loss was far higher than for soda, so excessive sulfur buildup was not a problem. In a modern mill, the recovery efficiency for sulfur is almost equal to that of soda, so a chemical supplying equal equivalents of soda and sulfur cannot be used. In mills where waste sulfuric acid from the chlorine dioxide plant is added to the liquor system, saltcake is usually not needed at all.

Smelt-Water Explosions

Following the wide-spread acceptance of the Tomlinson furnace in the 1930's, the kraft pulp industry has been plagued by occasional severe recovery boiler explosions. These incidents have been thoroughly investigated, and most were determined to be caused by explosive mixtures of water and smelt. Today, based on extensive consultations between equipment manufacturers and users, all recovery boilers are subject to prescribed emergency shutdown procedures in case of any water tube leaks.

As well, the general operating directives for recovery boilers have always underlined the safety aspects. Particular attention has been focused on the procedures for discharging smelt into the dissolving

tank. A representative system is illustrated in Figure 10-26. As the smelt pours over the water-cooled spout, the flow is first disrupted by a steam jet, and further shattered by a high-volume recirculating spray, with the objective of eliminating high concentrations of smelt that might lead to an explosion. Even so, the dissolving tank is fitted with an over-size vent to quickly relieve pressure in the event that an explosion occurs.

The Black Liquor Recovery Boiler Advisory Committee (BLRBAC), an international group representing researchers, manufacturers, operators and insurers, has developed a set of safety standards and requirements for recovery boilers. BLRBAC subcommittees routinely examine all reports of operating problems and mishaps, so that standards can be updated on a continuing basis (4).

10.4 RECAUSTICIZING

The function of the recausticizing plant is to convert sodium carbonate (Na_2CO_3) into active sodium hydroxide (NaOH) and remove various impurities introduced from the furnace and lime kiln. The operation starts with the dissolving of furnace smelt in weak liquor ("weak wash") to form green liquor (so-called because of its green color). The green liquor is then clarified to remove "dregs", and subsequently reacted with lime (CaO) to form white liquor. The white liquor is clarified to remove precipitated "lime mud" ($CaCO_3$), and is then ready to be used for cooking. Auxiliary operations include washing of both the dregs and lime mud for soda recovery, and the calcining ("reburning") of lime

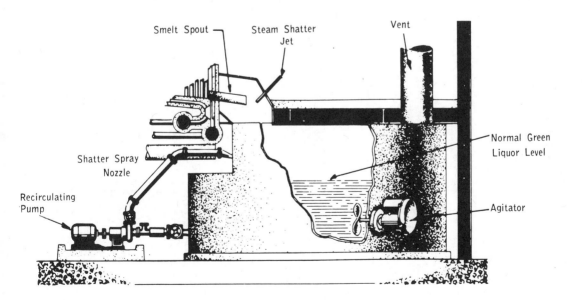

FIGURE 10-26. Smelt dissolving system.

mud to regenerate lime. A typical operating sequence is illustrated in Figure 10-27.

The causticizing reaction occurs in two steps. The lime first reacts with water ("slakes") to form calcium hydroxide $Ca(OH)_2$, which in turn reacts with sodium carbonate to form sodium hydroxide:

1) $CaO + H_2O \longrightarrow Ca(OH)_2 + Heat$

2) $Ca(OH)_2 + Na_2CO_3 \longrightarrow 2NaOH + CaCO_3$

The extent to which the second reaction is carried to completion is known as causticizing efficiency. A high percentage conversion is desirable to reduce the loading of inert sodium carbonate in the recovery cycle. The equilibrium causticizing efficiency is a function of liquor concentration and sulfidity as illustrated in Figure 10-28. Typical mill values are 3 to 6% below the corresponding equilibrium values.

Green Liquor Clarification and Dregs Washing

The strength of the green liquor is controlled by the amount of dilution added to the dissolving tank. Weak wash (from lime mud and dregs washing) is used instead of water to minimize dilution of the liquor system. The green liquor is then clarified to remove insoluble materials (dregs) consisting of unburned carbon and inorganic impurities (mostly calcium and iron compounds). The clarifier is usually of the type shown in Figure 10-29; the operation is basically of settling and decantation. The clear liquor may be pumped to a separate storage tank; but modern clarifiers usually have built-in storage capacity above the clarified zone.

The dregs are pumped out of the clarifier as a concentrated slurry, mixed with wash water, and are usually settled again in another sedimentation thickener known as the dregs washer (Figure 10-30). Often the dregs are further washed and thickened on

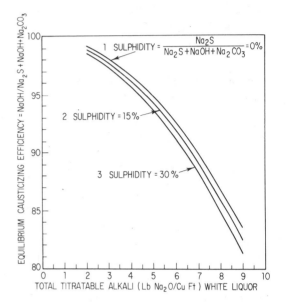

FIGURE 10-28. Equilibrium causticizing efficiency as a function of liquor alkali concentration and sulfidity.

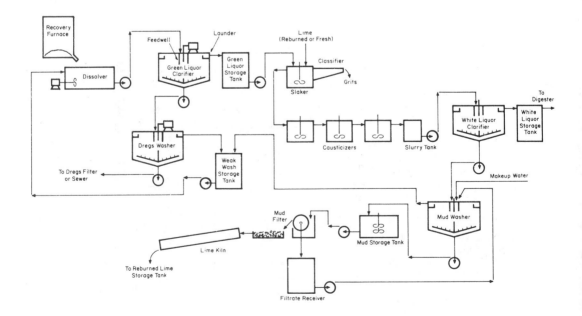

FIGURE 10-27. Typical causticizing system (Betz Laboratories Inc.).

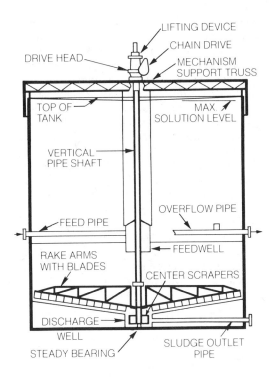

FIGURE 10-29. Typical liquor clarifier (Dorr-Oliver).

a vacuum filter (which has been precoated with lime mud) prior to disposal as landfill. The clear liquor containing the recovered soda chemicals is known as weak wash or weak liquor.

Slaking and Causticizing

Green liquor and reburned lime (from a surge bin) are fed continuously at a controlled rate to the slaker, where high temperature and violent agitation promote rapid conversion of quick lime (CaO) into milk-of-lime (slaked lime). Proper slaking is important to the subsequent operations of causticizing and lime mud settling. The slurry is channelled from the mixing section where unreacted particles (grit) settle to the bottom and are raked out for disposal as landfill (refer to Figure 10-31). A significant portion of the causticizing reaction also takes place in the slaker.

If both lime and green liquor are fed at relatively high temperature to the slaker, considerable steam generation can be expected due to the exothermic reactions. A high temperature is helpful in accelerating the reaction rate and ensuring a good causticizing efficiency. All slaking and causticizing is presently carried out in vented, atmospheric tanks. However, experimental data indicate that significant gains in process and thermal efficiency could be realized by utilizing a pressurized reaction vessel in series with pressurized filtration and storage of the white liquor.

Liquor continuously overflows from the slaker into a series of agitated tanks with about 2 hours of total retention, sufficient time for the causticizing reaction to be carried to completion. Two or more tanks in series (Figure 10-32) are used to minimize short-circuiting of the reaction mixture and ensure maximum conversion.

White Liquor Clarification and Lime Mud Washing

It is important to obtain a clear white liquor for cooking to avoid a coloration problem with bleached pulp and to eliminate inerts from the liquor cycle. The

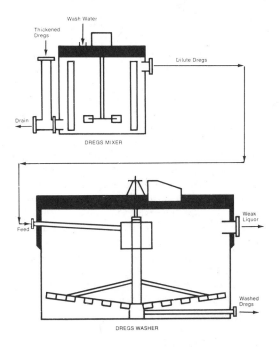

FIGURE 10-30. Dregs mixer and washer system (Dorr-Oliver).

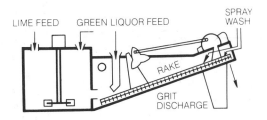

FIGURE 10-31. Lime slaker-classifier (Dorr-Oliver).

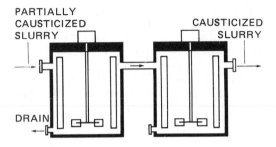

FIGURE 10-32. Typical series arrangement of causticizers (Dorr-Oliver).

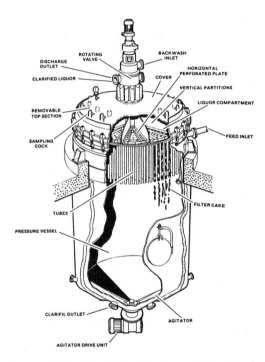

FIGURE 10-33. "Clarifil" white liquor filter (Dorr-Oliver).

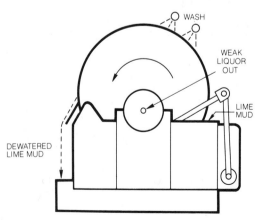

FIGURE 10-34. Lime mud precoat filter (Dorr-Oliver).

major part of the clarification operation is typically carried out in the same type of equipment as illustrated for green liquor (refer back to Figure 10-30). Some mills producing high-quality bleached pulps may utilize a "polishing filter" following sedimentation clarification to ensure maximum clarity of the cooking liquor. The current trend in the industry is to eliminate the sedimentation clarifier and carry out the entire clarification process by filtration utilizing equipment such as that shown in Figure 10-33.

The lime mud underflow from the thickener section of the clarifier (or from the filter) is pumped at high solids content (usually 35 - 40% solids) and is usually dilution-washed in sedimentation equipment. Sometimes, two stages of dilution washing and thickening are used. The lime mud system also includes a storage tank to provide for a uniform, uninterrupted flow to the lime mud filter (Figure 10-34) which dewaters the lime mud for calciner feed.

Variables of Operation and Operating Problems

Common problems that plague the causticizing operation are:
- poor or erratic causticizing efficiency,
- insufficient removal of dregs,
- poor settling or filtering characteristics of lime mud.

These problems are inter-related. Good control of the causticizing reaction requires a uniform flow and quality of green liquor to the slaker, along with careful metering of consistent-quality reburned lime. If the density or activity of the green liquor is varying, or if lime availability or reactivity is fluctuating, the causticizing efficiency will also vary. A slight excess (1 or 2%) of hydrated lime is necessary for a high causticizing efficiency, but a large excess will adversely affect the settling characteristics of the mud.

Recovery furnace operation can impact heavily on the causticizing plant. Variation in reduction efficiency will affect the activity of the green liquor and/or the amount of dregs produced. If dregs are carried into the causticizing reaction, they largely remain with the lime mud and can adversely affect the settling characteristics of the mud and the reactivity of the reburned lime.

The most important variables affecting causticizing plant operation are summarized in Table 10-3. For further information on the operation and control of the causticizing process, references 5, 6 and 7 are recommended.

10.5 CALCINING

The causticizing operation consumes quick lime (CaO) and produces lime mud ($CaCO_3$) as a by-product. The purpose of calcining (or lime "reburning") is to convert the $CaCO_3$ back into CaO for reuse in the causticizing process. The traditional lime kiln remains by far the most popular method of calcination, and this section will concentrate on lime kiln operation. However, a few mills are using fluidized-bed systems, and a flash calcining system is being promoted as another alternative.

Three phases are involved in a typical calcination operation:

1) drying the lime mud
2) raising the temperature of the lime mud to the level (about 800°C) required for the calcination reaction
3) maintaining a high temperature for sufficient time to complete the endothermic reaction.

The desired reaction is the chemical breakdown of calcium carbonate into quick lime and carbon dioxide:

$$CaCO_3 + Heat \longrightarrow CaO + CO_2$$

A well-controlled calcination will yield a product which is 90 - 94% CaO and reacts rapidly with the green liquor. Excessive temperature along with chemical impurities can promote the formation of non-reactive glass-like particles that are likely to be discharged as "grits" from the slaker.

The heat requirement for a modern calcining operation is detailed in Table 10-4 (8). Many existing plants require much higher inputs of energy because of design or operating limitations. These limitations are manifested by higher exit gas temperature, higher lime product temperature, and greater radiation losses.

Lime Kiln

A typical lime kiln is illustrated in Figure 10-35. Wet lime mud is fed into the high end of the kiln and the solid phase moves countercurrent to the flow of hot gases as the kiln rotates. The transfer of heat into the mud at the "cold end" is optimized by providing extended surface area, usually by means of steel chains

attached to the kiln shell and hanging in the hot gases (Figures 10-36 and 10-37). In the hotter zones of the kiln, the metal shell is lined with refractory brick.

As its temperature is raised, the lime mud material becomes plasticized and forms into pellets, aided by the rolling and lifting action of the kiln. Normally, the size of the aggregates ranges up to about 3 cm in diameter. Occasionally, the pellets keep on growing to form large "balls", or adhere to the brick to form "rings" (Figure 10-38). The soda content of the lime mud has a significant effect on its aggregating properties during the reburning operation, and is typically controlled to less than 1% Na_2O.

The "hot end" of the kiln is typically maintained at 1150 – 1250°C by firing oil or gas. Without reclaiming heat from the kiln product, the reburned lime would be discharged at a temperature of about 950°C. Most modern kilns are equipped with integral tube coolers to recover the major portion of this heat in direct contact with part of the entering air (Figure 10-39). These coolers are attached to the discharge end of the kiln in such a way that the calcined lime falls into one of the coolers; it then reverses direction and flows uphill (countercurrent to the air) to the opposite end of the cooler where it is discharged at a temperature of about 350°C. Air is supplied to the kiln by a forced draft fan, but the major work is done by the induced draft fan that pulls the combustion gases through the kiln.

The gases leaving the kiln are laden with lime mud dust and must be cleaned up before discharge. In the majority of cases, the dust is removed in a suitably-designed scrubber, most commonly a venturi-type (illustrated in Chapter 27). More recently, electrostatic precipitators have become the

TABLE 10-3. Variables affecting causticizing plant operation.

Green Liquor	• dregs carryover
	• strength (density) uniformity
	• activity and sulfidity
	• temperature- flow rate
Lime	• availability (%CaO)
	• reactivity ("causticizing power")
	• addition rate
	• method of calcining
Operating Factors	• method of agitation
	• retention time for causticizing reaction

TABLE 10-4. Heat requirement for lime calcination.

Conditions:	• 35% water in lime mud feed
	• 350°C product temperature
	• 90% CaO product
	• 10% excess air
	• 230°C exit gas temperature

Energy Requirement	1000's BTU/ton of product
Heat in exit gases	
• sensible/latent heat of water	2,410
• combustion gases	1,240
• CO_2 gas from breakdown of $CaCO_3$	140
Heat of dissociation	2,500
Radiation loss	1,000
Heat in lime product	300
Total	7,600

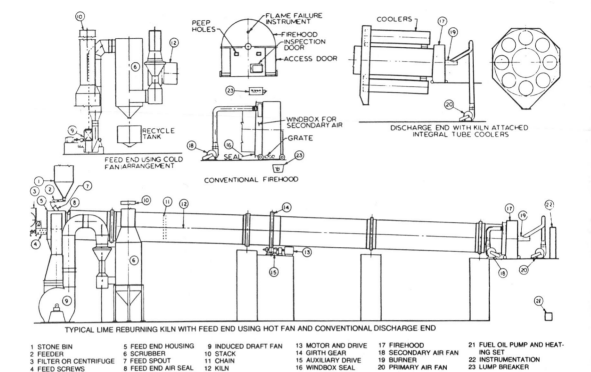

TYPICAL LIME REBURNING KILN WITH FEED END USING HOT FAN AND CONVENTIONAL DISCHARGE END

1 STONE BIN	5 FEED END HOUSING	9 INDUCED DRAFT FAN
2 FEEDER	6 SCRUBBER	10 STACK
3 FILTER OR CENTRIFUGE	7 FEED SPOUT	11 CHAIN
4 FEED SCREWS	8 FEED END AIR SEAL	12 KILN
13 MOTOR AND DRIVE	17 FIREHOOD	21 FUEL OIL PUMP AND HEATING SET
14 GIRTH GEAR	18 SECONDARY AIR FAN	22 INSTRUMENTATION
15 AUXILIARY DRIVE	19 BURNER	23 LUMP BREAKER
16 WINDBOX SEAL	20 PRIMARY AIR FAN	

FIGURE 10-35. Arrangement of kiln equipment (Fuller Co.).

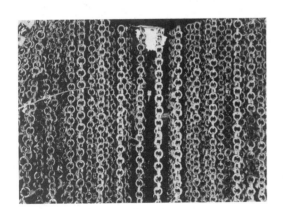

FIGURE 10-36. Curtain chain system in a lime kiln (Fuller Co.).

FIGURE 10-37. Garland chain system in a lime kiln (Fuller Co.).

equipment of choice for this service, especially for the largest-size kilns. The feasibility of using fabric filters for efficient dust removal has also been demonstrated, but this method has not caught on with the industry.

Generally, to achieve the best time/temperature balance and obtain the optimum thermal efficiency, a large kiln with a length-to-diameter ratio of over 30 is required. The retention time is controlled by the slope and rotational speed of the kiln; slope is usually designed between 1/4 and 1/2 inch per foot of length; speed varies from 0.5 to 1.5 rpm. Retention time should fall into the range of 2 to 3 hours. A good review of the factors affecting productivity and fuel economy, and steps for upgrading are given in references 8 and 9.

FIGURE 10-38. These large ring deposits were dislodged from the kiln by shotgunning (Weyerhaeuser Co.).

Many older kilns still in operation are handicapped by having a low length-to-diameter ratio. Often, these older units are severely overloaded as well. Fortunately, it is now possible to, in effect, lengthen an existing kiln by moving the evaporation zone outside of the shell and using the entire length of the kiln for calcining. Both a flash dryer and cage mill are employed in a typical system where the kiln exit gases are used to dry the lime sludge before it enters the kiln (Figure 10-40).

Fluid Bed Calciner

A fluidized bed reburning system is illustrated in Figure 10-41. The exhaust gases from the fluid bed are used to dry the lime mud, which is then collected in two cyclones and discharged into a surge bin. The dry lime mud is blown into the calcination bed where the fuel is being burned, and is almost instantaneously transformed into lime pellets. The heavy pellets fall into the cooling compartment, while the lighter dust is recycled with the air for drying the mud.

The system is compact, requiring much less space than a kiln of comparable capacity, and is capable of producing a high-quality lime of uniform size. Other advantages claimed are good thermal efficiency and ease of startup. Although these systems have been available for over 20 years, there are fewer than 20 installations within the pulp and paper industry. The economics of the fluidized bed system may be relatively more favorable for smaller-size plants.

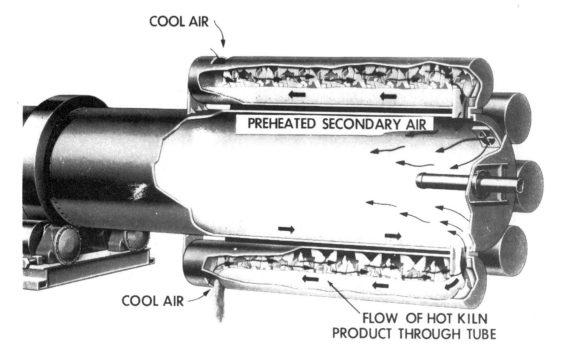

COOL AIR

PREHEATED SECONDARY AIR

COOL AIR

FLOW OF HOT KILN PRODUCT THROUGH TUBE

FIGURE 10-39. Integral tube coolers on lime kiln, used for recovering heat from the hot kiln product (Allis Chalmers).

Flash Calcining

Flash or gas suspension calcining is a recent development, and at least two manufacturers are promoting this technology as another alternative for

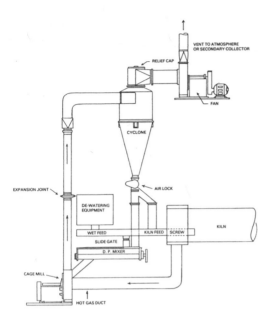

FIGURE 10-40. This flash drying system utilizes exit gases from the kiln to pre-dry the lime feed. (C-E Raymond).

pulp mill lime reburning. Fuller, for example, has over 25 flash calcining systems in operation on a variety of materials. The heart of the system is a swirl-type furnace in which the combustion of fuel is in intimate contact with the material being calcined. Fuller claims that the turbulent swirling mixture produces a uniform temperature profile, and that conventional flame temperatures are never encountered because the material absorbs the excess heat instantaneously.

10.6 BY-PRODUCT RECOVERY

Two by-products of alkaline pulping are economically important: turpentine and tall oil. Both are obtained in quantity from resinous woods such as the southern pines. When demand from chemical processors is high, these by-products make a significant contribution to the profitability of a kraft pulping operation.

Unfortunately, demand in recent years for both turpentine and tall oil has been extremely cyclical, and this volatility is reflected in the market prices for these commodities. The list prices have ranged from $0.60 to $2.00 per gallon for turpentine, and from $90 to $270 per ton for crude tall oil. During periods of slack demand it has often been more cost-effective for mills to use one or both of these by-products as fuel feedstocks rather than sell them to processors at depressed prices.

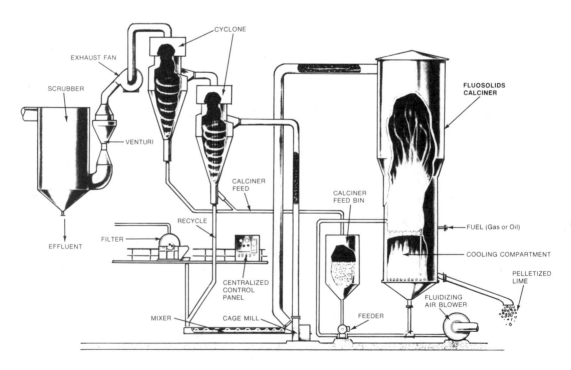

FIGURE 10-41. Fluidized bed system for lime mud reburning (Dorr-Oliver).

Turpentine

Turpentine is recovered primarily from digester relief vapors, as illustrated for a batch pulping operation in Figure 10-42. The vapors are conducted to a cyclone separator where liquor carryover is removed, then to a condenser where the steam and turpentine are condensed while giving up their heat to process water. The condensate drains to a decanter (Figure 10-43) where the turpentine and water separate into two liquid phases and overflow from top and bottom, respectively. The water is combined with other contaminated condensate streams for steam stripping, while the turpentine flows to a storage tank. During storage, additional water separates from the turpentine and is periodically removed.

Turpentine is a mixture of many cyclic chemical compounds, principally pinenes and turpenes. The raw turpentine is sold to chemical processors where it is fractionally distilled and converted into numerous products, including camphor, synthetic resins, solvents, flotation agents and insecticides.

Tall Oil

The resinous material in pines and other species is made up of fatty and resin acids, as well as sterols and related alcohols. During kraft cooking, the fatty and resin acids become saponified, i.e., are converted into sodium soaps, and are dissolved in the residual cooking liquor. (Generically, soaps are defined as the metal salts of complex organic acids.)

As black liquor is evaporated, tall oil soap solubility drops until it reaches a minimum at 25 to 30% solids. In practice, the black liquor is usually concentrated through 2 effects of evaporation and then pumped to a skimming tank. The skimming tank is suitably designed to minimize turbulence and provide sufficient retention time for good separation. Here, the soap (along with some "unsaponifiables") rises to the surface and is removed, while the skimmed liquor is passed on to the next evaporator effect. The recovered soap is allowed to settle in storage to remove additional entrained liquor.

For optimum soap separation, the use of injected

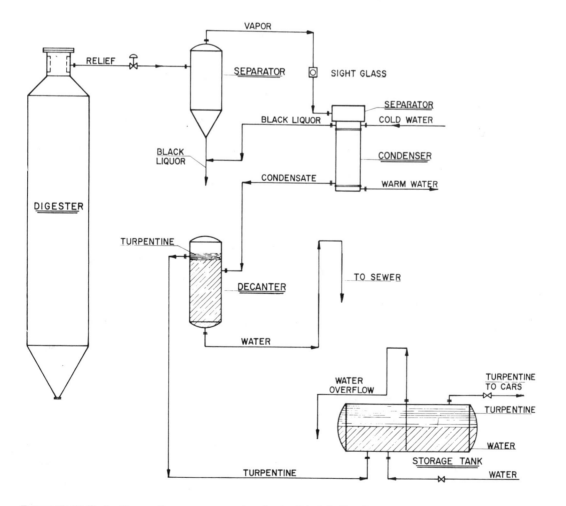

FIGURE 10-42. Typical turpentine recovery system for kraft batch digester.

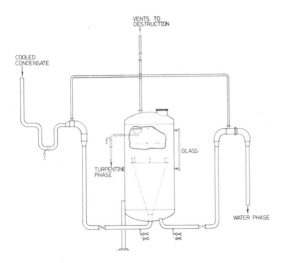

FIGURE 10-43. Turpentine decanting system (MoDo Chemetics).

air and chemical additives has proved effective (11). Air bubbles adhere to the soap particles and increase their buoyancy; adding just 0.5% to 1.0% air by volume to the inlet side of the skimmer pump can increase crude tall oil recovery by up to 10%. Additional recovery up to 13% is claimed using lignosulfonate additives.

Some mills sell the recovered soap to a chemical processor without further refinement. However, a large number of mills, especially where waste sulfuric acid is available from other operations (usually from the chlorine dioxide plant), carry out an acidulation process to convert the soap into raw tall oil and recover the soda. The acidulation reaction is the reverse of saponification; an excess of acidity causes hydrogen to displace the sodium on the organic molecules, and a portion of the sulfuric acid (H_2SO_4) is converted to salt cake (Na_2SO_4). The acidulation mixture is either allowed to separate in a settling tank or is processed through a continuous centrifugal separator. The acid phase (brine) containing sodium sulfate and residual lignin is added to the black liquor system. The crude tall oil is washed with hot water to remove free acid and is allowed to settle further in storage tanks prior to shipment. One processing system for tall oil is shown schematically in Figure 10-44.

The yield and chemical makeup of raw tall oil depends on the pulping raw material, i.e., the wood species, the relative maturity of the trees when harvested, and the method and duration of chip storage. Some representative data are given in Table 10-5. At the chemical processing plant, the raw tall oil is separated by vacuum distillation into resin oil

and a number of fatty acid fractions. These materials are used in the manufacture of a wide range of products, including soaps, lubricants, surface active agents and paper sizings.

10.7 RECOVERY OF SULFITE LIQUORS

Sulfite recovery technology has developed extensively since 1950. Depending on the process used, heat recovery, sulfur recovery, or recovery of both base chemical and sulfur may be possible.

For the combustion of calcium base liquors, only heat recovery is possible due to the formation of calcium sulfate ($CaSO_4$). With ammonium base liquors, heat recovery with and without sulfur recovery is practiced, but no base recovery is possible because the ammonia is converted into elemental nitrogen.

The chemical recovery of magnesium-based liquors is well established utilizing a relatively uncomplicated process. The liquor is concentrated and burned in a specially-designed recovery furnace. The process differs from kraft liquor burning in that no smelt is produced. Rather, the chemicals are swept through the boiler with the combustion gases in the form of magnesium oxide (MgO) ash and sulfur dioxide (SO_2) gas. About 80 - 90% of the ash is separated using a cyclonic collector; the remainder of the ash and the SO_2 are removed from the flue gas in a gas/liquid contactor. The scrubbing liquid used is magnesium hydroxide solution from the slaking of MgO, and the cooking liquor is thereby regenerated. A representative system is illustrated in Figure 10-45.

Soda Base Recovery

Several processes for chemical recovery from sodium-base sulfite waste liquors are based on the combustion of the concentrated liquor in a kraft-type recovery furnace. The resultant smelt is similar in composition to that produced by the combustion of kraft liquor. The significant difference is that the

TABLE 10-5. Typical tall oil yields and compositions from different areas of North America (Janes).

| | Yield lb/ton pulp | Percentage of Total | | |
		Resin Acids	Fatty Acids	Unsapon-ifiable
Southern States	90	46	40	14
Mid-Atlantic States	80	39	48	13
Canada	40	43	32	25
S. Western States	125	48	36	16
East of Cascades	50	30	41	29
West of Cascades	30	43	27	30

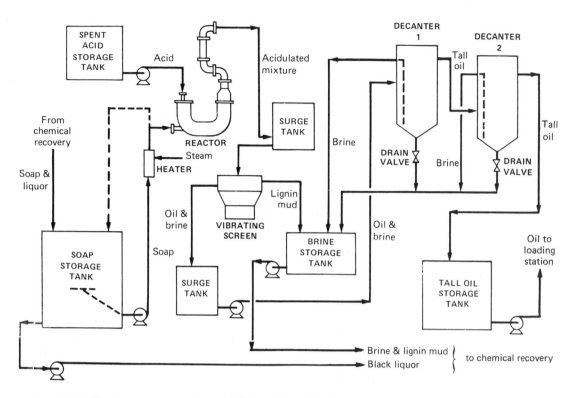

FIGURE 10-44. Continuous process for acidulation of raw tall oil.

sulfidity of the smelt and the SO_2 content of the flue gas are significantly higher for the sulfite liquor. The task remains to convert the smelt chemicals into a regenerated cooking liquor; and different approaches toward this objective have been taken commercially in the Rauma, Tampella and Stora processes (12).

The Rauma process for soda recovery is shown schematically in Figure 10-46. Green liquor containing Na_2S and Na_2CO_3 reacts with CO_2 in the carbonating tower, producing $NaHCO_3$ and liberating H_2S:

$$Na_2CO_3 + Na_2S + 3CO_2 + 3H_2O \longrightarrow 4NaHCO_3 + H_2S$$

The bicarbonate thus formed is then split thermally in a decomposer, producing Na_2CO_3 and CO_2 for the above carbonation. The major portion of the Na_2CO_3 reacts with $NaHSO_3$ in the reactor and produces Na_2SO_3 and CO_2. Portions of the Na_2SO_3 are then used for cooking liquor preparation and for "sulfitation". In the sulfitation step, Na_2SO_3 reacts with SO_2 to produce bisulfite ($NaHSO_3$). The SO_2 for sulfitation is obtained by burning the H_2S along with makeup sulfur.

The Tampella process for soda-base recovery is illustrated in Figure 10-47. Green liquor containing Na_2S and Na_2CO_3 is partially carbonated with flue gas to form $NaHS$ and $NaHSO_3$:

$$Na_2S + Na_2CO_3 + 2CO_2 + 2H_2O \longrightarrow NaHS + 3NaHCO_3$$

The precarbonated liquor then passes to a steam stripper where, at reduced pressure, sodium hydrosulfide ($NaHS$) reacts with sodium bicarbonate ($NaHCO_3$) to liberate H_2S gas:

$$NaHS + NaHCO_3 \longrightarrow H_2S + Na_2CO_3$$

The H_2S gas, along with makeup sulfur, is burned to provide sulfur dioxide (SO_2) which is then absorbed by the Na_2CO_3 to form bisulfite liquor. The Tampella process is flexible and can be modified to produce liquors for various cooking methods.

10.8 ALTERNATIVE KRAFT RECOVERY

Despite many attempts in recent decades to develop improved recovery technologies, there has been little change in the basic kraft recovery process over the last 50 years. While the existing process has many advantages (reliable proven technology, efficient chemical recovery, ability to meet emission

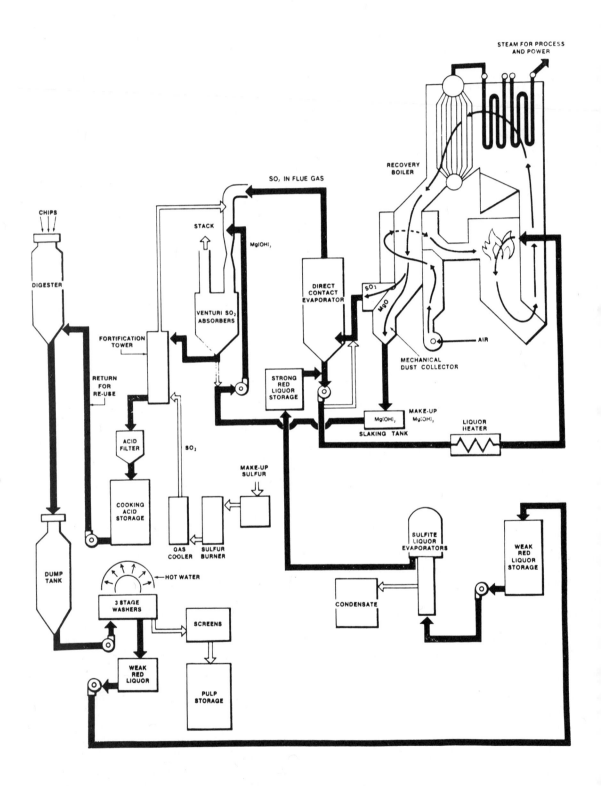

FIGURE 10-45. Magnesium bisulfite chemical recovery system.

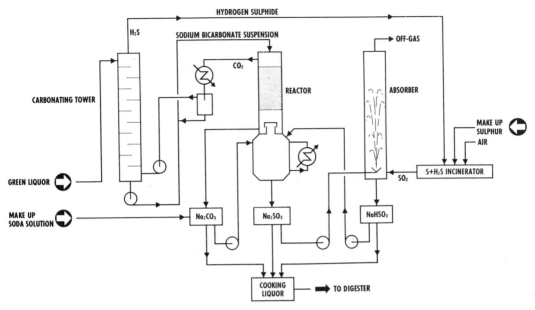

FIGURE 10-46. The Rauma process for soda base recovery.

requirements, good thermal efficiency), it also has a number of serious drawbacks (high capital cost, the need for large-scale equipment to keep unit costs down, the specific danger of smelt/water explosions).

One promising approach is via black liquor gasification. Presently, two processes (Chemrec and MTCI) are being brought to the pilot plant stage at pulp mill sites (13). A pressure gasification process is also under development in Finland.

Another approach is to use a process that generates caustic directly during the combustion step, and therefore, eliminates the need for subsequent causticizing and calcining operations. A number of agents are available which will react with sodium carbonate to drive off CO_2, so the resultant inorganic product self-causticizes when dissolved in water. One well-researched system is borate auto-causticizing which involves pulping with an alkaline borate solution, however, no commercial systems are in operation.

Thus far, none of the new technologies offers sufficient benefits to justify the risk of implementation on a mill scale. According to Grace (14), four incentives that could induce mills to try a new technology are:
1) Significantly lower capital cost
2) Elimination of causticizing/calcining
3) Improved energy values
4) Fully adaptable to changing pulping/bleaching technology

If all four advantages were present in a new process, the impetus for conversion would be overpowering. Perhaps this new process will be a combination of gasification and autocausticizing technologies.

REFERENCES

(1) MADDEN, D.J. **Advantages and Disadvantages of Different Types of Evaporators** *Pulp and Paper* (Buyer's Guide 1978)

(2) CHANDRA, S. **Blow Heat Recovery Reduces Liquor Evaporation Requirement 25% to 35%** *Pulp & Paper* (November 1987)

(3) MACCALLUM, C. and BLACKWELL, B.R. **Improved Air Systems for Modern Kraft Recovery Boilers** *P & P Canada* 88:10:T392-395 (October 1988)

(4) McGEE, T. **The Recovery Boiler and the Insurance Industry** *Tappi Journal* (October 1986)

(5) DORRIS, G.M. and ALLEN, L.H. **Operating Variables Affecting the Causticizing of Green Liquors with Reburned Lime** *Journal P & P Science* 13:3:J99-105 (May 1987)

(6) CAMPBELL, A.J. **The Effects of Lime Quality and Dosage on Causticizing and Lime Mud Settling Properties** *P & P Canada* 86:11:T320-324 (November 1985)

(7) LINDBERG, H. and ULMGREN, P. **The Chemistry of the Causticizing Reaction - Effects on the Operation of the Causticizing Department in a Kraft Mill** *Tappi Journal* (March 1986)

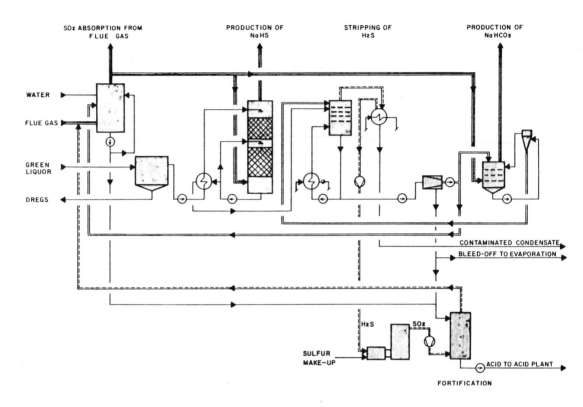

FIGURE 10-47. Tampella process for sodium base sulfite chemical recovery.

(8) KRAMM, D.J. **Update on Lime Kilns .... (3 Parts)** *Paper Trade Journal* (May 15, May 30, June 15, 1979)

(9) PUHR, F. **Pulp Mill Lime Kiln Improvement Projects Produce Rapid Payback** *Pulp & Paper* (October 1988)

(10) NORBOM, H.R. **Minor, Inexpensive Kiln Upgrades Yield Major Gains in Performance** *Pulp & Paper* (February 1985)

(11) GUPTA, J. **Increasing Crude Tall Oil Yield** *Tappi Journal 66*:10:41-43 (October 1983)

(12) PATRICK, K., et al **Sulphite Liquor Recovery for Canadian Newsprint Mills** *P & P Canada 84*:8:T195-201 (August 1983)

(13) EMPIE, H.J. **Alternative Kraft Recovery Processes** *Tappi Journal* (May 1991)

(14) GRACE, T.M. **New Recovery Technologies** *PIMA Magazine* (October 1991)

Chapter 11

Bleaching

The whiteness of pulp is measured by its ability to reflect monochromatic light in comparison to a known standard (usually magnesium oxide). The instrument most commonly used is the Zeiss Elrepho reflectance meter which provides a diffuse light source. Fully bleached sulfite pulps can test as high as 94, and unbleached kraft pulp as low as 15 Elrepho units.

Unbleached pulps exhibit a wide range of brightness values (Table 11-1). The sulfite process produces relatively bright chemical pulps, up to 65, whereas those produced by the kraft, soda and semichemical processes can be quite dark. Mechanical pulp brightness is mainly a function of the species and condition of the wood pulped. Generally, only those species providing brightness values over 55 are utilized for mechanical pulping.

TABLE 11-1. Approximate brightness ranges of unbleached pulps.

Brightness Range	Type of Pulp
15 – 30	kraft
40 – 50	NSSC, ammonium bisulfite
50 – 65	groundwood, bisulfite, sulfite

Cellulose and hemicellulose are inherently white and do not contribute to pulp color. It is generally agreed that "chromophoric groups" on the lignin are principally responsible for color; oxidative mechanisms are believed to convert part of the lignin's phenolic groups to quinone-like substances that are known to absorb light. Heavy metal ions (e.g., iron and copper) are also known to form colored complexes with the phenolic groups. Extractive materials can contribute to the color of mechanical pulps made from resinous woods.

Lignin Preserving or Lignin Removal?

Two approaches are used in the chemical bleaching of pulps. One approach is to utilize chemicals that selectively destroy some of the chromophoric groups, but do not materially attack lignin. The other approach is to almost totally remove residual lignin.

The selective approach (often referred to as "brightening" to distinguish it from true bleaching) is used for high-yield pulps with significant lignin content. This methodology is limited in many cases to achieving brightness values in the 70's; but under favorable circumstances, values above 80 can be obtained. Although the brightness gains can be substantial, no known method of selective brightening produces a permanent effect; exposure to light and atmospheric oxygen causes the lignin to rapidly discolor, as can be readily observed with old newspapers. Brightening techniques are discussed in Section 11.11.

To produce high-quality, stable paper pulps (as well as dissolving grades), bleaching methods that delignify the pulp must be used. The early stages of bleaching are usually considered as a continuation of the delignification process started in cooking. The later stages employ oxidizing agents to scavenge and destroy the residual color. The entire bleaching process must be carried out in such a way that strength characteristics and other papermaking properties are preserved.

11.1 BLEACHING SEQUENCES

Modern bleaching is achieved through a continuous sequence of process stages utilizing different chemicals and conditions in each stage, usually with washing between stages (e.g., Figure 11-1). The commonly applied (or proposed) chemical treatments and their shorthand designations are as follows:

Chlorination (C)
- Reaction with elemental chlorine in acidic medium

Alkaline Extraction (E)
- Dissolution of reaction products with NaOH

Chlorine Dioxide (D)
- Reaction with ClO_2 in acidic medium

Oxygen (O)
- Reaction with molecular oxygen at high pressure in alkaline medium

Hypochlorite (H)
- Reaction with hypochlorite in alkaline medium

Peroxide (P)
• Reaction with peroxide in alkaline medium
Ozone (Z)
• Reaction with ozone in acidic medium

The practice of designating bleaching stages and sequences using this symbolic shorthand has evolved informally over many years. However, the complexity of modern bleaching practices coupled with variable symbolism has caused misunderstandings regarding bleaching practices. Adherence to standardized guidelines are now necessary to facilitate clarity in technical communication. The following recommended conventions have been extracted from a comprehensive protocol submitted by the TAPPI Pulp Bleach Committee (1):

1. Using the above symbols, a conventional five-stage bleach sequence consisting of chlorination, alkaline extraction, chlorine dioxide, alkaline extraction, and chlorine dioxide would be designated as CEDED.

2. When two or more bleaching agents are added as a mixture or simultaneously, the symbol of the predominant chemical should be shown first, and the symbols depicting all of the added chemicals should be enclosed in parentheses and be separated by a plus sign. For example, when a lesser percentage of chlorine dioxide is added with chlorine in the chlorination stage, the sequence would be designated (C+D)EDED. If chlorine dioxide is the dominant chemical species in the chlorination stage, the designation should be (D+C)EDED.

3. When two or more chemicals are added sequentially with mixing in between points of addition, the symbols depicting the added chemicals should be shown in order of addition and should be shown in parentheses. For example, when chlorine dioxide is added before chlorine in the chlorination stage, the designation

should be (DC)EDED.

4. If the ratio of added chemicals is to be shown, the percentage number should immediately follow the symbol of the designated chemical and should be expressed in terms of oxidizing equivalence. For example, the term (D70C30) indicates sequential addition of 70% ClO_2 and 30% Cl_2, all expressed as active chlorine.

For softwood kraft pulps, a number of bleach sequences utilizing between four and six stages are commonly used to achieve "full-bleach" brightness levels of 89 - 91. Numerous CEHDED and CEDED full-bleach systems are in operation dating from the 1960's and 1970's; sequences more typical of modern mills are (D+C)(E+O)DED and O(D+C)(E+O)D. Lower brightness levels can be achieved with fewer stages. A level of 65 can be easily reached with a CEH or OH sequence. Intermediate levels of brightness can be achieved with CED, DED, OCED, CEHH, CEHD, or CEHP. In most cases, each bleaching stage is followed by a washing stage to remove reaction products; and subsequent bleach stages serve the same purpose, while creating additional reaction products. For this reason, the industry has often employed sequences with alternating acid/alkaline stages.

Sulfite pulps and hardwood kraft pulps are "easier bleaching" than softwood kraft pulps. Both have lower lignin content, and in the case of sulfite pulps the lignin residues are partially sulfonated and more readily solubilized. Consequently, a somewhat simpler process can be used to achieve a comparable brightness level.

The initial stages of a sequence are used primarily to delignify the pulp. This part of the bleaching sequence can be considered a continuation of the delignification process which starts with cooking. Typically, significant brightening occurs only after the removal of residual lignin.

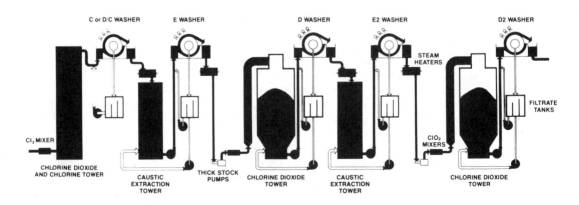

FIGURE 11-1. Flow diagram of CEDED bleach plant (Dorr-Oliver).

Short-Sequence Bleaching

Recent advancements in equipment design and process control, together with the use of oxygen in the alkaline extraction stage, have resulted in the lowering of chemical requirements in subsequent stages. This has led to the development of short-sequence bleaching, which produces 88 brightness softwood kraft pulp in only 3 or 4 stages. The most widely adopted short sequence is probably (D+C)(EO)D.

11.2 PREPARATION OF BLEACH CHEMICALS

All conventional bleaching chemicals can be produced from air, water and brine (NaCl solution) utilizing varying amounts of energy. Since the raw materials are cheap, it follows that manufacturing costs are mainly a function of how much energy is expended.

Oxygen

Two systems are utilized for oxygen generation: cryogenic air separation and adsorption-based separation. The former method is universally used by chemical processors who supply liquid oxygen in over-the-road tankers, and is sometimes employed for onsite generation in high-usage mills (i.e., over 80 ton/day), especially in situations where coproduct nitrogen can also be utilized. The adsorption-based oxygen plant yields a gas product and is suitable only for onsite generation.

Cryogenic plants compress air and remove water, CO_2 and other impurities in a pretreatment system. The dry air is then cooled and liquified at cryogenic temperatures using turbo expanders and/or refrigeration systems. Finally, the liquified air is distilled to separate it into oxygen, nitrogen and other coproducts.

By contrast, adsorption systems separate air at ambient temperatures using molecular sieve adsorbents. The adsorbent picks up nitrogen and other impurities at the system feed pressure, while the product oxygen passes straight through the bed. After the bed is saturated with nitrogen, it is regenerated by lowering the pressure to release the adsorbed components. Adsorption processes are cyclical, and multiple beds are used to maintain continuous oxygen production.

Chlorine and Caustic Production

Chlorine and sodium hydroxide are generally produced by electrolysis of brine according to the following net reactions:

$$2NaCl + \text{Electrical Energy} \longrightarrow 2Na + Cl_2$$

$$2Na + 2H_2O \longrightarrow 2NaOH + H_2$$

Three types of electrolytic cells are used commercially: diaphragm cells, mercury cells, and membrane cells. The operating principle of each is illustrated in Figure 11-2.

In the mercury cell process (Figure 11-3), metallic sodium forms an amalgam with mercury which is then removed from the cell before the sodium reacts with water. Therefore, the mercury cell system produces essentially pure NaOH. With the diaphragm cell, the caustic must be concentrated by evaporation to crystallize and separate NaCl which is present in the withdrawn cell liquor. The membrane cell is the most recent development and is now used in most new electrolysis plants because of greater operating flexibility and lower processing cost. It utilizes a cation exchange membrane that is impervious to chloride ion flow, and thus produces an essentially pure NaOH product.

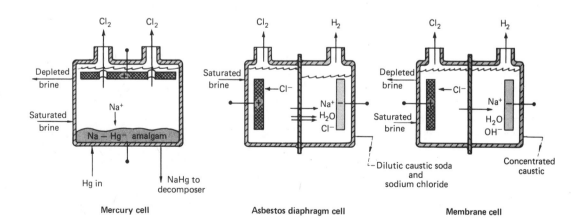

FIGURE 11-2. Basic electrolytic cells for producing chlorine and caustic (Olin Corp.).

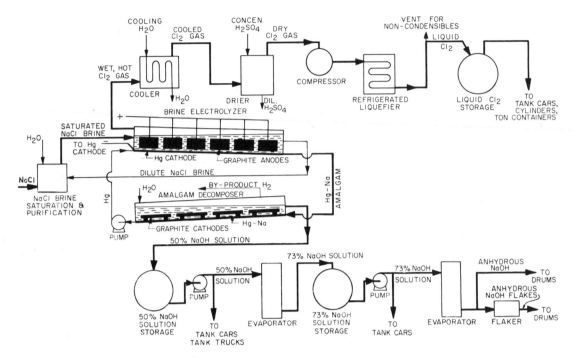

Figure 11-3. Mercury cell chlor-alkali process (CIL).

In all cases, chlorine gas forms at the anode and is vented from the cell. The gas is cooled, scrubbed with sulfuric acid to remove moisture, and usually liquified under compression and refrigeration.

Chlorine and caustic are generally supplied by the chemical industry. The chlorine is shipped as liquid, usually in tank cars. Caustic is most commonly shipped as 50% solution. Where justified by high consumption, some larger pulp mills have utilized on-site generation. Here, the chlorine can be handled as a gas and the caustic at lower concentration, thus eliminating some of the unit operations necessary for a market chemical plant.

There has been a dramatic shift away from chlorine bleaching in recent years due to environmental concerns. A severe chlor-alkali imbalance now exists in the industry. Alternative (i.e., non-electrolytic) methods for producing caustic are being developed, for example by causticizing sodium carbonate.

Handling of Chlorine and Caustic

A typical chlorine handling system is shown in Figure 11-4. Both liquid and gaseous chlorine are safely handled in carbon steel tankage and piping as long as the entire system is moisture-free. Tank cars are pressurized with dry air. Liquid chlorine flows under this pressure head to steam-jacketed evaporators for conversion to the gas phase, and then

the chlorine gas is either mixed with pulp stock or used for hypochlorite makeup. Barometric legs are commonly utilized before each point of use as a precaution against "suckbacks" of water into the dry chlorine system.

Caustic is usually stored as a 50% solution and diluted to 5% in a continuous dilution system. Measurement of either solution density or conductivity can be utilized as the basis for an automatic dilution system.

Preparation of Hypochlorite

Hypochlorite solution is prepared at the mill site by reacting chlorine with 5% caustic solution or milk-of-lime:

$$2NaOH + Cl_2 \longrightarrow NaOCl + NaCl + H_2O$$

$$2Ca(OH)_2 + 2Cl_2 \longrightarrow Ca(OCl)_2 + CaCl_2 + 2H_2O$$

These reactions are reversible. An excess of alkali is required to drive the reaction to the right and also prevent subsequent decomposition of the hypochlorite into chlorate. However, decomposition may occur at higher temperature even in the presence of alkali. Because the above reactions are exothermic, cooling may be required to maintain the solution below the critical level of 50°C (especially following lime slaking).

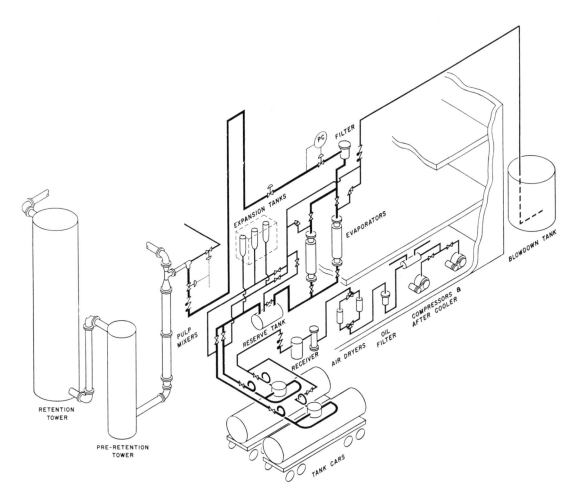

FIGURE 11-4. Typical chlorine unloading system.

Most hypochlorite makeup systems operate continuously and automatically utilizing an oxidation/reduction potential measurement, which depends on the relative amounts of hypochlorite and residual chlorine present.

Sodium Chlorate Production

Sodium chlorate ($NaClO_3$) is the principal feedstock for on-site generation of chlorine dioxide (ClO_2), and is therefore an important bleaching chemical. It is also produced by electrolysis of brine. In this process, the anodic and cathodic products are not separated (as is done for caustic and chlorine cells), but are allowed to react to form sodium hypochlorite ($NaOCl$). Excess alkali is not maintained to stabilize the hypochlorite (as is done for mill hypochlorite makeup), and formation of chlorate proceeds:

$$3NaOCl \longrightarrow NaClO_3 + 2NaCl$$

The solution is removed and evaporated; the NaCl crystallizes out to leave only $NaClO_3$ in solution. The solution is then further evaporated to crystallize out the $NaClO_3$. The material is shipped out in crystal form and is dissolved at the mill site for storage. The net reaction for the overall chlorate production process can be expressed:

$$NaCl + 3H_2O + \text{Electrical Energy} \longrightarrow NaClO_3 + 3H_2$$

Chlorine Dioxide Generation

Chlorine dioxide (ClO_2) is a gas at room temperature/pressure, and liquifies at 11°C. However, the chemical is unstable and potentially explosive in the pure form. In air mixtures, it is easily detonated by exposure to heat, light, mercury, and various organic substances. Rich mixtures can produce severe explosions; lower concentrations (below 20%) produce only mild detonations described as "puffs".

Chlorine dioxide is somewhat more soluble in water than chlorine (10 to 11 grams per liter at 4°C). Since it cannot be shipped either in pure form or as a concentrated solution, ClO_2 is always manufactured at the mill site. It is generated as a gas from chemical reduction of sodium chlorate in a highly acidic solution; the gas is then absorbed in cold water to produce ClO_2 solution at a concentration of about 7 g/liter.

Numerous methods of on-site ClO_2 generation have been developed. Originally, the methods differed primarily with respect to the reducing agent that was used to react with the chlorate. More recent developments have been more concerned with reducing the quantity of by-product chemicals.

Four seminal processes used in North America for generating chlorine dioxide are listed in Table 11-2. These processes are the bases for numerous

TABLE 11-2. Seminal processes for chlorine dioxide generation.

Process	Reducing Agent	Diluent for Generated Gas
Solvay	methanol	air
Mathieson	SO_2	air
R-2	NaCl	air
R-3/SVP	NaCl	water vapor

modifications and improvements that have been introduced over the years. In their original format, the Solvay and Mathieson systems (which date from the 1950's) are inherently difficult to control, and provide relatively poor conversion efficiency. Both of these processes, as well as the R2 process (which dates from the 1960's), operate with relatively high concentrations of sulfuric acid and produce large volumes of sulfuric acid/sodium sulfate solution as by-product. The newer, more widely-used R3/SVP process employs a lower concentration of sulfuric acid and produces only sodium sulfate crystals as by-product. In processes that use sodium chloride as the reducing agent (e.g., the R2 and R3/SVP processes), chlorine gas is generated along with the chlorine dioxide gas.

In a 1970's modification of the R2 and R3 processes, hydrochloric acid (HCl) is substituted for all of the NaCl and part of the H_2SO_4, thus reducing the amount of by-product Na_2SO_4. Two 1980's modifications of the R3/SVP process are the R7 and R8 processes. In the R7 process, by-product chlorine is reacted with sulphur dioxide and water in such a way as to form a mixture of hydrochloric and sulfuric acids, which is then used to replace a portion of the sodium chloride and sulfuric acid otherwise fed to the generator. The net effect is a significant reduction in the amount of sulfur and chlorine by-products. In the R8 process, methanol replaces NaCl

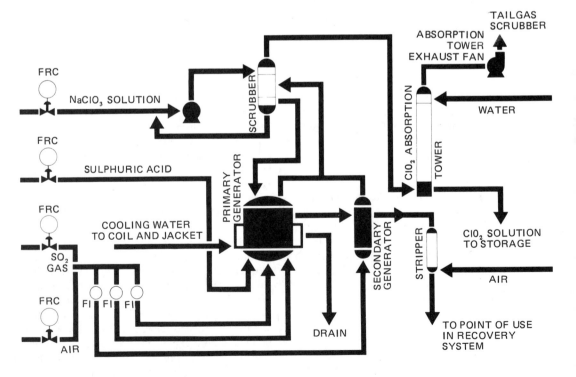

FIGURE 11-5. Modified Mathieson process for chlorine dioxide generation.

as the reducing agent, thus eliminating by-product chlorine; while higher acid concentration leads to the crystallization of sodium sesquisulfate [$Na_3H(SO_4)_2$] and lowers the amounts of by-product sulfur and sodium.

Examples of Generation Systems

The Mathieson process is illustrated in Figure 11-5. The reactants (sodium chlorate, sulfur dioxide and sulfuric acid) are fed to the primary generator along with sparged air which acts as a mixer and diluent for the generated ClO_2 gas. The generator operates with acid normality of 9 - 10 and chlorate concentration of 20 - 30 g/liter. The temperature is controlled in the 35 to 40°C range with jacket cooling. The contents overflow into a secondary reactor and then to a stripper. The principal net reaction is:

$$2NaClO_3 + SO_2 \longrightarrow 2ClO_2 + Na_2SO_4$$

The sulfuric acid acts only as a catalyst and goes through the reaction unchanged.

The R3 process is shown as a plant layout in Figure 11-6. The net reaction for this process is:

$$2NaClO_3 + 2NaCl + 2H_2SO_4 \longrightarrow 2ClO_2 + Cl_2 + 2Na_2SO_4 + H_2O$$

The reaction mixture of sodium chlorate, sodium chloride and sulfuric acid is controlled at higher temperature by continuous circulation through a heat exchanger. The generator is also maintained under a vacuum, thus producing water vapor as a carrier for the ClO_2 and Cl_2 gases. Because the generator acid solution is concentrated by evaporation, the sodium sulfate (saltcake) by-product crystallizes out of solution; the crystal slurry is continuously pumped to a filter where the saltcake is removed and washed while the acid is returned to the reactor.

Hydrogen Peroxide

Several processes are used commercially to produce hydrogen peroxide (H_2O_2), and literally dozens of others are being researched by chemical companies in North America, Europe and Japan. Many of these processes are proprietary and closely guarded. The most widely used method is based on the autooxidation of quinones. In this process, anthraquinone (which is dissolved in a multicomponent organic solution) is catalytically hydrogenated to hydroquinone; a subsequent aeration step produces H_2O_2 (which is water-extracted and concentrated) and regenerates the quinone.

A number of firms are pursuing an electrolytic route. For example, the new H-D Tech process, which utilizes a diaphragm cell, appears ideal for onsite generation. Dilute caustic is fed to the anode side where O_2 is generated. The generated O_2 is taken off the top of the anode side and, along with supplemental elemental oxygen, is fed to the high-surface-area cathode made of a carbon chip composite, where it forms H_2O_2 and caustic. The concentration of H_2O_2 produced is 30-80 g/L and the peroxide/caustic ratio can be adjusted to that needed in the bleaching process.

DuPont has unveiled a system for directly reacting hydrogen and oxygen into hydrogen peroxide, which is scheduled for commercial implementation in the 1990's. In this low-temperature, high-pressure process, oxygen and hydrogen are bubbled through a special aqueous medium containing a proprietary palladium catalyst that is loaded on activated carbon; single-pass conversion efficiencies of up to 13% are claimed.

11.3 CHLORINATION AND EXTRACTION

Chlorination and alkaline extraction have traditionally been employed in the first two stages of a pulp bleaching sequence as a means to effectively delignify the pulp. The chlorine reacts selectively with the non-carbohydrate constituents, rendering them water-soluble or soluble in alkaline media. Most of the chlorination reaction products are actually removed in the subsequent alkaline extraction stage. The industry is currently moving away from elemental chlorine as a pulp bleaching agent because of concerns about toxic chlorinated organic compounds in the environment. However, most mills will continue to use a modified chlorination/extraction for many years to come.

Chlorination

Chlorine reacts with lignin primarily by means of substitution and oxidation. In substitution, chlorine replaces a hydrogen on the organic molecule with the simultaneous formation of a hydrochloric acid molecule:

$$RH + Cl_2 \longrightarrow RCl + HCl$$

where R represents the organic molecule.

Oxidation can be pictured as forming elemental oxygen which reacts with the pulp:

$$H_2O + Cl_2 \longrightarrow O + 2HCl$$

The amount of HCl which is formed during chlorination is indicative of the reaction balance. Generally, more than 50% of the applied chlorine reacts by substitution, which is the preferred reaction mode. Oxidation-type reactions can have a relatively more degrading effect on the cellulose.

Traditionally, the high-density brownstock entering the bleach plant for chlorination was diluted with ambient fresh water to relatively low consistency (2.5 - 3.5%) to facilitate mixing and dissolution of the gaseous chlorine into the pulp stock and to dissipate the appreciable heat of reaction. Since all other bleach stages are operated at medium consistency (10 - 13%) or higher, the chlorination stage was conspicuous as a major consumer of water. Chlorination effluent is now commonly recycled for stock dilution in the same stage, and this practice has caused a sharp rise in chlorination temperature. Mills using only fresh water usually carry out chlorination at 20 - 30°C, while those mills utilizing significant recycle may operate as high as 60°C. Many mills are also

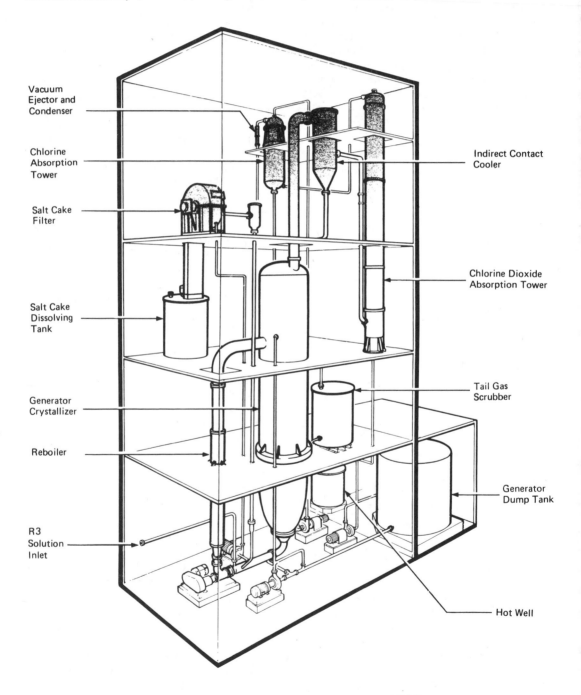

FIGURE 11-6. Conventional layout of R-3 chlorine dioxide generating plant (ERCO).

converting to a medium-consistency chlorination stage by utilizing the latest mixing technology.

The variables affecting chlorination are listed in Table 11-3. Time and temperature are directly interrelated as illustrated in Figure 11-7. Unfortunately, most existing chlorination towers are of fixed retention time, providing between 45 and 90 minutes under normal operating conditions. The chemical application must be sufficient to provide adequate driving force and ensure the desired degree

TABLE 11-3. Variables affecting chlorination.

Pulp Properties
- bleach demand (e.g., permanganate number)
- method of pulping
- CED viscosity

Operating Conditions
- time/temperature
- chemical application (as available Cl2)
- proportion of ClO2 in application
- sequence of chemical applications
- initial pH
- consistency

Process Conditions
- degree of mixing

Control Parameters
- stock color
- ORP level
- residual chemical concentration

of chlorination; the amount applied is usually about 75 - 80% of the full chlorine demand as indicated by the kappa number test. Typically, chlorine applications are in the range of 6 - 8% on oven-dry pulp for softwood kraft and 3 - 4% for sulfite pulps and hardwood kraft pulps. Measurement either of stock color or oxidation/reduction potential at a selected point downstream (after mixing and initial reaction) can be used as the basis for rapid feedback control of chemical application. Unfortunately, no automatic control method is capable of responding to large, rapid changes in lignin content. Manual kappa number testing of the incoming stock is still advisable, but the need can be minimized by stock blending prior to the bleach plant (refer back to section 9.8).

In the traditional chlorination stage, it was customary to maintain a small residual chlorine concentration in the retention tower discharge to ensure that the reaction had not stopped prematurely. However, with high-temperature chlorination, the chemical must be totally consumed in a short period; otherwise severe degradation of the cellulose results.

pH is extremely important in chlorine-water systems as the determinant of the chlorine species mix. In acid media, the following equilibrium exists:

1) $Cl_2 + H_2O \rightleftharpoons HOCl + H^+ + Cl^-$

When a base is present, a different equilibrium exists:

2) $HOCl + OH^- \rightleftharpoons OCl^- + H_2O$

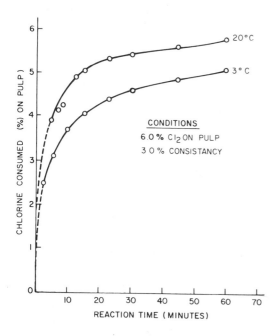

FIGURE 11-7. Effect of temperature on the rate of chlorination (from Duncan and Rapson).

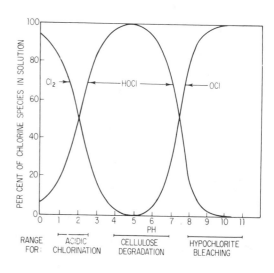

FIGURE 11-8. Chlorine equilibria in water at varying pH.

The relative proportion of molecular chlorine, hypochlorous acid (HOCl) and hypochlorite ion (OCl⁻) in solution depends on pH as illustrated in Figure 11-8. Each chlorine species is distinctive with respect to its action on lignin and cellulose. Hypochlorous acid can be quite destructive to cellulose, and therefore, bleaching within the pH range of 2-9 is usually avoided.

During traditional chlorination with fresh water dilution, enough hydrochloric acid is formed to keep the pH below 2.0 where the dominant species is molecular chlorine. Recycle of chlorination effluent serves to further reduce the pH of the chlorination stage and thereby almost eliminate the deleterious effect of the hypochlorous acid. A greater chloride ion concentration also serves to shift the equilibrium (of reaction #1 above) toward a higher relative concentration of molecular chlorine.

Over the past two decades it has become established practice in bleacheries utilizing chlorine dioxide to apply a small portion of ClO_2 in the chlorination stage, displacing perhaps 5 to 10% of the chlorine on a chemical equivalent basis. The ClO_2 serves to minimize degradation to the cellulose, especially where recycle of chlorination effluent is not practiced and pH is near 2.0. Other advantages claimed are higher final brightness, better brightness stability, and reduced effluent color. Numerous investigations were carried out to determine the most efficient level and method of application, with some conflicting claims. Most mills apply the ClO_2 prior to the Cl_2 addition to ensure quick reaction and no carryover of residual chemical into the washer system, thus avoiding corrosion problems.

The Importance of Mixing

A modern, versatile chlorination system is shown schematically in Figure 11-9. An example of an upflow, pump-through tower commonly used for chlorination is depicted in Figure 11-10. Thorough mixing of the chlorine gas with the dilution water and with the stock is essential to ensure uniformity of chemical treatment. Incremental chemical addition is also helpful toward that objective. Because of the low solubility of molecular chlorine in water (especially at higher temperature), the mixer must rapidly and uniformly combine the gas, liquid and pulp. A substantial input of mechanical energy is required. It is estimated that 80% or more of the chlorine is attached to the pulp within the initial 40 seconds.

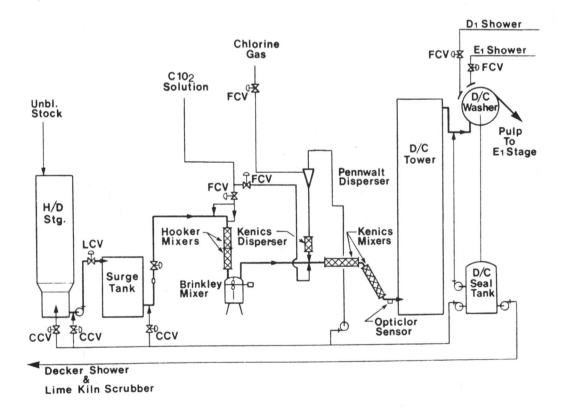

FIGURE 11-9. A modern, versatile chlorination system utilizing sequential ClO_2 addition.

FIGURE 11-10. Low-consistency pump-through up-flow chlorination tower.

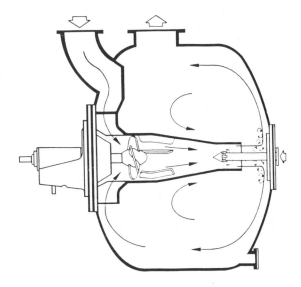

FIGURE 11-11. Chlorine mixer (Impco/Ingersoll-Rand).

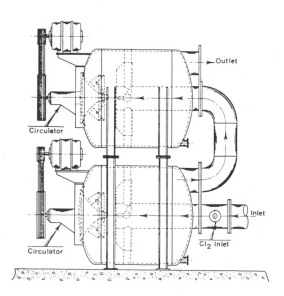

FIGURE 11-12. Double chlorine mixer (Sandy Hill Corp.).

FIGURE 11-13. In-line static mixer (Impco/Ingersoll Rand).

Since pulp can consume more chlorine than required to solubilize lignin, poor mixing will cause a portion of the pulp to be over-chlorinated while another portion is under-chlorinated. Many different methods of mixing are employed in modern chlorination systems. A series of small agitated tanks is advantageous from the standpoint of sequential chemical addition and location of sensors. A series of in-line static mixers with tanks in between provides the same advantages. Examples of chlorine mixers are shown in Figures 11-11 to 11-13.

The Dioxin Problem

In 1986, the production process for bleached chemical pulp was identified as a contributor of polychlorinated dioxins and dibenzofurans to the environment. Since these compounds (hereafter referred to simply as dioxins and furans) are recognized as powerful toxins and carcinogens, their detection in bleach plant effluents prompted a flurry of investigative activity in Europe and North America to identify point sources and corrective measures. The chlorination/extraction sequence was subsequently found to be the major source of these compounds.

The industry moved rapidly to implement measures to reduce or eliminate the formation of dioxins and furans (2). For example, most mills no longer utilize oil-based defoamers or tolerate wood chips containing preservatives because these chemical products are known to contain precursors of dioxins and furans. Unfortunately, the wood furnish itself also appears to contain precursors, and basic process changes are necessary to further reduce discharges to "non-detect" levels. The most effective actions appear to be either:

1) high-level (over 50%) substitution of chlorine with chlorine dioxide in the chlorination stage.
2) step-wise, incremental chlorination utilizing a low "chlorine multiple" (i.e., ratio of applied chlorine to kappa number), thus avoiding concentrations of chlorine that lead to the formation of highly substituted compounds, while moving a portion of the delignification load downstream.

Chlorinated Organic Compounds

In addition to dioxins and furans, a host of chlorinated organic compounds are formed during pulp chlorination, each possessing varying levels of toxicity. New regulations are being formulated to control these compounds, also known as "adsorbable organic halides" or AOX (3). The obvious way to conform with the new discharge limits is to reduce the amount of chlorine used in the pulp-bleaching process. New systems are moving away from molecular chlorine, with greater reliance being placed on oxygen, peroxide and chlorine dioxide.

Caustic Extraction

The purpose of the alkaline extraction stage following chlorination is to remove the chlorinated and oxidized lignin by solubilization. The extraction stage is carried out at medium consistency (12 - 15%) at temperatures between 60 - 80°C, and with retention times up to 2 hours. The final pH should be above 10.8, otherwise the solubilization will be incomplete. The required caustic application is directly dependent on the chlorine usage in the previous stage, and is generally equal to about 60% of the chlorine application.

Efficient washing following the chlorination stage is important; carryover of acid effluent will neutralize some of the caustic and thereby increase chemical costs. Uniform application of caustic to the pulp is also important. One arrangement that works well (where vacuum washers are used) is chemical addition from a shower pipe above the washer mould (Figure 11-14). Further blending is provided in a single- or double-shafted steam mixer used for temperature control (Figures 11-15 and 11-16).

Oxygen-Enriched Extraction

A significant bleaching development of the past decade is the oxygen-enriched extraction stage. A small amount of molecular oxygen (about 10 lb per ton of pulp) is mixed with the pulp under hydraulic head as it enters the extraction stage. The oxygen selectively reacts with residual lignin on the pulp, thus reducing the demand for chlorine dioxide in the

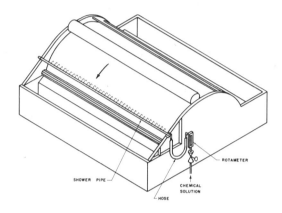

FIGURE 11-14. Installation of chemical shower pipe on backside of bleach stage washer.

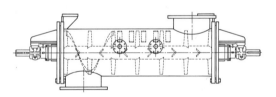

FIGURE 11-15. Single-shaft gravity steam mixer (Impco/Ingersoll Rand).

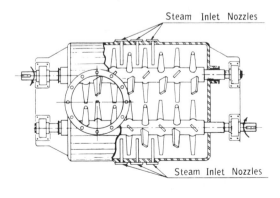

FIGURE 11-16. Double-shaft steam mixer.

subsequent stage. Since chlorine dioxide is a far more expensive bleaching agent, the economic incentive for using oxygen at this point in the process is very favorable. The oxygen-enriched extraction also provides some reduction in effluent color and appears to make shives more reactive to subsequent bleach treatments (4).

Perhaps the most popular method of oxygen-enriched extraction utilizes a medium-consistency mixer which fluidizes the stock and injected oxygen in the mixing zone thus promoting adherence of the oxygen to the fiber and rapid completion of the reaction. Typically, a mill having a downflow extraction tower (Figure 11-17) will install an upflow tube for retaining the pulp under pressure for 5 to 10 minutes. In fact, since the reaction is usually completed in 1 or 2 minutes, a large-diameter pipe between the mixer and the tower may be all that is required. Upflow extraction towers usually do not require a preretention tube.

Additional enhancement of the extraction stage can be obtained by adding small amounts of peroxide or

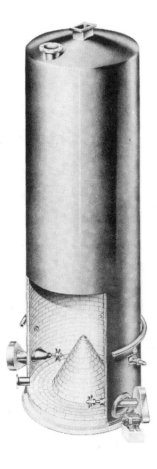

FIGURE 11-17. Downflow high-density bleach tower.

hypochlorite in combination with the oxygen. The effect is roughly additive, and the combined action is claimed to provide better control of the process.

2nd Extraction Stage

Alkaline extraction in the later stages of a multistage sequence serves to remove colored and degraded products and "open up" the fiber to more effective treatment in the subsequent oxidative stage.

11.4 OXYGEN BLEACHING

Oxygen bleaching (or delignification, or "prebleaching") was developed commercially during the late 1960's and early 1970's in Sweden and South Africa. The obstacle preventing earlier exploitation of oxygen as a bleaching agent had been its lack of selectivity. The breakthrough came with the discovery that a magnesium "protector" in the process successfully controlled cellulose degradation up to a certain degree of delignification.

Because the effluent from an alkaline oxygen prebleaching stage (i.e., prior to a conventional CEDED sequence) is totally compatible with the kraft chemical recovery system, much of the initial interest in the process stemmed from its inherent reduction in pulp mill pollution. Oxygen delignification prior to chlorination was rapidly adopted in Sweden in the 1970's, but the process has been slower to catch on in North America because of lingering doubts about its effect on pulp strength. Today, North American mills are showing greater interest, principally because of increased environmental concerns, but also because the application of medium-consistency equipment now provides more process options.

Typically, it is possible to reduce the lignin content by up to 50% in the oxygen prebleach stage; further delignification would cause excessive cellulose degradation. A commensurate reduction in discharge of pollutants is achieved by washing the dissolved solids from the oxygen-delignified stock and recycling them to the pulp mill recovery system. As a result, the total solids load to the recovery boiler will increase significantly, by about 4% with softwood pulp and 3% with hardwood pulp. Since these solids are already partially oxidized, steam generation will increase by only 1 to 2%.

Most kraft mills employing oxygen delignification systems use oxidized white liquor as the source of the alkali in order to maintain the sodium/sulfur balance in the chemical cycle. In most instances, air systems are used for white liquor oxidation because they are more economical to operate, even though the initial capital cost is higher than for oxygen systems. The use of oxidized white liquor increases the load on the causticizing plant and lime kiln by 3 to 5%.

Two types of systems are used commercially for oxygen delignification; these are generally characterized as high-consistency and medium-consistency systems (5). Low-consistency systems have been attempted, but have proven to be unsuccessful.

High-Consistency System

Figure 11-18 shows a representative system where pulp is handled in the 25% to 28% consistency range. Typically, a press is used to raise the feed consistency to about 30%. Fresh caustic or oxidized white liquor along with the magnesium protector is added to the pulp at the discharge of the press. A thick stock pump then transfers the pulp to a fluffer via a feed pipe in which a gas-tight plug is formed. The fluffed pulp flows down the pressurized reactor as a loose bed, while gaseous oxygen is continuously dissolved into the liquid phase and reacted with the pulp. In order for the oxygen to move freely from the gaseous to the liquid phase, a high specific surface is important. In some high-consistency reactors, a series of tray-like compartments are used to avoid excessive compaction of the pulp as it moves down the tower.

Steam is injected into the top of the reactor to maintain temperature in the 90 to 130°C range.

Oxygen gas is added to either top or bottom to maintain a partial pressure within the 90 to 130 psi range. Relief is taken from the head space at the top of the tower to remove combustibles and other noncondensible gases, and control overall reactor pressure. At the bottom of the tower, the reacted pulp is diluted with post-oxygen filtrate and pumped at about 6% consistency to a blow tank.

From the beginning, safety has been a major concern with the high-density systems. However, through the use of continuous monitoring systems coupled with a properly designed safety system, these reactors have proven to be safe and reliable.

Medium-Consistency System

A representative medium-consistency system is shown in Figure 11-19. Pulp from a brown stock washer or decker at about 10-14% consistency is charged with caustic or oxidized white liquor. It is then preheated in a low-pressure steam mixer and pumped through one or more medium-consistency gas mixers to an upflow pressurized reaction tower. Steam and oxygen are added upstream of the medium-consistency mixer or are added directly into it. The reactor bottom may be hemispherical or conical, and the top is equipped with a discharger. Often, post-oxygen filtrate is added at the top of the

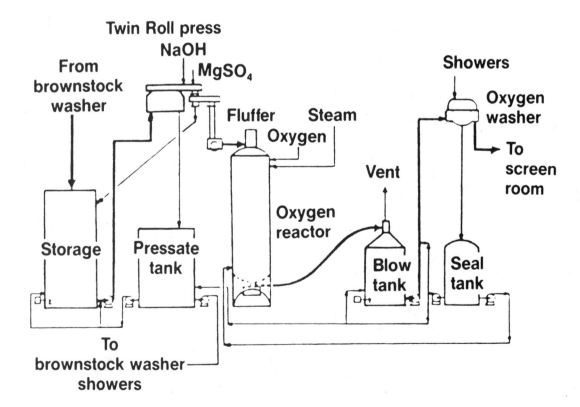

FIGURE 11-18. Representative high-consistency oxygen delignification system (E.P. Eddy Forest Products).

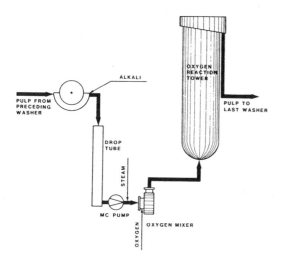

FIGURE 11-19. Medium-consistency oxygen delignification system (Rauma).

reactor to facilitate discharging the pulp. The stock is depressurized through a separator where steam is flashed and small amounts of by-product gases, oxygen and other noncondensibles are released.

In some medium-consistency systems, the top of the reactor is at atmospheric pressure. This type of system is sufficient for delignifying sulfite pulps and may be used for kraft pulps to obtain a 30-35% reduction in lignin content.

Although good results are obtained from both types of oxygen delignification systems, the trend is definitely toward medium-consistency systems because of their lower capital cost and inherently safe operation.

11.5   HYPOCHLORITE BLEACHING

Hypochlorite is a true bleaching agent. It preferentially destroys certain chromophoric groups of lignin, and consequently has been used in limited applications for lignin-preserving brightening. Subsequent to this initial reaction, the chemical attacks lignin vigorously. Unfortunately, depending on the accessibility of the lignin, the cellulose is also attacked to some extent.

Sulfite pulps are relatively easy to bleach with hypochlorite because the lignin residues are partially sulfonated and readily solubilized. With kraft pulp, the remaining lignin or lignin derivatives are less reactive, more condensed, and less accessible. Because high brightness cannot be attained for kraft pulp using hypochlorite without also causing severe cellulose degradation, application has been limited to the production of semi-bleached pulps up to 75 brightness, or to the intermediate stage of a full-bleach sequence (e.g., CEHDED or CEHDP).

However, sodium hypochlorite is now used sparingly in full-bleach sequences because of the greater cost-effectiveness of chlorine dioxide. The trend has been reinforced by the movement away from chlorine bleach chemicals.

The rate and extent of hypochlorite bleaching reactions are dependent on the nature of the pulp, chemical application, temperature, retention time, pH and consistency. Initially, the lignin "protects" the cellulose against degradation. As bleaching proceeds, the internal surfaces of the fibers are opened up and the cellulose becomes more reactive. In mills producing dissolving pulp grades, a relatively severe hypochlorite stage is often used for controlled reduction of pulp viscosity.

As in the other bleaching stages, time and temperature are related variables. Most hypochlorite bleaching is carried out at 35 - 40°C with a retention time of 1 to 2 hours. The hypochlorite application usually corresponds to the bleach chemical demand. A small residual is carried at the end of the bleaching stage to ensure that a driving force is operative during the entire retention time.

Usually, enough excess caustic is added with the hypochlorite solution to ensure that the terminal pH is 9.0 or higher. Hydrochloric acid and other acidic reaction products must be neutralized to prevent the pH from dropping into the range where hypochlorous acid becomes active (refer to chlorine water equilibria in Section 11.3). However, if too much alkali is present at the start of hypochlorite bleaching, the initial reaction rate can be severely retarded.

Hot Hypochlorite Stage

With the current emphasis on bleachery effluent recycle, oxidation-reinforced extraction, and short-sequence bleaching, some interest has been shown in hypochlorite bleaching at higher temperatures. It has been demonstrated (6) that, with good mixing, satisfactory hypochlorite bleaching can be carried out at 70°C in less than 10 minutes. Good control of chemical application is essential since the hypochlorite is totally consumed.

Calcium vs. Sodium Hypochlorite

Calcium hypochlorite has been used longer and more extensively than sodium hypochlorite for pulp bleaching, and it is still used successfully in some mills (7). The industry largely changed to sodium hypochlorite during the 1960's because mills were unwilling to contend with troublesome calcium hypochlorite makeup and handling systems in view of the modest cost differential. However, with the current escalating price for sodium hydroxide, it may be appropriate for mills that are using hypochlorite to review the economics.

11.6 CHLORINE DIOXIDE BLEACHING

Chlorine dioxide (ClO_2) was first used commercially for pulp bleaching in 1946, and by the late 1950's was an integral part of virtually all bleaching sequences for high-brightness kraft pulp. The industry rapidly embraced chlorine dioxide as a bleaching agent because of its high selectivity in destroying lignin without significantly degrading cellulose or hemicellulose; it thus preserves pulp strength while giving high, stable brightness.

The initial utilization of chlorine dioxide bleaching was in the later stages of a multi-stage bleaching sequence. The CEHDED sequence became dominant in the early 1960's for the production of full-bleach kraft pulp. As the cost of ClO_2 became more competitive with sodium hypochlorite in the late 1960's, the CEDED sequence became the industry standard. In the 1970's, attention was given to replacing part of the chlorine in the chlorination stage with chemically equivalent chlorine dioxide to improve pulp strength and color stability and reduce effluent color (refer back to Section 11.3).

A typical chlorine-dioxide bleach tower is illustrated in Figure 11-20. The stock at 11-14% consistency is heated to the requisite temperature of 70-75°C in a steam mixer, and flows by gravity into a thick stock pump. In older systems, the ClO_2 solution is added to the pressurized line through peripheral diffusers, and the stock goes through a single- or double-shaft mixer (e.g., Figure 11-21) to an up-flow tube. Newer systems or retrofitted systems are more likely to employ a modern medium-consistency mixer. Most of the chemical is consumed under hydraulic pressure in order to reduce flashing and "gas-off" of ClO_2 vapor. Although some towers are entirely upflow, most consist of an upflow leg followed by a larger downflow tower to provide controlled retention, usually 2 to 4 hours. The residual ClO_2 at the bottom of the tower is usually neutralized with SO_2 or NaOH to eliminate toxic fumes and reduce corrosion during the subsequent washing step. The SO_2 acts as a reducing agent, while the caustic causes the ClO_2 to dissociate into chlorate and chlorite, according to the following reaction:

$$2ClO_2 + 2NaOH \longrightarrow NaClO_2 + NaClO_3 + H_2O$$

ClO_2 is usually considered to have 2.5 times the oxidizing power of Cl_2 on a mole per mole basis (2.63 times on a pound per pound basis). Assuming complete reduction to chloride in both cases, the valence change is five for ClO_2 and two for molecular chlorine. The actual reaction mechanisms are fairly complex, and it has been shown that some portion of the ClO_2 is converted to chlorate and is

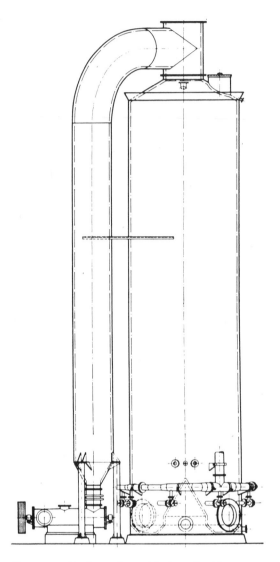

FIGURE 11-20. Downflow chlorine dioxide tower with upflow pre-retention tube.

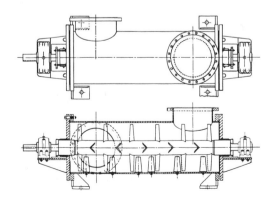

FIGURE 11-21. Single-shaft pressurized chemical mixer (Impco/Ingersoll Rand).

unreactive. A terminal pH of 3.5 - 4.0 is recommended for optimum chemical usage in the first dioxide stage. However, a somewhat higher pH (5.5 - 6.0) has been found to provide the highest brightness from the second ClO_2 stage.

The amount of ClO_2 added to the pulp depends on the final brightness desired. However, the brightness level reached also depends on the thoroughness of the previous treatments. The pulp will generally respond to higher chemical applications up to a certain brightness level; this level can only be exceeded by the addition of large, uneconomical excess chemical applications.

11.7 PEROXIDE BLEACHING

Peroxide (usually hydrogen peroxide) is used in the bleaching of both mechanical and chemical pulps. When applied at moderate temperatures up to 60°C, peroxide is an effective lignin-preserving bleaching agent, improving the brightness of mechanical and chemi-mechanical pulps without significant yield loss (refer also to Section 11.11). At slightly higher temperatures (i.e., 70 - 80°C), peroxide has traditionally been used as part of a chemical pulp full-bleach sequence to provide marginal increases in brightness and to improve brightness stability.

Peroxide is an efficient bleaching chemical when used in the second extraction stage of a CEDED or (DE)(EO)DED sequence. Studies have shown that for each chemical equivalent of peroxide used, a greater chemical equivalency of ClO_2 is saved in the subsequent stage. Peroxide bleaching is also useful as a terminal stage, which can either be carried out in a conventional bleach tower or in a high-density storage chest. As chemical pulp mills move away from their dependency on chlorine-containing chemicals, peroxide will become a more dominant bleaching agent.

From an environmental standpoint, bleaching without the use of any chlorine-containing chemical would be desirable. In this situation, all bleachery effluents could be recycled to the chemical recovery system, making the wastewater discharge relatively innocuous. Bleaching systems using only oxygen, ozone and hydrogen peroxide have been demonstrated on a laboratory scale. However, some outstanding technical problems must be solved before full commercial exploitation is possible.

Peroxide Chemistry

Peroxide bleaching is strongly affected by pH, which must be adjusted and buffered at about 10.5 for best results. The concentration of active perhydroxyl ion (HOO^-) increases with pH according to the following reaction:

$$H_2O_2 + OH^- \longrightarrow HOO^- + H_2O$$

At pH levels above 10.5, competition from undesirable side reactions detracts from the bleaching action. The pH is usually controlled by addition of sodium hydroxide and sodium silicate. The silicate (usually with added magnesium sulfate) acts as both a stabilizer and a buffering agent in the peroxide bleaching system. (Refer to Figure 11-22.)

Although higher temperature accelerates the peroxide bleaching reactions, undesirable side reactions can gain greater momentum. The stability of peroxide is adversely affected at higher temperature, especially in the presence of heavy metal ions. A prior acid bleach stage and the application of a chelating agent (most often DTPA) provide additional protection against the metal ions.

11.8 OZONE BLEACHING

Ozone (O_3) pulp bleaching has been extensively studied in laboratory and pilot plant operations. Since this process is most feasible in sequence with an oxygen stage, potential implementation of commercial ozone pulp bleaching in North America has been delayed by the slow adoption of oxygen delignification. Other problems that still need to be resolved are the relative inefficiency of ozone generating equipment and the non-selectivity of ozone, which can easily depolymerize cellulose. Given the current trend toward oxygen delignification, the advantages of ozone bleaching compared to other methods make it a viable process alternative (8).

The ozone bleaching process under greatest scrutiny is the high-consistency gas-phase process, which is carried out in the same manner as high-consistency oxygen bleaching (refer back to Section 11.4), and would utilize the same proven mechanical equipment. A major distinction is that ozone bleaching is carried out under acidic conditions, and therefore, the pulp to be bleached is pretreated with acid prior to fluffing and feeding the reactor.

Ozone gas is generally produced by the corona discharge method. In this process, a high voltage is applied across a discharge gap through which an oxygen-containing gas is passed. The discharge causes dissociation of the oxygen molecules, some of which recombine in the form of ozone. The economics of pulp bleaching dictate that the ozone be generated from oxygen gas and that the oxygen carrier gas be recovered and returned to the ozone generator.

The major impetus for the development of ozone bleaching has been the elimination or reduction of chlorine-containing chemicals. For example, an OZEP bleach sequence in a new mill, perhaps in combination with extended delignification, offers significant potential for environmental gains. The elimination of chlorides in the bleach plant would facilitate the recycle of all bleachery effluents back to the chemical recovery system.

11.9   BLEACHING EQUIPMENT

The vast majority of pulp mills operating in North America utilize downflow or upflow towers to provide retention time for the bleaching reactions, followed by vacuum washers to remove reaction products (refer to Sections 9.3 and 9.6). Major auxiliary equipment includes various types of pumps for moving the stock (refer to Section 9.7) and mixers for blending in steam and chemicals. In a few mills, diffusion washers are used at the top of upflow towers (Figure 11-23) in place of vacuum washers.

With a series of upflow towers, all stages must be operated simultaneously, since the stock continuously overflows the top at the same rate as the bottom feed. Thus, the feed to the chlorination tower automatically establishes the stock rate for all subsequent upflow towers. With downflow towers, the rates are usually adjusted to maintain constant levels; however, downflow towers have the advantage that retention times can be varied by changing the stock level. When they are operated at somewhat less than full capacity, downflow towers have both upstream and downstream surge capacity, so that problems in one stage do not necessarily shut down the other stages. Also, stock can more readily be pumped out of a downflow tower during shutdowns or for maintenance. Many bleach plants utilize a combination of upflow and downflow towers. All atmospheric stages with a potential "gas off" problem (e.g., the C, D, or E+O stages) must have at least an upflow leg to keep the chemical under hydraulic pressure during the initial reaction period.

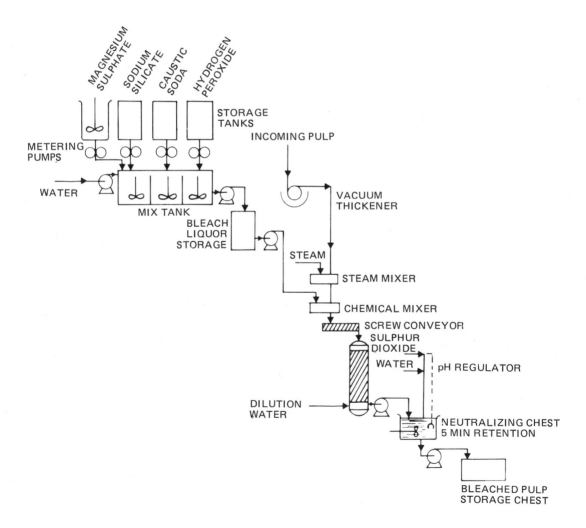

FIGURE 11-22. Peroxide bleaching system.

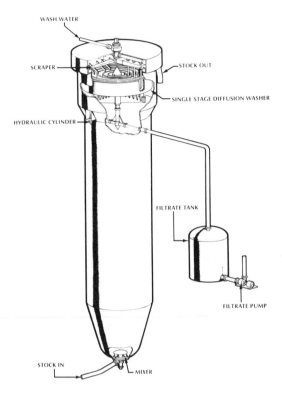

FIGURE 11-23. Kamyr diffusion washer mounted on upflow tower.

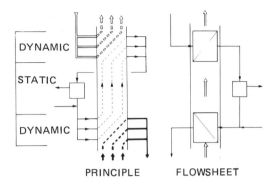

FIGURE 11-24. Displacement principle as applied in a continuous process and its corresponding flowsheet notation.

Displacement Bleaching

Rapson and Anderson showed in 1964 that rapid bleaching can be performed at conventional temperatures when the chemical agents are continuously displaced through the pulp mat rather than mixed into the pulp and retained in the conventional manner. The rapid technique was originally called "dynamic bleaching", but is now known commercially as "displacement bleaching". The reaction rates are accelerated because each fiber is continuously exposed to a concentrated chemical solution, rather than a progressively weaker solution as is the case with conventional bleaching. It was also found that with efficient displacement, washing between subsequent stages was unnecessary; most of the displaced solution is recycled along with fresh chemical makeup, and only a small portion of each displacement is sewered.

The principle of displacement bleaching is illustrated in Figure 11-24, along with its corresponding flowsheet notation. The development of the Kamyr atmospheric diffusion washer allowed the displacement principle to be applied commercially. Kamyr succeeded in mounting a series of diffusion units in a single tower to accomplish a multi-stage bleaching sequence.

Typically, the chlorination is carried out at medium consistency and the chlorinated stock is then pumped to the bottom of the displacement tower incorporating stages EDED (Figure 11-25). To ensure that maximum brightness is achieved, some systems now include a second tower which provides longer retention time between the second displacement with chlorine dioxide and the final wash (9). A number of displacement bleaching plants are in operation, generally meeting claims of significantly lower energy and water consumption and reduced space requirement. Chemical usages are somewhat higher, mainly because displacement falls short of the theoretical ideal.

11.10 RECYCLE OF FILTRATES

Prior to the 1970's, most bleach plants employed warm, fresh water showers on all the post-tower washers. This practice entailed high fresh water usage, the input of substantial energy to heat the water, and the discharge of large effluent volumes. As energy costs increased and environmental concerns became more paramount, various schemes for reducing water usage were investigated.

Initially, relatively "clean", 4th-, 5th- or 6th-stage washer filtrates were utilized as wash water on compatible earlier-stage washers. After it became evident that little risk was involved, mills became more adventurous in recycling their bleachery filtrates, thus taking advantage of chemical and energy savings, and reducing effluent volumes. The major problem to be overcome was corrosion of washers from the buildup of chlorides and other chlorine compounds, and from variable pH. Bleach plants are now usually provided with suitable metallurgy to withstand a more corrosive environment.

Countercurrent washing systems are necessary to achieve the greatest reduction in fresh water usage. With a true countercurrent system (i.e., where filtrate

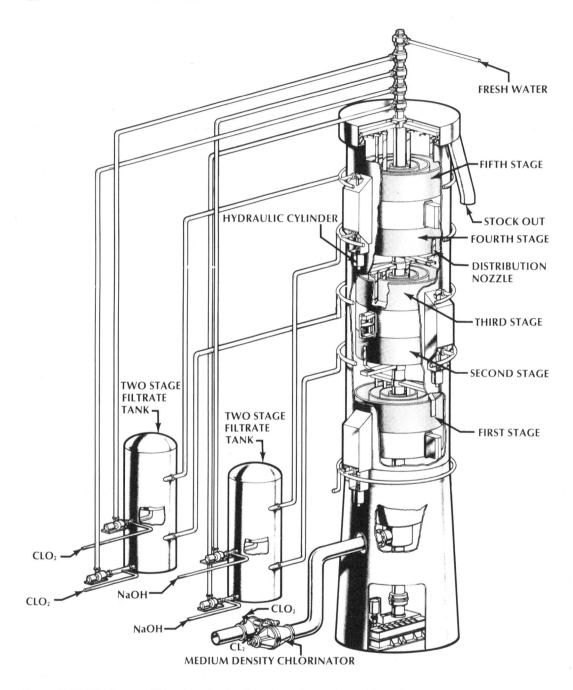

FIGURE 11-25. Displacement bleachery having five stages in one tower (Kamyr).

from each stage is used as shower water on the previous stage), it is essential that there be a net flow through the sheet and seal box system. In other words, the dilution factor must be greater than zero; and typically, a dilution factor of 2 is used. Otherwise, the residuals in the filtrate are carried back to the same stage, and impurities are allowed to build to an excessive level.

Complete countercurrent washing for a conventional CEDED sequence breaks down with respect to the application of alkaline extraction filtrate on the chlorination stage washer. Problems with foaming and precipitation of chlorolignin materials precludes any mixing of chlorination and extraction effluents in the bleach plant. The normal practice is to use fresh water on the lower bank of showers and extraction filtrate on the top bank of showers. The remaining extraction filtrate is sewered and represents the main bleed of contaminants from the entire bleach system. Most of the chlorination filtrate is re-used for stock dilution and chlorine injection.

Re-Use of Extraction Filtrate

An interesting recycle process has recently been developed which specifically reduces the amount of sodium hydroxide required to extract chlorinated pulp. In what is referred to as the Papricycle Process (10), extraction filtrate is applied in the usual way to the top bank of the chlorination washer showers. More filtrate is then applied at the C-washer repulper, replacing the normal addition of sodium hydroxide. The filtrate reacts almost instantaneously with the chlorinated pulp, and the pulp stock is then immediately taken to an additional washing stage where more extraction filtrate displaces the reaction products. The pulp mat then continues into a conventional (or oxygen-enriched) extraction stage. This procedure decreases by about one-third the amount of sodium hydroxide normally required during alkaline extraction by making use of alkalinity that previously was discarded.

Closed-Cycle Concept

The ultimate countercurrent washing and water recycle scheme for a conventional bleach kraft operation is embodied in the Rapson and Reeve concept of the closed-cycle mill. The complete system has been implemented in one Canadian mill, and although problems have been encountered, the objective of very low effluent discharge has largely been achieved (11).

The flowsheet for the closed-cycle mill is shown in Figure 11-26. The following processing steps are essential for achieving minimal discharge:
1) Countercurrent washing in the bleach plant.
2) Utilization of sufficient chlorine dioxide in the first bleach stage to achieve chlor-alkali balance, i.e., equal chemical equivalents of sodium and chlorine.
3) Use of bleach plant effluent for brown stock washing and white liquor dilution.
4) Utilization of evaporator condensates (that have been cleaned up by steam stripping) as bleach plant wash water.
5) Recovery of salt from concentrated white liquor by crystallization. The salt would preferably be used in a captive electrolysis plant for manufacturing bleach chemicals.

Some of the original precepts have been modified in light of actual operating experience, and a high degree of closure has been achieved. The main problems have stemmed from the high chloride concentrations in the recovery cycle. The benefits of the system, in addition to low effluent treatment cost, are substantial savings in chemical and energy costs. Savings in raw water treatment cost could also be important at some mill sites.

11.11 PULP BRIGHTENING (MECHANICAL PULP BLEACHING)

The term "brightening" is used here to denote lignin-preserving bleaching treatments. It must be noted that terminology is not standardized; in practice, the terms "bleaching" and "brightening" are often used interchangeably. Some technologists apply the term "brightening" only to those specific bleaching treatments that provide an increase in brightness.

In order to retain the advantage of high yield, mechanical and chemi-mechanical pulps must be decolored or brightened by methods that do not solubilize any appreciable amount of lignin. Chemicals used commercially for this purpose today are sodium hydrosulfite ($Na_2S_2O_4$) which has a reductive action, and the peroxides (usually H_2O_2) which have an oxidative action. Single-stage treatments are sufficient for improving the eye appeal of newsprint-type pulps. Brightness levels over 80 can often be achieved by a modern two-stage peroxide process. Unfortunately, no present treatment provides a brightness gain that is completely stable on exposure to light and air.

Hydrosulfite Brightening

Sodium hydrosulfite (or, more precisely, sodium dithionite) is inherently unstable. It can either be

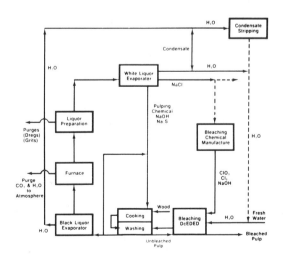

FIGURE 11-26. Schematic of closed-cycle bleached kraft mill.

supplied in a "stabilized" form or it can be generated on site by reacting SO_2 with a proprietary caustic solution of sodium borohydride. The brightening effect is obtained by the chemical reduction of colored quinoid lignin structures to colorless phenolic. The principal reaction is the oxidation of the hydrosulfite radical ($S_2O_4^=$) to sulfite ($SO_3^=$) or bisulfite (HSO_3^-), releasing reactive hydrogen:

$$S_2O_4^= + 2H_2O \longrightarrow 2HSO_3^- + 2H^+ + 2e^-$$

The hydrogen then combines with the pulp to decolor the "chromophores".

The variables affecting hydrosulfite brightening are listed in Table 11-4. The operation is usually carried out within the 4% to 8% consistency range to facilitate air-free mixing in economy-of-scale retention towers. A pH in the 5.5 - 6.5 range gives the best compromise between effective brightening and loss of chemical. Generally, a retention time of 1 to 2 hours at 60°C is recommended. If lower reaction temperature is used, the retention time should be increased. Chemical application is typically limited to 1% or less on pulp because higher applications provide only marginal brightness gain.

Undesirable side-reactions occur to a greater or lesser extent, depending on process conditions. Higher temperatures and lower pH levels, especially with inadequate mixing of chemical and stock, lead to wasteful losses of hydrosulfite. Exposure of the hydrosulfite solution to atmospheric air or entrained air will cause rapid oxidation. Unless the pulp stock is deaerated, some loss of chemical can be expected from air oxidation.

Iron, copper and other heavy metal ions are known to restrict brightness gains and adversely affect brightness stability. Sodium tripolyphosphate ($Na_5P_3O_{10}$) is widely used during hydrosulfite brightening as a chelating or sequestering agent to immobilize the metal ions and prevent the formation of undesirable colored salts. Other chelating agents, such as sodium silicate, are also effective.

The results achieved with hydrosulfite brightening are dependent on the characteristics and handling of the derivative pulpwood. For example, mechanical pulps from cedar and cypress are relatively unresponsive. Also, where pulpwood has been stored for a considerable period and/or contains large portions of rot or heartwood, the response of the pulps to hydrosulfite brightening is likely to be poor.

Peroxide (also see Section 11.7)

Traditionally, peroxide brightening has been a single-stage process, carried out over a range of consistencies up to 15%. In recent years, demand for higher-quality, brighter mechanical pulps has prompted the development of more sophisticated processes. The move is toward two-stage, high-consistency systems with recycle of chemical.

Single-stage systems are usually adequate for brightness gains up to 15 points. If only moderate gain is needed (up to 10 points), a medium-consistency system is adequate. For brightness increases above 10 points, a high-consistency system is generally required; these need shorter retention time and lower chemical dosage, but the equipment costs are higher. Presses are required to reach 25-40% consistency before the peroxide is added.

Maximum brightness gains (above 15 points) require substantial peroxide residuals (12). Thus, economics will usually dictate that residual chemicals are recycled. In single-stage systems, it is possible to reduce chemical costs by washing the pulp after the tower and recirculating some of the filtrate. However, because significant amounts of spent liquor cannot be recycled to a high-consistency tower, a two-stage process is usually needed, where

TABLE 11-4. Variables affecting hydrosulfite brightening.

Pulpwood Quality
- species
- condition (e.g., rot and heartwood content)
- period of storage
- moisture content

Contaminants
- entrained air
- heavy metal ions

Process Variables
- chemical application
- temperature/time
- pH
- use of chelating agent
- consistency

Process Conditions
- mixing

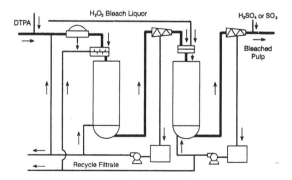

FIGURE 11-27. Two-stage peroxide brightening sequence (Interox).

the first tower is at medium consistency and the second tower at high consistency (Figure 11-27). A two-stage process also has an advantage in providing a cleaner pulp; the first stage acts as a wash step for extractives and organics which are pressed out between stages.

Mixed Sequence

Where the required brightness gain is slightly higher than that provided by a single medium-consistency treatment, consideration should also be given to the two-stage peroxide/hydrosulfite sequence. Because the effects of peroxide and hydrosulfite are largely additive, it may make economic sense to operate a two-stage system with low chemical applications in each stage. In a representative system (Figure 11-28) the first-stage alkaline peroxide treatment is carried out in a downflow tower. At the bottom, SO_2 solution is injected to consume the residual chemical and lower the pH. Hydrosulfite solution is then added and the second-stage reaction is carried out in an upflow tower to exclude air entrainment.

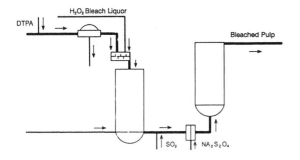

FIGURE 11-28. Two-stage peroxide/hydrosulfite brightening sequence (Interox).

REFERENCES

(1) **TAPPI Information Sheet, TIS 0606-21** (Issued 1988)

(2) BETTIS, J. **Bleach Plant Modifications, Controls Help Industry Limit Dioxin Formation** *Pulp & Paper* (June 1991)

(3) PRESLEY, J.R. **Bleach Plant Faces New Environmental Hurdle in Absorbable Organic Halides** *Pulp & Paper* (September 1990)

(4) VAN LIEROP, B., et al **Using Oxygen in the First Extraction Stage of a Bleaching Sequence** *Journal P&P Science 12*:5:J133-140 (September 1986)

(5) TENCH, L. and HARPER, S. **Oxygen-Bleaching Practices and Benefits: an Overview** *TAPPI Journal* (November 1987)

(6) MILNE, P.T. **High-Temperature, Short-Retention Hypo Stage Cuts Chemical, Water Use** *Pulp & Paper* (March 1981)

(7) ELLIOT, R.G. **Utilization of Calcium Hypochlorite in a CEHD Bleach Plant** *TAPPI54*:5:762-764 (May 1971)

(8) SHACKFORD, L.D. **Shifting Environmental Requirements Renew Interest in Ozone Bleaching** *Pulp & Paper* (October 1990)

(9) GULLICHSEN, J. **Displacement Bleaching - Past, Present, Future** *Tappi 62*:12:31-34 (December 1979)

(10) BERRY, R.M. and FLEMING, B.I. **Using Extraction-Stage Filtrate to Preextract Chlorinated Pulp** *Journal of P&P Science 14*:5:J121-124 (September 1988)

(11) ISBISTER, J.A., et al **The Closed-Cycle Bleached Kraft Pulp Mill at Great Lakes - An Advanced Status Report** *Pulp & Paper Canada 80*:6:T174-180 (June 1979)

(12) LACHENAL, D., et al **Bleaching of Mechanical Pulp to Very High Brightness** *TAPPI Journal* (March 1987)

Chapter 12

Overall Pulp Mill Operating Strategy

Assuming a stable market for its product, the profitability of a pulp mill depends on the quantity of production and the cost of production. When product demand is strong and pulp prices are high, operating emphasis is always on a high production rate. When the market is less buoyant, greater attention is usually devoted to controlling costs. Pulp markets are noted for being cyclical, with periods of extreme demand alternating with periods of slack demand.

The overall cost of production is mainly determined by the unit costs of capital, pulpwood/ chips, energy, chemicals, labor, and transportation. These costs vary significantly from country to country and from site to site. One comparison of bleached kraft market pulp costs (including transit to Western Europe) from "greenfield" sites in four different countries is shown in Figure 12-1. According to these 1989 estimates, Chile enjoyed a big cost advantage over Sweden, while Western Canada and the U.S. South were in between. Since wood cost differentials are the major factor, it is of interest to review (in Figures 12-2 and 12-3) further comparisons of wood growth rates and costs for softwood and hardwood. Countries such as Chile, Brazil, New Zealand and South Africa are blessed with high growth rates and relatively cheap wood.

In general, economy of scale in North America favors large kraft pulp mills, both with respect to relative capital cost per ton of production and with respect to operating costs. Figure 12-4 illustrates the effect of bleached kraft mill size on capital cost per ton and shows that 700 to 800 tons of daily capacity is a minimum size for a competitive mill. Indeed, some experts now believe that the minimum economically feasible size for a new bleached kraft mill is over 1000 tons per day. It should be noted, however, that other types of pulping operations, e.g., CTMP and NSSC are economically viable at outputs as low as 300 tons per day.

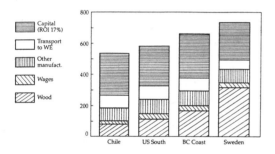

FIGURE 12-1. Relative costs for greenfield bleached kraft market pulp in U.S. dollars (R. Brandlinger of Sodra Skogsagarna).

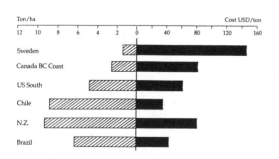

FIGURE 12-2. Relative softwood growth rates (on left) and costs (on right). (R. Brandlinger)

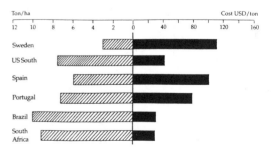

FIGURE 12-3. Relative hardwood growth rates (on left) and costs (on right). (R. Brandlinger)

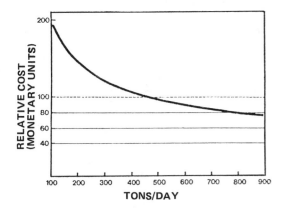

FIGURE 12-4. Variation of capital cost with size for a bleached kraft mill. The relative scale shows the capital investment per ton of production (Poyro).

12.1 SELECTION OF A MILL SITE

A number of factors are important in selecting the site for a pulp mill (1). Among the more basic requirements are the availability of desired pulpwood species, skilled labor, transportation facilities, adequate power at reasonable rates, and water supply. Other considerations are proximity to chemical suppliers, local tax rates, waste disposal requirements, and the provision of cultural and physical amenities. The "bottom-line" objective in site selection is to locate the plant where overall minimum cost is attained for the delivered product.

Typically, the site for a greenfield mill is considered in two phases. First, an initial screening is performed on a relatively large number of proposed sites to narrow the search down to a few promising locations. Then, a more detailed analysis is carried out before making the final decision. A representative schedule and list of factors for consideration are shown in Figure 12-5.

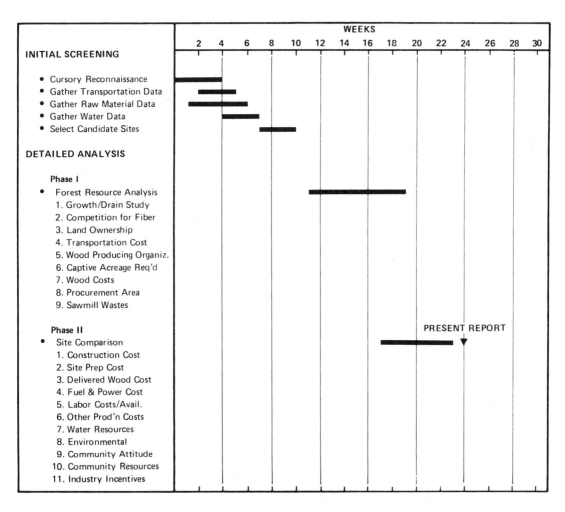

FIGURE 12-5. Typical site selection schedule (R.B. Ellerbee).

Traditionally, the criteria for plant selection were concerned only with the factors that significantly affected production and distribution costs. Today the situation is made far more complex by the necessity (as dictated by environmental legislation) to file permit applications, complete environmental impact assessments, and hold public hearings (2). In addition, all factors must be further assessed in relation to potential future changes in public attitudes, governmental regulations, and geopolitical relationships.

Because of corporate frustration in dealing with all the requirements of a greenfield project, it is generally felt that rehabilitation projects will be more dominant in the 1990's (3). In any case, it is probably correct to assume that most of the attractive greenfield sites in the United States (perhaps excluding Alaska) which are supported by adequate low-cost wood have been optioned or purchased. A better corporate strategy for growth in many instances is to expand and upgrade an existing facility or to purchase a facility from another company that is being under-utilized.

12.2 PROCESS DESIGN AND PLANT LAYOUT

Any industrial process is made up of a number of unit operations. In the design of a pulp mill, these unit operations are laid out in a logical sequence beginning with the handling of the fibrous raw material (e.g., chips) and ending with the handling of the pulp product either as a high-consistency stock, wet-lap, or in the form of bales for shipment.

Pulp manufacture is a mature industry. One result is that the manufacturing process design has evolved into a fairly standardized model. All kraft mills, with very few exceptions, incorporate the same processing steps. This is not to suggest that all plant layouts are the same. For example, some kraft mills are laid out in a straight-line sequence with an "operating corridor", thus having the advantage that additional operating lines can be installed parallel to the first during subsequent expansions. Other mills may use a "wraparound layout" for greater space economy and in order to cluster operating departments around a central operating room.

Regardless of its physical appearance, the layout should be planned to facilitate manufacturing operations, promote the effective utilization of operators and maintenance personnel, and provide for the convenience, safety and comfort of employees while they are carrying out operating and maintenance tasks. The layout may also be aimed at minimizing or facilitating the movement of material through the process. In the pulping process, the movement of fiber is characteristically accompanied by the movement of large quantities of process water or liquor.

Building requirements for kraft mills vary according to climatic conditions. In cold northern climes, mills tend to be housed in complete and well-insulated enclosures; whereas mills in warm climes tend toward outdoor installation of equipment and lightly-clad structures.

Kraft pulping is usually characterized as a continuous manufacturing operation. Strictly speaking, this concept infers that the process stream moves constantly from operation to operation without a controlled storage at any point in the process. In fact, many mills utilize batch cooking operations; and virtually all mills incorporate a number of "buffer storages" in order to isolate sections of the process from upsets or discontinuities in other sections. Figure 12-6 illustrates the major storage buffers for a typical bleached kraft pulping and recovery sequence of operations. Generally, each of these buffer capacities is equivalent to several hours of operation.

12.3 PRODUCTION CONTROL

The objective in mill operations is to process the available fibrous raw material through the various conversion and separation steps into a uniform, high-quality product in the most expeditious manner possible with regard both to the volume input of raw material and the yield of the product. During periods of trouble-free operation, the various unit operations are adjusted to process material at a coordinated rate (usually as set by digester production) and buffer storages are maintained at between 40 and 60% of capacity.

In the event of mechanical problems or other difficulties which curtail or shut down operations in one section of the mill, the departments directly upstream and downstream are able to continue operating while storage capacity or storage inventory is available. Initially their operating rates may be unaffected, but if the difficulties continue for several hours, it is ultimately necessary to reduce production rates and finally to shut down these departments when no more storage space is available for receipt of product or if no inventory is available to draw from. Any protracted problem in a single department will eventually cause the shutdown of all operating departments because of the "domino effect".

Following a period of operation where problems have been experienced and corrected, some storage tanks may be relatively full while others are virtually empty. It then becomes an objective of operating supervision to bring all storage inventories back into the desired 40 to 60% range, thus requiring that some departments operate at marginally higher rates of throughput than others.

An additional objective with respect to liquor storage levels is to maintain a reasonably constant

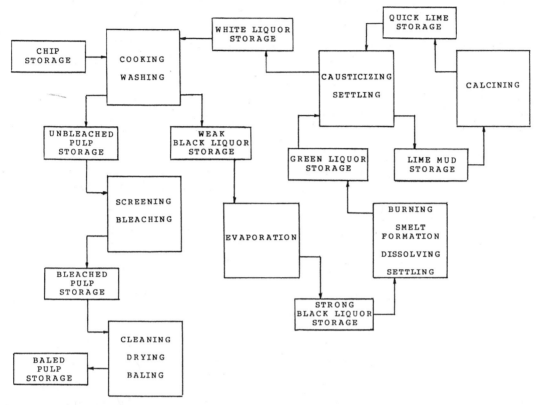

FIGURE 12-6. Kraft process schematic showing buffer storages.

inventory of active/activatable soda within the overall system. Toward this end, it is necessary that the levels in the various tanks be converted to equivalent soda and then tallied so that a running inventory is maintained. If the inventory is found to be decreasing, the rate of chemical makeup (usually in the form of saltcake, soda ash or caustic) is increased; if the soda inventory is building up, the makeup rate is reduced.

Lost Time Analysis

Systematic analysis of lost time toward the objective of identifying and quantifying reasons for production losses can be an important tool for improving production line performance. One well-described method of lost-time analysis (4) relies on records both of production interruptions and changes in the production rate through each department. The method identifies two key measures of production performance:

• Availability (a measure of reliability)—the portion of the total operating time during which the department was available for full production.
• Utilization—the portion of the total operating time

during which the department was utilized for full production (within the limit set by its availability). The ratio of utilization to availability, always less than 1.0, provides important information. A ratio close to 1.0 may indicate that the department is a bottleneck. A low ratio (say, below 0.8) may indicate that the production capacity of the department is not being fully utilized. The size of the buffer storages upstream and downstream from the department will influence results, and should be taken into consideration when the analysis is made.

Defining Production Bottlenecks

Within the mill, each process unit operates within a capacity limitation which usually is well defined by operating experience. Emphasis is given to modifying or replacing equipment within the process unit that acts as a restraint to overall mill productivity (5). After one bottleneck has been removed, the production rate can then be marginally improved until restrained by another process unit.

Most established kraft mills are restrained by recovery boiler capacity limitations. The recovery boiler is the highest capital cost item in the mill and is usually not over-designed. The boiler cannot be easily modified for higher capacity, and safety factors cannot be compromised. However, operating strategies can be followed to mitigate the impact of a recovery boiler limitation, which is usually defined in terms of liquor solids feed rate. For example, the loading of inorganic solids can be minimized by using the absolute minimum application of active chemical for cooking and by maintaining cooking liquor causticity at the highest practicable level to minimize the chemical cycle "dead load". A former strategy for reducing the organic solids loading was to increase the kappa number target from cooking; but it is no longer acceptable to push a greater lignin-removal loading into a bleachery utilizing chlorine-containing chemicals because of environmental ramifications.

Similar strategies are employed for other production bottlenecks. For example, a limitation in pulp drying capacity is usually circumvented by shipping the pulp at higher moisture content, even though the cost of shipping the additional water seriously detracts from the profitability of the increased tonnage. A limitation in digester capacity may be countered by reducing the impregnation time, but possibly at the penalty of higher screen rejects. The possible adverse ramifications of all operating strategies must be carefully evaluated against the benefits.

The general actions that could be considered to achieve a selected operating goal can be summarized as follows:
• adjustment of operating parameters
• adjustment of operating equipment
• adjustment of quality control targets
• adjustment of throughput rate
• scheduling of plant maintenance
• modification of the order mix
• adjustment of the wood input mix
• adjustment of slush pulp inventories (to make full use of storage capacity)

Because contingency factors can come into play, the use of a computer model has been suggested to provide evaluation of ongoing mill economic performance and assess the impact of various strategic options.

12.4 INVENTORY CONTROL

Inventory control with respect to any manufacturing process refers to the steps taken to maintain proper levels of raw materials and finished goods. For a typical market kraft pulp mill, the main inventory concerns are with respect to pulpwood/chips, pulping and bleaching chemicals, and bales of finished pulp.

In principle, pulpwood and chip inventories should be large enough to cover the anticipated fluctuations in raw material supply. Beyond this, it is difficult to generalize because of the many diverse arrangements that mills have with their pulpwood and chip suppliers. In the far north, for example, where logging is carried out on a seasonal basis, the overriding consideration is simply to provide sufficient pulpwood by the end of summer to sustain the mill through the winter until logging can resume. In these situations, the pulpwood inventory fluctuates markedly on a seasonal basis. (As a point of interest, pulp quality parameters may show a corresponding seasonal trend reflecting physical changes that occur to the roundwood or chips during storage.) In warmer climates where logging is carried out on a round-the-year basis, pulpwood/chip inventories are normally maintained within a much narrower range; the main requirement is for sufficient inventory to even out variations in pulpwood characteristics, to provide for blending of species, and to protect against short interruptions in supply. Many mills are associated with lumber mills and are duty-bound to take all wastewood chips regardless of specific inventory requirements.

In situations where purchased chips make up a significant percentage of the raw material supply, protection against a possible strike in the lumber and plywood industry is often a major consideration. This protection usually constitutes adequate chips to meet pulp mill demands for 60 to 90 days. Such mills may sometimes find it advantageous to inventory large volumes of chips in order to exploit deflated chip prices or to provide leverage when negotiating chip contracts. Another important point to consider is species variation in purchased chips; it may be necessary to stockpile certain species when supply exceeds demand and then reclaim these chips when the situation is reversed.

New mills sometimes find it desirable to stockpile large inventories of pulpwood or chips prior to startup while they consolidate their purchasing arrangements. These inventories can be three to four times the normal inventory. It is important in the design of a mill that adequate area be allocated for these exceptionally high inventories when they can be justified by mill management.

In some mills the production department is accountable for inventory control of pulpwood/chips and chemicals. Sometimes a single individual in the mill may be given primary responsibility for all aspects of pulpwood/chip acquisition and storage. In these instances, finished stock inventory (i.e., pulp bales) is generally controlled by the physical distribution manager. In other organizations, the acquisition, movement and storage of both raw materials and finished products may be the responsibility of a materials-management group.

For finished goods inventory, the management objective is to have sufficient material on hand to meet the needs of the customer while at the same time minimizing the amount of working capital invested in inventory and controlling the operating costs associated with carrying the inventory. Obviously, these two aspects must be balanced to achieve the best mix of customer satisfaction and low cost. Generally, when demand for pulp is high, the mill is producing to fill specific orders and no real impetus for inventory control exists. The need for inventory control is paramount when pulp demand is soft and the marketplace is more competitive. Here, the tendency is to ensure that sufficient supplies are available in a number of locations in order to respond as rapidly as possible to marketing opportunities. However, this strategy may not be cost effective if too much working capital is tied up in inventory.

12.5 ECONOMIC CONSIDERATIONS

Production costs can be divided into two categories:
- variable (or direct) costs
- fixed (or period) costs

Variable costs are those charges which are dependent on the volume of production, whereas fixed costs remain constant regardless of the production rate. The important variable cost items are wood chips, chemicals, and steam. Fixed cost items include operating labor, maintenance, salaries, taxes, insurance, and depreciation.

Although the profitability of "baseline pulp production" is dependent on both variable and fixed costs (Figure 12-7), additional tonnage (above the established base level) can be thought of as a function only of direct costs. The important point is that incremental production (i.e., production above baseline) is extremely profitable; and therefore a significant incentive always exists for increased mill productivity. The difference in profitability between baseline tonnage and incremental tonnage is illustrated in simplified form in Figure 12-8.

Cost-Effectiveness

The relative obsolescence (or modernity) of a mill facility relates to its profitability. However, the chronological age of equipment is only one factor affecting relative obsolescence. An older mill, well located in relation to its markets, can benefit from certain lower production costs and can compete successfully with newer mills.

Sometimes, external factors beyond a mill's control can be more significant. Raw material costs, environmental requirements, transportation costs, and purchased energy costs are examples of external factors that can change, favorably or unfavorably, to affect a mill's profit. The key criteria used by management to evaluate a mill's modernity are

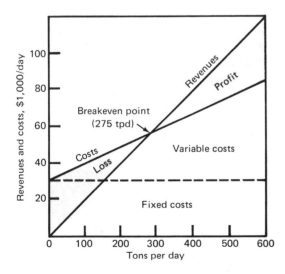

FIGURE 12-7. Production cost and profit for a pulp mill with break-even point at 275 tpd.

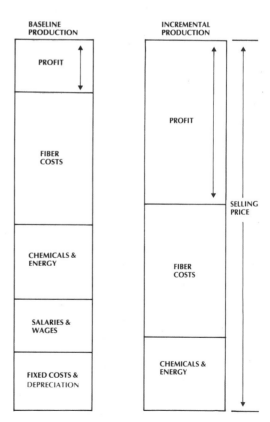

FIGURE 12-8. Comparison between profit margin between baseline ("normal") production and incremental production.

energy efficiency, raw material costs, manhours per unit of production, overhead costs and capital costs. How effectively a mill deals with these factors determines its cost effectiveness.

A mill may, in fact, be highly unprofitable even though it incorporates the latest technology if it has startup problems through unwise choices of equipment or design. The first few years have a disproportionate impact on the ultimate rate of return. Therefore, a mill must come up to a designated capacity rapidly in order for the project to be cost-effective.

Application of New Technology

Technical progress in mill pulping operations appears to evolve slowly in contrast to the seemingly continuous stream of developments being touted by research laboratories and equipment manufacturers. Although the amount of progress in any one year is hard to discern, the progress over several years viewed in retrospect is substantial.

The industry is anxious to adopt cost-saving techniques, but there are sound reasons why a cautious approach toward change is usually followed. The high capital investment in relation to profit margins has discouraged fast write-offs and high-risk ventures. The application to commercial scale from laboratory or pilot plant development involves unanswered questions; frequently, unforeseen problems are encountered that must be resolved, sometimes at considerable expense. It is

not uncommon for new mills, especially when utilizing unproven technology, to require extensive "shakedown" periods in order to bring operations to a desired level of efficiency. Awareness of the high cost of prolonged startups has contributed to the conservative approach taken by the industry in trying out innovations.

Value of Integration

Large integrated operations (for example, Figure 12-9) can obtain savings in costs for collective mill departments, such as power stations, wood handling, and water/effluent treatment, as well as administration overhead. The integration of wood processing with pulp production facilitates optimal utilization of wood residues and reduces transportation costs for wood handling. Further integration with papermaking reduces pulp processing costs and eliminates transport of the pulp.

In addition to "technical integration" (i.e., integration of manufacturing processes), many forest products companies are also cooperating at other levels, for example, through joint sales agencies and pooling of shipments.

12.6 COST CONTROL

A well-managed mill usually has an efficient measurement and control program for monitoring routine costs. Nonetheless, there is almost always a wide scope for ferreting out operational and design practices that are wasteful of wood, chemicals, energy and manpower.

The cost control system consists mainly of setting budgets and examining variances from the budget levels. The program is useful because it immediately focuses attention on the obvious high-cost areas. The process of cost monitoring is also valuable because it involves all levels of supervision and instills an across-the-board cost consciousness. However, budget levels that are initially set too high do not provide a challenging goal and may even encourage sloppy operation. The explanation of cost variance should concern itself with the "why" rather than the "what". In some instances, it may be observed that a reduction in unit cost offsets an increase in usage; unless the cost control format provides for variances in both unit cost and usage, the higher usage may go unnoticed.

As an example of a cost monitoring program, Figure 12-10 shows the development of unit cost figures for wood and steam. Wood cost is by far the most important single factor affecting pulp mill profitability. Where it is possible to exert some control on chip quality and species mix, these efforts are usually handsomely rewarded. By the same token, steam (or energy) costs are usually a significant portion of the overall per-ton cost.

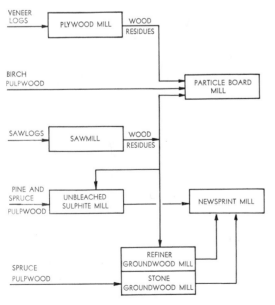

FIGURE 12-9. Schematic of integrated forest products mill.

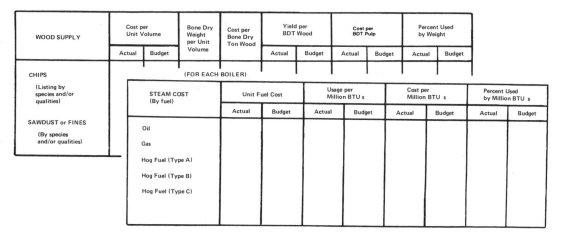

FIGURE 12-10. Unit cost sheets for wood and steam.

Generally, more effort is expended in controlling steam usage; but a 10% reduction is steam cost will have the same effect on pulp cost as a comparable reduction in steam usage.

Most mill managers encourage the initiation of cost reduction proposals by staff members. The proposal may involve a change in process, the application of a new control procedure or a new technology, or the utilization of improved equipment. Any proposal involving the expenditure of capital funds will have to be justified on the basis of standard return-on-investment criteria (6). The formalized request for capital expenditure should answer the following questions:
1) What is the present situation and in what way is it deficient?
2) What alternative actions could be employed to improve or correct these deficiencies?
3) What are the advantages and disadvantages of each alternative?
4) Which alternative is recommended and why?

When investment in new equipment is required to reach higher production levels, the additional capital costs must be weighed against the incremental improvement in profits. Since small increases in tonnage above the baseline can reap large profits, this type of investment generally appears to be attractive. However, "incremental investments" are not always straightforward. Too often, the true costs are underestimated by neglecting the side effects of the increased production. For example, if a bottleneck is overcome in the washing area to allow a higher production rate, what will be the effect of additional evaporation load on the pulp dryer? Or, what will be the effect of greater solids loading on the recovery furnace?

The Energy Balance

The energy balance is an effective tool for identifying areas of wastage or excessive usage of energy. Balances showing the inputs and outputs for different sections of the mill should be carried out regularly. Ideally, if sufficient instrumentation and measuring equipment is available, a computer can be programmed to display a reasonably complete balance on a routine basis.

The balances must be examined to determine if any waste heat streams can be recycled or if low-level heat can be utilized to replace fresh steam usage. In particular, the thermal efficiency of the various boilers should be reviewed critically. As an adjunct, systems to optimize the usage and thermal values of wood residual materials (e.g., hog fuel) need to be developed and improved.

REFERENCES
(1) STOVALL, J.H., et al **Considerations in Selecting a New Mill Site** *Tappi 63*:8:63-66 (August 1980)
(2) KINSTREY, R.B. **Greenfield Mill Site Permitting Can Take Years of Preparation** *Pulp & Paper* (February 1991)
(3) BASSETT, C.D. **Mill Construction Trends** *Tappi Journal 67*:3:46-48 (March 1984)
(4) AURELL, R. and ISACSON, C. **Lost-Time Analysis: Key to Improved Production Line Performance** *Pulp & Paper* (October 1982)
(5) FORBES, D.R. **Practical Aspects of Facilities Planning** *Tappi Journal* (September 1989)
(6) BLACKBURN, T. and LANE, J. **Project Management at the Mill Level** *Tappi Journal* (March 1986)

Chapter 13

Preparation of Papermaking Stock

Stock preparation is the interface between the pulp mill or pulp warehouse and the paper machine. In an integrated mill, stock preparation begins with dilution of the heavy stock at the discharge of the high-density pulp storage chests and ends with the blended papermaking furnish in the machine chest. In the independent paper mill, stock preparation begins by feeding pulp bales into the repulping system.

The basic objectives in stock preparation are to take the required fibrous raw materials (pulps) and non-fibrous components (additives), treat and modify each furnish constituent as required, and then combine all the ingredients continuously and uniformly into the papermaking stock. The primary concern is to produce a uniform papermaking furnish to ensure stable paper machine operation and a high standard of paper quality. The following operations are usually involved:

- **Pulping.** Baled pulp (or other fibrous raw material) is dispersed into water to form a slush or slurry. The operation can be either batch or continuous.
- **Refining (or Beating).** The fibers are subjected to mechanical action to develop their optimal papermaking properties with respect to the product being made. The operation is usually continuous, but some non-wood and specialty pulps are still treated batch-wise.
- **Utilization of Wet-End Additives.** A wide variety of mineral and chemical agents are added to the stock, either to impart specific properties to the paper product or to facilitate the papermaking process. Preparation is usually carried out batch-wise. The objectives, methodology, and chemical interactions (wet-end chemistry) are discussed in Chapter 15.
- **Metering and Blending.** The various fibrous and non-fibrous furnish components are continuously combined and blended to form the papermaking stock.

Additional operations may be carried out as part of stock preparation depending on the particular system requirements. For example, pulp screening and cleaning may be necessary for some high-quality furnishes. The special requirements of secondary stock preparation systems are covered in Chapter 14.

It should be noted that the paper machine white water and broke handling systems are sometimes considered to be part of stock preparation. However, it is more appropriate to consider these systems in Chapter 16 where their function and operation are easier to understand in the context of overall paper machine operation.

13.1 REPULPING (DISPERSION)

Repulping (or pulping) refers to any mechanical action that disperses dry, compacted pulp fibers into a water slush, slurry, or suspension. The extent of repulping can be just sufficient to enable the slurry to be pumped, or it can be adequate to totally separate and disperse all the fibers. In batch repulping operations, the defibering is usually completed within a single vessel. With continuous repulping, supplemental in-line treatment is commonly utilized following the pulper to ensure complete dispersion; typical devices for this purpose are called fiberizers, deflakers, or dispergers (fine deflakers).

The various fibrous raw materials (e.g., virgin pulps, broke, various reclaimed papers, etc.) have different energy requirements for repulping. Among virgin pulps, unbleached kraft is notably difficult to reslurry and defiber, especially after storage at low moisture levels. Usually, broke streams are fairly easy to repulp unless wet strength additives have been used. The greatest problems are experienced when utilizing waste papers and other reclaimed fibers (See Chapter 14). Generally, energy requirements can be minimized by operating at the highest practical consistency (up to 18%) and at temperatures above 50°C.

A common type of pulper adaptable for either batch or continuous operation is illustrated in Figure 13-1. Pulpers of this design employ one or more revolving elements that provide the turbulence and circulation necessary to disintegrate the fiber bundles, and they are open at the top to allow charging of pulp bales or sheets without danger of clogging. One manufacturer has now modified the conventional round pulper tub so that one side is flattened, and some older pulpers are being retrofitted with a baffle insert; this so-called D-shaped tub is reported to provide improved submergence and circulation.

FIGURE 13-1. The Black Clawson Hydrapulper is a vertical pulper equipped with under-rotor extraction through a perforated plate. Depending on defibering requirements, the perforations may be as small as 1/8 inch.

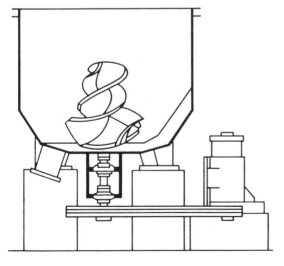

FIGURE 13-2. Schematic of Helico high-consistency pulper (Lodding, Div. of Thermo Electron).

Along with changes to the vat, the rotor design has undergone extensive development in recent years to allow repulping in the range of consistencies from 15 to 18%. As consistency is increased, defibering is improved due to greatly increased fiber-to-fiber friction, while energy consumption is significantly reduced. The new conical-shaped rotors protrude well into the vat. The proper sizing of the vat in relation to the rotor and the number and location of the guide vanes are extremely important in creating the fold-over action that is necessary when operating at higher consistency levels. The general configuration of a high-consistency pulper is shown in Figure 13-2.

Pulp bales (or slabs or sheets) are most easily fed into the pulper from a floor that is approximately level with the top of the vat. One system for handling incoming stacks of pulp bales prior to repulping is shown in Figure 13-3.

A representative device for in-line dispersion is illustrated in Figures 13-4 and 13-5. The stock suspension (up to 4.5% consistency) is split into streams by the slits in the rotor and stator rings, which mate with one another. The stock is therefore accelerated and decelerated repeatedly, and hydrodynamic shear forces are produced by the severe velocity gradients. The resulting forces serve to loosen the bonds between fibers and reduce "flakes" into constituent fibers.

13.2 REFINING

The terms "beating" and "refining" are often used interchangeably. More precisely, beating refers to the mechanical action of rotating bars opposing a stationary bedplate on a circulating fiber suspension where the individual fibers are oriented perpendicular to the bars. This batch operation is exemplified in the traditional Hollander beater which is still used in some older paper mills, especially for handling such difficult furnishes as jute, hemp, flax and cotton (see Figure 13-6). Refining refers to the mechanical action carried out in continuous conical or disk-type refiners where the fibers move parallel to the bar crossings. In all cases, the objective is to "develop" or modify the pulp fibers in an optimal manner for the demands of the particular papermaking furnish.

Although refining and defibering are usually considered as separate operations, considerable overlap may occur in practice. Some mechanical modification of the fibers takes place during repulping and dispersion operations. By the same token, a refiner can function as an effective defibering device.

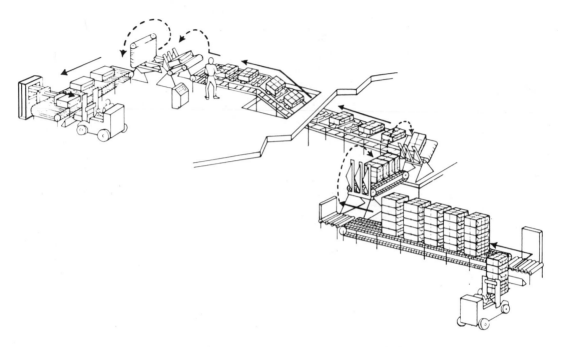

FIGURE 13-3. Two-storey transfer, destacking and dewiring system for pulp bales.

FIGURE 13-4. Representative deflaker (Escher Wyss).

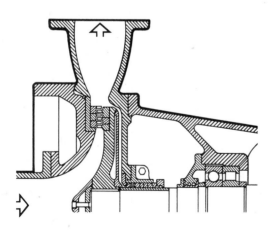

FIGURE 13-5. Operating principle of Escher Wyss deflaker.

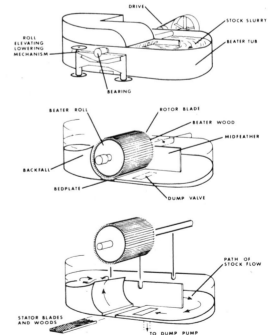

FIGURE 13-6. Simplified representation of a beater.

Mechanism of Refining

The mechanisms involved in refining are illustrated in Figure 13-7. Both mechanical and hydraulic forces are employed to alter the fiber characteristics. Shear stresses are imposed on the fibers by the rolling, twisting, and tensional actions occurring between the bars and in the grooves and channels of the refiner. Normal stresses (either tensional or compressive) are imposed by the bending, crushing, and pulling/pushing actions on the fiber clumps caught between the bar-to-bar surfaces.

The major effects of refining on the individual fibers are summarized in Table 13-1. The initial action is to partially remove the primary wall. Although the primary wall is permeable to water, it does not swell and prevents the fiber from swelling. Removal of the P-layer exposes the secondary wall and allows water to be absorbed into the molecular structure. The consequent loosening of the internal structure promotes fiber swelling and renders the fiber soft and flexible. This so-called "internal fibrillation" is generally regarded as the most important primary effect of refining following removal of the primary wall. The further action of (external) fibrillation involves loosening of the fibrils and raising of the finer microfibrils on the surfaces of the fibers, resulting in a very large increase in surface area for the beaten fibers. As the fibers become more flexible, the cell walls collapse into the lumens, thus creating ribbon-like elements of great conformability. The effects of refining on sheet structure can be seen in Figures 13-8 and 13-9. Fibrillation is well illustrated in Figure 13-10.

Fiber shortening (or cutting) always occurs to some extent during refining, mainly due to the shearing action of the bar crossings. Fiber cutting is often considered undesirable because it contributes to slower drainage and loss of strength. But, in some applications, a cutting action may be promoted to obtain good sheet formation from a long-fibered pulp furnish or to control sheet drainage on the paper machine. Refining also produces fines consisting of fragments of broken fibers and particles removed from the fiber walls.

One obvious effect of refining is the dramatic change in the drainage or dewatering properties of the pulp. Pulp drainability is rapidly reduced as refining proceeds, mainly due to the increased concentration of fines.

System Analysis

Two factors are of primary importance in analyzing refiner performance: the amount of effective energy applied per unit weight of pulp (net specific energy), and the rate at which the energy is applied (refining intensity). The first factor can be precisely measured; but it is more difficult to evaluate how intensively the fibers are hit. The term "specific edge load" is widely applied for this purpose (1), and is calculated by dividing the rate of net energy application (net power) by the total length of bar edges contacting the stock per unit time.

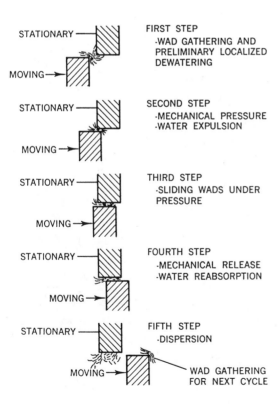

FIRST STEP
- WAD GATHERING AND PRELIMINARY LOCALIZED DEWATERING

SECOND STEP
- MECHANICAL PRESSURE
- WATER EXPULSION

THIRD STEP
- SLIDING WADS UNDER PRESSURE

FOURTH STEP
- MECHANICAL RELEASE
- WATER REABSORPTION

FIFTH STEP
- DISPERSION

WAD GATHERING FOR NEXT CYCLE

FIGURE 13-7. Illustration of refining method.

TABLE 13-1. What happens during refining.

Primary Effects
- removal of primary wall; formation of fiber debris or "fines".
- penetration of water into the cell wall (referred to as "bruising" and "swelling").
- breaking of some intra-fiber bonds; replacement by water-fiber hydrogen bonds ("hydration").
- increased fiber flexibility.
- external fibrillation and foliation.
- fiber shortening.

Secondary Effects
- fractures (cracks) in the cell wall.
- fiber stretching and/or compression.
- partial solubilization of surface hemicellulose into "gels".
- straightening of fiber (at low consistency).
- curling of fibers (at high consistency).

FIGURE 13-8. Photomicrographs of kraft softwood pulp before and after refining (courtesy of Institute of Paper Science and Technology).

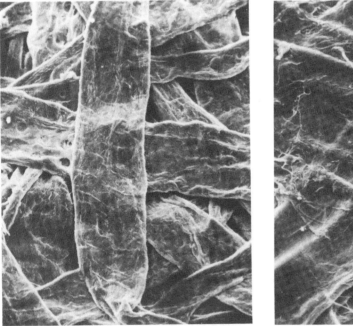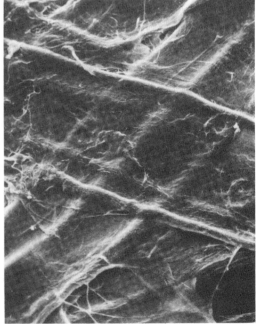

FIGURE 13-9. Scanning electron micrographs of southern pine earlywood fibers before and after refining (courtesy of Institute of Paper Science and Technology).

The problem is that the energy is actually transferred to the fibers in three phases and in three different ways (2). In the edge-to-edge phase, the fibers trapped between the bar edges get a strong hit directed over a short length of the fibers. In the edge-to-surface phase, the leading edges of both rotor and stator bars press the fibers against the flat bar surfaces, and their relative movement imparts a brushing treatment to the fibers. Finally, after the leading edges reach the trailing edges of the opposite bars, the fibers undergo a further gliding action during a surface-to-surface phase until the trailing edge of the rotor bar has cleared the trailing edge of the stator bar.

The energy split between the three phases, as well as the refining result, depends on the sharpness of the bars, the width of the bars and grooves, and the roughness of the bar surfaces. If the energy is consumed mainly in the first phase, the fibers will undergo significant cutting. If, however, most of the energy is utilized in the second and third phases, the fibers will become more fibrillated.

The net energy applied by the refiner is the difference between the measured refiner brake energy (gross energy) and the backed-off or no-load energy requirement. Some energy is necessary simply to rotate the unit against the stock flow; this varies with the type of refiner and is substantial for certain designs.

Obviously, the amount of energy absorbed by the pulp is the major factor affecting the change in pulp properties; but the manner in which the work is carried out is also a significant determinant. Refining carried out at low intensity will produce greater fibrillation, less cutting, and more satisfactory development of fiber properties. In other words, a gradual, step-by-step application of mechanical energy to the fibers provides the optimal treatment, in contrast to that produced by a more abrupt and concentrated application of the same amount of energy. Unfortunately, the attainment of optimal low-intensity refining on a mill scale often requires the installation of a prohibitively large number of refiners. Therefore, fiber development must be compromised to a degree in order to control the capital and operating costs of the refiners. Because of their greater refining intensity, mill refiners are not capable of duplicating the more idealized action of laboratory beaters.

Variables Affecting Refining

Some of the variables affecting refining which are related to fibers, equipment, and process are listed in Table 13-2. Various pulps respond in different ways to a given refining treatment. Generally, kraft pulps are more difficult to refine (i.e., require more energy) than sulfite pulps. Soda pulps are notably the easiest to refine. Unbleached pulps are more difficult to

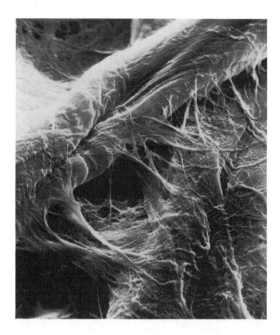

FIGURE 13-10. Scanning electron micrograph of refined fibers showing unravelling of the microfibrils (courtesy of Institute of Paper Science and Technology).

TABLE 13-2. Variables affecting refining.

Raw Materials
- fiber morphology (species derivation)
- pulping method
- degree of pulping
- bleaching treatment
- prior processing (e.g., drying, mechanical treatment)
- fiber length distribution
- fiber coarseness
- earlywood/latewood ratio
- chemical composition (lignin/cellulose/ hemicellulose)

Equipment Characteristics
- bar size and shape
- area of bars and grooves
- depth of grooves
- presence or absence of dams
- materials of construction
- wear patterns
- bar angles
- speed of rotation (peripheral speed)

Process Variables
- temperature
- pH
- consistency
- additives
- pretreatments
- production rate
- applied energy

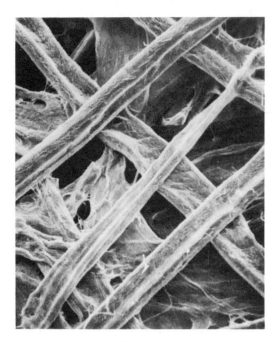

FIGURE 13-11. Refined high-yield softwood pulp (courtesy of Institute of Paper Science and Technology).

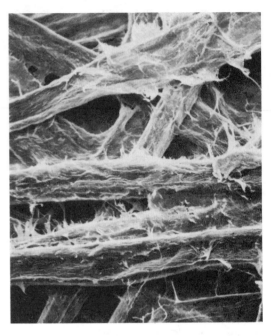

FIGURE 13-12. Refined low-yield softwood pulp (courtesy Institute of Paper Science and Technology).

refine than comparable bleached pulps, and those with higher lignin content are less responsive to beating because the lignin does not absorb water, and therefore the fibers do not "swell" as much (compare Figures 13-11 and 13-12). High-yield mechanical and chemi-mechanical pulps are generally not refined in the paper mill because their high fiber stiffness causes severe cutting. In some instances a light "post-refining" of mechanical pulps is carried out for drainage control on the paper machine, but not for the normal purpose of developing fiber properties.

As a general rule, pulps containing large percentages of hemicelluloses are easy to refine and respond well to the input of mechanical energy. The great affinity of hemicellulose for water promotes "swelling" and fibrillation. On the other hand, dissolving-type pulps which are high in alpha cellulose refine slowly and produce weak sheets.

Dried chemical pulps, including secondary fibers, do not absorb water as readily and are more difficult to refine than pulps which have never been dried. Over-drying or uneven drying of pulp may contribute to a lower strength paper product, owing to uneven strength development, if sufficient time is not allowed for re-wetting.

The "refinability" of mixed-furnish secondary pulps is mainly a function of the chemical pulp

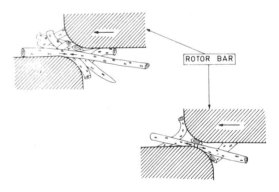

FIGURE 13-13. Illustrating the effect of consistency on the amount of fiber-to-fiber and fiber-to-metal contact during refining.

content. The higher the proportion of chemical fibers, the greater is the potential for the development of pulp properties through refining.

Higher pH levels (above 7) promote faster beating in accordance with the greater swelling of cellulose in alkaline media. A high consistency generally contributes to a good beating response because of increased fiber-to-fiber contact and less cutting action (Figure 13-13).

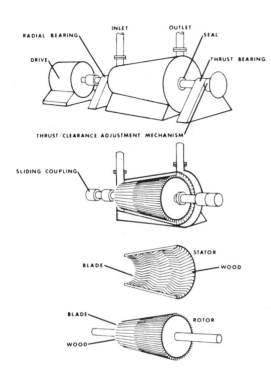

FIGURE 13-14. Simplified representation of a Jordan conical refiner.

Types of Refiners

Two major types of continuous refiners are used for stock preparation: disc refiners and conical refiners. The conical refiners can be further differentiated into low-angle types (Jordans) and high-angle types (Claflins). The conical types are illustrated in Figures 13-14 to 13-16.

In conical refiners, the rotating plug (rotor) and its housing (stator) are fitted with metal bars oriented lengthwise. The fibers flow parallel to the bars. The position of the plug determines the clearance of the bars and controls the amount of work done on the fibers for a constant stock throughput.

Disc refiners are a more recent development, and are available in a wide variety of designs and disc patterns. There are three basic types:

• rotating disc opposing stationary disc.
• two opposing rotating discs (Figure 13-17).
• rotating double-sided disc between two stationary discs (Figure 13-18).

For the last design, the stock flow can follow either a parallel arrangement (duoflow) or a series arrangement (monoflow). The two flow options are illustrated schematically in Figure 13-19. (Additional disc refiner designs are illustrated in Section 5.3.)

Disc refiners offer significant advantages over conical refiners, and it is difficult to understand why some papermakers still prefer to use conical types. The major advantages are:

• lower no-load energy consumption; this advantage becomes more significant with higher energy costs.

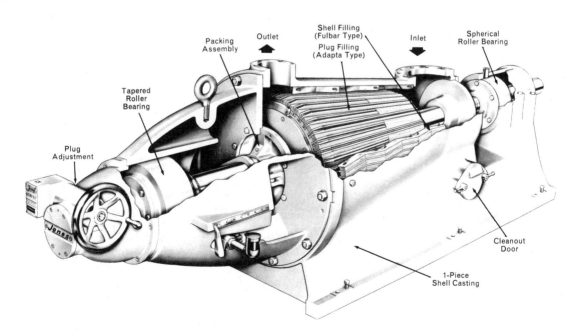

FIGURE 13-15. Jordan Refiner (Jones Div., Beloit Corp.).

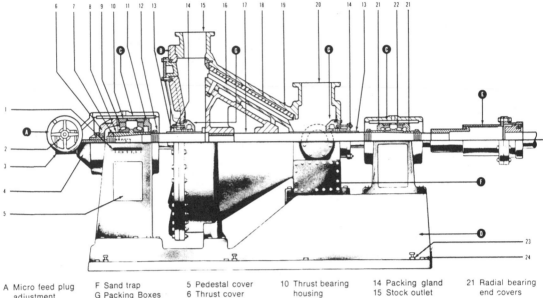

A Micro feed plug adjustment	F Sand trap	5 Pedestal cover	10 Thrust bearing housing	14 Packing gland
B Large head section	G Packing Boxes	6 Thrust cover	11 Thrust bearing end cover	15 Stock outlet
C Bearing assemblies	1 Plug indicating gauge	7 Lock nut and washer	12 Push nut	16 Plug liner
D Base	2 Dial indicator and lock screw	8 Thrust bearing sleeve	13 Packing box sleeve	17 Shaft
E Coupling	3 Jack	9 Seal ring		18 Shell liner
	4 Thrust bracket			19 Shell housing
				20 Stock inlet
				21 Radial bearing end covers
				22 Radial bearing housing
				23 Leveling jacks
				24 Anchor bolts

FIGURE 13-16. Claflin refiner (Bolton-Emerson).

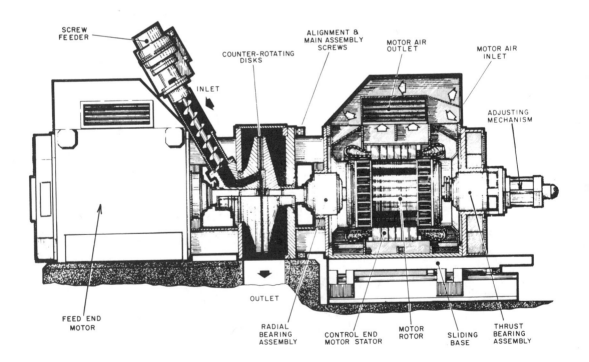

FIGURE 13-17. Disc refiner with two opposing rotating discs (Jones Div., Beloit Corp.).

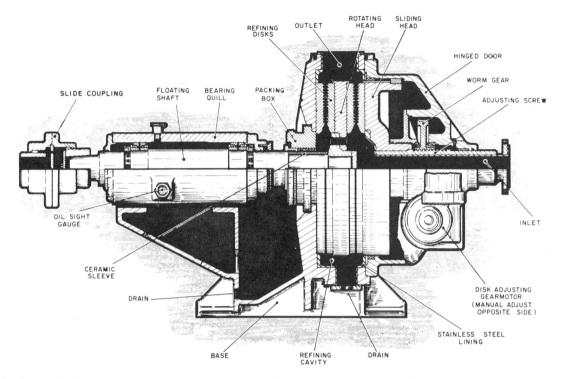

REFINING DISKS, OUTLET, ROTATING HEAD, SLIDING HEAD, HINGED DOOR, WORM GEAR, ADJUSTING SCREW, SLIDE COUPLING, FLOATING SHAFT, BEARING QUILL, PACKING BOX, OIL SIGHT GAUGE, CERAMIC SLEEVE, DRAIN, BASE, REFINING CAVITY, DRAIN, INLET, DISK ADJUSTING GEARMOTOR (MANUAL ADJUST. OPPOSITE SIDE), STAINLESS STEEL LINING

FIGURE 13-18. Disc refiner with double-sided rotating disc between two stationary discs, operated in "duoflo" configuration (Jones Div., Beloit Corp.).

• utilization of higher stock consistencies.
• application of higher loading and greater rotational speed.
• greater versatility of refiner plate design.
• self-correcting wear patterns; i.e., uniform mating of high and low spots, in contrast to conical refiners (refer to Figure 13-20).
• more compact design; smaller space requirement.
• lower capital investment per ton of production.

However, it must be noted that for the same operating conditions, there is little difference between conical and disc refiners regarding their ability to develop fibers.

Perhaps the most popular stock preparation refiners are those equipped with a two-sided rotating disc sandwiched between two stationary refiner plates. Because the pressure is equal on both sides, the rotating disc centers itself between the two non-rotating heads. This "floating disc" principle assures that the refining energy is split equally between the two sides and that the thrust loads developed (i.e., the opposing forces in the axial direction of the shaft) are equal in both directions, thus eliminating the need for thrust bearings.

In disc refiners, it is the gap between plate surfaces that determines the amount of work done on the pulp

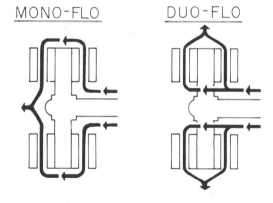

FIGURE 13-19. Alternate flow paths for refiner with double-sided rotating disc.

at constant throughput. This gap must be carefully controlled to maintain loading, and at the same time avoid plate clashing. A number of methods are used to measure and control the clearance, depending on equipment manufacturer.

Higher disc speeds provide lower refining intensity for the same throughput, and therefore provide better fiber development. However, higher rotational

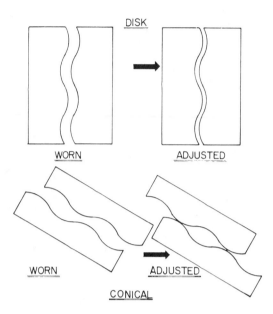

FIGURE 13-20. Wear patterns for disc and conical refiners. The gap between refining surfaces usually varies between 0.1 and 1.0 mm depending on stock consistency and throughput.

speeds waste more energy because the no-load energy requirement increases by the cube of the rotor speed. Most of the no-load energy is dissipated at the periphery where the disc velocity is greatest. Typical disc refiners operate with a maximum peripheral velocity in the 4700 to 5700 fpm range, and the maximum rotational speed of the refiner is dependent on the diameter of the disc. In order to circumvent the limitation of diameter on capacity, one manufacturer (Sunds) has added a conical refining section at the periphery of the disc, and their so-called conical disc refiners are claimed to be the largest refiners on the market.

Disc Refiner Plates

Plates for disc refiners consist of a variety of bars cast onto a base plate. The configuration of these bars is a significant factor in accomplishing specific refining effects. The plate patterns shown in Figure 13-21 are typical of those commonly used for stock preparation. The courser patterns provide a high-intensity action which is more suitable for cutting fibers; the finer patterns are more appropriate for strength development (3).

A number of alloys are used for the manufacture of refiner plates depending on specific requirements. Although pulp quality is a consideration, the choice of plate metallurgy is usually based primarily on cost-effectiveness. To date, the use of the more exotic metals such as titanium, or the use of jet age

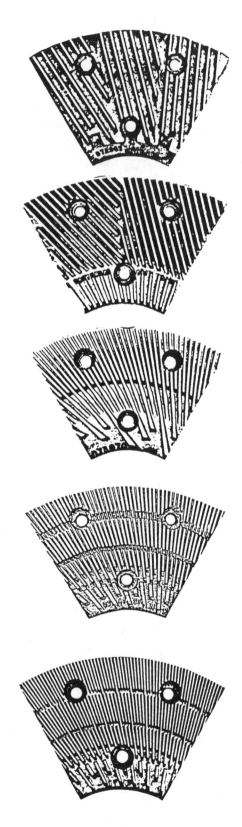

FIGURE 13-21. Common disc refiner plate patterns.

plastics has proven to be uneconomical. Perhaps the most widely used material is Ni-hard, an abrasion-resistant nickel chromium white iron. One advantage of Ni-hard is that the refiner bars retain relatively sharp leading edges as they wear, a factor that is compatible with long service life. However, Ni-hard is not suitable for corrosive stocks, and more expensive alloys may be required.

Plate wear occurs during refining as a result of normal abrasion, and is accelerated by the presence of foreign materials in the stock. Plate life is also directly dependent on the corrosiveness of the stock. A plate has reached the end of its life when either fiber quality or throughput falls below acceptable levels. Measures taken in the paper mill to provide cleaner stock and less corrosive refining conditions will ensure longer plate life.

Figure 13-22 illustrates the diverse refining requirements of a wide range of paper products, and graphically points out the advantage of the disc refiner with respect to energy requirements.

Effect of Refining on Paper Properties

Figure 13-23 shows the typical pattern of change for some common sheet strength properties as the stock is refined. The actual response to refining will depend on the type of pulp fibers, the equipment used, and the operating conditions. Tear strength always decreases with refining due to the strength attrition of individual fibers; other strength

parameters (e.g., burst, tensile, folding endurance) increase due to improved fiber-to-fiber bonding (as illustrated in Figure 13-24). The paper stock itself becomes slower (i.e., more difficult to drain) and the resultant paper sheets become denser (less bulky), with reduced porosity, lower opacity, and decreased dimensional stability.

The results of laboratory handsheet testing are generally presented as a function of beating time, beating energy, or freeness (drainability). However, for comparative purposes it is more advantageous to evaluate one important property as a function of another relevant property. This technique is most effective where the properties vary in divergent directions with degree of refining. The most common example of paired properties is tear as a function of either tensile or burst. Other relationships used are tear as a function of sheet density and porosity as a function of tensile.

Where it is possible to carry out refining or beating at two extreme levels of intensity, two sets of beating curves can be generated to characterize the papermaking potential of the pulp over the complete range from gentle to harsh beating (see Figure 13-25).

Control of the Refining Process

In most refining systems, there are generally two specific objectives:
• optimum strength development,
• control of stock drainability and sheet formation.

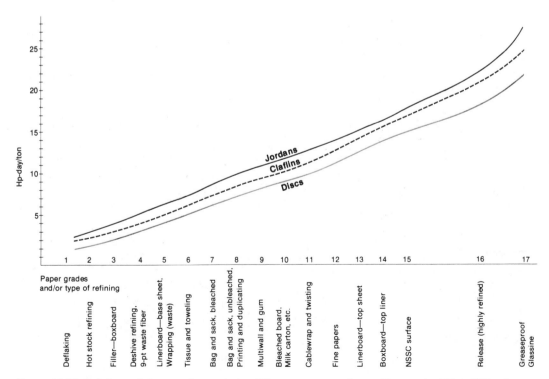

FIGURE 13-22. Relative refining requirements for different paper grades and types of refining (Gilbert).

For ease of control and minimum energy consumption, these two objectives are best met in a two-stage system.

In the primary refining stage, the virgin fiber is given a low-intensity treatment for optimum strength development. This stage is usually operated at a constant specific energy for the product being produced. Specific energy control relies on input signals from consistency and flow transmitters plus a further signal from a kilowatt meter. In order to minimize energy requirements, the refiner size and speed should not be greater than required to handle the anticipated throughput.

In the secondary stage, the entire pulp furnish (i.e., pre-refined virgin fiber, secondary fiber, broke, etc.) is refined immediately ahead of the paper machine. Here, optimum results are obtained with high-intensity refining and relatively low specific energy. This stage can be controlled to constant freeness by

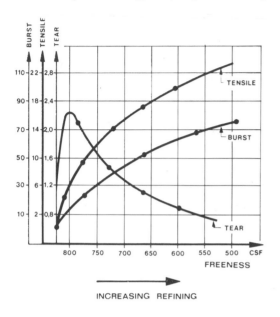

FIGURE 13-23. Typical strength development during refining.

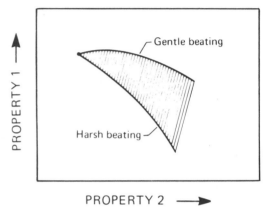

FIGURE 13-25. The papermaking potential of a pulp in respect to properties 1 and 2.

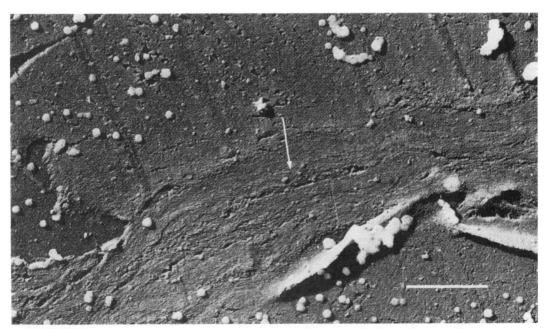

FIGURE 13-24. Scanning electron micrograph showing the bonded area (arrow) between two moderately refined pinus radiata kraft fibers. Note the wavy appearance of the bonded area and the gels of surface hemicellulose. (Dark color is due to staining. The marker corresponds to 1 micron.)

utilizing a sensor to provide feedback control of refiner loading, as for example, the system shown in Figure 13-26. The best results are usually obtained with an on-line instrument measuring either drainage time or wetness permeability, which is installed as close to the refiner as possible (4).

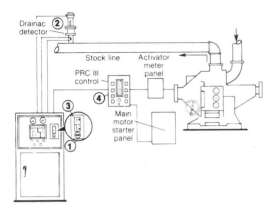

FIGURE 13-26. **Representative refiner control system utilizing a continuous drainage sensor. The set point (1) is dialed into the control unit. The detector (2) measures the actual drainage; a comparison with the set point generates an error signal to the controller (3). The controller then changes the set point on the refiner motor load controller.**

13.3 METERING AND BLENDING OF FURNISHES

Metering and blending of the various furnish components is probably the most under-rated function of the stock preparation system from the standpoint of how much harm can occur if not handled properly. It must be emphasized that variation in furnish composition will manifest itself as erratic behavior of stock on the forming wire, fluctuation of wet web response during subsequent processing steps on the paper machine, and variability in product quality.

Accurate proportioning of pulps and additives into a uniform blend depends on control of both consistency (or concentration) and flow rate for each component stream. When a pulp component is supplied from a high-density storage chest, a series of controlled dilution steps and mixing stages are necessary to achieve the desired level of control. Pulp components are usually metered into the machine stock chest at consistencies between 2.8 and 3.2%. Such modifying operations as deflaking and refining are usually carried out at intermediate consistency levels.

Although batch systems of metering and blending are still being used for small specialty papermaking operations, the trend is toward continuous, automated systems. These systems rely on modern consistency control loops and magnetic flow meters to regulate control valves. In some instances, the

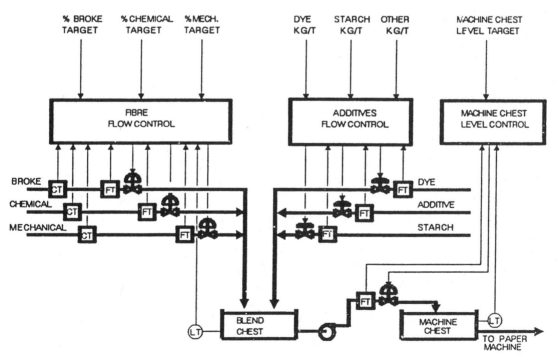

FIGURE 13-27. **Typical stock proportioning system.**

various pulp and broke stocks are blended first, followed by addition of mineral fillers and chemicals. The basic control principle is that component streams are ratioed to an overall flow demand.

Figure 13-27 shows a typical stock proportioning system of medium complexity in which the various pulp and broke stocks are delivered to the blend chest together with dye, starch and other additives. The ingredients are mixed by an agitator in the blend chest to assure furnish uniformity. The blended furnish is then delivered to the machine stock chest. While not shown in the figure, it must be noted that each stock flow is individually controlled for consistency, and a "fine tuning" of consistency is generally made on the combined flow from the blend chest to the machine chest.

For relatively simple furnishes such as for newsprint, the blend chest is often omitted, and the final mixing occurs in the machine chest.

The machine chest typically contains the final furnish mixture, although in some instances, small concentrations of additives may be added just prior to the headbox. The machine chest stock (at 2.6 - 2.8% consistency) is continuously circulated to a constant head tank (stuff box) which feeds the stock through a control valve (called the "basis weight valve") into the paper machine approach system. Here, the stock is combined with circulating white water from the wire pit, and the consistency is dramatically reduced to the level (typically 1% or less) required at the headbox.

REFERENCES

(1) STEVENS, W.V. **Refiner Analysis Offers Improved System Efficiency, Horsepower Usage** *Pulp & Paper* (March 1981)

(2) LUMIAINEN, J. **Refining Intensity at Low Consistency - Critical Factors** *Paper Technology* (November 1991)

(3) SHARPE, P.E. and RODARMEL, J.L. **Low Consistency Refiner Plate Design and Selection** *P&P Canada 89*:2:T57 (February 1988)

(4) SHAW, P. **Do You Drive Your Car, or Run Your Refiner Blindfolded?** *Paper Technology* (June 1991)

Chapter 14

Secondary Fiber

Secondary fiber is defined as any fibrous material that has already undergone a manufacturing process and is being recycled as the raw material for another manufactured product. Strictly speaking, broke from the dry end of the paper machine, finishing room trimmings, and repulped rolls could be considered as secondary fiber; but in practice, internal recycling is not included as part of secondary fiber utilization.

Secondary fiber utilization is currently increasing at a rapid pace in North America. The driving force is governmental legislation which seeks to reduce landfill loadings and lessen dependency on forest resources by mandating minimum secondary fiber content levels for certain paper grades, notably newsprint. Traditionally, the supply of waste paper has been market-driven; but recently the emphasis has been on source separation, where paper and other recyclables are often collected irrespective of market demand. While significant imbalances in supply and demand have occurred in the past, the present over-supply of certain wastepaper grades (e.g., old newspapers) is large by historic standards. It now appears that enough new recycling capacity is being brought on line in the early 1990's to return to a demand-driven marketplace. In the meantime, excess wastepaper is being sold at bargain prices in the overseas market.

Secondary fiber plants are usually located in areas of high population density, where a dependable supply of waste material can be more easily collected and transported. It is generally less economical to gather materials outside a fifty-mile radius of the mill site, due to high transportation costs for the bulky wastepaper bales. However, it can be worthwhile to bring in wastepaper from considerable distances when there are significant price differentials.

Wastepaper Grades

Five basic grades of wastepaper are defined by the U.S. Dept. of Commerce, and these definitions are generally accepted by the paper industry:

- **Mixed Paper.** Paper of varied quality; included in this category are office waste, boxboard cuttings and mill wrappers.
- **Old Newspapers (ONP).**

- **Old Corrugated Containers (OCC).**
- **Pulp Substitutes.** Mainly consisting of unprinted paper and board that has not been coated or adulterated in any way; included in this category are tabulating cards, white and semibleached sheets, cuttings, shavings or trim.
- **High-grade Deinked.**

The Paper Stock Institute further subdivides these grades into 80 definitive groups.

Pulp substitute grades command top prices because they can be utilized directly in the papermaking process for certain products. (These grades have always been fully recovered for recycle.) Other wastepapers must be cleaned up in the secondary fiber pulping system to remove contaminants. Separation and/or dispersion of such contraries as plastic laminates, adhesives, glues, waxes, etc. is sufficient for secondary pulps that are used for the inner plies of a multi-ply paperboard or for corrugating medium. But for printing grades, more selective removal of contaminants, including deinking, is necessary to prepare a suitable papermaking stock.

The relative usages by wastepaper grade are shown in Table 14-1 for 1977 and 1988. OCC accounts for about 50% of the total. Approximately 75% of all secondary fiber in North America is presently used for multiply paperboard and corrugating medium production, but an increasing proportion will be deinked and utilized for newsprint and other printing grades over the next decade.

14.1 WASTEPAPER PROCUREMENT

In order for wastepaper to be efficiently utilized as secondary fiber, it is necessary that the collected material be sorted and classified into appropriate quality grades. Sorting may occur at the source or be carried out by the collecting agency. Where many sources are involved, a dealer or broker usually assumes the tasks of collecting, sorting, baling, and supplying to the mill. Normally, the utilizing mill does not sort the incoming wastepaper, but monitors the material to ensure that minimum quality standards are being met.

Wastepaper sources are generally categorized as pre-consumer or post-consumer. Pre-consumer

TABLE 14-1. Relative Wastepaper Usage (API Data).

	1977	1988
Mixed Paper	19%	11%
ONP	16%	16%
OCC	43%	49%
Pulp Substitutes	15%	15%
High-grade Deinked	7%	9%

sources are converting plants and printers where wastepaper, in the form of cull rolls, clippings, off-quality product, or over-issue, is generally clean and well sorted. Typical post-consumer sources are homes, offices, and retail outlets from which the waste must be collected, sorted and baled. Wastepaper from post-consumer sources is considered less desirable because it is relatively less sorted and higher in contamination.

Wastepaper prices around the world are mainly determined by demand, and also to some extent by the ease with which supplies can be obtained (1). The relationship is hard to quantify and varies with the grade of wastepaper. When shortages arise, mills are usually prepared to pay higher prices to generate additional supplies. Wastepaper prices are highly cyclical, and their main characteristic is instability. Except for the highest quality grades, wastepaper prices have been depressed in recent years because of excess supply.

Wastepaper recovery methods have changed markedly in the United States over the past few years, especially for ONP and OCC. About 60% of the ONP now comes from curbside collection and other city-operated recycling programs. Municipal governments, not wastepaper dealers, are controlling the flow of ONP into the mills. In many instances, the wastepaper dealer is being bypassed as the municipality sorts, bales, and sells the ONP directly to the mills.

A similar pattern can be seen for OCC. Approximately 60% of this recycled material now goes directly from chain stores to the mills, many of whom have signed long-term contracts to take all the available OCC. Municipal curbside collection may prove to be another important source of OCC.

Raw Material Storage and Handling

Bales of wastepaper are typically delivered to the mill by truck or rail, and are inspected during off-loading to insure that they conform to mill requirements. It is good warehousing practice to segregate both by grade and by specific supplier, so that any quality problems can be traced back to the secondary fiber source. Outside storage should be avoided, if possible, because exposure to sunlight and weathering causes deterioration in pulp properties. Also, relatively "fresh" secondary fiber is more easily deinked than older stock. Generally, it is desirable to operate a storage facility on a "first-in, first-out" basis.

14.2 DEGREE OF RECYCLING

Two primary indices are used to compare the level of recycling in various countries (2). Recovery rate is the amount of wastepaper recovered for reuse compared with paper consumed. Utilization rate is the amount of secondary fiber used in paper/board production compared with the total fiber used. The recovery rate in the United States has been setting new records in recent years and reached 33% in 1990. By contrast, such countries as Japan and Taiwan without a forest resource base have long had recovery rates approaching 50%. The average recovery rate in Europe is about 30%, and in Canada about 25%.

Secondary fiber utilization in the United States is about 25% and is climbing slowly. In Japan, it is about 50%, and the average utilization rate in Europe is about 40%. In the leading paper exporting countries, the utilization rate is understandably less; for example, about 10% in Sweden and Canada.

It is generally considered that 50% represents a practical maximum for an overall utilization rate . Significant losses of both fiber substance and strength occur during each recycling. (Some investigators have suggested that a fiber can be recycled only four times before the loss in quality becomes too great for effective reuse.) At a sustained 50% level, it is apparent that half of the material being recycled at any time has already been through one or more previous cycles.

The effects of multiple recycling operations on fiber characteristics and sheet properties have been investigated to some extent in the laboratory. Most of these studies were undertaken in the decade from the late 1960's to the late 1970's (3). However, it is not clear how well a laboratory can simulate the fiber attrition that takes place from actual paper usage and recycle. Figure 14-1 summarizes results from one study on a never-dried unbleached softwood kraft pulp (3) that had been refined down to 285 CSF. Over the course of four recycles, the fines content increased from 16% to 35% and caused a severe reduction in stock drainage. Presumably, for the test handsheets, fines were removed after each recycle to maintain the same freeness level. As expected, the fibers exhibit progressively lower strength and bonding potential; the initial increase in tear is due to the effect of drying on fiber stiffness.

Figure 14-2 illustrates the relative effects of repeated recycling of newsprint on individual fiber

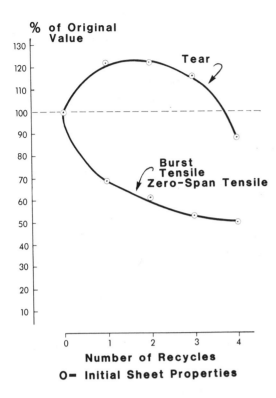

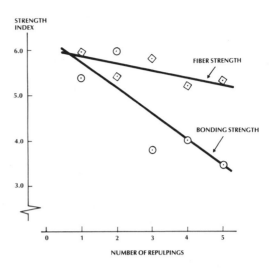

FIGURE 14-2. Effect of multiple recycling on fiber and bonding strengths of newsprint (P. Howarth).

FIGURE 14-1. Effect of multiple recycling on strength properties of unbleached kraft pulp handsheets.

strength and on the bonding strength between fibers. In this instance, both strength indices decrease, but the bonding between fibers shows the more dramatic loss. With each drying and slushing cycle, the fibers become less flexible and less permeable to water, and therefore do not conform as well as virgin fibers. Cumulative loss of hemicelluloses from the fiber surfaces also contributes toward reduced bonding.

It should be emphasized that various fibers may react in different ways to recycling. Investigative findings for one specific type of fiber cannot necessarily be applied to another type of pulp.

14.3 REPULPING OF WASTEPAPER

Wastepaper bales are conveyed to a pulper where the secondary fiber is dispersed into a wet pulp slurry. Depending on the papermaking requirements, the pulper may also cause ink and coating particles to separate from the fibers. Such variables as temperature, consistency, retention time, and chemistry must be controlled to optimize the operation.

At the heart of the repulping operation is the pulper itself. The most common design, illustrated schematically in Figure 14-3, is the low-consistency pulper with ragger and junker, which is usually run

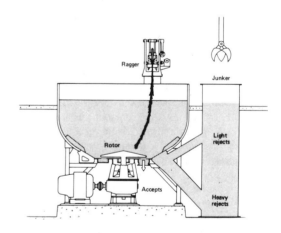

FIGURE 14-3. Schematic of continuous pulper with ragger and junker (Lodding, Div. of Thermo Electron).

at 5-8% consistency, and may be operated in either continuous or batch mode. Strings, wires and rags are continuously withdrawn from the stock as a "debris rope". Initially, a few "primer wires" are rotated in the stock to start the rope, after which the rope builds on itself. Heavy objects are thrown into a recess at the side of the pulper by centrifugal force; this material is usually removed from the "junking tower" by a bucket elevator or grapple.

The high-consistency pulper (as introduced and illustrated in Section 13.1) has also been applied to secondary pulping systems. It has the advantage of maintaining contaminants, such as plastics and wet

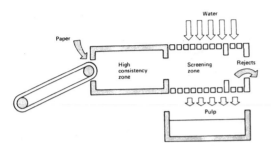

FIGURE 14-4. Schematic of drum-type pulper (Ahlstrom).

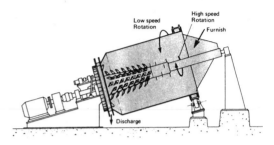

FIGURE 14-5. Betonniere high-consistency pulper (Lamort-Aikawa-Lodding).

strength flakes, in a larger form to facilitate their later removal; it also provides better ink removal and has a relatively gentle action on the fibers. This type of pulper must be operated batch-wise, but batch times are generally shorter than for low-consistency pulpers.

The drum pulper, illustrated schematically in Figure 14-4, consists of a slowly rotating sectionalized horizontal drum that is inclined toward the discharge end. Bales of wastepaper (which have previously been dewired and broken apart) and water enter the feed end where the drum rotation produces a tumbling action which causes the bales to break down into individual fibers. The pulp at 15-20% consistency then moves from the pulping zone into the screening zone where showers wash the fibers through perforations while contaminants are retained within the drum and are discharged at the end. Although its capital cost is relatively high, the drum pulper requires less energy than other designs and provides a gentle pulping action.

A special design of pulper, utilizing both a high-speed disintegrating rotor and low-speed drum rotation, is illustrated in Figure 14-5. The Betonniere, which loosely translates into "cement mixer", is suitable for operation in the 35-45% consistency range. It is said to be effective for the most difficult furnishes, including wet-strength papers, but can be used for any application where a high consistency is required following pulping.

Many pulping systems also utilize a secondary in-line repulping device, one type of which is illustrated in Figure 14-6. In addition to more defibering, this equipment has the ability to further separate both heavy and light-floating trash. In-line units generally consist of a conical housing containing the rotating defibering element. The accepted stock passes through a perforated stainless steel plate that is positioned immediately behind the rotor and is collected in an annular channel. Depending on their design or function, secondary pulping devices may be designated as dispergers, fiberizers, or deflakers.

Other than equipment design and applied energy,

the major variables affecting degree of defibering are stock temperature and consistency. In general, a higher temperature will facilitate defibering, while power requirements are reduced because of increased fluidity. Consistency should be maintained at the highest level that still gives good circulation.

A basic pulping system for relatively clean wastepaper stock is shown in Figure 14-7. The stock passes through the small-hole extraction plate of the pulper, then to a liquid cyclone for removal of small high-density contaminants, and finally through a fine screen. The screen rejects go through a deflaker or fiberizer to provide more complete fiber separation, and this stock is reclaimed through a secondary screening system back to the pulper.

Another pulping system which utilizes a side mounted low-consistency pulper and a secondary pulping device is well illustrated in Figure 14-8. This system features three stages of screening, and is said to be effective for wastepapers containing a high percentage of dirt.

14.4 CONTAMINANT REMOVAL

A prime objective in processing wastepaper stocks is to remove enough contaminants and/or upgrade

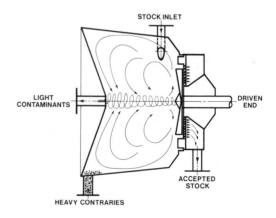

FIGURE 14-6. Secondary pulper for wastepaper processing (Escher-Wyss, Impco).

the material so that the secondary fiber is suitable to make the finished product within specifications. The major process steps that may be used are screening, cleaning, washing, bleaching, dispersion and deinking.

Even when stringent standards are applied to incoming wastepapers, contaminants cannot be totally avoided and must be dealt with in the process system. All extraneous constituents are considered contaminants, including dirt, rocks, sand, and tramp metal. Some of the more difficult product-related contaminants and their sources are listed in Table 14-2. In practice, such contaminants as glues, hot melts and latexes are lumped together into the category of "stickies" or "tackies".

As already noted, most of the primary pulping equipment and secondary in-line pulpers and/or deflakers have the ability to remove much of the gross contamination. Some are augmented with a so-called flotation purge system to help remove the lighter weight contaminants. Beyond this, all systems have suitable screens for removing oversize particles and centrifugal separators (forward and reverse cleaners) for removing all types of fine particle contamination. Each of the major suppliers of secondary pulping equipment specifies or recommends an assembly of screens and cleaners which is compatible with their own pulping equipment and with the requirements of a particular wastepaper furnish.

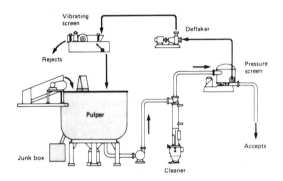

FIGURE 14-7. Simplified conventional pulping system for recycled fiber (Lodding, Div. Thermo Electron).

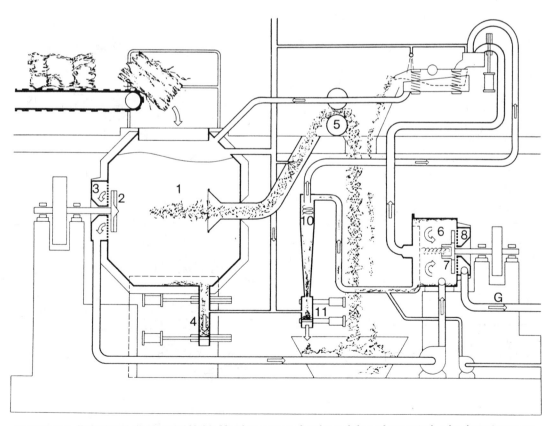

FIGURE 14-8. Functional diagram of Voith-Morden system for the pulping of unsorted, mixed wastepapers. (1) pulper, (2) impeller, (3) & (8) screen plate, (4) & (11) dirt trap, (5) ragger, (6) secondary pulper, (7) pulping assembly, (9) lightweight contraries screen, (10) forward cleaner.

TABLE 14-2. Common contaminants in wastepaper repulping systems.

Type of Contaminant	Typical Sources	In-Mill Problems
Hot Melts	Adhesives and coatings	Cannot be handled in conventional systems; fouls equipment; causes defects in products.
Polystyrene Foam	Blocks and beads used in packaging	Difficult to remove; sticks to rolls; causes sheet indentations and "pickouts".
Dense Plastic Chips (e.g., polystyrene)	Blister packs and see-through packages	Breaks into small pieces; hard to remove; causes "shiner" in product.
Wet Strength Resins	Laminated paper products, tramp material	Slows down pulper process; causes product defects.
Latex	Adhesives and coatings; flying pasters; rubber bands	Difficult to remove; causes product defects.
Pressure Sensitives	Roll splices; case seals; miscellaneous	Sticks to fabrics and felts; causes sheet defects and web breaks.
Waxes	Coatings and laminates	Difficult to disperse; fouls equipment and degrades products.
Asphalt	Laminated products	Agglomerates in pulper; sticks to fabrics; causes black spots in products.
Foreign Fibers	Vegetable and synthetic rope fibers	Causes product defects and web breaks.

A dispersion system is sometimes incorporated into a secondary pulping system to control certain stickies. The objective is to disperse the contaminants thoroughly over the surfaces of the fibers and thereby nullify any adverse effects. Both hot and cold systems are available; the hot systems generally yield a cleaner appearance, but the cold systems are adequate for stocks used as filler plies in multiply products. Figure 14-9 shows a modern dispersion system suitable for either atmospheric or pressurized operation. In this sequence, the incoming dilute stock is thickened to 35% consistency by a two-stage screw press arrangement and fed through a plug screw (to prevent steam blow-back during pressurized operation) to a vertical shredder which fluffs up the stock. The fluffed pulp is then mixed with steam in a preheater before being fed to the disperser, a device similar in operation to a disc refiner.

Relatively elaborate systems are used to remove the more troublesome contaminants. These treatments include bleaching, deinking (to be discussed in the next section), hot melt extraction, and solvent extraction. However, some contaminants (e.g., vinyl acetates, polypropylene) have defied all efforts to effectively handle them in the wastepaper system, and they are best excluded from the furnish.

14.5 DEINKING

Deinking of pulp fibers is essentially a laundering or cleaning process where the ink is considered to be the dirt. Chemicals, along with heat and mechanical

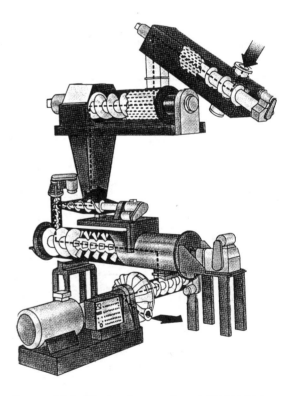

FIGURE 14-9. Dispersion system suitable for atmospheric or pressurized operation. (Krima, Cellwood Machinery).

energy, are used during repulping to dislodge the ink particles from the fibers and disperse them in the stock suspension. The ink particles are then separated from the so-called "grey stock" by a series of washing or flotation steps, or by applying a hybrid process that utilizes both separation techniques.

The key chemicals used for stock deinking are surface active agents ("surfactants"), which affect the surface tension of liquids and solids. Typically, these agents are chemically modified mineral oils, where hydrophillic groups have been added to the molecular structures to make them partly soluble. Three specific types of surfactants are important in deinking applications:
• detergents—to remove ink from the fiber
• dispersants—to keep the ink particles dispersed and prevent re-deposition onto the fibers
• foaming agents—to reduce the surface tension of water and to promote foam formation
Other chemicals such as caustic soda, sodium silicate, and borax are also used to enhance the action of the surfactants.

Washing Process

In the washing process, detergents and dispersants are utilized in the pulper to remove the ink constituents from the fibers, break them down, and disperse them into very fine particles. The ink dispersion is subsequently separated from the pulp, typically by a multistage dilution/thickening washing sequence. A representative washing process is illustrated in Figure 14-10.

The separation of ink in the washing process corresponds to a stock thickening process, whether accomplished by washing equipment or by screens. If the ink particles are extremely small (less than 15 microns), the amount removed is theoretically proportional to the amount of water removed (Figure 14-11 and Table 14-3). In practice, the fiber network acts as a filter to reduce the actual removal efficiency.

The filtering effect during washing is minimized by operating at low consistency and by utilizing thickening equipment that does not involve mat formation, such as the sidehill thickener (Figure 14-12), slusher, or Celleco screen. An incoming consistency of 1.0 to 1.5% would be ideal because the filtering effect is low; however, older plants utilizing this consistency range had large space requirements. As a consequence, an incoming consistency of about 3% in conjunction with a countercurrent washing sequence is more common, but thorough reduction in particle size is more critical. Pulp is diluted with clarified water only ahead of the last washing stage.

Recently, washing machines, such as the double-nip thickener (Figure 14-13), have been developed that can thicken large volumes of low-consistency stock up to 8-12% consistency in a single stage. Equipment of this type operates in the manner of a low-retention papermachine wet end. The stock at low consistency is introduced via a headbox onto an endless fabric belt where water is removed by the action of the fabric as it moves rapidly around a turning roll.

Flotation Process

In the flotation process, chemicals are introduced during the repulping operation to promote flocculation of the ink particles and the formation of foam. The grey stock is subsequently aerated at low consistency (typically 0.8 to 1.2%) in a series of flotation cells, causing the light flocs of ink particles to rise to the surface where they are skimmed off. A representative flotation process is illustrated in Figure 14-14.

At the heart of the flotation process is the flotation cell, of which several designs are available (Figures 14-15 to 14-17). Here, air in the form of small bubbles is blended with the grey stock. The air bubbles become attached to ink and dirt particles, causing them to rise to the surface of the cell where they are removed as a dirt-laden layer of froth. Typically, 6 to 10 flotation cells in series are required for efficient ink removal depending on the level of dirt in the stock. The froth is subsequently cleaned in a secondary stage (usually two cells) to recover good fiber.

For reasons that are not clear, flotation deinking is far more effective when the wastepaper has a

TABLE 14-3. Theoretical ink removal.

Washer	Consistency, % Inlet	Consistency, % Discharge	Theoretical Ink Removal, % One-stage	Theoretical Ink Removal, % Three-stage	Dilution Water Required, gal/ton
Sidehill screen	0.8	3.0	74.0	98.2	26,740
Gravity decker	0.9	6.0	85.5	99.7	23,740
Inclined screw	3.0	10.0	72.2	97.8	6,950
Horizontal press	4.0	28.0	89.2	99.9	5,150

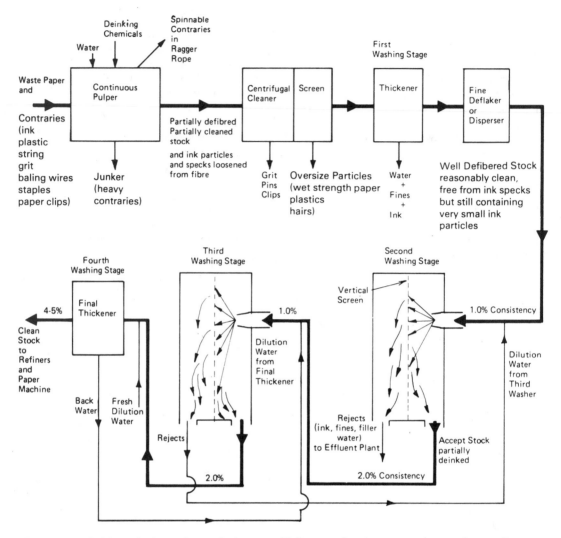

FIGURE 14-10. Deink wash plant using vertical screens (Celleco-type) and countercurrent wash water flow sequence.

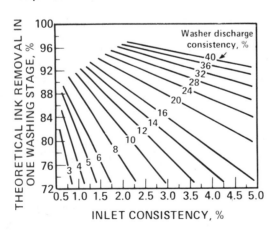

FIGURE 14-11. Theoretical ink removal in one washing stage as a function of washer inlet and discharge consistencies.

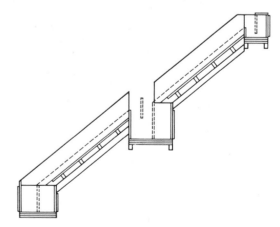

FIGURE 14-12. Two-stage sidehill screen/thickener.

significant ash content. A clay content of 8 to 10% is considered a minimum requirement, and 12 to 14% is preferable (5). The requisite level can usually be maintained by blending in coated waste with the wastepaper. Approximately 25 to 30% of the clay is removed with the flotation cell rejects.

The flotation method is generally favored in Europe and Japan. The washing method has traditionally been more popular in North America because of lower capital cost and lower space requirements. The flotation method is more selective in removing ink particles and therefore provides a significantly higher yield than the washing system. On the other

hand, the washing process removes a high proportion of fiber fines and fillers along with the ink particles which has the effect of upgrading the pulp. Flotation-deinked stock requires little washing because the

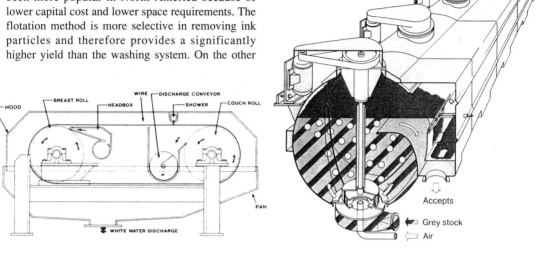

FIGURE 14-13. Double-Nip Thickener (Black-Clawson).

FIGURE 14-15. Voith flotation cell.

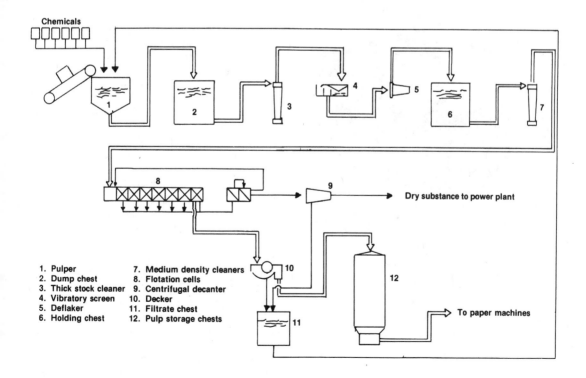

1. Pulper
2. Dump chest
3. Thick stock cleaner
4. Vibratory screen
5. Deflaker
6. Holding chest
7. Medium density cleaners
8. Flotation cells
9. Centrifugal decanter
10. Decker
11. Filtrate chest
12. Pulp storage chests

FIGURE 14-14. Representative flotation deinking process.

chemicals are concentrated in the withdrawn froth. Effluent treatment for the flotation process is usually simpler than for the washing process.

The size of the ink particles to be removed should be a primary basis for selecting the appropriate removal process. Inks that are amenable to being

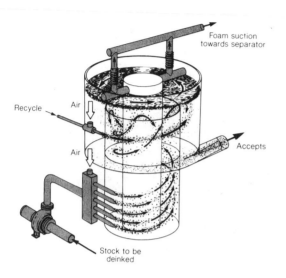

FIGURE 14-16. Vertical flotation cell (Lodding, Div. of Thermo Electron).

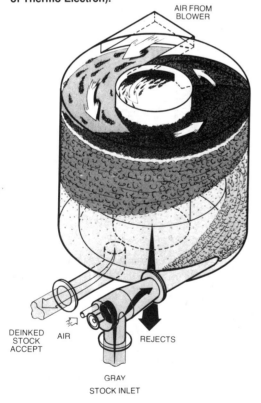

FIGURE 14-17. Swemac Hellberg flotation cell.

broken down during the repulping operation are more easily removed by the washing process than those that withstand treatment.

Combined Washing and Flotation

Since the washing and flotation methods have respective advantages, the notion of utilizing a hybrid process is attractive. However, the principles involved and the chemicals utilized have been quite different, and this lack of compatibility has been a stumbling block. The objective in washing is to break the ink down into particles under 15 microns, render them hydrophillic, and keep them finely dispersed. For effective flotation removal, the ink particles must form hydrophobic flocs, ideally in the size range from 30 to 60 microns.

The problem of process incompatibility has been overcome by the development of a new class of surfactant called a "displector" (coined from dispersant and collector). These chemicals provide enough hydrophillicity for the ink particles to remain dispersed during washing operations, while retaining enough adhesion between ink particles and air bubbles for flotation to be effective (6).

For maximum operating flexibility and improved secondary fiber quality, a modern combined or hybrid system utilizing both washing and flotation technology is now the preferred choice. A representative combined system is depicted schematically in figure 14-18. In this two-stage system, washing serves to remove fines and fillers along with the smaller ink particles. Washing also appears to enhance the subsequent flotation stage by removing some contaminant elements from the furnish which inhibit attachment of ink particles to bubbles. The subsequent flotation step is then effective in handling the more difficult-to-disperse inks while also removing other lightweight contaminants. It must be noted that, depending on system objectives, some hybrid process designers prefer to carry out flotation first followed by washing.

14.6 SECONDARY FIBER UTILIZATION

Secondary fibers have distinctly different characteristics when compared to virgin pulps. Therefore, papermakers should consider each secondary pulp as a distinct furnish component with separate targets for cleanliness, freeness, degree of refining, etc.

Regardless of how they have been processed, secondary pulps always contain some residual contamination that will have an impact on papermachine operation. Waxes, glues, inks, and a variety of dissolved solids tend to build up in the system and precipitate or agglomerate at various points along the papermachine. For example, stickies and hot melts are prone to precipitate on fabrics and

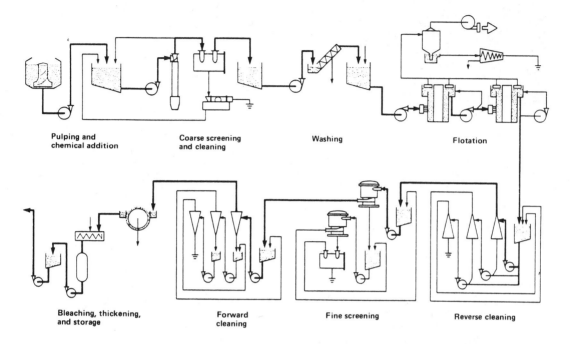

Pulping and chemical addition

Coarse screening and cleaning

Washing

Flotation

Bleaching, thickening, and storage

Forward cleaning

Fine screening

Reverse cleaning

FIGURE 14-18. Combined washing and flotation deinking system (Lodding, Div. of Thermo Electron)

dryer surfaces. Ink and clay may agglomerate into larger particles in the white water system and ultimately end up as visible dirt in the paper product. Mills that convert to a partial secondary pulp furnish find that a more stringent program of fabric cleaning and/or system boilout is required.

Generally, pressrooms have found that recycled newsprint and other printing papers that contain secondary fiber are not significantly different with respect to runnability. Differences in printability and appearance are more apparent. Recycled sheets tend to be more absorbent because of higher sheet porosity; therefore, more ink is required in printing which causes poorer printing resolution and higher rub-off. However, the biggest concern for white grades is appearance. Ink spots and other dirt specks often cannot be tolerated, and brightness may be significantly lower. Secondary fiber usually cannot be used as a furnish component in the most critical grades.

REFERENCES

(1) GARCIA, D.A. **Recycling Capacity to Increase at Record Rates as Laws Proliferate** *Pulp & Paper* (May 1990)

(2) BROEREN, L.A. **New Technology, Economic Benefits Give Boost to Secondary Fiber Use** *Pulp & Paper* (November 1989)

(3) HOWARD, R.C. **The Effects of Recycling on Paper Quality** *Journal of P&P Science* 16:5:J143-149 (September 1990)

(4) HORN, R.A. **...the Effects of Recycling on Fiber and Paper Properties** *Paper Trade Journal* (February 17/24, 1975)

(5) SCHRIVER, K.E. **Mill Chemistry Must Be Considered Before Making Deink Line Decision** *Pulp & Paper* (March 1990)

(6) SHRINATH, A. et al **A Review of Ink-Removal Techniques in Current Deinking Technology** *Tappi Journal* (July 1991)

Chapter 15

Non-fibrous Additives to Papermaking Stock

A wide range of chemicals is utilized in the papermaking stock furnish to impart or enhance specific sheet properties or to serve other necessary purposes. A general classification of wet-end chemical and mineral additives is given in Table 15-1. Such additives as alum, sizing agents, mineral fillers, starches and dyes are commonly used. Chemicals for control purposes such as drainage aids, defoamers, retention aids, pitch dispersants, slimicides, and corrosion inhibitors are added as required. The order of addition must be taken into account to prevent interaction at the wrong time and enhance retention in the paper sheet.

Not all papermaking chemicals are added to the wet stock. Sizing solution is often applied to the dried sheet at a later stage in the process (e.g., at the size press); and pigment coatings are used for the better quality publication grades. Increased papermill chemical and mineral consumption is anticipated mainly for coatings. The highest tonnage additive is clay, over half of which is used as part of surface coating formulations.

It is of interest to put an economic perspective on the chemical and mineral contribution to papermaking. Perhaps, on average, 10% of the cost of making paper can be attributed to chemicals. Figuring the value of annual North American paper and board shipments at $80 billion, the industry probably uses some $8 billion worth of additives per year.

15.1 RETENTION ON THE PAPER MACHINE

Two parameters are used to measure the retention of fibers and additives during paper forming:

1) Overall Retention (%) $= \dfrac{\text{amount retained in sheet}}{\text{amount added with stock}}$

2) Single-Pass Retention (%) $= \dfrac{\text{amount retained in sheet}}{\text{amount from headbox}}$

These formulas apply to the overall furnish or to any single component of the furnish. The cost of utilization for various additives is mainly related to their overall retention, because the portion not retained with the sheet is lost with the white water

TABLE 15-1. Classification of wet-end chemical and mineral additives.

Additive	Application
Acids and bases	Control pH
Alum	Control pH; fix additives onto fibers; improve retention.
Sizing agents (e.g., rosin)	Control penetration of liquids.
Dry-strength adhesives (e.g., starches, gums)	Improve burst and tensile; add stiffness and pick resistance.
Wet-strength resins	Add wet strength to such grades as towelling and wrapping.
Fillers (e.g., clay, talc, TiO_2)	Improve optical and surface properties.
Coloring materials (dyes and pigments)	Impart desired color.
Retention aids	Improve retention of fines and fillers.
Fiber flocculants	Improve sheet formation.
Defoamers	Improve drainage and sheet formation.
Drainage aids	Increase water removal on wire.
Optical brighteners	Improve apparent brightness.
Pitch control chemicals	Prevent deposit/accumulation of pitch.
Slimicides	Control slime growths and other microorganisms.
Specialty chemicals	Corrosion inhibitors, flame-proofing and antitarnish chemicals, etc.

overflow from the system. Even though modern papermaking systems have a high degree of "closure" to reduce the volume of effluent, losses of certain constituents can still be substantial.

Paper quality and paper machine operation are more affected by single-pass retention (1). A low level of single-pass retention indicates a high recycle rate of furnish materials with the recirculating white water; it gives rise to non-uniform distribution in the cross-section of the sheet and may contribute to two-sidedness (i.e., different surface properties on the two sides) in fourdrinier-made paper. The accumulation of fines and additives in the headbox loop retards drainage, and the fines fraction absorbs a disproportionate amount of certain additives by virtue of its high specific surface. Also, pitch and slime have a greater propensity for buildups and agglomerations, and are generally more difficult to control.

The major factors that affect the retention of such additives as rosin sizes, starches, resins and fillers during the sheet forming process are listed in Table 15-2. Retention of nonfibrous additives occurs through the mechanisms of filtration, chemical bonding, colloidal phenomena, and adsorption. Filtration (i.e., mechanical interception) is important for retaining larger particles, but smaller particles must be retained by other means (see Figure 15-1). For example, it is estimated that only about 2% of titanium dioxide pigment particles (average size 0.2 microns) are retained by the mechanism of filtration.

A number of retention aid chemicals are available to the papermaker (see also next section). Since

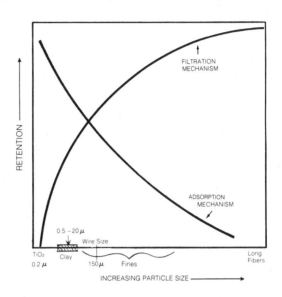

FIGURE 15-1. Effect of particle size on retention mechanism.

these chemicals act mainly through flocculation and entanglement, some care should be exercised in their utilization. Some effects on stock drainage and sheet formation should be anticipated. Primarily, the dispersive action of the headbox system must be adequate to avoid overflocculated stock that would be detrimental to sheet formation.

TABLE 15-2. Factors affecting retention during the sheet forming process.

Stock Factors	- pH
	- consistency
	- temperature
	- fiber characteristics
	- degree of system closure
Conditions on Wire	- sheet grammage
	- sheet formation
	- fabric characteristics
	- type of dewatering elements
	- machine speed
	- shake (if used)
Additives	- types and amounts of fillers
	- shape and density of mineral particles
	- types and amounts of other additives
	- order of addition
	- ionic balance
	- level of anionic trash

15.2 WET END CHEMISTRY

Wet end chemistry deals with all the interactions between furnish materials and the chemical/physical processes occurring at the wet end of the paper machine. While the subject is complex, it is possible to gain a basic understanding of the major concepts without delving too deeply into the technical aspects.

The major interactions at the molecular and colloidal level are surface charge, flocculation, coagulation, hydrolysis, time-dependent chemical reactions and microbiological activity. These interactions are fundamental to the papermaking process. For example, to achieve effective retention, drainage, sheet formation, and sheet properties, it is necessary that the filler particles, fiber fines, size and starch be flocculated and/or adsorbed onto the large fibers with minimal flocculation between the large fibers themselves. There is a wide range of phenomena (as listed in Table 15-3) which can influence these fundamental interactions (2).

TABLE 15-3. Factors affecting molecular and colloidal interactions.

Chemical concentrations
Electrokinetics
Polymer molecular weight, charge density, structure, conformation
Hydrodynamic shear
Residence and mixing times
Electrolyte concentrations and valences
Floc strength and reversibility
Specific surface area
Surface charge density
Particle sizes and morphology
Entanglement and filtration
Thermodynamics and kinetics of adsorption/bulk solution reactions

There are three major groups involved in wet-end chemistry: solids, colloids and solubles. Most attention is focused on the solids and their retention. In order to maximize retention, it is important to cause the fines and fillers to approach each other and form bonds or aggregates which are stable to the shear forces encountered in the paper machine headbox and approach system. In modern papermaking, this is usually accomplished by using synthetic polymers.

Certain colloidal materials derived from cellulose and hemicelluloses are released from pulps or wastepapers or added deliberately. The pulping process breaks down the cellulosic structures into smaller molecules which are potentially soluble. These smaller molecules have a negative impact on process control and runnability, and their natural retention is effectively zero. They concentrate within the system, consume chemicals and are generally a nuisance.

Control of wet-end chemistry is vital to ensure that a uniform paper product is manufactured. If the system is allowed to get out of balance (say, by over-use of cationic polymers), the fibers themselves will become flocculated and sheet formation will suffer. Also, functional additives (e.g., sizes, wet-strength agents) are often added at the wet end; if the chemistry is not under control, the functionality may not be adequately imparted and the product will be off-quality.

Unfortunately, wet-end chemistry is made more complex because of soluble materials. Whether they are organic or inorganic, added deliberately or inadvertently, they change the action of the polymers. In particular, wastepaper furnishes contain a relatively high concentration of solubles and a greater variety of chemical species which have an adverse impact on the controllability of the wet end (3).

Electrokinetics

The term, zeta potential, applies to the electrical charges existing in fine dispersions. A solid particle (e.g., fiber, starch, mineral) suspended in a papermaking stock is surrounded by a dense layer of ions having a specific electrical charge. This layer is surrounded by another layer, more diffuse than the first, that has an electrical charge of its own. The bulk of the suspended liquid also has its own electrical charge (see Figure 15-2). The difference in electrical charge between the dense layer of ions surrounding the particle and the bulk of the suspended liquid is the zeta potential, usually measured in millivolts.

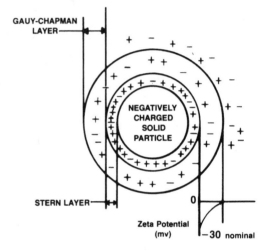

FIGURE 15-2. Pictorial representation of zeta potential.

The best retention of fine particles and colloids in the papermaking system normally occurs when the zeta potential is near zero. Pulp fibers, filler and size particles usually carry a negative charge, but the zeta potential can be controlled by absorbing positive ions from solution. Polyvalent cations such as aluminum (Al^{+++}) and ferric (Fe^{+++}) are most effective.

Papermakers alum, $Al_2(SO_4)_3$, is still a commonly used agent for wet end chemistry because it effectively neutralizes the negatively-charged fiber and pigment particles to zero zeta potential. At the proper pH, it also hydrolyzes to form an ionic polymer:

$$Al_2(SO_4)_3 + 6H_2O \longrightarrow 3H_2SO_4 + 2Al(OH)_3$$

This aluminum polymer has a significant flocculating effect by bridging from particle to particle and thereby forming large ionically-attracted flocs. However, the retention effect is sensitive to shearing forces or strong agitation. With higher paper machine speed, alum has become less effective.

Fortunately, synthetic polymers have been developed with good shear resistance. These polyelectrolytes are available either as cationic or anionic retention aids. The retention mechanism is a combination of ionic charges and long molecular chains linking fibers and particles together. Synthetic polymers have less pH dependence than alum and are used in very dilute form. Instruments are now available to measure both zeta potential and single-pass retention. Therefore, the economics of utilizing polyelectrolytic polymers to optimize zeta potential and retention can be monitored continuously and evaluated under commercial conditions.

Some researchers have found that simple adjustment of the papermaking system close to zero zeta potential will lead to optimum results (e.g., Figure 15-3). Others have found the optimum zeta potential to be approximately -9 mv (4). However, in commercial practice these findings have not always been confirmed. It appears that zeta potential is an indirect measure of a number of interacting factors, each of which could be dominant under certain conditions. As such, it cannot be relied upon to provide unequivocal information for operation of the papermaking system. Perhaps the best role for an on-line zeta potential measurement is to characterize a well-operated system, and then flag upsets by showing deviations from the norm.

15.3 APPLICATIONS OF NON-FIBROUS ADDITIVES

Sizing

The purpose of sizing is to enable paper products to resist penetration by fluids. Sizing can be achieved either by using wet-end additives or by applying a suitable coating to the surface of the dried paper. Sometimes a combination of treatments is required. The action of a wet-end sizing agent also imparts other desirable properties to the paper; however, the sheet remains porous. For products that require a vapor barrier, a surface coating must be used (see also Section 18.1).

A fundamental factor influencing the rate of liquid penetration is the contact angle formed between the impinging liquid and the sheet surface, as illustrated in Figure 15-4. The action of a wet-end sizing agent is to provide the fiber surfaces with a hydrophobic, "low-energy" coating that discourages aqueous liquids from moving extensively.

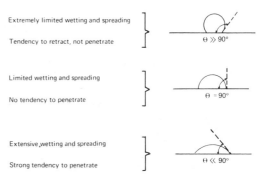

FIGURE 15-4. Illustrating the dependence of liquid movement on contact angle.

The traditional wet-end sizing agent is a modified rosin, most often in a saponified form to make it water soluble. Rosin size is usually shipped to the paper mill as a high-solids thick paste and is diluted through an "emulsifier" for metering into the stock. Natural rosin is the amber-colored resin obtained from southern pines. Formerly, it was tapped from growing trees or extracted from stumps. Now, more commonly, it is processed from tall oil (see also Section 10.6). Rosin is an amphipathic material, meaning that it has both hydrophillic and hydrophobic parts. To provide good sizing, it is essential that the hydrophobic parts are oriented outward, as shown in Figure 15-5. In practice, the rosin is precipitated onto the fibers by the action of alum as an oriented monolayer of aluminum resinate molecules.

Rosin, along with wax emulsions, are sometimes categorized as bulk or nonreactive sizes. Their retention is dependent primarily on precipitate

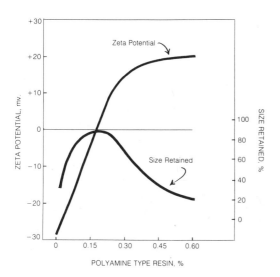

FIGURE 15-3. Effect of cationic retention aid on zeta potential and retention of synthetic size (E. Strazdins).

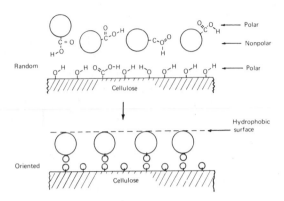

FIGURE 15-5. Illustrating the required orientation of amphipathic molecules to produce a well-sized paper.

particle size and electrostatic attraction to cellulose. Also, they depend on the drying process to promote flow and coverage of the fiber surface. But, they do not react chemically with the fiber.

Corresponding to the movement from traditional acidic papermaking toward a neutral or alkaline wet end (see Section 15.4), there is a trend toward greater use of synthetic sizing agents which react chemically with the cellulose hydroxyl groups to form stable ester linkages. These chemicals were introduced to the paper industry in the 1950's and provided the first opportunity to manufacture sized paper in an alum-free environment. Alkyl ketene dimer was the first commercially available reactive size and is still the most widely used. Acrylic stearic anhydride sizes (derived from fatty acids) and alkenyl succinic anhydride (derived from petroleum) are more reactive and have more selective applications.

Internal Strength

A number of natural and synthetic polymeric substances (refer to Table 15-4) may be admixed with the stock at the wet end to improve the physical properties of the dry paper sheet. Their action is to reinforce fiber-to-fiber bonds and thereby improve the burst and tensile strength, provide greater resistance to erasure, reduce "fuzz" or lint on the paper surface, and reduce the rate of water penetration.

The traditional internal strength additives are natural and modified starches and gums. Starches are polymers of glucose and are derived from various plants, principally corn, tapioca, potato and wheat. Gums are polymers of mannose and galactose and are derived from locust bean and guar seeds. Both starches and gums are usually cooked at low concentration prior to use to promote swelling and dispersion. (Refer to Section 18.1 for more information on starch.)

TABLE 15-4. Internal strength additives.

Natural Polymers
starches
- natural or unmodified
- chemically modified:
  - cationic starches
  - anionic starches
  - oxidized starches
  - dextrin

gums
- natural
- chemically modified

Cellulose Derivatives
carboxymethylcellulose
methyl cellulose
hemicellulose

Synthetic Polymers
phenolics
latices (latexes)
polyamines
polyacrylamides
* urea-formaldehyde
* melamine- formaldehyde
* polyamids

*Used mainly for wet strength retention

The trend today is toward increased use of such synthetic polymers as latexes and polyacrylamides either alone or in combinations with starches and gums. By means of copolymerizing or cross-linking, these products have evolved over the past two decades to meet a wide range of specific requirements for greater paper strength with different degrees of stiffness and stretch.

Wet Strength Resins

Ordinary paper will retain a significant portion of its strength when immersed in most oils or solvents (see Table 15-5). But because of the special interaction between water and cellulose, the normal fiber-to-fiber bonds are destroyed in aqueous media. The action of wet-strength resins is to tie fibers and fines together with additional bonds that are not taken apart by water. Wet-strength paper is defined as such if it retains more than 15% of its tensile strength when wet. Some papers actually retain up to 50%. Wet strength develops during aging; this effect is more pronounced with treated papers, but is also true for untreated papers.

The most common wet-strength agents are urea-formaldehyde, melamine-formaldehyde, and polyamide resins; they are water-soluble and available in both anionic and cationic forms. These agents are applied at an intermediate degree of polymerization, so that the final "cure" is obtained in the dryers. Since wet-strength resins are water-

TABLE 15-5. Comparative tensile strength of ordinary paper when immersed in various liquids (K.W. Britt, 1963).

Dry Paper	13.7
Petroleum	12.1
Xylene	10.9
Hexanol	9.2
Amyl Alcohol	8.6
Butanol	7.9
Dioxane	7.2
Propanol	6.7
Acetone	6.3
Ethanol	3.6
Methanol	1.4
Water	0.4

soluble, they must be fixed onto the fibers. Retention can be poor under some conditions. Anionic resins are best added with alum, but only after rosin size has been precipitated. The best retention on the stock is achieved over a relatively long period of contact.

Fillers or Loadings

Finely-divided white mineral fillers are added to papermaking furnishes to improve the optical and physical properties of the sheet. The particles serve to fill in the spaces and crevices between the fibers, thus producing a denser, softer, brighter, smoother and more opaque sheet. In some instances the paper can also be made cheaper because the fillers are often less costly than fiber.

The proportion of filler in the sheet is limited by the resultant reduction in strength, bulk and sizing quality. The majority of filled papers contain between 5 and 15% of sheet weight, but some heavily loaded grades exceed 30% (refer to Figure 15-6). An important innovation of the 1980's was the development of the multicomponent wet-end chemical system, described as a systems approach toward optimizing the balance between retention, dewatering and product strength (5, 6). A number of proprietary two-component systems are currently being offered by chemical suppliers which are purported to allow a higher level of loading without compromising machine productivity or product quality (see Figure 15-7).

The common papermaking fillers are clay (kaolin, bentonite), calcium carbonate, talc (magnesium silicate), and titanium dioxide. Clay is the most popular filler because it is cheap, plentiful, stable, and provides generally good performance. Calcium carbonate is used only in neutral or alkaline systems because of its solubility at lower pH levels. It is available at a higher brightness level than clay and is a better opacifier; it is especially useful for "permanent paper" because it neutralizes the acids which form during aging and cause deterioration.

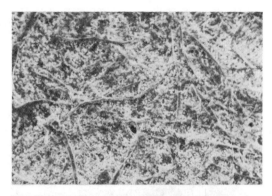

FIGURE 15-6. Scanning electron micrograph of cigarette paper at two different magnifications. Note particles of CaCO₃ filler (courtesy of Institute of Paper Science and Technology).

Titanium dioxide is the brightest and most effective opacifier; however, its relatively high cost limits its use to those applications where high whiteness and opacity must be obtained at low filler levels (2 - 3%) without loss of strength. Talc is notable as a "soft" filler, imparting a soft, silky feel to the paper product. Talc also has an affinity for pitch particles and is effective in preventing pitch deposits in the papermaking system.

Chemical Dyes

The absorption of dye by pulp fibers depends on the chemical nature of the dye, the capillary pore structure of the fiber, and the nature and polarity of the fiber surface. The principal types of water-soluble dyes are known as acid, basic, or direct. Chemically, acid and direct dyes are similar, both being the sodium salts of colored acids. The difference is in their affinity (or substantivity) for cellulose fibers. While direct dyes are readily absorbed by cellulosic fibers, the acid dyes can only be retained by adding rosin size and alum. Acid dyes are more soluble in water than other classes of dyes and have the advantage over basic dyes that they do not mottle in mixed fiber furnishes. As a class, direct

10μm = |———————|

FIGURE 15-7. Scanning electron micrograph of sheet formed with two-component system, showing fibers, modified bentonite clay filler, and fines (Hydrocol System, Allied Colloids).

dyes are less soluble than acid dyes, tending to form colloidal systems; they are often duller than basic dyes and more expensive for producing a given shade.

Basic dyes are the salts of color bases, and generally appear as the chlorides, hydrochlorides, sulfates, or oxalates. They are the most important class used in coloring paper. They have the advantages of low cost, high tinctorial strength, and great brilliance. Sometimes basic dyes are used in small quantities to improve the brilliance of acid or direct dyes, in this way producing a small amount of insoluble color "lake". Because there are several chemical groupings of basic dyes, considerable variation is found in the physical and dyeing properties among individual members. However, as a class, basic dyes possess relatively poor fastness to light, acids, alkalis, and chlorine.

All dyes are specific chemical compounds. The actual color produced in the paper from adding one or more of these compounds is affected by various processing conditions, such as the nature of the pulps used, the degree of refining, and the chemical balance. Therefore, color matching of products and control of color uniformity is a difficult task requiring experience and judgement. Often, consultation with a specialist in the field is required to solve a specific problem.

Control of Pitch

A common problem in paper mills is the depositing of pitch particles within the papermaking system. These particles accumulate in the openings of the forming fabric, thereby producing holes in the finished product. They also collect within the structure of press felts, thus reducing the felt's permeability, and on roll surfaces causing pickouts and non-uniform peeling of the sheet. Once deposited, pitch can only be removed by scrubbing with solvents or special cleaning compounds.

Pitch is composed of low-molecular-weight oleophilic materials (mainly fatty acids, resin acids and esters) which are released from wood fibers during chemical and mechanical processing, and precipitated as calcium and magnesium salts. The concern in a papermaking system is with "depositable pitch", rather than the total pitch content. Pitch that is well dispersed usually causes no problems.

In acid systems, alum is used for cationic fixation of pitch on to the fibers. Nonionic wetting agents are commonly used in alkaline systems to disperse the pitch. Since pitch is precipitated by calcium or magnesium, any step to reduce the amount or impact of these ions will be helpful. Where possible, use of hard water should be avoided. In some cases, chelating agents to inactivate the metallic ions are useful. To be effective, all chemicals used for pitch control must be added before agglomeration occurs.

15.4 ALKALINE PAPERMAKING

Traditionally, sized papers were manufactured only under acid conditions. The capability to produce papers unders neutral or alkaline conditions has now existed for over 30 years, and about 75% of European fine paper mills have converted to this technology. North American papermakers initially were more reluctant to change, but interest has surged since 1980. The original impetus came mainly from customer demands for higher brightness and opacity. Another significant factor was that calcium carbonate fillers became readily available and more price-competitive. Approximately 30% of North American fine paper is now produced by the alkaline process.

As mills have changed over, a wide range of benefits have been documented. Now, the major driving force toward conversion is the greater strength of the alkaline sheet which permits higher levels of clay and calcium carbonate filler. Typical ash content with alkaline sizing ranges from 18% to 25% vs 7% to 12% with an acid system (7). Substitution of filler for fiber provides significant economic advantages. Since calcium carbonate serves to stabilize the papermaking system in the pH range from 7.2 to 8.0, other means of pH control are no longer necessary. Additionally, the alkaline system is less prone to corrosion, so maintenance cost is lowered. Because of reduced chemical loading, alkaline systems are more easily closed than acid systems.

Along with synthetic reactive sizes, calcium carbonate fillers are at the heart of alkaline papermaking. Two forms of calcium carbonate having quite different morphologies are commonly used. Ground calcium carbonate is made from natural chalk deposits using sophisticated wet grinding technology. Precipitated calcium carbonate is a manufactured product made by passing flue gas through a milk of lime solution. While either form is often used as the sole filler, blends can be used to optimize the opacity and production goals for a particular grade.

So-called alkaline papermaking systems actually operate in the pH range from 7.0 to 8.0, just barely into the alkaline region. It is understandable that many industry people prefer to use the term "neutral papermaking".

REFERENCES

(1) BRITT, K.W. **Why Bother About First-Pass Retention of Solids on the Paper Machine?** *Paper Trade Journal* (April 15, 1977)

(2) NAZIR, B.A. and CARNEGIE-JONES, J. **Optimising Wet-End Chemistry - the Practicalities** *Paper Technology* (December 1991)

(3) GUEST, D. **The Effects of Secondary Fibre on Wet-End Chemistry** *Paper Technology* (June 1990)

(4) GRIGGS, W.H. and CROUSE, B.W. **Wet End Sizing - An Overview** *TAPPI 63:6:49-54* (June 1980)

(5) HAYES, A.J. **40% Filler Loaded Paper - Dream or Reality?** *Paper Tech. Ind.* (April/May 1985)

(6) BROWN, R. **Review of Methods for Increasing Filler Loading** *Paper Tech. Ind.* (October, 1985)

(7) WILEY, T. **Converting to Alkaline: U.S. Fine Paper and Board Mills Step Up Pace** *Pulp & Paper* (Buyers Guide 1989)

Chapter 16

Paper Manufacture - Wet End Operations

The art of true papermaking had its origin in China around 100 A.D. based on the discovery that a dilute suspension of fibers (obtained by macerating the inner bark portion of bamboo) could be formed into a mat by filtering the suspension through a fine screen. The mat was then made suitable for writing and drawing by pressing, drying and sizing.

Paper continued to be made by hand up to the beginning of the 19th century. A Frenchman, Louis Robert, was the first to patent a design for a continuous paper machine in 1799. The first machine ever to be built and successfully operated was started up in England in 1803. The Fourdrinier brothers took over development in 1804 and by 1807 had acquired all patent rights. The machine eventually became known as the Fourdrinier machine. The early paper machines consisted of a headbox adding paper stock to a moving wire supported between two rolls. The wet sheet was pressed once on the wire and then taken to a felt and run through another press nip before being accumulated on a roll for eventual drying in sheeted form.

16.1 INTRODUCTION TO THE PAPER MACHINE

Since the days of Louis Robert, the paper machine has undergone continual development, making it possible to produce a wider web of paper at ever increasing speed and to more exacting standards of quality. A basic fourdrinier-type machine is illustrated in Figure 16-1, and the following gives a brief functional explanation for each major component:

- the flowspreader takes the incoming pipeline stock flow and distributes it evenly across the machine from back to front.

- the pressurized headbox discharges a uniform jet of papermaking stock onto the moving forming fabric.

- the endless, moving fourdrinier fabric forms the fibers into a continuous matted web while the fourdrinier table drains the water by suction forces. (On other types of paper machines, the sheet may be formed by other means, e.g., between two fabrics, or on a fabric-covered cylinder.)

- the sheet is conveyed through a series of roll presses where additional water is removed and the web structure is consolidated (i.e., the fibers are forced into intimate contact).

- most of the remaining water is evaporated and fiber-to-fiber bonds are developed as the paper contacts a series of steam-heated cylinders in the dryer section.

- the sheet is calendered through a series of roll nips to reduce thickness and smooth the surface.

- the dried, calendered sheet is accumulated by winding onto a reel.

The fourdrinier machine, as described, can be considered a basic design suitable for a wide range of grades. Many variations, modifications, and

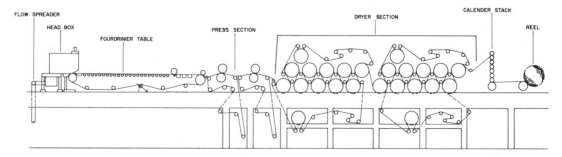

FIGURE 16-1. Fourdrinier paper machine (courtesy Beloit Corp.).

auxiliary on-machine operations have been developed for certain types of papers or special grades. For example, a large number of machines incorporate surface sizing, surface coating, and/or special calendering treatments. Some of the modifications are discussed in subsequent chapters (see Chapters 18 through 20).

16.2 APPROACH SYSTEM

The term "approach system" refers specifically to the fan pump loop wherein the papermaking furnish is metered, diluted, mixed with any necessary additives, and finally screened and cleaned before being discharged onto the paper machine forming fabric. The approach system extends from the machine chest to the headbox slice. Occasionally, certain stock chests and refiners may also be considered part of the approach system. While the headbox is integral with the approach system, it is also a component of the paper machine, and will be discussed in the next section of this chapter.

Although the machine chest stock should be reasonably free from impurities (assuming good control in the pulp mill or secondary fiber preparation system), most paper machine approach systems utilize screens and cleaners as insurance against foreign contamination. Traditionally, screens have functioned mainly to remove gross contamination and defloc the fibers, while the cleaners are designed to remove some debris. Pressure screens operating in this mode require only a small reject stream. Relatively large-diameter centrifugal cleaners are normally used to optimize removal of shives and slivers (refer to Sections 9.4 and 9.5). The simplified approach system shown in Figure 16-2 is a single-pump system where the fan pump is the sole source of pumping energy for the entire thin-stock circuit.

For products such as fine papers and coated grades, the modern requirement at the headbox is for virtually contaminant-free stock. One strategy to upgrade furnishes is to use fine slotted screens within the approach system, and at the same time improve the efficiency of the secondary and tertiary screens. This methodology is now commonly used in Europe and Japan, and is an economical way to achieve improved cleanliness. However, it must be kept in mind that modifications which provide greater efficiency of debris removal also tend to compromise the requisite ability of the approach screen to operate trouble-free without plugging for days or weeks at a time. (Another more-costly strategy is to install an independent fine slotted screening system between the blend chest and the machine chest; papermakers refer to this alternative as a thick-stock screening system.)

The workhorse of the approach system is the fan pump which serves to mix the stock with the white water and deliver the blend to the headbox. The fan pump is the largest pump in the paper machine system, and the demands made on it are very exacting. Both flow rate and pressure must be stable, without pulsations or surges, and yet capable of being varied over the entire range of machine operation. Usually, a single fan pump is used, but occasionally, two fan pumps are used in a series arrangement. A two-pump system is a requirement when air is removed from the stock under vacuum.

To ensure a uniform dispersion to the headbox, the stock is fed from a constant head tank (called the "stuff box"), through a control valve (called the "basis weight valve"), and is usually introduced axially into the suction line of the fan pump as illustrated in Figure 16-3 (1).

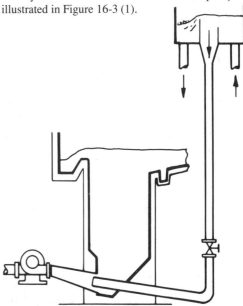

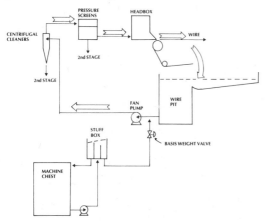

FIGURE 16-2. Simplified approach system to paper machine.

FIGURE 16-3. Recommended stuff box arrangement and stock entry into the machine approach system (Beloit Corp.).

Air Entrainment

Air bubbles entrained with the stock at the headbox can cause blemishes in the sheet (holes, thin spots), poor drainage on the wire, and instabilities in the approach system. Typically, small amounts of air are mixed into the stock during agitation and cascading of the flow. However, the major problem is usually with the wire-pit water. The free-draining water from the forming section of the paper machine falls into collecting trays, which

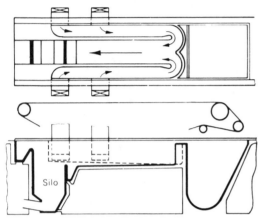

FIGURE 16-4. Representative wire pit located underneath the fourdrinier section (Beloit Corp.).

funnel the white water into a pit either underneath the forming fabric or at the backside of the machine (Figure 16-4). Inevitably, the white water picks up large concentrations of free air, and a relatively long period of open channel flow is desirable to release air and dissipate turbulence before the wire-pit water is recombined with the stock flow. Depending on what additives are present in the wire pit, it may be worthwhile to add defoaming agents to help release the air.

In spite of the best efforts to control air intake by design and operation, deaeration of the stock may be necessary in certain instances because of product quality requirements or to compensate for design faults. Deaeration is accomplished by spraying stock (usually centrifugal cleaner accepts) into a vacuum compartment where the air is boiled off, as in the Deculator system depicted in Figure 16-5 and 16-6 (2).

16.3 FLOWSPREADER AND HEADBOX

The function of the headbox (or flowbox) is to take the stock delivered by the fan pump and transform the pipeline flow into an even, rectangular discharge equal in width to the paper machine and at uniform velocity in the machine direction. Since the formation and uniformity of the final paper product are dependent on the even dispersion of fibers and fillers, the design and operation of the headbox

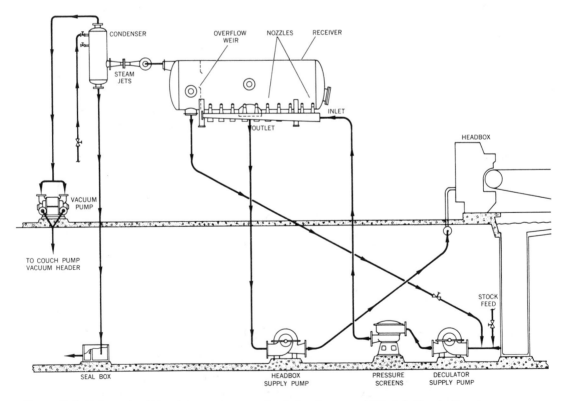

FIGURE 16-5. Schematic of Deculator system, used only for air removal (Clark & Vicario Corp.).

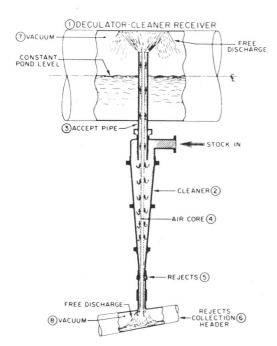

FIGURE 16-6. Detail of stock being sprayed into the Deculator vacuum compartment (Clark & Vicario Corp.).

system is absolutely critical to a successful papermaking system.

The following are specific operational objectives of the headbox system:
1) Spread stock evenly across the width of the machine.
2) Level out cross-currents and consistency variations.
3) Level out machine direction velocity gradients.
4) Create controlled turbulence to eliminate fiber floccing.
5) Discharge evenly from the slice opening and impinge on the forming fabric at the correct location and angle.

In addition, the following design objectives are important to the operator:
1) Inherent cleanliness.
2) Ease of operation.
3) Accuracy in measurement of headbox settings.
4) Ruggedness of construction to sustain constant slice opening shape.

Headboxes can be categorized, depending on the required speed of stock delivery, as open or pressurized types. Pressurized headboxes can be further divided into air-cushioned and hydraulic designs. In the hydraulic design, the discharge velocity from the slice depends directly on the feeding pump pressure. In the air-cushioned type the discharge energy is also derived from the feeding

pump pressure, but a pond level is maintained and the discharge head is attenuated by air pressure in the space above the pond.

The total head (pressure) within the box determines the slice jet speed. According to Bernoulli's equation:

$$v = \sqrt{2gh}$$

where v = jet velocity (m/s)
        h = head of liquid (m)
        g = acceleration due to gravity
             (9.81 m/s^2)

The jet of stock emerging from a typical headbox slice contracts in thickness and deflects downward as a result of slice geometry. The jet thickness, together with the jet velocity, determines the volumetric discharge rate from the headbox. Because the slice opening does not directly indicate jet thickness, the contraction of the jet must be considered if the flow rate is being calculated (3).

Open headboxes were employed with the early paper machines where a gravity head of stock was used to give the correct discharge velocity (Figure 16-7). As paper machine speeds increased, it became impractical to increase the height of stock by a corresponding amount, and the pressure headbox was developed. Open headboxes are today found mainly on slow-speed pulp machines.

Two current designs of air-cushioned headbox are illustrated in Figures 16-8 and 16-9. These boxes depend on the action of the rotating, perforated rolls to provide turbulence and level out velocity gradients. Shower sprays are arranged to cover the pond and sides of the box. Some older designs (e.g., Figure 19-10) incorporate a reception chamber (also called "swirl chamber" or "explosion chamber") into which the distribution system feeds, followed by an

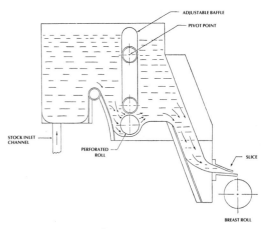

FIGURE 16-7. Schematic of open headbox cross-section.

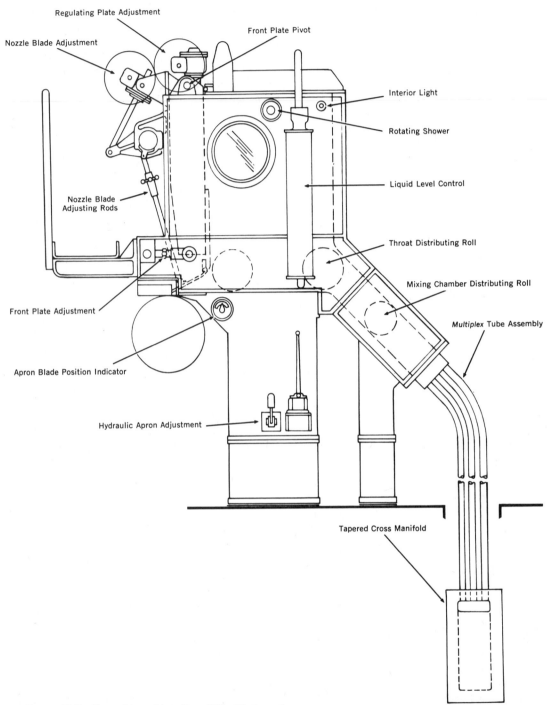

Regulating Plate Adjustment

Nozzle Blade Adjustment

Front Plate Pivot

Interior Light

Rotating Shower

Nozzle Blade
Adjusting Rods

Liquid Level Control

Throat Distributing Roll

Mixing Chamber Distributing Roll

Front Plate Adjustment

Multiplex Tube Assembly

Apron Blade Position Indicator

Hydraulic Apron Adjustment

Tapered Cross Manifold

FIGURE 16-8. Air cushioned headbox (Allis-Chalmers).

adjustable gap or throat. The gap is important; if too large, the high velocity streams from the inlet pipes will persist; if too small, troublesome vortices (eddy currents) can be created. An inside view of an air-cushioned headbox is shown in Figure 16-11.

Figures 16-12 and 16-13 illustrate two designs of hydraulic box. Typically, perforated rolls are used only sparingly; shear forces to provide turbulence and break up the fiber network are developed mainly in the small tubes and channels.

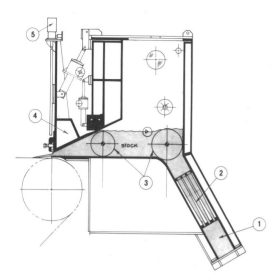

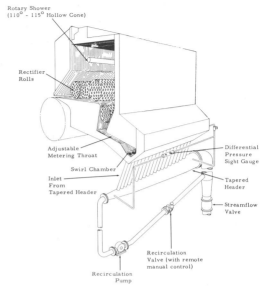

FIGURE 16-9. Air-cushioned headbox illustrating (1) rectangular tapered flow spreader with (2) tapered laterals, (3) rectifier rolls, (4) slice assembly, and (5) slice profile adjustment (LG Industries Ltd.).

FIGURE 16-10. Air-cushioned headbox, circa 1965 (Beloit Corp.).

FIGURE 16-11. View of the inside of an air-cushioned headbox. All surfaces are polished to reduce deposits (KMW).

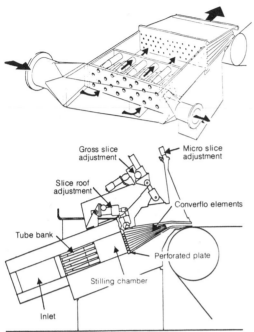

FIGURE 16-12. Two views of "Converflow" headbox (Beloit Corp.).

Flowspreader (Manifold Distributor)

Although the modern flowspreader is integral with the headbox, it is sometimes considered as a separate component because its function is so critical on high speed machines, and because its performance can be evaluated apart from the rest of the headbox (4). The design problem of taking a pipeline flow and uniformly distributing it across the width of the paper machine was not solved until the concept of the multitube, tapered inlet manifold with recirculation was introduced by J. Mardon in the mid-1950's. Today, this design of flowspreader (as illustrated in Figure 16-14) is a common feature on all modern headboxes.

The cross-section of the tapered inlet manifold can be circular or rectangular. Most designs utilize a thick plate at the top which can be precisely drilled for correct positioning of the small distributor tubes. The operation of the manifold can be understood from reference to Figure 16-14. The recirculation rate is controlled to maintain the same pressure at points A and B as evidenced by an absence of flow in the sight glass. The result of too little recirculation or too much recirculation is illustrated in Figure 16-15.

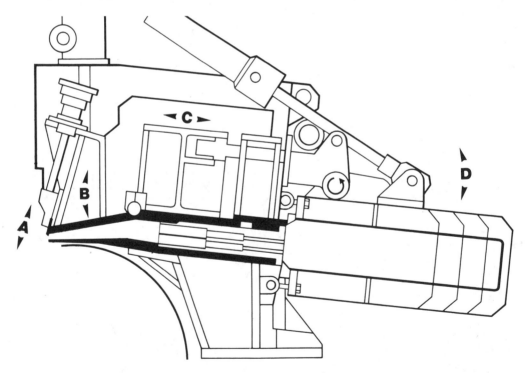

A Setting of slice and fine adjustment

B Adjustment of top lip, depending on flow rate and machine speed

C Horizontal movement of the top lip affects the point of impingement of the jet on the wire which in turn affects sheet formation

D The distributor can be swung clear for cleaning or inspection purposes

FIGURE 16-13. Hydraulic headbox (Escher-Wyss).

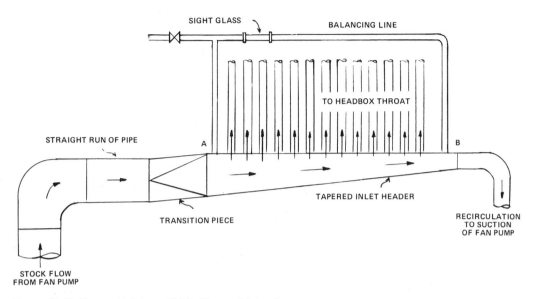

FIGURE 16-14. Tapered inlet manifold with small laterals.

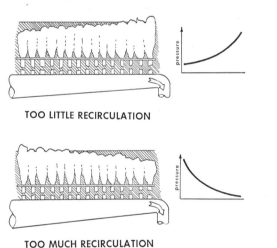

TOO LITTLE RECIRCULATION

TOO MUCH RECIRCULATION

FIGURE 16-15. Results of incorrect recirculation for tapered inlet manifold (J. Mardon).

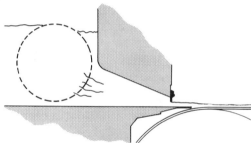

FIGURE 16-16. Illustrating action of slice rectifier roll (courtesy of Beloit Corp.).

Perforated Rolls (Rectifier Rolls)

If the stock distribution system (flowspreader) is ineffective for any reason, then the burden for dampening out flow variations is pushed nearer to the slice. In air-cushioned headboxes, hollow perforated rolls (also called rectifier rolls or "holey rolls") are commonly used, both to even out flow irregularities and to create turbulence to keep the fibers deflocculated (as per Figure 16-16).

The major design and operational variables for perforated rolls are hole diameter, % open area, wall thickness, direction of rotation, and rotational speed. Typically, hole diameter varies from 2 to 4 cm, open

area from 35 to 50%, and rotational speed from 6 to 15 rpm. The number of rolls and their relative placement has also been a design consideration. For example, the headbox illustrated in Figure 16-10 was manufactured in both 3-roll and 5-roll configurations. Perforated rolls must be installed to close tolerances within the headbox; any stock bypassing underneath or around the ends of the roll will not be affected by the action of the roll.

Slice

The headbox slice is a full-width orifice or nozzle with a completely adjustable opening to give the desired rate of flow. The slice geometry and opening determine the thickness of the slice jet, while the headbox pressure determines the velocity.

Every slice has a top lip and an apron (bottom lip), both constructed of suitable alloy materials to resist corrosion. The top lip is adjustable up or down as a unit (the "main slice") and also in local areas by use of individual micro-adjusters (hand-operated or motor-driven). Either the top lip or the apron

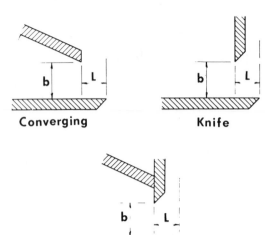

FIGURE 16-17. Various slice designs.

extension is also adjustable in the horizontal direction in order to change the impingement angle of the stock jet onto the forming fabric. The apron is usually slightly sloped toward the opening.

Slices may be broadly categorized as knife (also called vertical or straight), converging, or combination types as illustrated in Figure 16-17. However, many special designs are utilized that defy such a simplistic grouping. A representative combination slice is shown in Figure 16-18. The

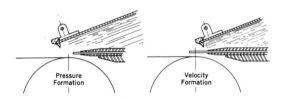

FIGURE 16-19. Two extremes of jet delivery.

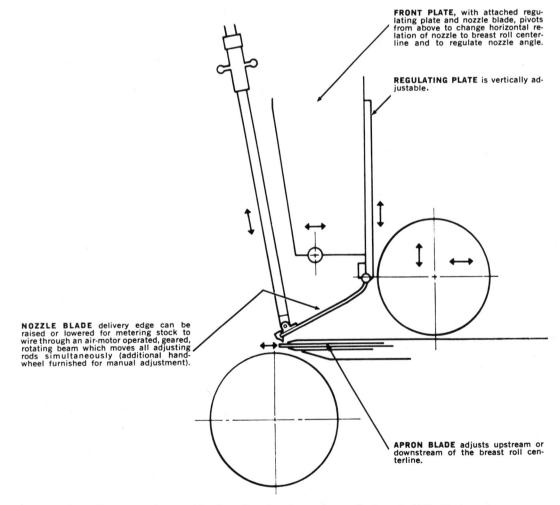

FRONT PLATE, with attached regulating plate and nozzle blade, pivots from above to change horizontal relation of nozzle to breast roll centerline and to regulate nozzle angle.

REGULATING PLATE is vertically adjustable.

NOZZLE BLADE delivery edge can be raised or lowered for metering stock to wire through an air-motor operated, geared, rotating beam which moves all adjusting rods simultaneously (additional handwheel furnished for manual adjustment).

APRON BLADE adjusts upstream or downstream of the breast roll centerline.

FIGURE 16-18. Representative combination slice showing various adjustments (Allis Chalmers).

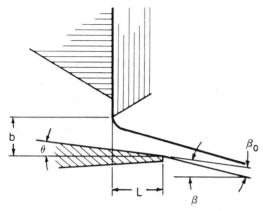

FIGURE 16-20. Terminology for defining slice geometry and jet angle (Nelson).

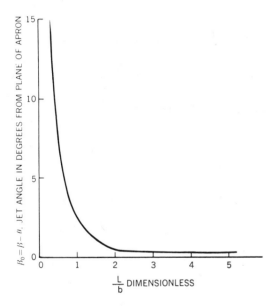

FIGURE 16-21. Effect of L/b ratio on jet angle (Nelson).

converging section is used to accelerate the stock, and a vertical section is used at the tip to provide a fine-scale turbulence to the jet.

Two types of jet delivery are possible, as illustrated in Figure 16-19. Most fourdrinier applications utilize velocity formation. Generally, pressure formation results in excessive discharge at the breast roll with low retention of fines and fillers and poor sheet structure.

Common notation for describing slice geometry is shown in Figure 16-20 (and also in Figure 16-17). The dimensions which determine the jet angle are:

L = projection of the apron beyond the inner vertical surface

b = slice opening between vertical knife and end of apron

The effect of L/b ratio on the impingement angle is shown in Figure 16-21. Both the point of impingement and the impingement angle onto the wire are important for achieving satisfactory formation and drainage on the forming fabric (see also Section 16.5).

Definition of the Dry Line

When viewed from above, a definite line of demarcation is visible on the fourdrinier (usually in the vicinity of the second dry box), which corresponds to the point where a glassy layer of water is no longer present on top of the stock (see Figure 16-22). The machine tender interprets the position and shape of the "dry line" as indicators of wet end operation. For example, a change in dry line position may be indicative of a shift in stock drainability, which can be compensated for by a change in either headbox consistency or dry box vacuums.

An irregular shape in the dry line may be due to a number of factors; but most frequently, it can be traced to non-uniformity of headbox delivery, as illustrated in Figure 16-23. Irregularities seen in the dry line can often be matched to cross-machine variations in sheet basis weight at the reel.

Slice Edge Bleeds

Most pressure headboxes are equipped with slice edge "bleeds" to give more control of edge sheet formation. By removing a small flow at the edge of the box, the natural tendency for the jet to "fan out" at the extreme edge is offset. The effect of primary and secondary bleed adjustments on the dry line (as seen from above the wire) is illustrated in Figure 16-24.

Operation of the Headbox

The main operating variables for the headbox are stock consistency and temperature, and jet-to-wire speed ratio. Typically, the consistency is set low enough to achieve good sheet formation, without compromising first-pass retention or exceeding the drainage capability of the forming section. Since higher temperature improves stock drainage, temperature and consistency are interrelated variables. Consistency is varied by raising or lowering the slice opening. Since the stock addition rate is controlled only by the basis weight valve, a change in slice opening will mainly affect the amount of white water circulated from the wire pit.

The ratio of jet velocity to wire velocity is usually adjusted near unity to achieve best sheet formation. If the jet velocity lags the wire, the sheet is said to be "dragged"; if the jet velocity exceeds the wire speed, the sheet is said to be "rushed". Sometimes, it is

FIGURE 16-22. View of fourdrinier wire (from above the headbox) showing the dry line.

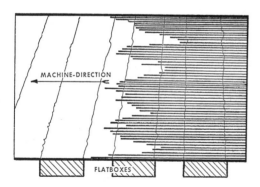

FIGURE 16-23. View of dry line, as seen from along-side the fourdrinier wire.

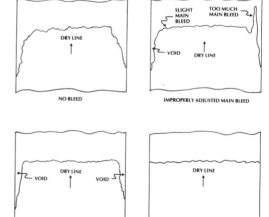

FIGURE 16-24. Effect of primary and secondary bleeds on dry line shape.

necessary to rush or drag the sheet slightly to improve drainage or change fiber orientation. Of course, the jet speed is not actually measured, but is inferred from the headbox pressure.

16.4 SHEET-FORMING PROCESS

Laboratory sheet forming (or making paper by hand) is carried out simply by draining a very dilute suspension of fibers on a fine-mesh wire screen. The sheet-forming process on a fourdrinier wire is far more complicated, involving, in addition to drainage, such effects as the generation and decay of turbulence, formation and breakdown of fiber networks, retention and transport of fine particles in the mat, compaction of the mat, and shear forces between the mat and free suspension. Chemical and colloidal effects are relatively less important.

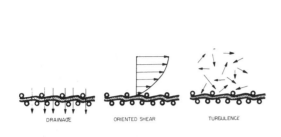

FIGURE 16-25. Hydrodynamic processes in sheet forming (J.D. Parker).

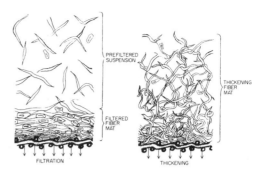

FIGURE 16-26. Mechanisms of fiber deposition by drainage (J.D. Parker).

In his classic monograph (5), Parker conceptualized sheet forming as a composite of three basic hydrodynamic processes: drainage, oriented shear, and turbulence, as illustrated in Figure 16-25. All these processes occur simultaneously and are not wholly independent of each other.

The most important effect of the drainage process is the dewatering of the fiber suspension to form the mat. When the fibers are free to move independently of one another, drainage proceeds by the mechanism of filtration, and the fibers are deposited in discrete layers (Figure 16-26). Filtration is the dominant mechanism in most fourdrinier forming applications, as shown by the layered structure and relatively uniform formation of the sheets. When the fibers in suspension are immobilized, they floc together in coherent networks; drainage then occurs by thickening, and a more felted and floccy sheet structure results (also Figure 16-26).

Papermaking suspensions spontaneously form networks during drainage unless sufficient dilution is used (as in laboratory sheet forming) or supplemental mixing energy (e.g., turbulence) is provided. Dilution is a powerful mechanism for dispersion; but the level required to adequately control floccing on paper machines is not economically feasible. Additional dispersion must be generated during drainage by the turbulence-inducing effects of drainage elements below the forming fabric, or by shear-inducing devices above the fabric (e.g., dandy roll or top former). In each design of commercial sheet-forming machine (whether fourdrinier or twin wire), the three elementary forming effects of dilution, turbulence and oriented shear are applied to different degrees in an attempt to optimize sheet quality.

16.5 WIRE PART (FOURDRINIER)

A representative fourdrinier section is illustrated in Figure 16-27. The forming medium is an endless, finely woven belt. Until the late 1960's, only woven metal wires (usually phosphor bronze) were used; today, plastic mesh fabrics are used almost exclusively because they provide much longer service life (up to ten times longer). However, the old terminology persists, and all woven forming media are usually called wires by papermakers whether of plastic or metal construction.

The fabric travels between two large rolls, the breast roll near the headbox and the couch roll at the other end. Typically, the breast roll is solid and serves only to support the fabric. (In some papermaking systems, the breast roll acts as a suction former.) The couch roll is a hollow, perforated shell containing one or two stationary high-vacuum suction boxes for dewatering the sheet. Most of the drive power for turning the wire section is applied to the couch roll and to the wire-turning roll.

The various elements between the breast roll and couch roll serve the dual functions of wire support and water removal. A number of different arrangements can be used depending on the particular requirements. Most machines today utilize a forming board immediately after the breast roll, followed by a number of foil assemblies. The wire then passes over a series of vacuum-augmented devices, from low vacuum (wet boxes) to high vacuum (dry boxes), and finally over the high-vacuum couch.

A system of wire return rolls carries the belt back to the breast roll. Stretch rolls and guide rolls are used to automatically maintain correct tension and eliminate lateral movement. A series of showers keeps the wire clean and free of buildups.

Slower paper machines (say, up to 400 m/min) are often equipped with a shake mechanism which imparts a cross-oscillation motion to the wire to improve sheet formation. These slower machines are also equipped with rubber deckles along the edge of the wire to contain the stock during the initial formation phase. Shakes and deckles are unnecessary on higher-speed machines because the sheet is set almost instantaneously.

Some fourdriniers are equipped with a dandy roll mounted above the wire and riding on the stock in

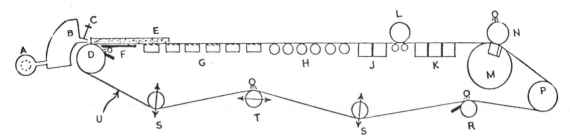

A. Tapered inlet manifold
B. Headbox
C. Slice and adjustment mechanism
D. Breast roll
E. Rubber deckle (only on slower machines)
F. Forming board
G. Hydrofoil assemblies
H. Table rolls
J. Vacuum boxes (to control drainage before dandy roll)
K. Vacuum boxes (to remove as much water as possible)

L. Dandy roll, with two small table rolls under wire for support
M. Suction couch roll
N. Lump breaker roll (to improve water-removal at couch and consolidate sheet)
P. Wire-turning roll or forward-drive roll (helping couch roll to drive wire)
R. First return roll with water spray and doctor.
S. Stretch rolls, having adjustment up and down to control wire tension
T. Guide roll, having adjustment backwards and forwards (at one end)
U. Forming fabric or wire

FIGURE 16-27. Diagram of fourdrinier wet end.

the suction box area. The skeletal roll has a wire cloth covering and serves to compact the sheet and improve formation in the top portion. Some dandy rolls carry a pattern in the wire facing that is transferred to the sheet to provide water marks or other special effects.

It is customary to trim off a narrow strip from each edge as the sheet leaves the couch. The edges are usually weak due to lower fiber weight and more erratic formation, and could be a source of breaks during web transfers. The cutting is performed by high-pressure water jets, called squirts, located over the wire just prior to the couch roll. The trimmed edges are carried around the wire-turning roll and washed into the couch pit, where they are re-slurried and later added to the broke stream.

Drainage Elements

Drainage on the fourdrinier wire is accomplished by hydraulic pressure gradients. The gradients, other than those caused by the weight of stock on the wire and the inertial impingement of stock from the slice, are induced by the drainage elements along the wire. In the free drainage area, the pressure gradients are produced by hydrodynamic suction forces as the wire passes over either rotating table rolls or stationary contoured supports. Further down the wire, various vacuum devices are used to remove additional water; these are assisted by pressure from a dandy roll or lump-breaker roll on some machines.

The first static element under the wire is the forming board (Figure 16-28). This element supports the wire at the point of jet impingement

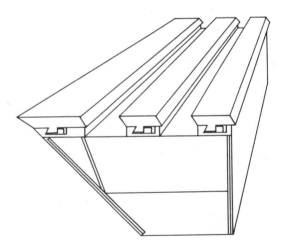

FIGURE 16-28. Representative forming board design (Albany International Company).

and is therefore under high loading. While its shape is restricted by the circumference of the breast roll, horizontal deflection must be kept at a low level to prevent wrinkles in the forming fabric. The lead blade must be designed so that the forming board can be set in close proximity to the breast roll centerline, thus enabling the angle of jet delivery to be optimized for best sheet formation. Generally, the forming board serves to retard initial drainage, so that fines and fillers are not washed through the sheet; however, the length and open area of the assembly must be compatible with overall drainage capability for the machine speeds and paper grades being produced. In view of these requirements, the forming board must be fabricated to exacting standards.

Until the mid-1960's, table rolls were used almost exclusively within the free draining section of the fourdrinier for removing water and setting the sheet. These rolls were originally used as wire support elements on the earliest fourdriniers. But as machine speeds were increased, it became apparent that a vacuum pulse was being created in the diverging angle between the wire and the roll by the pumping action of the water. The amount of induced drainage was found to be a function of wire speed, roll diameter and wire tension, and could be attenuated by cutting circumferential grooves in the rolls. Along with the vacuum pulse, a pressure pulse was being generated at the ingoing nip by the action of water carried on the underside of the wire and ringing the table roll. Although a small pressure pulse was generally considered beneficial to "activate" the sheet, the sharpness of both the pressure and vacuum pulses at higher speeds disrupted sheet formation and adversely affected retention of fines and fillers. In order to remove water that was pulled beneath the wire surface, it was necessary to install a stationary deflector following each table roll along the wire table. The space demands of the rolls and deflectors were large in relation to the small intervals along the wire where vacuum forces were operative.

The suction produced by the table roll is due to the separating of the roll and wire surfaces, both of which are moving. When the wire passes over a specially-designed angled stationary surface (called a hydrofoil or foil), the suction produced is less because only one surface is moving. However, the concept of using a fixed surface is attractive because the divergent angle can be totally adjusted and the leading edge can act as a deflector. Modern foils usually employ angles between 0.5 and 3.0 degrees; the larger the angle, the greater is the induced vacuum. The basic foil design is illustrated in Figure 16-29. A number of blade designs are available with adjustable angles, step angles, curved surfaces, or ramps in the divergent section (for example, Figure 16-30).

The basic foil concept was well known in the 1950's, but successful application was delayed until suitable materials of construction were available to minimize sliding friction and ensure reasonable service life for both wires and foils. Blades are now commonly available with ceramic inserts at the point of maximum wear (e.g., Figure 16-31). Foil assemblies generally hold 3 to 6 blades as illustrated in Figure 16-32.

The comparative actions of table rolls, foils and vacuum-assisted foils are illustrated in Figure 16-33.

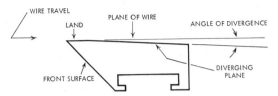

FIGURE 16-29. Basic hydrofoil blade with T-bar support slot.

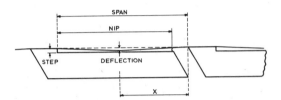

FIGURE 16-30. The so-called "Unfoil" utilizes a step instead of a divergent angle (Albany International Company).

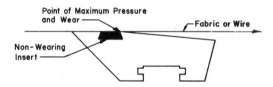

FIGURE 16-31. Foil blade with ceramic insert.

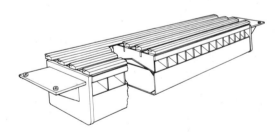

FIGURE 16-32. Assembly of foil blade elements (Beloit Corp.).

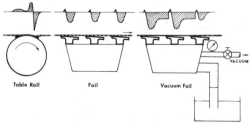

FIGURE 16-33. Examples of vacuum profiles achieved with three dewatering elements.

The action of foils (in relation to table rolls) is characterized by lower nip vacuum, longer-sustained vacuum, and closer spacing to obtain significantly greater drainage. Typically, little positive pressure is generated at the blade's leading edge, but some designs are available with a radiused nose to provide more wire-side activity.

Further down the table, vacuum-assisted foils or "wet boxes" are used to sustain higher drainage. Finally, high-vacuum "flat boxes" or "dry boxes" are used just prior to the suction couch. In the case of the vacuum-assisted element, the body of the unit serves as a seal leg to provide up to 20 inches of water vacuum (Figures 16-34). The water which enters at the top is discharged across a full-width slot (integral trap) at the bottom. The air drawn in with the water is evacuated through a small vacuum line.

In the case of the high-vacuum flat boxes, the water and air mixture is sucked from the box into a separator at the back side of the machine. The water flows down a barometric leg to a seal pit in the basement, while the air is removed by a vacuum pump (Figure 16-35). The boxes themselves are small and rugged in construction to withstand the high-vacuum conditions. The box covers are generally about 5 cm thick, with slotted openings (Figure 16-36). As the wet sheet moves over successive flat boxes, the water becomes increasingly more difficult to extract. Therefore, progressively higher vacuum levels must be utilized for diminished water removal. A typical profile of vacuums and water-removal rates for the flat boxes is shown in Figure 16-37.

Forming Fabric

A paper machine forming fabric is basically a cloth woven from polyester monofilaments, made endless by a seam to form a continuous belt. It is composed of machine-direction or lengthwise (warp) filaments and cross-filaments (weft, shute or filling). The meshes of the fabric permit the drainage of water while retaining the fibers. The traditional weaving variables (e.g., pattern, mesh, yarn diameter, degree of crimping, etc.) have provided a wide range of fabric constructions to suit the various paper grades being produced.

With today's double- and triple-layer constructions, the range of choices has been expanded by another order of magnitude. In 1983, about 90% of the fabrics used were of single-layer construction and 10% were double-layer. By 1990, single-layer fabrics were used in only 50% of applications; double-layer constructions accounted for about 40%, and triple-layer about 10%.

The functions of a forming fabric are best described under the headings of drainage, sheet support and mechanical stability. Fabric drainage can be characterized by such measurements as open

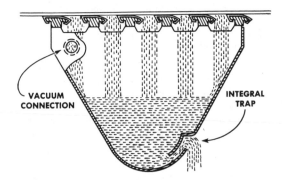

FIGURE 16-34. Low-head (up to 10 inches water) vacuum-assisted foil assembly (Huyck Corp.).

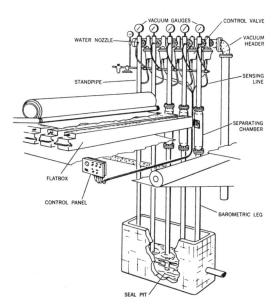

FIGURE 16-35. Schematic of a typical flatbox installation (Broughton Corp.).

area, permeability, void volume and void volume distribution. However, depending on the paper grade requirements, either rapid or slow drainage may be desired, and the fabric must be compatible with the drainage equipment on the paper machine. Sheet support refers to the ability to retain fibers and other furnish components on the fabric surface. One measure of fabric sheet support is the number of support points per unit area. Mechanical stability refers to a fabric's ability to run for long periods on the paper machine without excessive stretching or wrinkling. Obviously, all these fabric functions are interrelated, and optimization of one function may be at the expense of another.

Every fabric has two distinct surfaces, the forming surface (on top) and the wearing surface (the

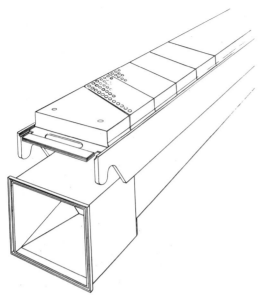

FIGURE 16-36. Dry suction box or flatbox (Beloit Corp.).

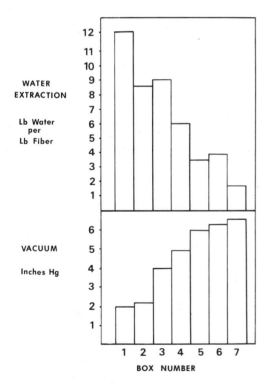

FIGURE 16-37. Profile of flatbox vacuum and water removal.

underside). The traditional metal wire was usually woven in the so-called three-shed design, with the warp wire going over one shute and under two (refer

to Figure 16-38). Thus, two-thirds of the warp wire was woven into the bottom or wearing plane of the wire cloth. Today's single layer fabrics are most commonly of four- and five-shed design, i.e., over one shute filament and under three or four (see Figure 16-39). The double- and triple-layer constructions basically provide two constructions within a single fabric, one designed for optimum sheet forming, while the underside is designed for wear and power transmittal characteristics (see Figure 16-40).

FIGURE 16-38. Retention of fiber during the early stage of drainage is illustrated for a metal wire. Note the difference in dimension between wire strand and fiber (courtesy MacMillan Bloedel Research).

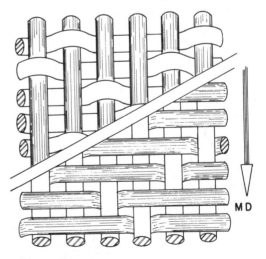

FIGURE 16-39. Four-shed forming fabric construction (Huyck Canada Ltd.).

Fourdrinier Drainage Profile

In general, water removal should take place along the wire in a fairly uniform manner, without either quiescent zones or abrupt, high-volume drainage. A water-removal profile on the order of that shown in Figure 16-41 might be considered ideal.

The drainage action on the fourdrinier depends on the paper making system, i.e., the fabric structure, arrangement of drainage elements, wire speed and tension; as well as on the stock temperature, consistency, thickness on the wire, and characteristics (e.g., freeness, additives, etc.). To achieve a given objective, the arrangement of drainage elements must be compatible with the stock and fabric characteristics.

As a general rule, the stock on the fourdrinier should be drained to the highest practicable extent with static and low-vacuum elements before applying higher vacuum forces. Dewatering under low-vacuum conditions is desirable for reducing wire drag and fourdrinier power requirements, and it also minimizes sheet wire marking (see Figure 16-42). Drainage with foils alone seems to depend mainly on stock freeness and thickness. For slow newsprint stock, it appears that water removal up to 2.0% consistency is the most that can be achieved with these static elements. An even greater limitation applies to high-caliper sheets such as food board and corrugating; it is difficult to get above 1.5 - 1.7% consistency. However, for a free-draining fine paper stock, a consistency above 3% can be reached without vacuum assistance.

For analytical purposes, the drainage table can be divided into four distinct zones, as illustrated in Figure 16-43. Initial drainage in the entry or blend zone is critical to good sheet properties, and is primarily a function of jet angle and placement of the forming board. Generally, drainage is by gravity and therefore very gentle. However, more powerful dewatering can be promoted on the forming board, if required, by proper selection of blades.

Gradual and positive water removal is combined with microturbulence to give good primary formation in the second zone. By properly controlling the rate of drainage, the removal of fines from the bottom of the sheet can be minimized. The microturbulence prevents premature "sealing" of the sheet against the wire, thus keeping the sheet open for drainage and minimizing wire marking.

After completing the actual formation of the web, the task remains to remove as much water as possible, but without damaging the sheet. This is done by utilizing increasingly higher suction forces, first by means of higher-angled foils, and then vacuum-assisted elements (wet boxes). Finally, in the last zone, increasingly higher vacuum is applied in step-wise fashion to obtain maximum consistency off the couch.

Generation of Turbulence and Oriented Shear on the Wire

A degree of agitation is necessary on the wire to keep fibers from floccing prematurely and causing poor sheet formation. It must be stressed that turbulence and shear on the wire cannot replace good dispersed delivery from the slice; however, good delivery can be adversely affected by an inactive wire section. Some agitation is achieved with combinations of foil blade angles, but this does not always provide enough control.

One method of providing additional hydraulic shear is to create controlled ridges in the flow from the headbox. When a ridge hits a drainage element, it splits into two smaller ridges, thus creating shear which helps to keep fibers dispersed. If these ridges are of the correct size, spacing and energy level, they can be quite helpful in maintaining good sheet formation. The two accepted methods of creating ridges are by means of a serrated top slice lip or by using small shower jets positioned close to the wire.

The so-called Sheraton Roll (Figure 16-44) is a new tool for introducing turbulence into the draining stock. This is essentially a table roll with a sawtooth surface which pulses the wire. The initial installations had spotty success, but a recent investigation (6) showed that the roll can be extremely successful if driven at a speed that generates the resonant frequency of the length of free wire above the roll.

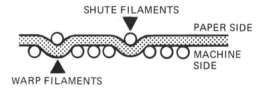

FIGURE 16-40. Cross sections of single- and double-layer forming fabrics (Capital Wire).

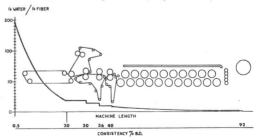

FIGURE 16-41 . Water removal profile down the fourdrinier (followed by pressing and drying).

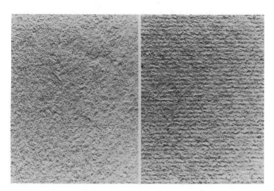

FIGURE 16-42. Top and wire-side surfaces of newsprint as viewed by reflected light under magnification. Note the fabric pattern impressed into the wire-side surface. (MacMillan Bloedel Research).

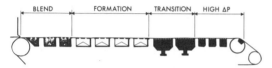

FIGURE 16-43. Drainage on the wire can be divided into four zones (Beloit Corp.).

FIGURE 16-44. Sheraton Roll installed on a board machine.

The traditional formation tool on older slow-speed machines was the dandy roll. The dandy roll was placed in the middle of the suction box section where it exerted a certain pressure on the sheet and was driven by the sheet and wire. These rolls were very effective in reworking the top section of the sheet to break up flocs. As machines became wider and faster, the diameter and stiffness of the dandy was increased and the rolls had to be driven so that the dandy speed matched the wire speed. On some modern installations, overspeeding the dandy by up

to 5% has provided improvements in formation and surface characteristics.

Today's larger dandy rolls require a different arrangement to operate properly. The dandy is now placed further back on the wire where the consistency is about 2%, and the dandy acts both as a water removal device and a structure-altering device. At the ingoing side, there is compression of the sheet between the dandy and the wire. Water is driven up through the sheet into the dandy and downward through the wire; the redistribution of water causes shear on the fibers and improves sheet formation. A modern roll typically has a surface covering of 35-40 mesh fabric, and the water held in the mesh is flung into a collecting pan on the upstream side.

16.6 TWIN-WIRE FORMING

Up to the 1950's, all paper and paperboard products were formed on conventional fourdrinier and cylinder vat machines, with the exception of some tissue grades. The inherent limitations of these methods in achieving higher operating speeds and improved product quality (especially for multiply boards) provided the impetus for further development work. Since the invention of the first modern twin-wire former by D. Webster in 1953, a number of two-wire designs have been introduced and utilized for various applications.

Modern twin-wire formers can be broadly categorized by application into the following groups:
• fourdrinier replacements (gap formers)
• fourdrinier modifications (hybrid formers)
• multiply formers
• tissue formers
Some formers can be placed in more than one grouping. This section is concerned with fourdrinier replacements and modifications. Multiply formers are considered in Chapter 19, and tissue formers in Chapter 20.

It must be noted that the conventional fourdrinier system has also undergone evolution and improvement during the four decades of twin wire former development. The fourdrinier wet end remains a viable forming system, and is still the best choice for producing certain paper grades.

Fourdrinier Replacements (Gap Formers)

In true twin-wire forming, the headbox slice jet impinges into the converging gap between two wires. Depending on the particular gap former design, the initial drainage can take place in one direction or in both directions. The dewatering action is due to pressure set up by the tension in the two wires and by water drainage elements outside of the wires. As fiber mats build up on both wire surfaces, the drainage resistance increases and the pressure in the fiber suspension also increases. The

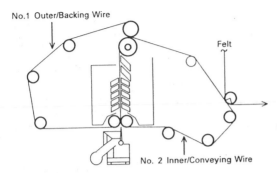

FIGURE 16-45. Gap blade former (Black Clawson).

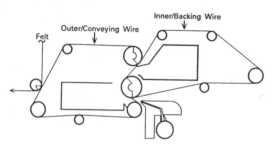

FIGURE 16-46. Gap roll former (Voith Corp.).

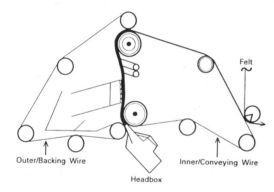

FIGURE 16-47. Gap roll/blade former (Valmet Corp.).

length of the forming zone is dependent on machine speed, basis weight, and stock freeness, as well as the tension in the two wires and the position of the dewatering elements. An even dispersion from the headbox is critical for gap formers because the sheet is set almost immediately and formation is provided by the jet. The jet impingement angle is also critical, more so than for a fourdrinier.

There are three basic configurations of gap formers:
• blade formers
• roll formers
• roll/blade formers

In a blade former, deflectors are used on either side of the converging wires to control the rate of

convergence and also direct the expressed water to saveall trays. A representative gap blade former is illustrated in Figure 16-45. In a roll former, it is the convergence of the two wires as they wrap one or more rolls which is the primary water removal mechanism. A representative roll former is shown in Figure 44-46. A roll/blade former is one in which both roll wrap and blade action are water removal mechanisms. The first commercial roll/blade gap former is illustrated in Figure 44-47.

The various configurations appear to have inherent advantages and disadvantages. For example, blade formers as a group seem to have better formation than roll formers, but lower retention and higher drive requirements. This is because the drainage forces are more severe in blade formers; these forces have a positive effect on formation due to shear introduced on the flocs, but they have a negative effect on first-pass retention. Improvements have been made with each successive generation of former design, and it is likely that the differences will become less significant.

Fourdrinier Modifications (Hybrid Formers)

The hybrid former (also called preformer or on-top former) has become a popular lower-cost alternative to a gap former for increasing speed, improving formation and reducing two-sidedness. Most hybrid formers are adaptations of existing fourdriniers, and for this reason, they are sometimes referred to as retrofit formers; however, hybrid former is a better term because not all installations are retrofits. All formers of this type are characterized by a preforming zone of "open wire" papermaking before the onset of twin-wire forming. In the twin-wire forming section, the slurry is subjected to drainage pressures generated by wire tension over rolls, curved surfaces, or blades, or by vacuum.

Some paper machine manufacturers now offer two or three different designs of hybrid former, each satisfying a set of requirements with respect to product basis weight, machine speed, and special needs. Two Beloit designs are illustrated as an example of this diversity. The Bel-Bond shown in Figure 16-48 is a vacuum former which utilizes a converging wire wedge over a turning radius to move surface water to the inside of the top wire where it is removed by vacuum. The Bel Bond is capable of removing in excess of 50% of the headbox flow, but typically the split is 40% up and 60% down. The Bel-Roll shown in Figure 16-49 is a roll former which extracts large quantities of water from the sheet as the converging wire wedge moves through a roll module which provides dewatering in both directions. Subsequent wrapping of a solid roll forces out additional water by centrifugal force. The Bel-Roll is placed a little further from the headbox (compared with the Bel-Bond) and can handle 40%

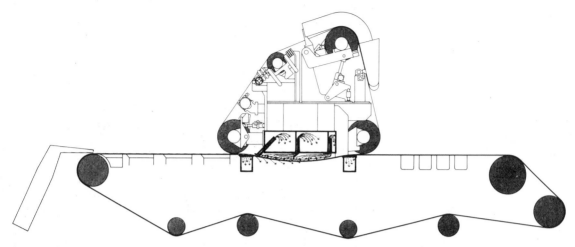

FIGURE 16-48. Bel Bond hybrid former (Beloit Corp.).

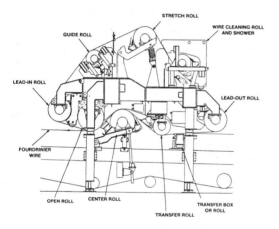

FIGURE 16-49. Bel Roll hybrid former (Beloit Corp.).

of the headbox flow.

Hybrid formers have generally equalled or exceeded expectations and have provided new life for old fourdriniers. Retrofits have given management and operators a method of trying twin-wire forming without the risk of an outright fourdrinier replacement. Since the sheet forming process on a fourdrinier is a stepwise filtration process, a partially formed web has a high consistency near the wire and remains close to headbox consistency on the top of the sheet. In the absence of high table turbulence or shear, the reflocculation of stock near the top surface is a major papermaking problem. Most top formers improve formation by creating high shear at the nip where the second wire is introduced.

16.7 WHITE WATER SYSTEM

White water is the drainage from wet stock, regardless of color, in pulping and papermaking operations. Besides fiber, white water may contain a variety of other furnish-derived materials. White waters may be classified as either rich or lean depending on their fiber content. This classification is an arbitrary one, but in most instances a consistency above 0.01 or 0.02% would qualify as being rich. In any papermaking process, the white water is richest from the area where the jet impinges on the wire and becomes progressively leaner as a fiber mat builds up on the forming fabric.

Figure 16-50 shows a conventional, well-closed white water system for a fourdrinier machine. The rich white waters drained from the static and low-vacuum elements are collected in the wire pit and immediately recirculated in the approach loop. The seal pit is divided into sections to facilitate usage of the richer seal pit water as makeup in the wire pit or dilution in the stock preparation area. Only the leanest water from the seal pit is removed from the system, and it is first taken to a fiber recovery unit. Although not shown on the figure, the complete white water system also includes such uses for clarified white water as wire and felt showers, press section waters, vacuum system waters, cooling water, etc.

The white water system of a modern paper machine must achieve a number of important objectives:

- provide sufficient quantities and qualities of white water for the various paper machine services during both regular operation and upset conditions.
- ensure that the fines content of headbox stock does not vary appreciably with time in order to minimize operating upsets and maintain uniform product quality.

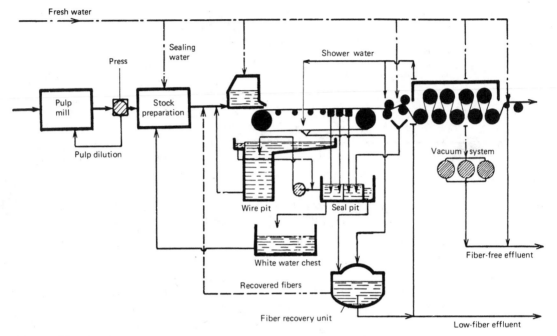

FIGURE 16-50. Well-closed white water system.

• maximize re-use of white waters in order to reduce fiber and energy losses and minimize effluent loadings.

In support of the above objectives, the following operating guidelines have been recommended as the basis for successful paper machine white water management (7):

• return fines to the headbox via the shortest possible route.
• maintain totally separate broke handling and fines recovery systems.
• provide sufficient white water storage to accommodate swings in stock chest levels.
• sewer only excess clear (filtered) white water.
• minimize load fluctuations to the saveall.
• provide independent white water systems for each paper machine in the mill.

The modern trend is toward greater "closure" of the white water system to reduce the fresh water introduced into the system (8). This objective can be partly achieved by using equipment that requires less fresh water and partly by re-using filtered white water. Also, fiber-free cooling waters can be collected separately and either re-used as fresh water or piped directly to the receiving water.

During the 1980's, a number of paper mills succeeded in reducing fresh water requirements by 80-90%. This degree of closure results in buildup of dissolved solids within the white water system. Some mills have reported loss of drainage, higher corrosion rates and deposition problems within the tightly-closed systems. However, the savings in effluent treatment, raw water treatment, and energy conservation provide strong incentives for the highest practicable closure level.

Fiber Recovery From White Water

To meet the requirements of both ecology and economy, the reusable fibers and fillers in paper machine white water overflow streams must be reclaimed into the machine furnish. Although drum-type filters are sometimes used for this service (refer back to Section 9.6), the two types of equipment most often employed are flotation savealls and disc filters. The flotation unit accomplishes recovery by attaching fine air bubbles to the fiber and filler, and then floating these materials to the surface for removal by a skimming device (Figure 16-51). The disc filter consists of a series of discs mounted on a center shaft (refer back to Figure 9-46). The water flows through the filtering medium into each sector and then through the core of the rotating shaft. A representative disc filter system is shown in Figure 16-52. The operating sequence of a typical disc saveall is shown in Figure 16-53.

16.8 BROKE SYSTEM

Broke is defined as partly or completely manufactured paper or paperboard that is discarded from any point in the manufacturing or finishing process. The term also applies to the furnish made by repulping these materials. Wet broke is taken

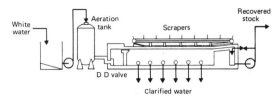

FIGURE 16-51. Flotation saveall system for recovering fiber from white water.

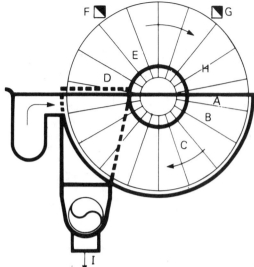

FIGURE 16-53. The disc filter cycle: (A) sector begins to gravity-fill from core outward; (B) vacuum on, cloudy filtrate collected; (C) clear filtrate obtained; (D) vacuum off; (E) atmospheric port opens; (F) knock-off shower peels mat; (G) wire washing starts; (H) atmospheric port closes; (I) thickened pulp discharged.

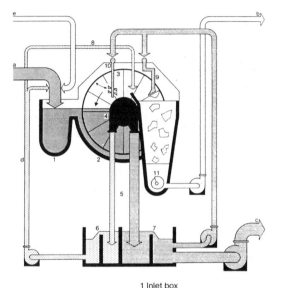

Legend

a White water
b High-consistency stock
c Clear filtrate
d Cloudy filtrate
e Sweetener stock

1 Inlet box
2 Filter vat
3 Filter discs
4 Suction and control head
5 Droplegs
6 Cloudy filtrate tank
7 Clear filtrate tank
8 Stock dilution
9 Cake removal shower
10 Cleaning shower
11 Pulping screw

FIGURE 16-52. Voith disc filter system.

from the forming and pressing sections, while dry broke emanates from the dryers, calenders, reel, winder, and finishing operations.

A broke handling and repulping system is an essential feature on any paper machine. As a minimum, a wet-end pulper (couch pit) is necessary after the forming section and presses, and a dry-end pulper is required to take waste from the dryers, calenders, reel and winder. During threading and machine breaks, both systems must be capable of handling maximum tonnage from the machine. At the same time, both systems are required to handle small amounts on a continuous basis (e.g., couch

trim at the wet end; winder trim and "slab-off" returns at the dry end). Another feature of the system is sufficient broke storage capacity to sustain long periods of upset operation. From the storage tank, a controlled flow is reintroduced through the blending system into the machine furnish.

The task is comparatively simple at the wet end, where the sheet disintegrates easily and minimal retention time is required. The system shown in Figure 16-54 utilizes direct discharge from both wire and presses. (With some press arrangements, the sheet must be doctored off the press roll and screw-conveyed to the side of the machine before it can enter the chute to the couch pit.) Any of several types of agitators can be used, with extraction generally through a perforated plate (Figure 16-55). Low-volume showers and a small pump are used during normal operation. When a break occurs, high-volume (deluge) showers automatically come on to slush the stock. When the couch pit reaches a pre-determined level, a high-volume pump kicks in to remove the slurry at the same rate that it is formed, and transfers it to storage.

The system at the dry end (e.g., Figures 16-56 and 16-57) functions according to the same principles. However, heavier-duty agitation and deflaking equipment is used (Figure 16-58) and retention time in the repulping unit is greater. Most of these units also utilize recirculation of the slurry to help break up the fiber bundles and prevent "festooning" (i.e., buildup on the surface) and plugging.

16.9 PRESSING

The primary objectives of paper machine pressing are to remove water from the sheet and consolidate the web. Other objectives, depending on product requirements, may be to provide surface smoothness, reduce bulk, and promote higher wet web strength for good runnability in the dryer section. The rather fragile paper web is transferred from the forming

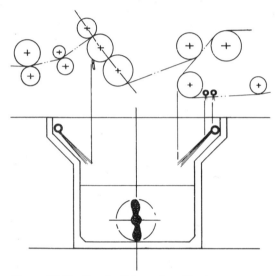

FIGURE 16-54. Couch pit for wet-end broke handling (Black Clawson).

FIGURE 16-55. Representative disintegrator for couch pit (Black Clawson).

section and conveyed on specially-constructed felts through a series of roll press nips and into the dryer section.

The pressing operation may be considered an extension of the water-removal process that was started on the wire. It is far more economical to remove water by mechanical means than by evaporation, so the papermaker is always looking for methods to improve pressing efficiency and reduce the evaporative load into the dryer section. Water removal should be uniform across the machine so that the pressed sheet has a level moisture profile entering the dryer section.

Sheet consolidation is a crucial phase of the papermaking process. It is here that the fibers are forced into intimate contact so that fiber-to-fiber bonding develops during drying. If the consolidation were to take place instead following drying, the product would invariably be weak and bulky. The cross-section photo-micrographs in Figure 16-59 illustrate the consolidation occurring between the 2nd and 3rd presses, as compared to the calendered sheet.

The sheet moisture content during pressing is usually expressed as consistency or % dryness. The reader should bear in mind that changes in % dryness are not a good indication of differences in moisture content; a better indicator is the water-to-fiber ratio. For example, a change in sheet dryness from 40 to 41% leaving the presses might be calculated as a 2.5% increase in dryness. In fact, the water-to-fiber ratio at 40% dryness is 1.50 and at 41% is 1.44, so the real difference in dryness is 0.06 or 4.0%. A value of 4.0% is quite different than 2.5%, and indicates the true importance of a 1% dryness change during pressing.

What Happens in the Press Nip?

Virtually all continuous wet pressing of paper webs is carried out in a two-roll press nip. The sheet is carried into the nip and supported during pressing by a specially constructed "felt". The basic concepts

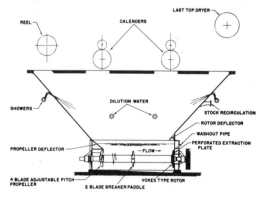

FIGURE 16-56. Dry-end broke handling arrangement (Black-Clawson).

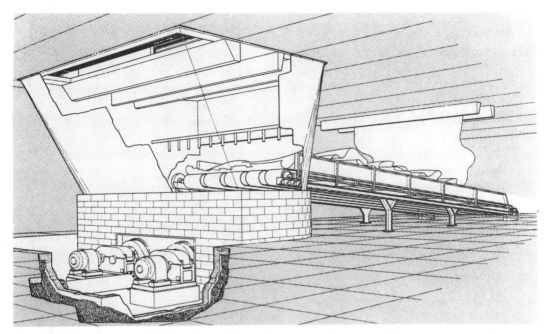

FIGURE 16-57. Some dry-end broke handling systems utilize a conveyor to bring broke from the size press or from other points within the dryer system.

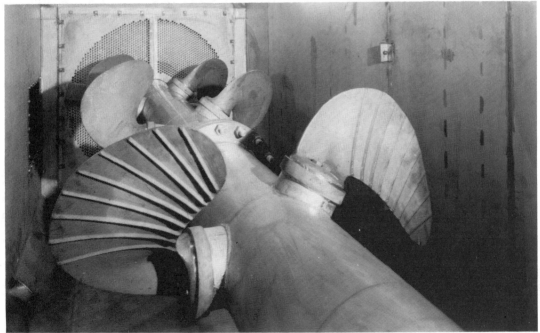

FIGURE 16-58. Typical paper machine dry-end broke pulper (Impco/Ingersoll Rand).

of pressing, first formulated by Wahlstrom, have been progressively refined by Wahlstrom (9,10) and others. Pressing is now considered to be basically a flow phenomenon controlled by the flow of water between fibers and from the fiber wall.

It is convenient to consider the pressing process as occurring in four phases, as illustrated in Figure 16-60:

• In phase 1, compression of sheet and felt begins; air flows out of both structures until the sheet is

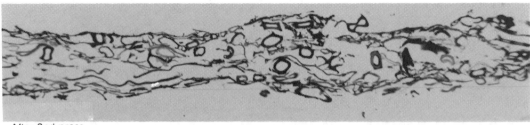

After 2nd press

After 3rd press

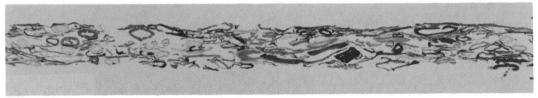

After calendering

FIGURE 16-59. Photo-micrographs of newsprint cross-sections illustrating sheet consolidation and compaction (courtesy MacMillan Bloedel Research).

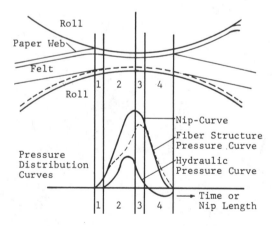

FIGURE 16-60. Four phases in the press nip.

saturated; no hydraulic pressure is built up (and therefore, no driving force for dewatering).

- In phase 2, the sheet is saturated and hydraulic pressure within the sheet structure causes water to move from the paper into the felt. If and when the felt also becomes saturated, water moves out of the felt. Phase 2 continues up to the mid-nip

where total pressure reaches maximum. It has been shown that the hydraulic pressure reaches maximum just prior to mid-nip.

- In phase 3, the nip expands until the hydraulic pressure in the paper is zero, corresponding to the point of maximum paper dryness.
- In phase 4, both paper and felt expand and the paper becomes unsaturated. Although a negative pressure is created in both structures, a number of factors cause water to return from the felt to the paper.

This depiction of pressing leads to insights regarding important factors in pressing. For example, the felt must have enough permeability when compressed to allow passage of water squeezed out of the paper. As the felt itself becomes saturated, some receptacles are needed to receive the water squeezed out of the felt. Addressing these concerns, the "clothing" manufacturers have designed felts which retain permeability at high nip loading; machine builders have provided receptacles of various types in the rolls and auxiliary equipment.

The reabsorption of water from the felt in phase 4 is recognized as a serious limitation to idealized water removal. Various mechanisms for this behavior are postulated, including capillary absorption and mechanical absorption. It is known that additional "rewetting" occurs following the nip unless sheet and felt are immediately separated; all modern press configurations provide for this separation.

Limitations to Pressing

Every press nip is limited in effectiveness either by how rapidly water can move out of the sheet or nip ("flow-limited" case) or by the extent of possible sheet compression ("pressure-limited" case). Generally, if

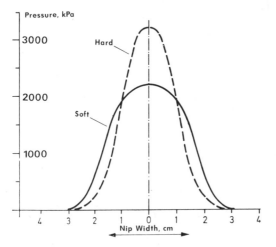

FIGURE 16-61. Press nip pressure distribution, utilizing hard and soft roll covers (45 kN/m loading).

the sheet readily loses its water and there is no impediment to water movement from the nip, or if the sheet has a low water content, a relatively high pressure can be applied to the web without reservation. Excessive pressure applied in a flow-limited situation is referred to as "sheet crushing". In its severest manifestation, sheet crushing causes disintegration of the sheet. More usually, the symptoms are localized washing out of fines, realignment of fibers, and washing out of fibers into the felt.

Pressing is always carried out in a series of nips (typically three) with increasing pressure in each nip. The first nip is almost always flow-limited to a certain extent. The effectiveness of the last nip is limited only by the magnitude of the pressure that can be applied. In theory at least, if the pressure could be increased by an order of magnitude, a significant increase in water removal could be achieved. However, as machine speeds increase, a higher pressure will have a diminished effect because of the brief residence time in a conventional roll press nip. Water removal is actually a logarithmic function of the nip impulse (area under the time/pressure curve).

Generalities About Presses

The sheet and felt are pressed in the nip between two rotating rolls. Generally, the top roll is mechanically loaded to create the desired pressure within the nip. In vertical press arrangements the lineal nip pressure is the sum of the mechanical loading and the weight of the top roll, divided by the length of the contacting face at the press nip. (The weight of the mating roll is not a factor in some inverse and cluster press arrangements.) The actual pressure developed in the nip follows a curve as illustrated in Figure 16-61. The maximum value depends on several factors, including lineal pressure, roll diameter, roll hardness, and the characteristics of the felt.

In single-felted nips, one side of the sheet is usually in contact with a smooth, hard roll, usually granite or an artificial stone material. Sometimes, a solid metal roll with or without a hard rubber covering is used. The roll in contact with the felt is usually covered with an elastomer material to control roll hardness. A somewhat resilient cover may be used to "soften" the nip on heavyweight grades or a bone-hard cover may be used with lightweight grades to obtain a narrow nip.

Because press rolls deflect under loading, it is necessary to camber or "crown" one or both of the rolls to achieve a uniform pressure profile across the contacting face. The amount of crown required is a function of roll diameter and length, material of construction, and applied load. A fixed crown (obtained by grinding the roll) is suitable for only one loading. The effect of correct and incorrect crowning on the pressure distribution is illustrated in Figure 16-62.

A machine manufacturing many different weights and grades of paper usually employs a controlled-crown roll to compensate for variable loading. Several designs of variable-crown rolls are available, two of which are illustrated in Figures 16-63 and 16-64. Most designs utilize hydraulic pressure to shift deflection from the rotating shell to a stationary shaft.

In general, uniform pressing across the machine is desired to achieve a reasonably level moisture profile. Occasionally, because of non-uniform drainage on the wire, or other problem, it may be desirable to compensate by uneven pressing action. In this situation, a controlled-crown roll with variable deflection across the machine may be employed, one design of which is illustrated in Figures 16-65 and 16-66. The width of the "streak" that can be corrected is limited by the bending characteristics of the roll shell. (Steam showers in the press section may be a better method of profile correction. Refer to subsequent discussion.)

Types of Presses

Fundamental research in wet pressing has shown that the most important requirement in press design is to provide the shortest path for the water to follow in escaping from the nip. The shortest distance is equal to the felt thickness (usually referred to as the vertical direction). Therefore, the main water flow

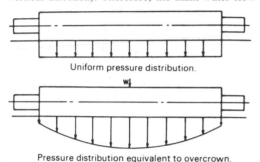

Uniform pressure distribution.

Pressure distribution equivalent to overcrown.

Pressure distribution equivalent to undercrown.

FIGURE 16-62. Although the amount of crown may be only 1.3 mm (0.05 inches) at the midpoint of the roll, small amounts of overcrowning or undercrowning will affect the pressure distribution.

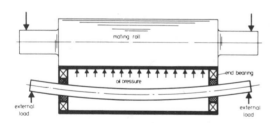

FIGURE 16-63. The "swimming roll" was the first design of controlled-crown roll; it utilizes oil pressure in the upper shell cavity to shift deflection from the rotating shell to the fixed shaft.

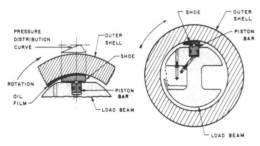

FIGURE 16-64. The Beloit controlled-crown roll utilizes a hydraulically-loaded, lubricated shoe to bear on the rotating shell.

should be perpendicular to the felt, and lateral flow should be minimized. Presses with dominant vertical flow have become known as transverse-flow presses.

The original presses had plain rolls. These presses were severely flow-limited because the water could only leave on the entering side of the nip by lateral movement (refer to Figure 16-67).

The suction press developed in the early 1900's was the first approach toward transverse-flow pressing. This design utilizes a suction roll (a perforated shell that rotates around a stationary suction box) as a receptor for the expressed water. The holes in the shell (Figure 16-68) provide an easier escape route for the water (as compared to the plain roll nip). Some water is able to flow directly to the holes, but (as can be seen in the figure) most of the expressed water also must travel laterally to escape the nip. Suction presses are still utilized on many paper machines; however, the hollow shell construction with perforations limits the amount of pressure that can be applied. Also, stress fatigue failure of suction rolls has been a continuing problem over the years.

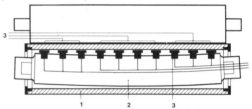

FIGURE 16-65. The Escher-Wyss Nipco Roll controls the force distribution by means of individually-controlled hydrostatic elements. (1) rotating shell; (2) stationary girder; (3) piston elements.

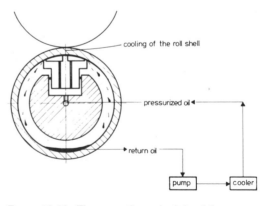

FIGURE 16-66. The operating principle of the Escher-Wyss Nipco Roll is illustrated. Oil under pressure is supplied to the underside of the elements, and through capillary ducts to the top side. As it escapes over the edge of the hydrostatic element, the oil provides a lubricating film between the shell and element.

FIGURE 16-67. Illustration of plain press nip. The expressed water from the sheet and felt can only leave at the entering side of the nip by lateral movement. (Drawing courtesy of Beloit Corp.)

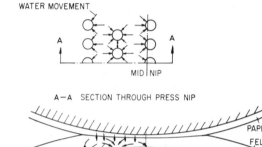

FIGURE 16-68. Lateral movement of water in a suction press is viewed in two planes.

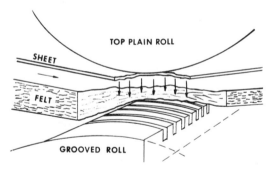

FIGURE 16-69. Illustration of grooved-roll press nip showing how the grooves provide an outlet for expressed water.

The grooved-roll transverse-flow press was introduced in 1963. As illustrated in Figure 16-69, the grooves in the roll cover provide easily-accessible receptacles for expelled water. The helically-cut grooves are typically 2.5 mm (0.1 inch) in depth, 0.5 mm (0.02 inch) wide on 3.2 mm (0.125 inch) centers (i.e., 8 grooves per inch). The maximum lateral distance for water travel in the grooved press is, therefore, only 1.3 mm (0.05 inch),

as compared to typical figures of 5 mm and 20 mm, respectively, for the suction and plain presses. Since the grooved roll is solid, higher pressures can be applied. The water caught in the grooves is thrown off by centrifugal force at high roll surface speeds, and the roll is cleaned by the action of sprays and doctor blades. Grooved roll covers must be fairly hard to maintain groove integrity.

Another innovation toward true transverse-pressing is the utilization of a blind-drilled receptor roll. Only the cover of the solid roll is drilled with small, closely-spaced holes. In comparison to the grooved roll (Figure 16-70), the blind-drilled roll has greater void volume, and since there is less tendency to close holes than grooves, blind-drilled patterns can be installed in softer roll covers. The wells tend to self-clean by the action of centrifugal force. A typical blind-drilled press arrangement is shown in Figure 16-71.

The fabric press is another development of transverse press which is used more in Europe than in North America. The concept is illustrated in Figure 16-72. A multiple-weave, non-compressible fabric belt passes through the nip between the felt and the rubber-covered roll to provide void volume to receive the expressed water. Water is removed from the fabric by passing it over a suction box on the return run. The shrink-sleeve press is a simplified modification of the fabric press, utilizing only a non-compressible fabric jacket or sleeve which is shrunk over the press roll to provide void volume. Modern press felt designs with high void volumes have reduced the benefits of a separate non-compressible fabric or sleeve.

One of the most effective strategies in the first press position (i.e., for flow-limited presses) is to use a double-felted and double-vented nip (i.e., with grooved or blind-drilled rolls in both positions) as illustrated in Figure 16-73. In this arrangement, dewatering takes place in both directions, thus reducing the distance for vertical travel. Double-felting is also useful at the second and third presses for sheets above 130 g/m^2 basis weight.

A totally new type of press was introduced in 1981. The "Extended-Nip Press" features a very wide nip to give the sheet a long dwell time at high pressure. When used as the last nip, this press provides not only a much drier sheet, but also a stronger sheet due to improved consolidation of the web structure. The configuration used to obtain the wide nip is shown in Figure 16-74. Key components are the stationary pressure shoe and the impervious elastomer belt, which form the bottom portion of the double-felted nip. The shoe is continuously lubricated by oil to act as a slip bearing for the belt.

The average loading of 4100 kPa (600psi) along the 25 cm (10 inch) length of shoe is equivalent to 1050 kN/m (6000 pli) by the normal method of

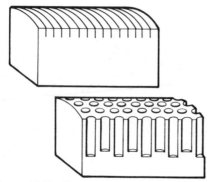

FIGURE 16-70. Comparison of roll segments showing the greater void volume of blind-drilled holes as compared to grooves.

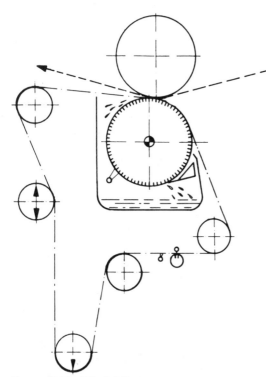

FIGURE 16-71. Blind-drilled press arrangement.

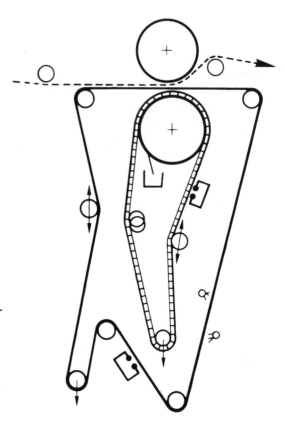

FIGURE 16-72. Fabric press arrangement (Albany International Corp.).

specifying press loading. The relatively low pressure level, together with a controlled rate of compression, reduces the crushing potential. A dramatic improvement in dewatering is accomplished because the pressure is sustained for a long period of time, up to eight times that obtained with conventional roll presses (refer to Figure 16-75). Wide nip presses have thus far only been applied to heavyweight sheets.

Since the introduction of the Extended-Nip Press, other suppliers have developed their own wide-nip shoe-press designs which also provide high sheet

dryness and good web consolidation. One design (shown in Figure 16-76) utilizes a concave shoe inside a flexible roll shell in the bottom position and a controlled crown roll in the top position.

Sheet Transfer from Wire to Press Section

Originally, all paper machines employed an open-draw sheet transfer from the forming section to the press section. One type of open-draw configuration is illustrated in Figure 16-77. The tension needed to pull the wet web off the forming fabric is provided by a speed differential between the press and forming sections. (Wet paper develops tension when stretched, but as with other visco-elastic materials, the tension decays rapidly with time. For each subsequent open draw, sheet tension must be re-imposed by an additional speed differential between sections.) Open draws are commonly used for heavyweight sheets and are still found on older lightweight machines operating at speeds below 600 m/min (2000 ft/min). Runnability problems are encountered at higher speeds because the required tension for removing the sheet from the forming fabric increases exponentially with speed.

The suction pickup concept, illustrated in Figure

16-78, was first applied commercially in 1954, and is now an integral part of most press arrangements for lightweight sheets. With this system, the sheet is picked up off the wire by a felt which wraps a suction roll at the point of contact. The sheet is then carried underneath the felt either to a transfer press or directly into the first press nip.

With a suction pickup, the first open draw on the paper machine occurs in the press section or the dryers. The first open draw is always a critical point with respect to machine operation, because the wet sheet is relatively weak and subject to stresses from

a variety of causes. At greater speeds, the need to provide support for lightweight sheets throughout the pressing operation and into the drying section has become paramount. A number of "no-draw" press configurations are offered by equipment manufacturers.

Types of Press Arrangements

The straight-through press is the oldest and simplest press arrangement, and is still used on pulp

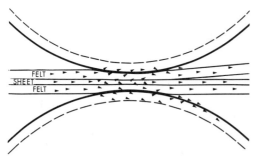

FIGURE 16-73. Double-felted nip with grooved rolls used in both positions (courtesy of Beloit Corp.).

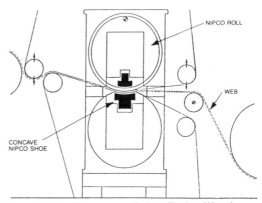

FIGURE 16-76. Shoe press (Sulzer Escher Wyss).

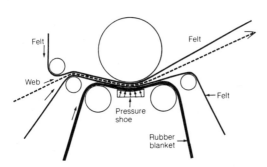

FIGURE 16-74. Arrangement of Extended-Nip Press (Beloit Corp.).

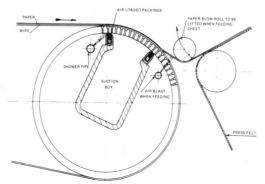

FIGURE 16-77. Open-draw sheet transfer. The position of the sheet takeoff in relation to the couch roll suction box is typical.

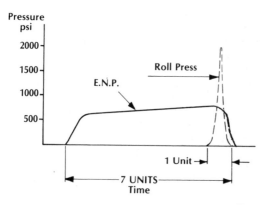

FIGURE 16-75. Comparison of nip pressure profiles (Beloit Corp.).

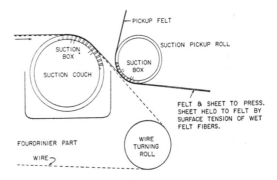

FIGURE 16-78. Suction pickup arrangement.

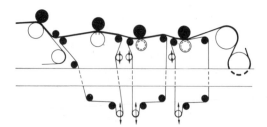

FIGURE 16-79. Straight-through press (Beloit Corp.)

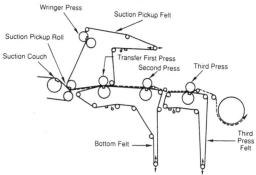

FIGURE 16-80. Suction transfer press arrangement.

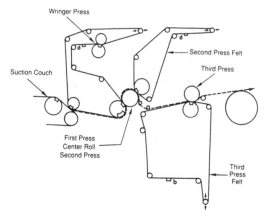

FIGURE 16-81. Twinver press arrangement.

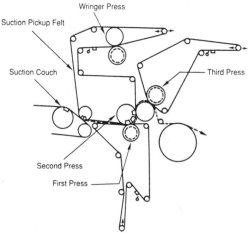

FIGURE 16-82. Modern three-nip no-draw press section.

and board machines. Originally, each press within the series arrangement consisted of a smooth top roll and a bottom felted roll (Figure 16-79), so that only the top surface of the sheet received smooth roll contact. Later, an inverse second press (smooth roll on bottom) was utilized so that the wire side of the sheet also contacted a smooth surface.

A number of transfer press arrangements have been used on suction pickup machines to transfer the sheet from the pickup felt to the first press felt without an open draw. One early arrangement utilizing a double-felted transfer nip is shown in Figure 16-80. The original suction transfer presses were not designed for dewatering, but modern transfer presses with double-felted, double-vented nips are very effective.

A number of arrangements have been developed as an alternative to the transfer press, the best known being the twinver press as illustrated in Figure 16-81. This arrangement has a number of advantages; the draw is eliminated between the first and second presses; the wire side contacts the smooth roll through two nips; and only one smooth roll is required for those two nips.

The demands of high-speed operation led to the development of no-draw presses which provide support to lightweight sheets throughout the pressing operation. A representative arrangement utilizing a double-felted nip is shown in Figure 16-82. Sheet marking problems and suction roll failures on some wide machines prompted the development of

alternate no-draw configurations using separate rolls for the first and second press nips.

Most of the water is removed from the top side of the sheet in triple-nip no-draw press arrangements. Single-sided dewatering can cause problems on certain paper grades because of differences in coating and inking receptivity between the two sides. This concern has led to the installation of separate open-draw fourth presses on some machines as illustrated in Figure 16-83. The fourth press serves to increase web dryness to the dryer section as well as reduce two-sidedness.

Hot Pressing

Pressing at sheet temperatures in the 60 to 90° C range is a modern technique for increased press dewatering and improved sheet consolidation. A higher sheet temperature softens the cellulose fibers and makes the sheet more compressible; it also increases water fluidity, enabling the water to leave the nip more rapidly. Operating data indicate that each 10° C increase in temperature will raise sheet dryness by about one percentage point from the last press nip.

The common method of increasing sheet temperature is by means of steam showers (11), both on the fourdrinier and in the press section. Typically, the steam is applied to one side of the sheet while suction is applied through the fabric support on the opposite side; the steam is pulled into the paper structure where it condenses and gives up its heat. Many modern steam shower designs are equipped with segmented control across the machine to provide effective profile correction.

An exciting new development in hot pressing technology is embodied in the TEM-SEC press which was invented by a small Spanish paper machinery company, DG International. This press consists basically of a large-diameter (up to 3 m) center roll which is internally heated by steam injection (up to 3 bar). Acting on this center roll are two or more press rolls which are either grooved or blind-drilled. Modest nip loadings in the 80 to 280 kN/m range are applied to these nips. A representative layout is illustrated in Figure 16-84 (12). The dryness levels achieved by these TEM-SEC installations have generally been higher than expected. Because of the absence of high pressure, this approach requires a lower capital investment than wide-nip presses. As with other methods that produce higher press dryness, improved sheet strength is an added benefit. The paper surface in contact with the hot metal receives a smoothing finish.

Press Clothing

The press fabric (still commonly called the "press felt") plays a dual role in pressing operations. It supports and conveys the paper web through the various operations and assists in paper web dewatering. It also acts as a transmission belt to drive other components of the press section. To perform these functions consistently and economically, the felt must be fabricated to exacting standards of durability, dimensional stability, permeability, cleanability, pliability, and surface finish.

The evolution of press clothing from "conventional felts" to engineered fabrics has paralleled changes in paper machine design (13). Originally, all felts were woven from spun yarns and were mechanically felted or "fulled" to provide bulk and stability. These felts, manufactured mainly from wool, had low vertical permeability and poor service life. The needling process developed in the 1950's made possible the introduction of the "batt-on-base" felt. This construction involves the vertical needling of a fibrous web (batt) onto a woven base fabric.

This change allowed higher percentages of synthetic fibers to be used.

As synthetic monofilament and multifilament yarns became available, many more options became possible for the base structure. The resultant fabrics became known as "batt-on-mesh" constructions because the base somewhat resembled a screen. Single-layer (see Figure 16-85) or multiple layer weave patterns may be used depending on requirements. The type of batt selected is determined by the particular press position being considered and by the finish requirements of the sheet.

Conventional multi-layered fabric constructions sometimes have pronounced "knuckles" at yarn

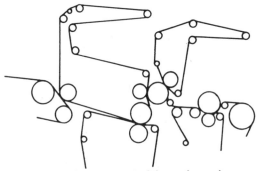

FIGURE 16-83. Arrangement of three-nip no-draw presses with added open-draw fourth press.

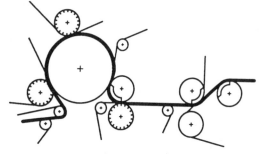

FIGURE 16-84. Press layout showing a conventional first press followed by three TEM-SEC press nips.

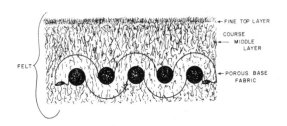

FIGURE 16-85. Single-layer batt-on-mesh press fabric construction.

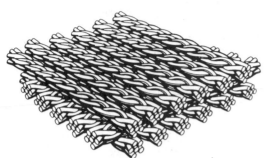

FIGURE 16-86. This triple-layer planar base fabric is composed of plied monofilament yarns in both directions.

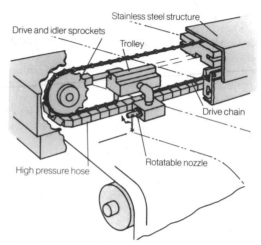

FIGURE 16-88. This single high-pressure traversing nozzle delivers 1.0 to 1.5 gpm of water at 1000 psig (Lodding Engineering Corp.).

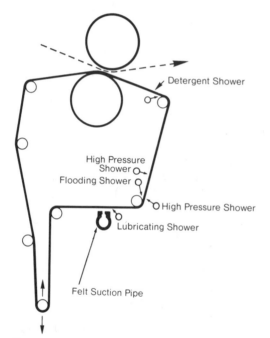

FIGURE 16-87. Conditioning system for a modern press fabric (Albany International Corp.).

crossover points which cause marking on certain grades of paper. To overcome this problem, alternate base constructions were developed in which a grid of warp (machine-direction) and shute (cross-direction) yarns were laid perpendicular to each other without the benefit of any interweaving. These "planar" press fabrics (Figure 16-86) offer greater sheet-side support because the top-side warp yarns can be closely grouped. Different types of base constructions can also be needled together into a multi-layer base to provide further options. As a group, multilayer fabrics have the advantage of a large void volume in the non-compressible base structure. If they are properly dewatered with suction boxes before entering the nip,

these fabrics often have sufficient capacity to retain all the water expressed from the sheet in the nip.

Specifying a press fabric for a particular position of the paper machine is usually a joint effort by the supplier and the papermaker. Often, the specification for a fabric evolves from a program of machine trials and design modifications. Ultimately, press fabric selection depends on the grade of paper being produced, its weight and finish requirements, the machine type, the severity of the conditions under which the fabric will be used, and the fabric conditioning equipment available. The manufacturer can affect the properties of the fabric by modifying yarn count, yarn form and weight, weave patterns, batt distribution and fineness, chemical treatments, overall synthetic content, and other manufacturing processes where applicable.

Felt Run

The typical press fabric run includes tensioning and positioning rolls, and provides a means to condition and dewater the felt to keep it permeable and open. A variety of mechanical and/or chemical conditioning treatments are used depending on the requirements.

Most systems utilize hydraulic energy (in the form of high- and low-pressure showers) as the primary means to loosen and flush out fines and fillers from the fabric structure. A representative system is shown in Figure 16-87, and one type of traversing high-pressure shower is shown in Figure 16-88. Following the showering treatments, the fabric is dewatered and the loosened filling material is removed either by a suction box or a "wringer press". If hydraulic energy by itself is insufficient, detergents and/or chemicals can be added to provide

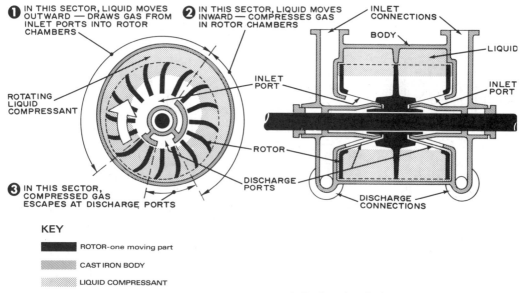

❶ IN THIS SECTOR, LIQUID MOVES OUTWARD — DRAWS GAS FROM INLET PORTS INTO ROTOR CHAMBERS

❷ IN THIS SECTOR, LIQUID MOVES INWARD — COMPRESSES GAS IN ROTOR CHAMBERS

INLET CONNECTIONS

BODY

LIQUID

INLET PORT

INLET PORT

ROTATING LIQUID COMPRESSANT

ROTOR

❸ IN THIS SECTOR, COMPRESSED GAS ESCAPES AT DISCHARGE PORTS

DISCHARGE PORTS

DISCHARGE CONNECTIONS

KEY

ROTOR-one moving part

CAST IRON BODY

LIQUID COMPRESSANT

FIGURE 16-89. **Schematic of water-ring vacuum pump (Nash Engineering Co.).**

a more specific action. Occasionally, a stationary or rotating "brush" contacting the forming side can be helpful in dislodging contaminants.

Variables of Operation

There are a large number of factors affecting press dewatering from the standpoints of press layout, nip configuration, press roll coverings and crowns, stock and sheet characteristics, felt clothing selection and conditioning, and operational variables. In most installations, nip pressure is the most significant operating parameter, and it should be maintained at the highest level consistent with quality requirements.

The selection of nip width and dwell time can be related to sheet basis weight and speed. For lightweight grades, maximum dewatering generally occurs with narrow hard nips, while heavyweight sheets are more responsive to wide, soft nips. Press performance deteriorates with increased machine speed because of lower residence time in the nip. Sheet rewetting is a particular problem with lightweight sheets; the felt and sheet should always be separated immediately unless serious runnability problems are caused.

16.10 VACUUM SYSTEM

A properly designed and installed vacuum system is a vital component in a successful paper machine system. Adequate vacuum pumping capability must be provided for all vacuum-assisted drainage elements, suction boxes, and felt-conditioning boxes in the forming and press sections. Several types of vacuum generating equipment are effectively used in

paper machine systems, including centrifugal exhausters and rotary lobe pumps. The most commonly used equipment is the positive-displacement, liquid-ring vacuum pump as illustrated in Figure 16-89. Pumps of this type provide full-range vacuum capability and easily handle mixtures of water and air.

A representative vacuum system with eight pumps is illustrated schematically in Figure 16-90. Figure 16-91 shows a row of pumps in the basement of a machine room. The pumping load is distributed to the respective units using a vacuum header which is flanged at intervals to isolate different services (i.e., vacuum levels). The use of a header gives the system flexibility for shared service (in case one pump is shut down for any reason) or future modifications.

REFERENCES

(1) BYERS, D. **Proper Component Fit Is Key Design Factor for Headbox Approach System** *Pulp & Paper* (November 1991)

(2) STEWARD, J. **The Design and Operation of Paper Machine Stock Deaeration Systems** *Pulp & Paper* (March 1981)

(3) KEREKES, R.J. and KOLLER, E.B. **Equations for Calculating Headbox Jet Contraction and Angle of Outflow** *TAPPI 64:1:95* (January 1981)

(4) TRUFITT, A.D. **Design Aspects of Manifold-Type Flowspreaders** Joint Textbook Committee (1975)

(5) PARKER, J.D. **The Sheet Forming Process** TAPPI STAP No. 9 (1972)

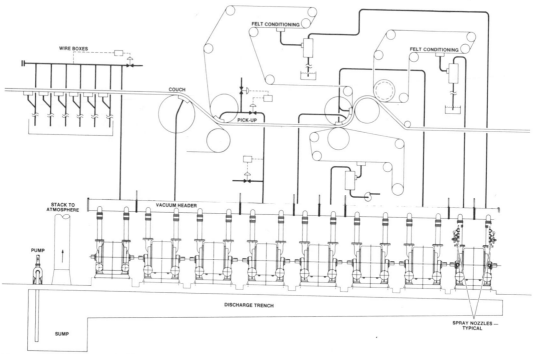

FIGURE 16-90. Schematic diagram of paper machine vacuum system (Nash Engineering Co.).

FIGURE 16-91. Typical vacuum pump installation (Nash Engineering Co.).

(6) KALLMES, O., ET AL. **A Novel Approach To Optimizing Sheet Formation on the Fourdrinier** *Tappi Journal* (April 1989)

(7) MERRIMAN, K. **Paper Machine White Water Management** CPPA Tech 90 Papermaking Course Notes (November 1990)

(8) WAHREN, D. **Water Conservation in Papermaking** *P&P Canada 86:3:T88-92* (March 1985)

(9) WAHLSTROM, B.P. **Our Present Understanding of the Fundamentals of Pressing** *P&P Mag Canada 70:19:T349* (December 20, 1969)

(10) WAHLSTROM, B.P. **Pressing - the State of the Art and Future Possibilities** *Paper Tech* (February 1991)

(11) BELL, N.E. **Hot Pressing Using Steam Showers Can Effectively Boost Production** *Pulp & Paper* (September 1987)

(12) WALKER, K. **Advances in Hot Pressing Technology** *Tappi Journal* (August 1990)

(13) HANSEN, R.A. **Steady-State Press Felts Optimize Paper Machine Output, Performance** *Pulp & Paper* (December 1991)

Chapter 17

Paper Manufacture - Dry End Operations

17.1 PAPER DRYING

After pressing, the sheet is conveyed through the dryer section where the residual water is removed by evaporation. On conventional paper machines, the thermal energy for drying is transferred to the paper by wrapping a series of large-diameter, rotating, steam-filled cylinders. A representative steam-cylinder drying system for lightweight papers is shown in Figure 17-1.

The massive dryer section is the most expensive part of a paper machine in terms of capital cost. It is also the most costly to operate because of the high energy consumption. Therefore, efforts to increase evaporation rate (to reduce the number of dryers) and conserve energy (to reduce steam usage) are usually well justified. Unfortunately, the drying operation does not appear to receive as much attention as other parts of the process, and many opportunities to improve efficiency are lost.

Criteria of Performance

Two indices are important in assessing the performance of a dryer section: evaporation rate and steam economy. Evaporation rate is measured as pounds of water evaporated per hour per square foot of dryer surface contacted (or equivalent metric units). A high evaporation rate is desirable with respect to equipment requirements, but drying must always be carried out within the constraints of the paper grade being produced. For example, a relatively low drying rate is required on some grades to ensure product quality. Generally, uniform evaporation across the machine is desired. However, if it is necessary to compensate for moisture streaks or other profile problems, techniques are available to raise or lower the evaporation rate at selected positions across the machine.

The evaporation drying rate is greatly influenced by the steam pressure used inside the drying cylinders. When assessing the operation of different machines making the same type of paper, comparisons should be made on the basis of "lines of equivalent performance" as illustrated in Figure 17-2. Actual paper machine evaporation rate data for different products are provided in a series of TAPPI data sheets (e.g., Figure 17-3).

Steam economy is measured as thousands of BTU's per pound of water evaporated (or kJ per kg) or as mass of steam per unit mass of water evaporated. Obviously, a low steam usage is desirable for the most economical operation. A value of 1.3 kg steam per kg water evaporated is typical for a modern well-designed, well-maintained system, but many machines have significantly higher usage. The steam/condensate and air handling systems have the greatest impact on energy consumption.

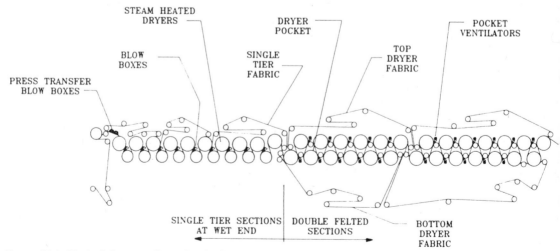

FIGURE 17-1. Typical dryer configuration for lightweight papers.

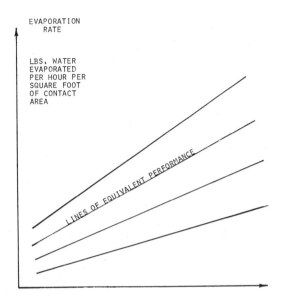

FIGURE 17-2. Effect of steam temperature on cylinder drying evaporation rate.

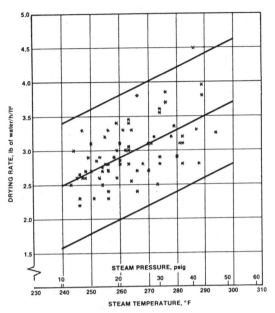

FIGURE 17-3. Evaporation drying rates for paper machines producing newsprint at speeds over 2000 ft/min. (TAPPI TIS 0404-15, 1986)

Zones of Evaporation

The drying rate varies along the machine. The first two or three cylinders into the drying section serve principally to raise the temperature of the sheet (the "warm-up zone"). Evaporation then quickly reaches a peak rate which is maintained as long as water is present on the fiber surfaces or within the large capillaries ("constant rate zone"). At the point where the remaining free moisture is concentrated in the smaller capillaries, the rate begins to decrease ("falling rate zone"). Finally, at about 9% moisture, the residual water within the sheet is more tightly held by physicochemical forces, and the evaporation rate is further reduced ("bound water zone"). The various zones are illustrated in Figure 17-4.

All factors being equal, the more evaporation which takes place in the constant rate zone, the higher will be the average evaporation rate. By the same token, the average rate will be lower if a significant amount of time is spent in the bound water zone. Unfortunately, some machines are forced to "over-dry" to compensate for poor drying uniformity. A natural levelling effect in the profile occurs at lower moisture contents because the physicochemical bonds become progressively more difficult to break.

Description of Drying Process

The wet web from the press section containing 55-60% moisture (40-45% dryness) is passed over a series of rotating steam-heated cylinders (usually 60 or 72 inches in diameter) where water is evaporated

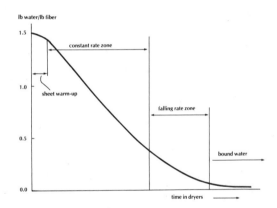

FIGURE 17-4. Drying curve.

and carried away by ventilation air. The wet web is held tightly against the cylinders by a synthetic, permeable fabric called a dryer felt. The fabric also serves to support and guide the sheet through the dryer section. (In some cases, it also aids in controlling cross-direction shrinkage and keeps the sheet flat, i.e., prevents sheet cockling.)

Most paper machines have three to five independently-felted dryer sections, each with independent speed control to maintain sheet tension between sections and adjust for any sheet shrinkage that occurs. All top and bottom felt runs are equipped with tensioning and positioning rolls. Usually, three to five sections are also grouped for

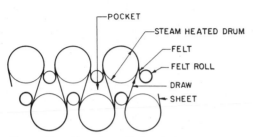

FIGURE 17-5. Two-tier cylinder drying configuration.

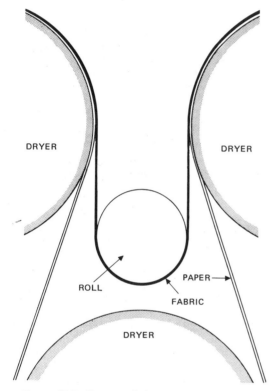

FIGURE 17-6. Dryer pocket.

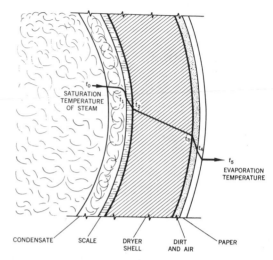

FIGURE 17-7. Temperature profile through dryer cylinder illustrating effect of thermal resistances.

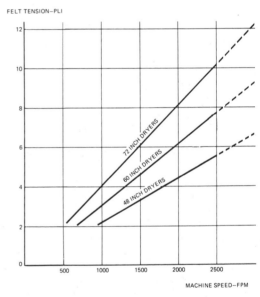

FIGURE 17-8. Recommended felt tensions for newsprint machines (TAPPI TIS 0404-04).

independent steam pressure control; these may be the same or different from the felt groupings. A typical configuration for conventional two-tier drying is illustrated in Figure 17-5.

Attention should now be directed to a "dryer pocket" (Figure 17-6). Paper drying can best be visualized as a repetitive two-phase process. In phase 1, the sheet picks up sensible heat while in contact with the steam cylinder. In phase 2, the sheet flashes off steam in the open draw between the top and bottom cylinders, thus causing the sheet to spontaneously cool and become ready to pick up sensible heat again.

A typical temperature profile between the steam inside the dryer cylinder and the paper wrapping the cylinder is illustrated in Figure 17-7. The major

thermal resistances are usually provided by the condensate layer inside the cylinder and the air layer between sheet and cylinder. The dirt film can also be significant for certain machines and/or paper products, and doctors may be required to keep the surfaces clean.

The air layer is minimized by utilizing adequate felt tension to keep the paper web firmly against the dryer cylinder surface. It has been found that increasing felt tension beyond a certain point does not further reduce the air layer (1). The tension needed is directly proportional to machine speed and varies with cylinder diameter to the 1.5 power. Figure 17-8 provides recommended felt tensions for newsprint machines as a function of machine speed and cylinder diameter.

The condensate layer is probably the most significant resistance to heat transfer on high-speed machines; this aspect will be covered later in this section. If non-condensibles are allowed to accumulate within the steam cylinders, they can adversely affect heat transfer and can also cause nonuniform drying.

The major resistance to steam flashing off in the pocket is the buildup of humidity, causing a lower differential of partial pressures. The problem of achieving adequate pocket ventilation will also be covered later in this section.

Steam and Condensate System

The heat energy for paper drying comes from steam as it condenses inside the dryer cylinders. This type of heat is referred to as "latent heat". The temperature at which steam condenses and the relative amount of latent heat depend on the steam pressure, as illustrated by the steam table data in Table 17-1. Steam always condenses at the "saturation temperature" as defined by the pressure in the system; this is important with respect to controlling drying uniformity across the machine. Steam is usually transported at a temperature considerably above the saturation level (i.e., it is "superheated") to prevent condensation within pipe lines.

TABLE 17-1. Properties of saturated steam.

Pressure, psig	Temperature, °F	Latent Heat, BTU/lb
0	212	971
5	227	961
10	239	953
15	250	946
20	259	940
30	274	928

As steam pressure increases, the condensing (saturation) temperature increases and the latent heat decreases. Therefore, the heat transfer rate increases with pressure, but more steam must be condensed for a given amount of heat transferred.

The condensate that forms in the dryer cylinders is removed by a specially-designed syphon assembly (2). On slow machines, the condensate collects in a puddle at the bottom of the cylinder; a stationary syphon angled into the puddle is often used on these machines. With increasing speed, the puddle begins to cascade; and finally a true rimming condition is reached where the condensate covers the entire inside surface due to centrifugal force (Figure 17-9).

For high-speed machines (and some slower machines), rotating syphons are used where the syphon assembly is fixed to the dryer shell to minimize the syphon-to-shell clearance (Figure 17-10).

On slow machines, puddling condensate within the cylinder has a positive effect on heat transfer rates. With the onset of rimming conditions, there is still considerable turbulence in the condensate layer which helps to maintain heat transfer. But with increasing machine speed, the condensate layer becomes more immobilized (due to greater centrifugal force) and it becomes more important to reduce syphon clearances to minimize the thickness of the condensate film. However, at the high speeds of modern paper machines, the condensate layer has become so immobilized that even a very thin layer is a significant impediment to heat flow.

Fortunately, the condensate layer still has some peripheral motion at high machine speed. This motion can be converted into a wave action by means of "dryer bars" attached axially to the interior surface of the dryer cylinder (Figure 17-11). This wave action (called "sloshing") greatly decreases the thermal resistance of the rimming condensate, and thus allows an increased rate of heat transfer (3). Dryer bar segments can also be installed in certain dryer cylinders to selectively improve evaporation

SLOW-SPEED PUDDLE   CLIMBING PUDDLE   CASCADING   RIMMING

FIGURE 17-9. Behavior of condensate in a dryer cylinder with increasing speed is illustrated from left to right. The transition between puddle and rimming usually occurs at a machine speed between 1200 and 1400 ft/min.

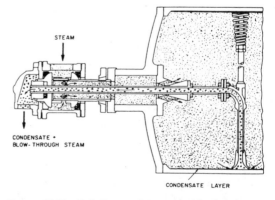

FIGURE 17-10. Rotating syphon assembly showing details of the rotary pressure joint and the action of blow-through steam (Johnson Corp.).

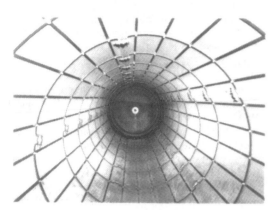

FIGURE 17-11. Representative installation of 25 dryer bars (Beloit Corp.).

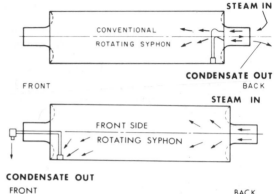

FIGURE 17-12. Older dryer design (top) utilizes steam entry and condensate extraction at the same end. In more recent design (bottom), the steam enters at the backside while condensate is removed at the frontside.

rates at cross-machine locations for moisture profile correction.

On most existing paper machines, the steam feed and the condensate removal both take place through the same side, usually the front journal. On wider machines, this arrangement may favor accumulation of non-condensibles. In more recent designs, steam is introduced at the backside of the cylinder and condensate is removed at the frontside (tending side) as shown in Figure 17-12. The current trend for wide paper machines is to use two syphons, one at each end of the dryer cylinder to ensure more even condensate removal.

The pressure differential required to "pump" pure condensate from the dryer shell to the centerline journal against centrifugal force for a high-speed machine is very large (e.g., 20 psi at 3500 ft/min). In practice, the differential is considerably reduced by allowing steam to entrain the condensate and thereby reduce the effective density (as illustrated in Figure 17-10). This so-called "blow-through steam" also serves to evacuate the non-condensible gases from the dryer cylinder.

To ensure that condensate is removed continuously, the differential pressure (as measured between the steam and condensate lines) is usually controlled at a level somewhat above the minimum required. The main effect of higher differential pressure is to increase the blow-through steam, which typically amounts to about 15-20% mass fraction of the condensation rate. Although a substantial amount of blow-through steam can be tolerated, excessive rates will erode the piping components, and reduce the flexibility of the blow-through steam handling system.

Blow-Through Steam Handling Systems

There are two common systems for handling blow-through steam. In the cascade system, the steam is separated from the condensate and then re-used in the lower-pressure dryer sections, as illustrated in

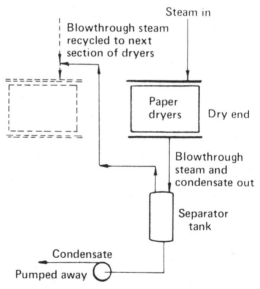

FIGURE 17-13. Cascade system for utilizing blow-through steam.

Figure 17-13. The main disadvantage of this system is the interdependency of sections; only the highest pressure section can be varied independently. The second approach, which is now the most popular, is the "thermo-compressor" system, as illustrated in Figure 17-14. Here, the lower-pressure blow-through steam is "boosted" in pressure by mixing with high-pressure motive steam and then re-used, usually in the same section. This system provides totally independent pressure control for each section and freedom in partitioning of cylinders, but at the cost of losing some electrical generation from the high-pressure motive steam. The thermo-compressor nozzle is illustrated in Figure 17-15.

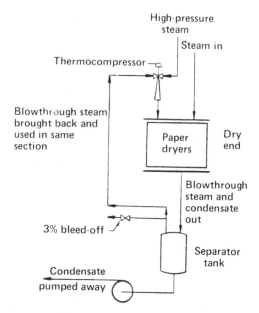

FIGURE 17-14. Thermo-compressor system for utilizing blow-through steam.

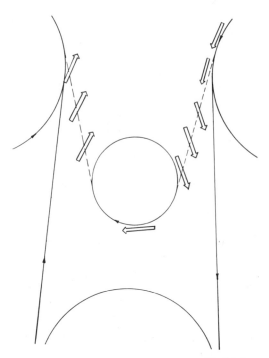

FIGURE 17-16. Illustrating how a permeable fabric provides displacement of air in a dryer pocket.

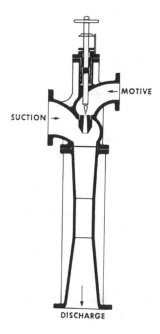

FIGURE 17-15. Thermo-compressor.

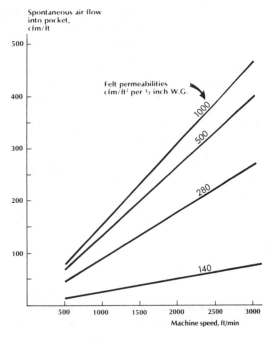

FIGURE 17-17. Spontaneous air flow into a dryer pocket per unit width as a function of machine speed and fabric permeability.

Pocket Ventilation

Prior to 1960, dryer felts were woven from wool and were virtually non-permeable. When dryers were clothed with these "conventional" felts, the pockets were essentially sealed except at the ends, and ventilation to remove humid air was difficult. Typically, high-velocity air was injected into one end of the pocket, which induced a flow of air across the pocket to flush out the humidity. This system was rather ineffective, especially on wider machines.

In the early 1960's, clothing manufacturers began to

introduce synthetic fabrics of more permeable construction. These "open" fabrics were found to provide ventilation by spontaneously carrying air into and out of the pocket, as illustrated in Figure 17-16. Investigators found that the amount of air displacement was primarily a function of felt permeability and machine speed, as shown in Figure 17-17.

The permeable fabric was further exploited by supplying hot, dry air expressly into the pocket. Two general methods are used: through the felt roll as illustrated in Figure 17-18, and through an exterior duct as illustrated in Figure 17-19. There are many variations to these basic designs. By providing variable supply air across the machine, these systems can also be utilized for moisture profile correction.

The improved pocket ventilation achieved with modern clothing and air supply systems has been responsible for greatly increased evaporation rates. The impact on the dryer pocket humidity profile is quite dramatic, as shown in Figure 17-20.

Sheet Flutter

After synthetic fabrics were introduced, the permeabilities were soon raised to take advantage of the improved ventilation. However, the stronger air currents induced by these fabrics compounded an existing problem with sheet flutter on faster lightweight machines. (Sheet flutter refers to the general billowing and flapping of the paper sheet in the open draws of the dryer section. Flutter at the sheet edges leads to creases and breaks.) The sheet flutter problem prompted the clothing manufacturers to offer dryer fabric constructions covering the complete range of permeabilities, including some with reduced permeability at the edges where flutter problems are most severe. In the meantime, the ever-present trend toward increased machine speed was also exacerbating the flutter problem.

A number of design approaches were taken to control the sheet flutter problem, including changes in the pocket geometry and the use of air doctors to divert the more damaging air currents. However, the most widely utilized stratagem on existing machines was to switch to a single-felt configuration for the first dryer section where the sheet is most prone toward wrinkling (4). In the so-called serpentine felt run (Figure 17-21), the bottom felt is eliminated and the top felt follows the sheet and wraps both the top and bottom dryers. Thus, the felt supports the sheet in the draws between top and bottom dryer cylinders, but also runs between the sheet and the bottom dryer surfaces. The serpentine felt run is effective in controlling sheet flutter up to a point, but instabilities are still evident at higher speeds. Unfortunately, with this arrangement, the bottom tier of dryer cylinders becomes redundant.

On a typical modern-design, high-speed machine producing lightweight sheets, the bottom tier of

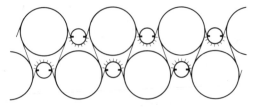

FIGURE 17-18. Pocket ventilation through felt rolls (courtesy of Beloit Corp).

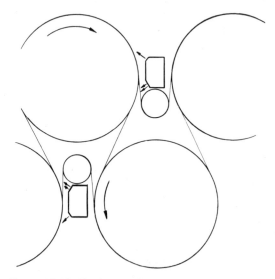

FIGURE 17-19. Pocket ventilation with ducts (courtesy of Beloit Corp.).

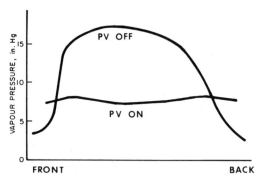

FIGURE 17-20. Example of the vapor pressure profile across a dryer pocket with and without pocket ventilation.

dryers has now been eliminated in the wet-end sections; and the sheet wraps a suction roll between dryer cylinders for greater stability (refer back to Figure 17-1). Blow boxes may also be used to control the boundary air layer. For the next generation of high-speed machines, some designers are proposing a complete single-tier arrangement with a number of "reverse sections" to provide drying from both sides of the sheet.

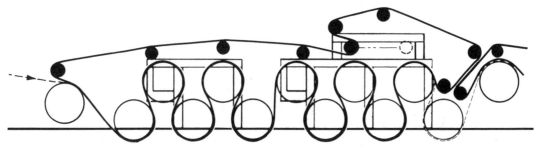

FIGURE 17-21. Serpentine felt run (Beloit Corp.).

Hood Ventilation

Air is an important part of the paper drying process. Depending on what type of hood arrangement is used, from 7 to 20 pounds of air are utilized for each pound of water evaporated. To prevent drips, buildups and corrosion within the hood, the volume and temperature of the exhaust air must be sufficient to avoid localized condensation. The supply air should be strategically introduced into the drying process for best utilization.

The early dryer hoods consisted of little more than a false ceiling with exhaust fans. All the process air was sucked from the machine room and was not effectively utilized. Partially-enclosed hoods were an improvement, but totally-enclosed hoods provide much better control of supply and exhaust air flows and ensure a more comfortable working environment (refer to Figure 17-22). The modern generation of hoods (the so-called "high-dew point hoods") are well sealed and insulated. Diffusion air is totally eliminated, and the amount of fresh makeup air is sharply reduced by operating at high temperature with partial recycle.

Steam Economy and Heat Recovery

Although some steam may be wasted by a poorly operated steam/condensate system (5), most of the heat energy expended for paper drying ends up with the exhaust air. Steam economy is, therefore, strongly affected by the amount of air used and the extent to which heat can be recovered into the supply air. All modern hoods are equipped with a heat recovery system similar to that pictured in Figure 17-23. The primary element is the air-to-air heat exchanger for transferring heat from the hot, humid exhaust air into the fresh, ambient supply air.

Steam economy can be optimized in the case of a high-dew point hood because less air is used and a higher level of heat recovery is possible from condensing vapors. Unfortunately, the amount that can be recovered in the supply air is typically limited to 10-15% (see Figure 17-24). Heating of both machine room supply air and process water is often incorporated into the heat recovery system;

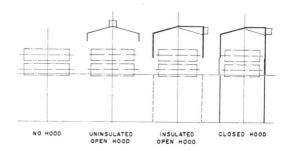

FIGURE 17-22. Different types of dryer hoods.

although useful energy is recovered, this "low-level heat" is normally not credited against steam economy.

Alternate Methods of Drying

Other methods of paper drying (besides steam cylinder drying) are used for certain applications. Air-impingement drying and infra-red drying are commonly used for coatings where contact with a steam cylinder would cause adhesion problems. Airborne drying, a technique commonly used for pulp drying, is also utilized in the production of extensible papers; since little tension is applied in the machine direction, the paper is not pre-stretched as it is during conventional drying (refer to Section 20.3). Air-through drying has found application for the drying of lightweight porous papers, mainly tissues (refer to Section 20.7).

Microwave or dielectric drying would appear to have application at the dry end of the machine. Free water (as opposed to bound water) preferentially absorbs the microwave energy to provide self-compensating profile correction, as illustrated in Figure 17-25. However, this technique has not caught on, perhaps because the economics are not favorable.

Press Drying and Impulse Drying

Press drying refers to any process that utilizes the combined application of pressing and drying under restraint. A range of technologies are under active development, but none is known to be used

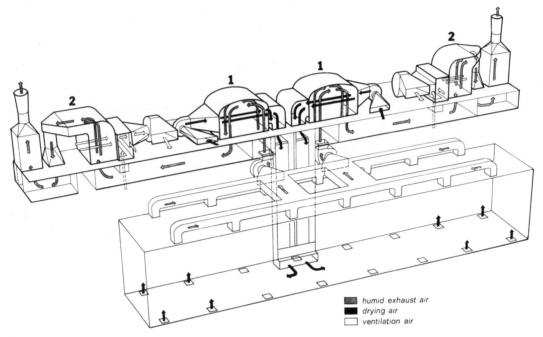

humid exhaust air
drying air
ventilation air

FIGURE 17-23. Typical arrangement of dryer section heat-recovery equipment: (1) heat exchanger for heating supply air to the dryer; (2) heat exchanger for heating machine room ventilation air.

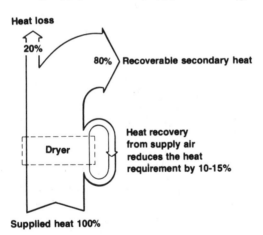

FIGURE 17-24. Sankey diagram of heat flow in the paper drying process.

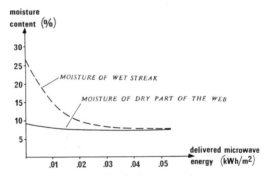

FIGURE 17-25. The selective evaporative action of microwave energy is shown for the drying of wet streaks at the dry end.

commercially as yet. This approach holds the promise of more economical water removal along with improved product quality.

Impulse drying is a form of press drying, but the conditions of temperature and pressure are more intense. The basic idea is to provide an additional driving force within the press nip; this force is created by generating a high-pressure steam front at one surface of the sheet (in contact with a very hot press surface) that literally pushes out water toward the cold press surface. Drying rates two or three orders of magnitude greater than those for conventional drying are claimed; but again, the technique has not yet been applied commercially.

17.2 CALENDERING

Calendering is a general term meaning pressing with a roll. Most, but not all, paper grades are calendered with the principal objective being to obtain a smooth surface for printing. Some sheet compaction always occurs during calendering; in some cases (as for example, with foodboards and tissues) this may be seen as a disadvantage. But thickness reduction is an objective with newsprint-type sheets where it is important to obtain a specified length of paper in a standard-diameter reel. Generally, calendering is performed on dry paper, but some calendering treatments may be carried out on partially-dried paper. (It has been suggested that a smoothing press or fourdrinier lump breaker roll are types of calendering equipment, since their primary role is to smooth the sheet.)

Another common objective of calendering is to improve the cross-direction (CD) uniformity of certain properties, particularly of thickness, which is important for reel-building and converting. It must be noted that where calendering pressure is varied to compensate for non-uniformity in one CD property (e.g., caliper), then the CD profiles of other properties (e.g., density, smoothness) are made less uniform. Generally, the induced non-uniformities are small, and caliper control remains an important aspect of calendering. (Cross-direction profile control methods will be discussed in Section 17.3.)

Calendering changes the surface and interior properties of the sheet by passing the web through one or more two-roll nips where the rollers may or may not be of equal hardness. The pressures are extreme and the time that any section of the web actually spends in the nip is infinitesimally small. The basic objective is to press the paper against the smooth surface with sufficient force to deform the paper plastically and replicate the calender roll surface onto the paper (6). This replication process can be enhanced by the application of greater pressure and/or shear forces and by heating or moistening the fibers to make them more pliable. At one time, calendering was likened to flat-ironing a cotton shirt, but, in fact, virtually no burnishing action occurs with roll-nip calendering. (The only type of calender still in common use that works by burnishing is the brush calender, in which the surface of paperboard is contacted by a rapidly rotating brush.)

On-Machine vs. Off-Machine Calendering

For reasons of economy and efficacy, most calendering operations are carried out on-machine. Calendering that is concerned with reel-building must, of necessity, be part of the paper machine process. Off-machine calendering is a relatively costly process, and is only resorted to when adequate paper surface finishing cannot be obtained on-machine. The classic off-machine operation using alternating iron and compressed fiber rolls is known as supercalendering (refer to Section 18.3 for discussion).

Up to the mid-1970's, on-machine calendering was limited to the traditional operation where paper was passed through one or more nips formed by a set of iron rolls. The difference in printing qualities between conventional machine-calendered and supercalendered papers are very substantial. However, during the late 1970's and 1980's, thanks to new elastomeric roll coverings and innovative equipment designs, the disparity between on-machine and off-machine capability has been narrowed. The new on-machine technology has enabled paper producers to provide a range of machine-calendered finishes to serve more market niches. Some mill references have even suggested that on-machine soft-nip calenders can replace supercalenders for certain less critical grades.

Types of Machine Calenders

In traditional hard-nip machine calenders, the web is calendered to a uniform thickness, as illustrated in Figure 17-26. Small-scale fiber concentrations (flocs) are forced to occupy the same thickness as light-weight spots because the hard rolls concentrate pressure on the high points. Hence, basis weight variations also become density variations; these in turn become variations in surface properties, which for example, are visible as print mottle.

In supercalenders, the nips are formed by mating a hard roll with a soft roll. The covers of the soft rolls are made of compressed paper, rubber or cast polymers, which have approximately the same hardness as paper under compression in the nip. The soft cover serves to distribute the pressure in the nip more evenly over the hills and valleys in the paper web and produces a product with more constant density rather than constant caliper as illustrated in Figure 17-26. Consequently, the surface properties will be more uniform, but basis weight variations will show up as thickness variations.

Although many different arrangements and configurations of hard-nip machine calenders are available, the equipment depicted in Figure 17-27 is reasonably typical. The up-to-date features of this "stack" are the heated rolls and the variable-crown rolls. Calendering at high temperature is desirable because the paper becomes more pliable and can be

UNCALENDERED PAPER

FIGURE 17-26. Illustrating the calendering action of hard nips and soft nips. In hard-nip calendering the basis weight variations also become density variations. In soft-nip calendering, the high basis weight areas pass through without being crushed, and become visible as thickness variations. (from Reference 6)

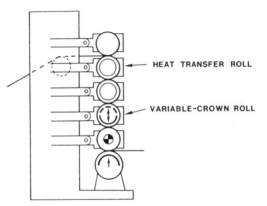

FIGURE 17-27. **Modern versatile hard-nip calender with facilities to heat the web and adjust loads in the bottom nips. (from Reference 6)**

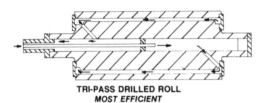

TRI-PASS DRILLED ROLL
MOST EFFICIENT

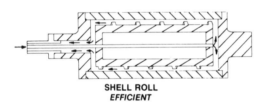

SHELL ROLL
EFFICIENT

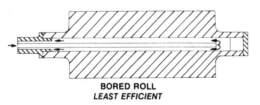

BORED ROLL
LEAST EFFICIENT

FIGURE 17-28. **Rolls used for temperature control showing three levels of drilling sophistication. All these rolls utilize hot water circulation.**

calendered at lower pressure. Therefore, the first two wrapped rolls of the calender stack should be heated (Figure 17-28). The king roll (bottom roll) must have a variable-crown, but it is also desirable to have one intermediate roll with a variable-crown to facilitate changing nip loads. The queen roll (second from bottom) is usually the driven roll.

On-machine soft-nip calendering is an attempt to get some of the benefits of supercalendering without paying the price. This technique was widely adopted in the 1980's as a means of improving paper surface qualities. A number of configurations are used, but the arrangement shown in Figure 17-29 is probably

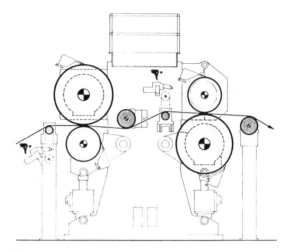

FIGURE 17-29. **On-line soft calender (Kusters).**

the most common. Since the side contacting the metal roll receives a much better finish than the side contacting the resilient roll, it is necessary to have two nips for equal finish. The resilient roll is usually water-cooled to remove heat generated in the cover. The hard roll is equipped with variable crown. An important design feature of these soft-nip calenders is that all rolls are individually driven; this is to enable the nips to be opened and closed at machine speed to facilitate threading of the web without damaging the roll covers.

Calendering Variables

Clearly, the objectives in calendering are to control the thickness and surface properties of the paper, without adversely affecting strength properties. The major variables which bear on the capability of the calender to achieve these objectives are listed in Table 17-2 under the categories of paper properties and operating parameters. Obviously, the properties of the paper entering the calender

TABLE 17-2. **Calendering variables.**

Paper Properties:	thickness
	bulk
	surface properties
	strength properties
	furnish
	temperature
	moisture content
Operating Parameters:	machine speed
	nip loads
	roll diameters
	roll temperatures
	roll hardnesses
	number of nips

influence the calendering operation and the properties of the calendered paper. The initial values will have an influence on the final values, as will the characteristics of the furnish.

The moisture content of paper has a profound effect on its compressibility. Consequently, various methods to utilize moisture for greater calendering effect have been utilized. On some machines, the web leaving the dryer section is passed over a cooled "sweat roll" where moisture is condensed and transferred to the surface of the paper. Steam showers or mist showers are also used to increase the surface moisture content. For certain paperboard grades, moisture is sometimes added to the sheet surface by means of water boxes prior to the first calender nip. In some instances, one or two calender

nips may be installed between the last two dryer sections to provide calendering action at higher sheet moisture content, in which case the equipment is called a "breaker stack".

The reductions in sheet roughness and caliper are inter-related, and are both functions of nip pressure, retention time and number of nips. Generally, there is limited scope for independent control of either caliper or smoothness when using conventional machine calenders; however, temperature is known to have a somewhat disproportionate effect on roughness as illustrated in Figure 17-30. Of course, the smoothing effect also depends on the finish of the rolls; most calender rolls are "superfinished" following grinding by utilizing belt polishers.

17.3 PROFILE CONTROL

Paper quality variability is measured and controlled in two dimensions: the machine direction (MD) and cross direction (CD). During the production process, the former varies with time and the latter mainly with respect to cross-machine position. The ultimate goal in papermaking is, of course, to produce a uniform product, and this requires that all variations be reduced to negligible levels.

Virtually all MD variations can be traced back to high-frequency or low-frequency pulsations in the headbox approach system. Significant progress has been made in recent years in identifying the sources of these disturbances, and by means of process or equipment modifications either eliminating the cause or dampening the signal. For example, by properly positioning the rotating foils on some approach system pressure screens, not only can reinforcement of potentially troublesome pressure pulses be avoided, but in some cases, nullification of one pulse by another can be achieved.

The present discussion will be concerned only with CD profile control. However, it must be recognized that when an on-machine CD sensor is

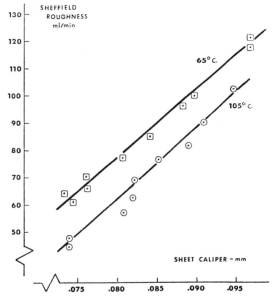

FIGURE 17-30. Effect of calendering temperature on the roughness/caliper relationship for newsprint.

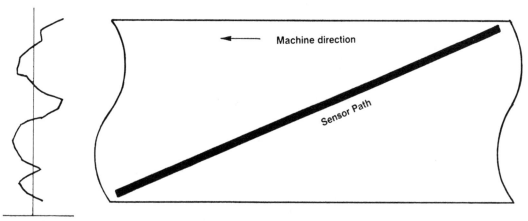

Figure 17-31. **Sensor path across the sheet and typical CD profile.**

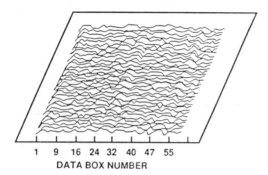

1   9   16 24 32 40 47 55
DATA BOX NUMBER

FIGURE 17-32. Series of CD profiles using an on-machine high-speed scanning gauge.

scanning the sheet, it traces a diagonal path, measuring a "now" profile that includes both MD and CD variations, as illustrated in Figure 17-31. A typical series of scan profiles is shown in Figure 17-32. Until the superimposed MD variability is eliminated from the measured profiles, a "pure" CD profile cannot be identified and successfully controlled. The conventional procedure for doing this, called exponential multiple-scan trending, weights the current measurement at each CD position to the long-term historical value rather than to the most recently measured value. Using this method, approximately 10 scans are needed to register 90% of a step-wise profile change. Various techniques designed to provide more rapid CD measurement resolution are under development (7).

Sensors

Many different types of sensors are used to measure paper properties on-line. The sensor usually does not measure the desired property directly, but rather a related characteristic that is well matched in sensitivity. The sensing methods include nuclear, infrared, microwave, visible light, magnetic reluctance, and ultrasonics. In a few cases, two or more sensors are combined for a particular measurement.

Typically, a cluster of sensors measuring different properties are mounted on a traversing platform which scans the sheet. Each sensor will generally consist of a radiation source (e.g., nuclear isotope, light bulb, etc.), radiation detector, analog and digital electronics, measurement algorithm, software routines and power supplies. The packaging and design of each sensor must provide for accurate measurement in environments which include high variable temperature and humidity, dirt and dust.

The most common on-line measurements are for grammage, moisture content, and sheet caliper. Grammage is typically measured by beta-particle attenuation, where the paper between the source and the detector attenuates the radiation according to its mass per unit area. The detector is usually an ionization chamber, and a suitable beta-particle emitting isotope is selected based on the paper or paperboard grammage. The moisture content of lightweight low-moisture paper is generally measured by infrared radiation absorption. For higher weight papers and high moisture contents, microwave sensors are usually preferred. Caliper measurement is usually done with contacting or semi-contacting sensors utilizing a magnetic reluctance principle.

Control

The first scanning gauges were installed in the 1960's, and the industry discovered the extent of CD variation. The early gauges were mainly monitoring tools for trouble-shooting machine faults.

The traditional method of manually controlling the CD caliper was by means of air jet cooling of one or two calender rolls. The individually controlled jets (see Figure 17-33) dissipate frictional heat produced at the high spots of the rolls; the high spots contract slightly, yielding a more uniform roll contour. The amounts of roll contraction are extremely small, but are sufficient to alter the caliper profile of the sheet as it leaves the stack. The earliest automated system for caliper control, introduced in 1973, was based on this traditional method. Successive developments of hot and cold showers, induction heating, and very hot confined air showers have provided increased control capability.

A number of methods have evolved since the late 1960's for manual control or attenuation of the CD moisture profile, including steam sprays during forming and pressing, zonal controlled-crown press rolls, localized pocket ventilation, sectionalized dryer bars, and moisturizing sprays. The earliest automated system started in 1977 and utilized water sprays and steam boxes as control elements. During the 1980's, moisture control strategies evolved rapidly toward greater sophistication and reliability,

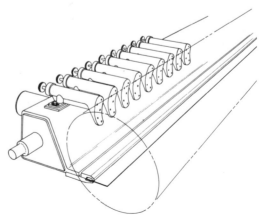

FIGURE 17-33 . Calender cooling air (Beloit Corp.).

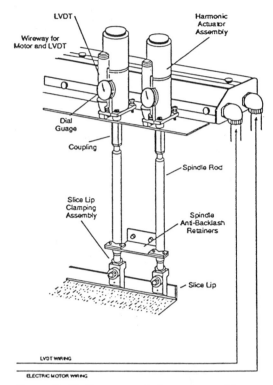

FIGURE 17-34. Beam-mounted arrangement for slice lip actuators.

PRIOR TO CROSS DIRECTION CONTROL

38% IMPROVEMENT
NEW SLICE LIP AND APRON

61% IMPROVEMENT
REMOTE MANUAL CONTROL

93% IMPROVEMENT
CLOSED LOOP CROSS DIRECTION CONTROL

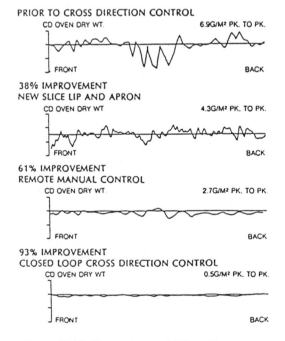

FIGURE 17-35. Three stages of CD profile improvement from implementation of automatic CD grammage control (Cutshall).

based on a combination of wet-end and dry-end correction. Steam showers for hot pressing are still commonly used for wet-end correction. Dry-end correction can be achieved by adding more energy to the wet areas (e.g., by infrared radiation) or by adding more moisture to the dry areas by means of fine water sprays.

Manual setting of the slice micro-adjuster screws has been the traditional means of CD grammage (basis weight) control. The first efforts toward automatic CD grammage control date back to the late 1960's, with mixed results. Further effort was subdued until the early 1980's when advanced slice micro-adjuster actuators and advanced control strategies became available (8). In one arrangement shown in Figure 17-34, the actuator consists of an enclosed synchronous electric motor driving through a harmonic 160:1 gear reducer and then a worm screw to translate rotational movement to vertical movement. The actuator is connected to the slice lip with a rigid spindle rod, which is restrained from horizontal movement. The spindle rod is attached to the slice lip with a specially designed non-slip clamp. The purpose of all this hardware is to move the slice lip in vertical increments as small as 13 microns in a precise and repeatable manner. The linear variable differential transformer (LVDT) is the device that delineates the slice lip position in relation to each point of attachment.

A graphic picture of what can be accomplished with modern headbox equipment and control methods is shown in Figure 17-35 where improvements in the CD grammage profile occurred in three increments. Installing a new slice lip and apron provided a 38% improvement. Installing slice lip actuators for remote manual control increased overall improvement to 61%. Finally, putting the system on computer control provided a total improvement of 91%.

17.4 REELING

After drying and calendering, the paper product must be collected in a convenient form for subsequent processing off-machine. Typically, a paper machine is equipped with a drum reel of the type illustrated in Figure 17-36 which collects the product to a specified diameter.

The reel drum (called a "pope reel" on some models) is motor-driven under sufficient load (amperage) to ensure adequate tension on the sheet from the calenders. During normal operation, the web wraps around the reel drum and feeds into the nip formed between the drum and the collecting reel, which is held by the secondary arms. While the reel builds up, an empty spool is positioned on the primary arms.

Just before the reel has built up to the required diameter, the new spool is accelerated to machine speed by a rubber wheel and then loaded by the primary arms against the reel drum. When the reel build is complete, the secondary arms release their pressure against the drum, causing the paper reel to slow down; a loop of paper then billows out between the reel and the drum, which is blown upward by air jets. At the proper moment, the backtender breaks the sheet at the loop so that it wraps the new spool. The full paper reel is then removed from the reel rails by a crane.

Once paper has begun winding on the new spool, the primary arms are lowered so that the spool rests

on the reel rails, and the secondary arms engage the reel spool to maintain the driving pressure of the reel against the drum. The primary arms are disengaged and returned to the upright position to receive a new spool.

On some older machines, it may still be common practice during reel building for the backtender to run his hand across the face of the reel or "sound" the reel at various locations with a "billy stick" to check for any significant cross-machine variations in sheet caliper or density. This type of monitoring should be unnecessary on modern machines with automatic CD profile control.

17.5 PAPER MACHINE DRIVES

The paper machine drive system must have the capability to independently control the speed of each section over a wide range and within narrow limits. As the paper web is transferred between sections of the paper machine, it is necessary to "draw" the sheet out in order to impose a degree of tension required for web control. Failure to correlate the speed of the various sections will either cause the sheet to be pulled apart or will produce a slack sheet that may fold over on itself, again causing breakage.

There are two basic types of drive systems used on

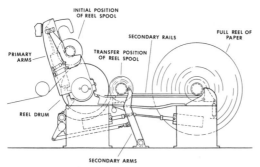

FIGURE 17-36. Schematic of modern drum reel (courtesy of Beloit Corp.).

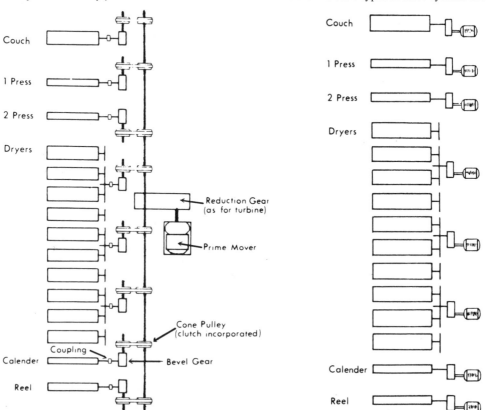

FIGURE 17-37. Illustrating basic types of machine drives. The mechanical drive is on the left, and the sectional electrical drive is on the right.

paper machines, the mechanical drive and the sectional electrical drive, as illustrated in Figure 17-37. In the mechanical drive system, a single motor or steam turbine supplies power to the entire machine by means of a line shaft that runs parallel to the machine, and the indrive for each section is transmitted mechanically by various means to provide speed adjustment. Mechanical drive systems were commonly installed up to the 1960's, but are now found only on older machines.

Displacing the mechanical drive was the adjustable-voltage, direct-current (dc) sectional

electrical drive, generally made up of a series of direct-current motors connected together electrically so that all sections are driven at correlated speeds. This type of drive operates on the principle that a dc motor's speed varies almost proportionally with the applied voltage. Up until about 1982, dc drives utilized analog speed regulators consisting of motor power supply (e.g., thyristor), speed feedback tachometer and regulator as illustrated in Figure 17-38.

Application of digital regulation to the paper machine electric drive embodies the latest state-of-the-art control. Direct digital control using microprocessors combines high resolution, near-absolute accuracy in settings, and freedom from drift. Typically, the accuracy of the digital regulator under steady-state conditions is within 0.01% of top speed as compared to 0.1% for the analog regulator. Figure 17-39 illustrates a five-section paper machine drive system with microprocessor control.

In the early 1980's, adjustable-frequency alternating-current (ac) drives were successfully developed in Europe and Japan; and this design is expected to be the drive of choice in the 1990's. Alternating-current drives operate on the principle that an ac motor's speed varies almost proportionally with the applied frequency. The variable frequency is accomplished by first converting mill power to direct current, and then the dc power is switched on

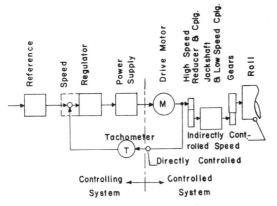

FIGURE 17-38. Analog speed regulator for direct-current sectional electrical drive.

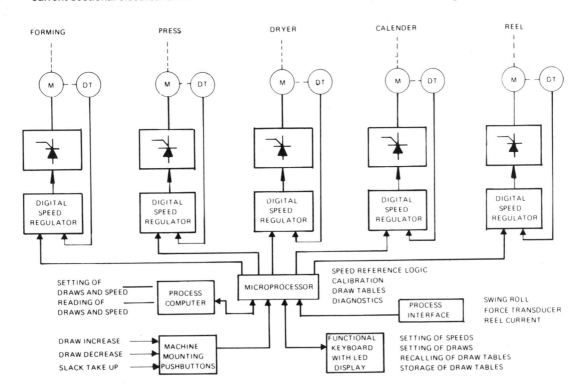

FIGURE 17-39. Paper machine drive section control by microprocessor.

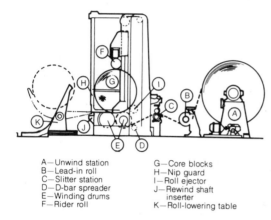

A—Unwind station
B—Lead-in roll
C—Slitter station
D—D-bar spreader
E—Winding drums
F—Rider roll
G—Core blocks
H—Nip guard
I—Roll ejector
J—Rewind shaft
    inserter
K—Roll-lowering table

FIGURE 17-40. Cross-section of typical two-drum winder.

and off, and in opposite directions, by semiconductor devices under the control of numerical digital regulators. The prime advantage of the ac drive is that ac motors can be used which are far more reliable and cost-effective.

17.6 WINDING

Winding is usually defined as a process which changes a material manufactured in web form into roll form for further processing. The true winding process occurs on the paper machine during reeling. Subsequent winding processes are roll-to-roll, and could more accurately be called rewinding operations. However, terminology is already well entrenched; the first roll-to-roll operation is called winding, and the assembly of equipment for carrying out this operation is the winder.

Depending on how deeply one gets involved with winding, the technology can become quite complicated. An in-depth treatment is obviously beyond the scope of this book. Fortunately, an excellent book has recently been published devoted entirely to the process and equipment of winding (9).

The purpose of the winder is to cut and wind the full-width, large-diameter paper reel into suitable-size rolls. These rolls may then be wrapped and sent directly to the customer (e.g., newsprint) or they may be processed through subsequent coating, calendering or sheeting operations. During winding, the two edges of the reel (typically 3-5 cm) are trimmed off and conveyed back to the dry-end pulper.

The most common design of winder is shown in Figure 17-40. While other winders vary in design and arrangement, they all utilize the basic functional elements listed in Table 17-3. The action of a spreader unit with two bars is illustrated in Figure 17-41.

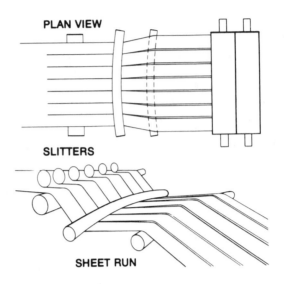

PLAN VIEW

SLITTERS

SHEET RUN

FIGURE 17-41. Schematic of spreader unit with two bowed bars. (Beloit Corp.).

TABLE 17-3. The major components of winders and their functions.

Unwind Stand
- supports the reel
- provides braking to control acceleration and deceleration

Tensioning Roll
- provides feedback to unwind stand brake to control tension

Slitting Station (with guide rolls)
- slits the web and removes trim

Spreader Bar (or Spreader Roll)
- spreads the web following slitting to prevent interweaving of web edges

Rewind Unit
- drives the winder
- controls roll structure

The full-width machine reel is transferred from the reel stand to the unwind stand by an overhead hoist. From the unwind stand, the paper is threaded through the web-tensioning rolls, the adjustable slitters, adjustable spreader bar (or roll), and onto fiber or plastic cores. Commonly, a steel shaft is inserted through the cores to provide a locking arrangement, but some of the newer winders operate "shaftless" by providing a retaining surface on one side to prevent cross-machine wandering.

On most winders, a rider roll is used during initial winding to provide drive friction against the rewind rolls, and is lifted off when the roll weight is

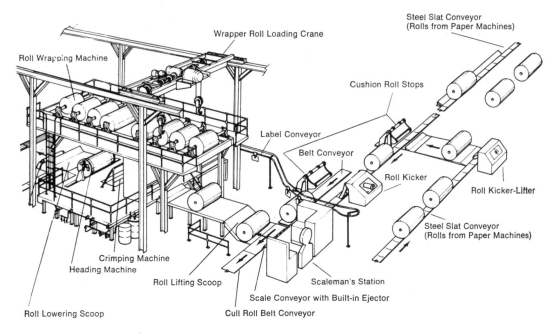

FIGURE 17-42. Roll handling and finishing system at Bowaters Newfoundland.

sufficient to maintain nip pressure. With some newer winders, the paper rolls are oriented horizontally with the rewind rolls, and drive friction is provided by pneumatic loading; this method provides uniform nip pressure throughout the entire winding process. The two rolls of the rewind unit are driven at slightly different speeds to provide "wound-in tension" within the rolls.

The winder drive must be capable of speeds $2\frac{1}{2}$ to 3 times faster than the paper machine in order to have time to change rolls, change reels, make splices (to repair breaks and remove defective paper), set up the slitting arrangement, and adjust the spreader bar. A single run of small-diameter rolls is called a set, and some reels are of sufficient size to produce six or seven sets. Obviously, a winder can more easily keep pace with the paper machine when large-diameter reels are provided and standard-diameter rolls are produced. Other factors that will affect the ability of the winder to keep ahead of the machine are the condition of the reels, web breaks at the reel and winder, and the frequency of order changes that necessitate repositioning the slitters and spreader bar.

17.7 ROLL FINISHING

The steps in roll finishing are scaling, wrapping, crimping (i.e., folding over the wrapper overlap), heading (i.e., a circular piece is glued over the crimped overlap), and labelling. At one time, all these functions were carried out manually and roll finishing was a relatively labor-intensive operation. Today, most of the wrapping operations are carried out semi-automatically, and the labelling function is handled by a data processing print unit.

A modern roll finishing system for a newsprint mill is illustrated in Figure 17-42. This system easily handles 2000 rolls per day with a basic three-person crew.

REFERENCES

(1) **Recommended Tensions in Dryer Felts** TAPPI TIS 0404-04 (1984)

(2) ROUNDS, D.A. and GARVIN, S.P. **Rotating and Stationary Siphons in High-Speed Dryer Sections** *Tappi Journal* (October 1991)

(3) PULKOWSKI, J.H. and WEDEL, G.L. **The Effects of Spoiler Bars on Dryer Heat Transfer** *P&P Canada 89:8:T258* (August 1988)

(4) EDGAR, C.B. **Sheet Flutter Can Be Reduced Through Use of Single Felting...** *Paper Trade Journal* (January 15, 1977)

(5) REESE, R.A. **Process Monitoring Can Upgrade Performance at the Dryer Section** *Pulp & Paper* (January 1979)

(6) CROTOGINO, R.H. and GRATTON, M.F. **Hard-Nip and Soft-Nip Calendering of Uncoated Groundwood Papers** *P&P Canada 88:12: T461-469* (December 1987)

(7) TAYLOR, B.F. **Optimum Separation of Machine-Direction and Cross-Direction Product Variations** *Tappi Journal* (February 1991)

(8) CUTSHALL, K.A. **Cross Direction Control, a Major Factor in Paper Uniformity** *Canadian Mill Product News* (November 1990)

(9) FRYE, K.G. **Winding** *TAPPI Press* 1990

Chapter 18

Surface Treatments

18.1 SIZING

Sizing operations are carried out primarily to provide paper with resistance to penetration by aqueous solutions. The treatment also provides better surface characteristics and improves certain physical properties of the paper sheet, such as surface strength and internal bond.

Two basic methods of sizing are available to the papermaker: internal sizing and surface sizing. These are used either as sole treatments or in combination. Internal sizing utilizes rosin or other chemicals to reduce the rate of water penetration by affecting the contact angle. Surface sizing typically utilizes starch particles to fill in the surface voids in the sheet, reducing pore radius and therefore the rate of liquid penetration. The relative advantages and disadvantages of surface sizing in relation to internal sizing are summarized in Table 18-1. Internal sizing was considered in Section 15.3 and will not form part of the present discussion.

TABLE 18-1. Advantages and disadvantages of surface sizing, in relation to internal sizing.

Advantages:
- more specific action; optimum control
- less sensitive to changes in wet-end operations
- 100% retention of additive
- reduction in wet-end deposits
- increased press clothing life due to lower deposits
- improved paper quality

Disadvantages:
- additional capital investment required for sizing and drying
- additional energy required for drying
- problems at size press can shut down machine

Surface sizing is most commonly applied on-machine at a station between dryer sections, which is referred to as the "size press". For board grades, sizing solutions may be applied at the machine calender stack. For the highest-quality grade papers, an off-machine operation known as tub sizing may be utilized. The most common material used in surface sizing solutions is starch, either cooked or in a modified form (e.g., oxidized or enzyme-

converted). Often wax emulsions or special resins are added to the starch solution. Other agents may be used as well to provide specific strength and optical improvements.

Conventional Size Press

Sizing solution is commonly applied within a two-roll nip; hence the term, size press. The traditional size press configurations are categorized as vertical, horizontal, or inclined as illustrated in Figure 18-1. In each case, the objective is to flood the entering nip with sizing solution; the paper absorbs some of the solution and the balance is removed in the nip. The overflow solution is collected in a pan below the press and recirculated back to the nip. The vertical configuration provides the easiest sheet run, but the pond depth of solution in each nip is unequal. The horizontal size press arrangement solves the problem of unequal top- and bottom-side absorption by providing identical pond forms on either side of the sheet. The inclined configuration is a compromise, and was developed to avoid the rather awkward vertical run of the sheet in the horizontal size press.

The retention time of the sheet in the pond and nip of the size press is very brief, and consequently, the

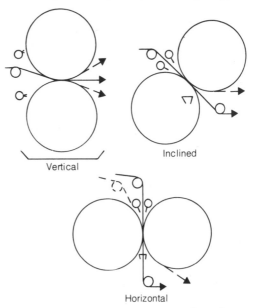

FIGURE 18-1. Traditional size press configurations.

operation must be carefully controlled to ensure that the requisite amount of solids is absorbed uniformly across the sheet. At the same time, the amount of water absorption should be minimized so that the steam requirement for subsequent drying is maintained at the lowest level. The main variables affecting size press performance are summarized in Table 18-2. Each size press application is unique and must be evaluated and optimized with respect to its own peculiar demands and conditions.

TABLE 18-2. Variables affecting solids pickup at the size press.

Sheet Characteristics
- paper substrate (grammage, density, smoothness, capillary structure, void size, etc.)
- level of internal sizing
- moisture content

Sizing Solution
- solids content
- temperature
- viscosity (fluidity)
- composition (type of starch, additives)

Design and Operational
- machine speed
- sizing solution pond depth
- nip pressure
- nip width (affected by roll hardness and diameter)

There are two basic mechanisms for incorporating starch solutions into the sheet at the size press. The first is the ability of the sheet to absorb the size solution; the second is the amount of solution film passing through the nip and the manner in which the paper and roll surfaces separate. Factors that favor greater absorption are low solution viscosity (higher solution temperature), low machine speed, high sheet moisture, high sheet porosity, and low level of internal sizing. The factors favoring greater film thickness include high sheet roughness and low nip pressure.

The sheet moisture content has a significant effect on solution pickup. Although higher sheet moisture promotes absorption, the level is typically controlled at 4 to 5% or less to ensure an even moisture profile and to keep the sizing agent nearer to the surface.

Of the factors affecting pickup, solution viscosity and solids content are the most easily manipulated. The solids concentration, however, is usually maintained at the highest manageable level consistent with the desired viscosity in order to minimize the amount of water that must be subsequently evaporated.

Conventional Size Press Limitations

At higher paper machine speed, the pond in a conventional size press begins to absorb kinetic energy from the converging web and roll surfaces. As the sheet moves more rapidly toward the nip, the nip pressure causes excess solution to flow backward and/or upward with greater force. Eventually the hydrodynamic forces become sufficient to cause the solution to break the surface of the pond and splash out of the nip. This turbulence results in uneven pickup of solids across the machine.

Also at higher speed, a greater quantity of solution remains on the surfaces between the paper and each roll at the exit of the size press nip. As the sheet leaves the nip, each film of solution is split unevenly into two layers, part remaining with the paper and part with the roll. Again, the result is uneven pickup of solids.

Another difficulty with the conventional size press, which becomes more severe at higher machine speed, is matching the speeds of the two rolls. Minute speed differences can lead to paper surface marking and increased web breaks.

Size Press Modifications

Slight modifications in design have allowed conventional size presses to operate successfully at somewhat higher speeds. For example, larger-diameter rolls keep the pond turbulence more manageable. Plastic baffles are sometimes used to isolate the sizing pond from the high-speed surfaces of the paper and rolls.

To avoid pond problems altogether, many modern paper machines are now utilizing some version of the gate roll size press. This design, illustrated in Figure 18-2, has an offset pond on each side which is not in contact with the sheet. The offset pond feeds a metering nip which controls the amount of solution going to the second nip. The second nip controls the uniformity of the film which is finally transferred to the size press roll. All the rolls in the gate roll train are run at different speeds to minimize film split patterning.

With the gate roll design, it is possible to increase the starch solution concentration which reduces the evaporative load on the "after dryers" and keeps the starch on the sheet surface. Nevertheless, problems associated with film splitting are still present, and

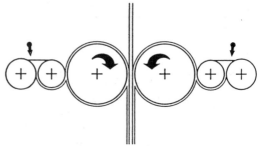

FIGURE 18-2. Gate roll size press.

may be amplified by the higher starch solution concentration (1). Also, supplanting a conventional size press with a gate roll design involves replacing two rolls with six, thus contributing toward higher initial investment and increased maintenance costs.

The roll maintenance and film-split pattern problems of the gate roll press led to the development of blade or rod metering size presses. In most designs, short-dwell coater heads are used to supply the sizing material and either a bent blade or rod is incorporated into the head to control the wet film thickness (Figure 18-3). (Refer to next Section for discussion of short-dwell applicators and blade/rod metering equipment.)

Coating Applications

Although its original purpose and primary application is for applying sizing solutions to paper, the functionality of the size press has been greatly expanded. Today, this equipment is also used for pigmented coatings and other specialized surface applications; however, the descriptive term, size press, has been retained.

Some producers of printing papers are now including pigment in the starch application as a means of adding value to their products. With this method, some of the pigment penetrates into the sheet, and overall pigment pickup is limited to about 3 g/m² per side. Nonetheless, the treatment is sufficient to significantly improve printing properties, and these so-called pigmented papers are finding a market with customers who are reluctant to pay a high premium for coated or supercalendered grades.

Nature of Starch

Starch is a carbohydrate synthesized in corn, tapioca, potato and other plants by polymerization of dextrose units. The polymer exists in two forms: a linear structure of about 500 units and a branched structure of several thousand units. The linear polymer, called "amylose", constitutes 27% of normal corn starch, while the branched polymer, called "amylopectin" makes up the remaining 73%. Fractionated starches are available for special uses.

Starch is supplied as a white, granular powder which is insoluble in cold water because of the polymeric structure and hydrogen bonding between adjacent chains. However, when an aqueous suspension is heated, the water is able to penetrate the granules and causes them to swell, producing a "gelatinized" solution or paste, depending on the concentration. Cooling this hot solution causes thickening, which is called "setback". Starch for application at the size press is "cooked" using either batch or continuous systems (2).

Unfractionated and unmodified starch, called "pearl starch", is "thick-boiling" (i.e., viscous) and has a tendency toward gelling or setback, even without cooling. Setback is avoided by using 100% amylopectin ("waxy starch") which forms a clearer paste and is non-gelling; however, a loss of sizing efficiency results because the linear fraction contributes more toward film formation.

Lower viscosity and setback resistance are achieved by using chemically- or thermally-modified starches. For example, a "thin-boiling", low-viscosity starch can be prepared at the mill by enzyme conversion, with film formation properties and setback resistance unaffected. Depending on the properties desired, a number of chemical methods may be utilized by the supplier to modify the dry starch product. Or alternatively, the user can employ a suitable conversion process (e.g., enzyme, thermal, thermal/chemical) when preparing the sizing solution.

In practice, a low-viscosity starch solution of 4 to 10% solids is used at the traditional size press in order to achieve starch pickup of 70 - 110 lbs per ton of product. Somewhat higher concentrations are used with gate roll presses. A higher viscosity starch can be used at the calender stack, where a more limited penetration is desired and a pickup rate of 10 lbs per ton is more typical.

Starch Preparation

Starch cooking may be either batch or continuous. The batch cooking operation (Figure 18-4) is always carried out in an open agitated vessel where the pre-measured mixture of water and dry starch is heated by direct steam injection or by circulation through a heat exchanger. The solution is heated to 88 - 93° C and held at this temperature for 20 to 30 minutes. Where live steam is used, it is essential that turbulent mixing conditions are maintained at the point of injection to provide even heating. The starch slurry should be kept below the boiling point to avoid foaming and spatter. Since cooking temperatures are limited, premodified starches are normally used in batch systems.

At the heart of most continuous systems is an

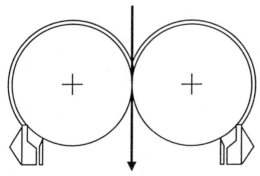

FIGURE 18-3. Size press equipped with short-dwell applicators.

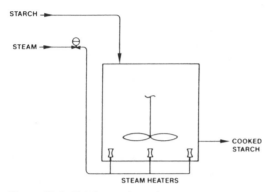

FIGURE 18-4. Batch starch cooker.

eductor which utilizes high-velocity steam energy for both mixing and heating. In the so-called thermal jet system, the steam is totally dispersed into the slurry and then retained in a pipeline long enough for the steam to fully condense. The temperature rise is extremely rapid, and steam pressure is controlled to provide the required final temperature. Since the jet cooker system is pressurized and can operate at elevated temperatures, it can be used for thermal conversion of unmodified starch.

Calender Size Application

Heavier paper and boards are commonly surface-sized at the calender stack in order to get improved calendering action and obtain a smooth, scuff-resistant surface for printing. Application of the size solution is usually by means of a water box with a reinforced rubber lip contacting the calender roll (refer to Figures 18-5 and 18-6). The solution is carried on the roll surface into the calender nip where it is applied to the board. One or more boxes may be fitted to one side of the stack, and the same number to the other side if equal absorption is desired. The pickup of solution is controlled in part by the roughness and absorbency of the sheet; therefore, more material will be picked up from the boxes installed at the top of the stack where the sheet is less compacted. After-calender drying is required to evaporate the water that is added with the size solution. Often, an additional calendering treatment is given to the sized sheet after it is dried.

Tub Sizing

The ultimate in sizing treatments is achieved by tub sizing. Here the sheet is run through a shallow bath containing a solution of starch and other additives, and the excess solution is removed by passing the sheet through a light nip. Initial drying of the sized sheet is usually accomplished by hot air impingement to avoid disturbing the size film.

In tub sizing, the objective is not only to improve surface properties, but also to impregnate the sheet

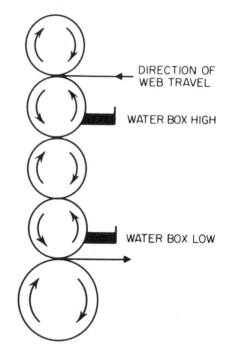

FIGURE 18-5. Calender stack equipped with water boxes.

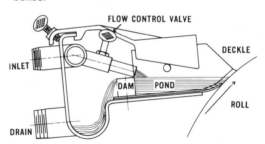

FIGURE 18-6. Water box arrangement (Lodding Eng. Corp.).

sufficiently to improve such properties as ply bond, burst, stiffness and tensile strength. Tub sizing is sometimes carried out on-machine, but better results are obtained with an off-machine operation.

18.2 PIGMENT COATING

The advancing technologies of printing and packaging have placed greater demands on the surface of the paper sheet. To meet the more stringent requirements, many paper surfaces are coated with suitable pigment-rich formulations to provide improved gloss, slickness, color, printing detail, and brilliance. The coating can be applied either on-machine or off-machine, depending on product requirements and operating philosophy.

Coating processes can be generally categorized as either pigment coating or functional coating.

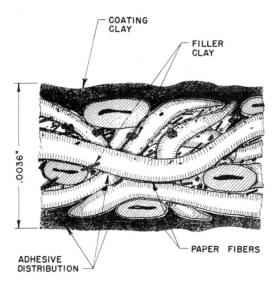

COATING CLAY

FILLER CLAY

.0036"

ADHESIVE DISTRIBUTION

PAPER FIBERS

FIGURE 18-7. **Cross-section of coated paper (N. Bearce).**

Without qualification, the term coating usually refers to pigment coating. Functional or barrier coatings (e.g., lacquer, varnish, waxes, resins, etc.) are more often applied off-machine as part of a converting operation (see Section 23.2). Although the application of starch solution is, strictly speaking, a functional coating process, this operation is always referred to as sizing, rather than coating.

The mineral pigment used in coatings is similar to a filler (see Section 15.3), but is usually somewhat finer; it is mixed with adhesives and other components to hold it onto the paper surface and provide suitable finish and rub-resistance.

The applied coating tends to fill in the void areas (hollows) on the surface of the paper sheet (Figure 18-7). After drying and calendering, the coating provides a smooth, even surface for printing. However, a satisfactory coated sheet can be produced only if the base sheet ("coating raw stock") is well-formed and free from defects. No surface treatment can compensate for poor raw stock, the quality requirements of which are usually more stringent than for uncoated papers. The base sheet is usually sized prior to the coating operation to control the receptivity of the surface to a particular coating mix.

Coating Formulations

The variety of coating formulations is awesome. It is not unusual for a coating mixture to contain more than ten ingredients, and some blends could contain more than fifteen different constituents. In many cases, the formulations have evolved and developed in response to more stringent requirements with respect to both the paper coating itself and the handling properties of the coating dispersion

(referred to simply as "color" in the mills). While the list of possible ingredients is large, the components can be conveniently grouped into three general categories: pigments, binders, and additives, as summarized in Table 18-3.

Coatings are aqueous dispersions ranging in total solids from 50% to more than 70%. Typically, 80% to 90% of the dry formulation weight is composed of pigment. China clay is the most common pigment, and several grades are used according to brightness and particle size. Other pigments include barium sulfate (mainly for photographic papers), calcium carbonate, synthetic silicates, titanium dioxide (TiO_2), and plastic pigments. Because of its high cost, TiO_2 is used only for specialty papers. The plastic pigments (polystyrene) are used in combination with other pigments to provide high gloss. Satin white (prepared from slaked lime and alum) is little used in North America because a glossy-coated sheet can more easily be obtained using less troublesome alternative pigment combinations.

The purpose of the binder (or adhesive) is to cement the pigment particles firmly to the paper surface and to each other. Binders are used at the lowest level consistent with the end-use requirements of the product. It must be kept in mind that the final dried coating is not a continuous film, but rather a porous structure of pigment particles cemented together at their points of contact. If too much binder is used, the voids begin to fill in and some light scattering capability is lost.

Coating binders fall into three classifications: starches, proteins, and synthetics. Starch and protein binders continue to be part of many formulations, although they are tending to be replaced by synthetics, mainly latexes (based on styrene-butadiene, acrylic, or vinylacetate polymers). Protein binders are either casein (isolated from acidified skim milk), soya extract, or animal glues.

Prior to 1958, coating additives tended to fall into the area of "black art". Since then, a more scientific approach has been followed and additives are now employed to meet specific requirements of the coating equipment and product end-use.

Coating Kitchen

Since the different components of a coating mixture vary greatly in physical and chemical characteristics, it is common practice to disperse and store each component separately and then mix together in the desired proportions. These operations are usually carried out in a centralized facility known as the "coating kitchen".

Some of the components must be prepared for use. For example, if pigments and binders are purchased and stored in dry form, dispersions or solutions must be made up. Assuming that all components are ready

TABLE 18-3.   Representative Coating Components.

Component	Examples	Function
Pigments	clay CaCO₃ (precipitated) TiO₂ plastic pigments (polystyrene)	Builds a fine porous structure. Provides a light scattering surface.
Adhesives (Binders)	water soluble adhesives (glues, starches, gums, casein, soya protein, etc.) polymer emulsions (latexes, acrylics, polyvinyl acetate.)	Binds pigment particles together. Binds coating to paper. Reinforces the base sheet. Fills the pores of the pigment structure.
Additives - Insolubilizers (Waterproofing Agents)	formaldehyde donors glyoxal latices	Makes the coating less sensitive to water.
- Plasticizers	stearates wax emulsions azite	Improves the flexibility of coating films.
- Rheology Control Agents (Thickeners)	natural polymers cellulose derivatives synthetic polymers	Controls coating viscosity and water retention properties.
- Dispersants	polyphosphates lignosulfonates silicates	Optimizes pigment dispersion.
- Preservatives	formaldehyde beta-naphthol	Prevents spoilage of formulation between runs.
- Defoamers	proprietary agents	Controls foam problems. Eliminates air bubbles.
- Dyes	lakes direct dyes acid dyes	Used for tinted formulations.

for use, the sequence of operations is to meter each component from storage into a high-viscosity mixer according to a preset order for each formulation, mix thoroughly, and extract into agitated holding tanks. In a modern system, the components are added by pushing a series of buttons or inserting punch cards into a computer console. The ingredients are then automatically metered either by weight or volume. Flowsheets for batch and continuous systems are shown in Figures 18-8 and 18-9.

Strainers are used at nearly every stage in the preparation and use of coating color to ensure that foreign material and over-size solids cannot be applied to the product or build up on equipment surfaces. Undispersed or oversize pigment, flocculated pigment, or undissolved adhesives can arise at various steps in the process and must be removed before they cause problems.

Rheology of Coating Suspensions

Rheology is the science of the deformation and flow properties of matter. Fluid mechanics usually considers only the behavior of Newtonian fluids, which are characterized by constant viscosity. However, starch solutions and pigment suspensions are among a host of rheologically interesting materials that exhibit strain-rate dependency, i.e., in which the viscosity is nonlinear and varies with the flow. The rheology of coating suspensions must be controlled so that the coating color can be easily pumped and also perform adequately under the high-shear conditions of the coating application system.

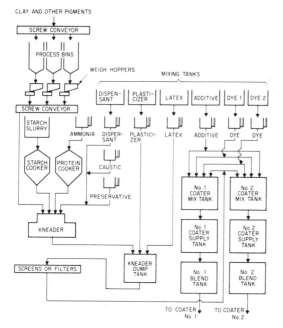

FIGURE 18-8. Batch high-solids coating preparation system.

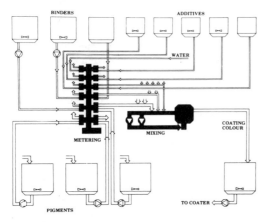

FIGURE 18-9. Flowchart of continuous coating preparation system (KMW).

Because of their potential impact on coating system design and development, it has been necessary to quantify the flow behavior and properties of a wide class of dispersions. This has led to the accumulation of a bewildering amount of experimental data, and to a framework of practical fundamentals for the coating specialist (3); but a full understanding of the behavior of interacting dispersions is still lacking.

On-Machine vs. Off-Machine

Coating color can be applied on-machine or off-machine. Strong arguments can be made for each method, and debate continues within the industry as

to the relative merits and cost-effectiveness of the two approaches. The course of action in a particular case should be based on a reasoned judgement. The relative advantages and disadvantages are summarized in Table 18-4.

TABLE 18-4. On-machine vs. off-machine coating.

Advantages of On-Machine Coating
- eliminates storage and transport of coating raw stock rolls, thus reducing handling and roll damage,
- minimizes space requirements,
- minimizes operator requirements,
- encourages early identification of raw stock quality problems.

Advantages of Off-Machine Coating
- provides great flexibility
- facilitates grade changes
- conditions can be tailored for the coating independent of paper machine conditions
- can handle tonnage from different paper machines
- eliminates interdependence between operations (i.e., downtime on either coater or paper machine does not impact on the other operation),
- provides superior coating quality.

Generally, on-machine coating is most feasible where the application is light to moderate and the quality requirements are not too exacting. Most types of coating equipment are suitable for either on-machine or off-machine operation, but a few of the more specialized techniques are best utilized off-machine.

Coaters

A wide variety of single-sided coaters, as well as a good selection of equipment for simultaneous two-sided application, is offered by equipment manufacturers. Some paper products require coating only on one side of the sheet; a one-sided application is generally easier to control and simplifies the subsequent drying operation. Many double-sided coated grades are produced utilizing two coating stations with dryers in between. Most coater designs incorporate the following features:

1) uniform application of color to the entire paper surface.
2) metering or attenuating the coating layer to control its weight or thickness.
3) smoothing and evening the surface.

The first on-machine coating was carried out using equipment adapted from the size press. In the Massey print roll coater (Figure 18-10), the color is supplied by a series of metering or transfer rolls that smooth out the coating color and spread it evenly by the time it reaches the two large applicator rolls. Typically, one or more of the metering rolls oscillate. The pressure is varied between rolls to control the amount of coating transferred.

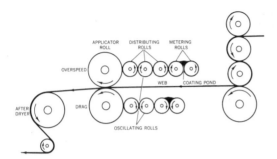

FIGURE 18-10. Massey print roll coater.

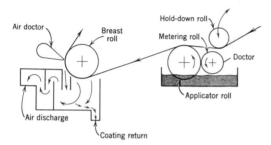

FIGURE 18-11. Air-knife coater.

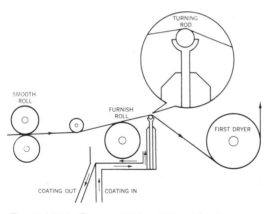

FIGURE 18-12. Representative design of rod coater (Champion).

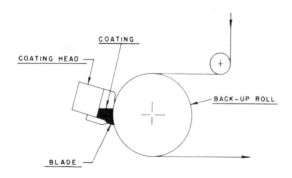

FIGURE 18-13. Pond-type blade coater.

In the air-knife coater (Figure 18-11), the sheet picks up the coating mixture from an applicator roll running in a trough of color. The sheet then passes over a backing roll, where a sharp jet of air impinges on the sheet, evening out the coating layer and blowing off the excess. In this design, it is vital that the air jet be oriented at the correct angle and be of uniform intensity across the entire width of the machine.

The principle of the rod coater is illustrated in Figure 18-12. Again, the sheet picks up the coating mixture from an applicator roll. Here, the doctoring and smoothing function is performed by a small-diameter roll or wire-wound rod which rotates in the opposite direction to the travel of the web. Since web tension is used to maintain pressure against the rod, this coating method is generally limited to heavier weight products that can withstand the tensile loading.

Since their introduction in the 1950's, blade coaters have undergone extensive development, and a large number of designs are currently being employed. In all cases, the web is given a generous application of color, and the excess is removed by using a metal blade. In some designs, the blade tip is bevelled at the same angle as the blade orientation; and the tip rides on a thin film of coating and performs the metering and smoothing function. In other designs, the blade is very thin and is flexed against the web. Generally, the so-called bent-blade designs allow higher coat weights and are less prone toward scratching. In all cases, the angle and pressure of the blade against the metal or rubber-covered backing roll determine the weight of coating retained by the sheet.

One of the earliest blade coaters was the pond or puddle-coater as illustrated in Figure 18-13. In more recent designs, the blade is separate from the applicator, leading to the designation of trailing blade coater. Recent variants are the inverted blade coater (Figure 18-14) and the vacuum blade coater (Figure 18-15). Designs which coat both sides at the same time are the Billblade coater (Figure 18-16 and 18-17) and the opposed-blade coater (Figure 18-18).

A specialized off-machine technique known as cast coating is used to produce paper of exceptional gloss and smoothness. Here, the wet coated paper is pressed into contact with a large-diameter, highly-glazed cylinder (called a Yankee cylinder or machine-glazed cylinder) during the drying phase (Figure 18-19). Great care must be taken at all steps of the process. The base sheet must be fairly porous so that water vapor given off during the drying of the coating can pass through it. The binder must have the ability to adhere to the hot chrome surface when wet, and then when dry, to separate without picking or plucking.

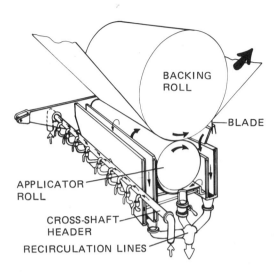

FIGURE 18-14. Flooded nip inverted blade coater (Beloit Corp.).

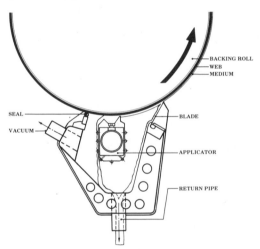

FIGURE 18-15. Coating under vacuum with fountain-type applicator and trailing blade (KMW).

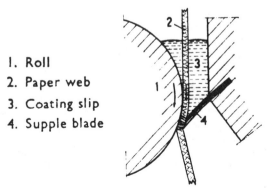

1. Roll
2. Paper web
3. Coating slip
4. Supple blade

FIGURE 18-16. Principle of Billblade two-sided coating system.

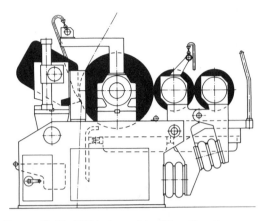

FIGURE 18-17. Billblade coater with optional metering rolls (Inventing SA).

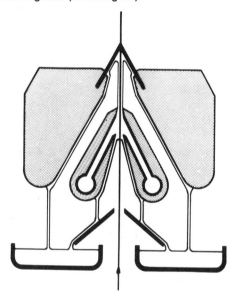

FIGURE 18-18. Opposed blade coater, showing circulation system (Inventing SA).

Short-Dwell Concept

In the coater designs reviewed thus far, a measurable time lag occurs between the application of color and the metering/smoothing operation, whether carried out by an air knife, rotating rod, or blade. Although the elapsed time is short, it is sufficiently long for some of the water and binder to migrate into the sheet; and thus the composition of the color that is doctored from the surface is slightly different from that retained on the sheet. This situation leads to inconsistent coating composition, which has been recognized for some time as a significant problem.

Although the impetus for development of a short-dwell coater design existed for some time, it was not until the early 1980's that a breakthrough design was

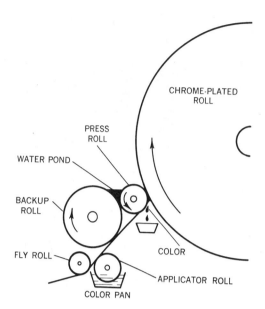

FIGURE 18-19. Cast coating.

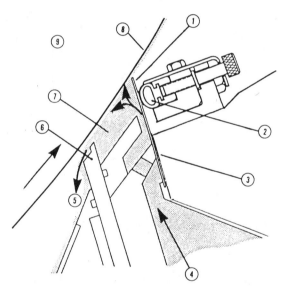

FIGURE 18-20. Short-dwell coating device (Consolidated Paper). 1) trailing blade; 2) loading and profiling tube; 3) blade clamp backing; 4) inflowing color; 5) excess color outflow; 6) adjustable weir; 7) color under pressure; 8) coated paper to dryer; 9) backing roll.

introduced by Consolidated Paper as illustrated in Figure 18-20. Turbulence is created in the pond by the incoming fluid, and pressure is maintained by the supply pump and the adjustment of the exit weir. Following disclosure of this new design, equipment manufacturers soon developed their own short-dwell applicator systems.

Drying of Coatings

Sometimes, conventional steam cylinder methods are used for the drying of coatings (refer back to Section 17.1). More often, other methods must be used to avoid disturbing the coating film. The two methods employed most often are hot air impingement and infra-red drying (4).

High-velocity convective hoods placed over conventional steam cylinders are a popular method of drying single-sided coatings. Tunnel-drying is another approach, suitable for both single- and double-coated sheets. In this method, the air temperature is controlled to suit the drying requirements and speed of the machine, while the paper is carried through the tunnel on rollers, supported on foils, or held up by air-impingement. A complete on-machine coating system consisting of two single-sided coating stations followed by their respective dryers is depicted in Figure 18-21. A similar system for off-machine coating is shown in Figure 18-22.

An infra-red emitter (usually gas-fired) provides a compact, high-intensity heat source which transfers its energy without any physical contact, ideal for the drying of coatings. However, since the infra-red radiation unit supplies only a source of heat, air must also be provided to carry away the moisture evaporated from the coating. Some drying units, therefore, combine infra-red and air-impingement principles for more efficient operation.

Factors Affecting Coated Sheet Properties

Five general factors are of importance in determining the nature and uniformity of the coating layer (5):
1) surface properties of the raw stock.
2) composition of the coating.
3) method of coating.
4) method of drying.
5) extent of supercalendering.

The surface properties of the raw stock influence the formation of the coated layer in two ways. Surface roughness has a significant impact on the coating thickness uniformity; while surface absorptivity determines the composition of the actual coating layer. When the coating first contacts the paper surface, capillary forces within the sheet structure cause a movement of water-soluble components into the smaller pores of the sheet, leaving behind at the surface (by filtration action) a formulation richer in pigment particles.

The type and amount of binder in the coating formulation has a pronounced effect on coating structure because it influences the rate of fluid penetration into the raw stock, the degree of filling between pigment particles, and the rate of drying. The basic structure of the coating layer is more fundamentally related to the size and shape of the pigment particles and the degree of packing.

TWO PASS FOUR ZONE DRYER

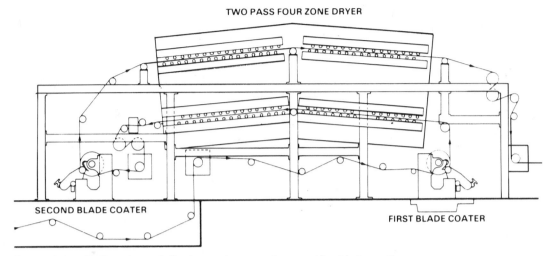

SECOND BLADE COATER   FIRST BLADE COATER

FIGURE 18-21. Air-float dryers following each stage of on-machine blade coating.

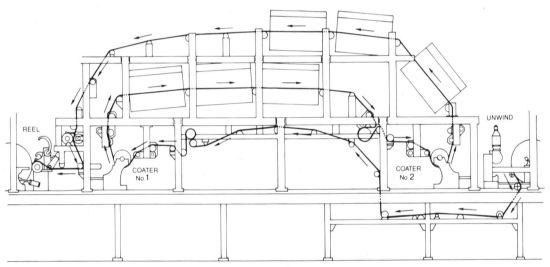

FIGURE 18-22. Air-float dryer installation for an off-machine coater.

Additives in the coating mix will normally determine the flexibility of the dried adhesive and the subsequent reorientation of pigment particles during supercalendering.

Figure 18-23 depicts magnified cross-sections of raw stock coated by three common techniques. The air-knife coater tends to deposit a uniform layer that follows the contours of the base sheet (i.e., good coverage). The roll coater provides good coverage, but patterning defects are introduced from the film-splitting. Blade coating results in good filling in of the surface valleys, but the uniformity of coating layer thickness is sacrificed to obtain increased smoothness. Each coating process forms a somewhat different layer; hence, some coating operations employ multiple coating steps to combine the advantages of two or more methods.

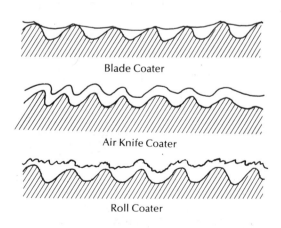

Blade Coater

Air Knife Coater

Roll Coater

FIGURE 18-23. Cross-sections of paper coated by three different methods.

The effects of drying conditions on coating structure can be significant. If the coating dries too quickly, those areas with a thicker deposit or a slower rate of absorption will retain a higher proportion of adhesive, and thereby yield a different coating structure. As the speed of the coating operation increases, and the time interval between application and drying becomes shorter, this problem can be more severe.

During drying, a considerable amount of shrinkage occurs in the thickness of the coating layer. The extent of shrinkage is mainly a function of the solids content of the initial dispersion, but is also affected by the shape of the pigment, the degree of dispersion, and the physical properties of the binder. The shrinkage is undesirable because a portion of the original sheet roughness returns, as shown in Figure 18-24; this problem is most severe with low coat weights.

Supercalendering is often carried out on coated sheets to compact the coating structure and develop a greater level of smoothness. If the coating structure is not uniform, it is likely that the supercalendering will further emphasize the non-uniformity. Areas of the coating structure relatively rich in adhesive will not develop as high a gloss as adjacent areas, and the sheet will exhibit a finely-mottled appearance.

Coating clays are a laminar form of clay (kaolin). When the coating is initially applied, the clay

"platelets" are randomly oriented, as shown in the top diagram in Figure 18-25. To produce a glossy appearance, a substantial number of these tiny platelets must be oriented more nearly parallel to the plane of the sheet. At high supercalendering pressures, this reorientation occurs while the binder is squeezed through the pigment particles. Light reflection from the supercalendered surface is more specular, thus improving gloss, as shown in the bottom diagram of Figure 18-25; however, low areas are not affected. The appearance of a coated, supercalendered sheet surface under magnification is shown in Figure 18-26.

18.3 SUPERCALENDERING

The typical stand-alone supercalender consists of a series of rolls arranged vertically, with alternating hard metal rolls and soft rolls made from compressed fibrous material. The web of paper is fed from an unwind stand into the top of the stack, through each nip, and out the bottom into a rewind unit. Quite often, the web is fed around lead rolls into each nip to prevent air entrapment that could cause creasing. A typical arrangement is shown in Figure 18-27. This equipment is used to develop smoothness and gloss in such products as coated and uncoated high-quality printing papers.

Machine calendering has been discussed previously (see Section 17.2). For many grades of paper, the required degree of smoothness can be achieved with hard steel nips. However, if a smooth, highly-glazed surface is required without over-compaction, then a different type of calendering action is required. On-machine soft-nip calenders are now being successfully utilized for intermediate-quality grades. However, for production of the highest-quality printing papers, there is no satisfactory substitute for supercalendering.

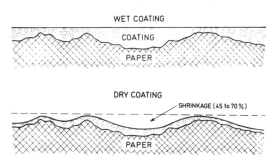

FIGURE 18-24. When a coating is dried, paper roughness reappears.

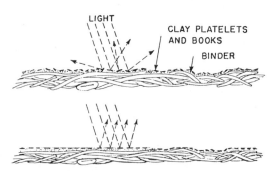

FIGURE 18-25. Schematic cross-section of clay-coated paper before and after supercalendering.

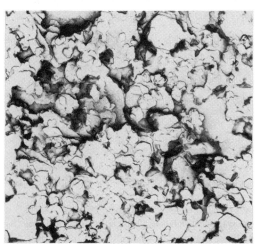

FIGURE 18-26. Scanning electron micrograph of clay-coated, supercalendered paper surface.

The unique results achieved with supercalendering are due to the intermediate fibrous rolls (sometimes referred to as bowls or filled rolls) which possess elastic or plastic properties. With the application of load on the nips, the metal rolls cause a depression or deformation on the fibrous rolls at the point of contact, and the deformation spreads out on either side of the nip (Figure 18-28). When rotated, this spread area will creep because of the constant effort of the material to return to its normal shape. This flow causes a relative motion of the filled roll surface against the metal roll surface, thus producing rolling friction, which helps to give the polishing and smoothing effects. The intensity of action is governed by the amount of plastic flow and by the nip contact pressure.

In contrast to the hard-nip machine calender, the supercalender characteristically converts part of the applied energy into heat through deformation of the soft rolls. Typically, about 25% of the drive power goes into roll heating. Until recently, this heating was viewed as undesirable because it reduced the life of conventional filled rolls. However, the development of modern "controlled temperature rolls" (refer back to Section 17.2) has made possible the utilization of heat for better control, particularly when optimizing production (6).

When the rolls alternate all the way down the stack, a high finish is imparted only to the surface in contact with the steel rolls. Such an arrangement is known as a single-finishing supercalender. The placement of two consecutive steel or fiber rolls in the middle of the stack has the effect of reversing the nips and giving a similar finish to both sides of the sheet. A stack with a reversing nip (Figure 18-29) is known as a double-finishing supercalender.

An important aspect of supercalender operation is to select the proper material for the filled rolls, which are made to a specified hardness. The composition is a mixture of cotton and other cellulosic fibers, sometimes containing up to 25% wool fibers. The traditional method of bowl preparation is to cut composition material into donut-shaped discs which are assembled and compressed onto a shaft, and then fitted onto a lathe and ground to a fine polish. Cotton-filled rolls are labor-intensive and sensitive to marking.

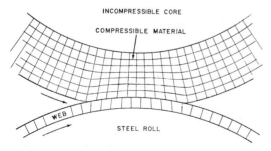

FIGURE 18-27. Representative supercalender arrangement.

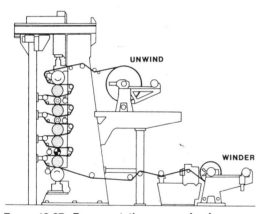

FIGURE 18-28. Diagram of the internal strain pattern developed by a filled roll running against a web and steel roll. (Nip width and penetration are exaggerated.)

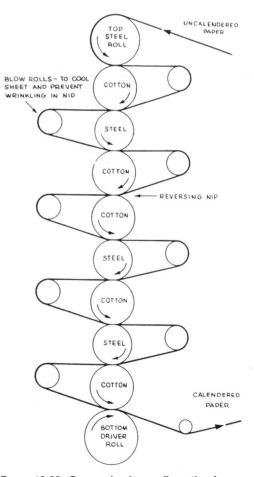

FIGURE 18-29. Supercalender configuration for double finishing.

The North American paper industry has been actively looking for alternate soft roll materials for supercalenders. In recent years, some polymer cover rolls, which do not generate heat and which are mark resistant, have replaced cotton-filled rolls in less critical applications. New soft-roll technology is expected to proceed along with the development of on-line soft calendering.

Supercalendering is almost always carried out as a separate (i.e., off-machine) operation because of the delicate nature of the filled rolls. These rolls are easily damaged or dented by torn paper or lumps from any source getting into the nips. To prevent marking of the paper surface, any damaged rolls must be immediately replaced. The frequent shutdowns and delays would be intolerable on a papermachine or coater.

REFERENCES

(1) KLASS, C.P. **Trends and Developments in Size Press Technology** *Tappi Journal* (December 1990)

(2) KLEM, R.E. **...Selecting the Optimum Starch Binder Preparation System** *Pulp & Paper* (May 1981)

(3) KLINE, J.E. **Rheology of High-Solids Dispersions** *Tappi Journal* (February 1984)

(4) GRANT, R.L. **Drying Pigment Coated Papers and Boards - A Review** *Paper Tech* (March 1991)

(5) LEPOUTRE, P. **Paper Coatings: Structure-Property Relationships** *TAPPI 59:12:70-75* (December 1976)

(6) SCHMIDT, S. and KIRBIE, R. **Recent Supercalender Developments...** *Pulp & Paper* (April 1981)

Chapter 19

Multiply Paperboard Manufacture

Paperboard can be loosely defined as "stiff and thick paper". The line of demarcation between paper and paperboard is somewhat vague, but has been set by the I.S.O. at a grammage of 224 g/m^2. Therefore, material above 224 g/m^2 is termed board while lighter weights fall into the category of paper. While no standard has been set for caliper, the 224 g/m^2 basis weight corresponds roughly to a caliper of about 0.010 inch or 10 "points". Boards can have a single-ply or multi-ply structure; they can be manufactured on a single fourdrinier wire, or single cylinder former, or on a series of formers of the same type or combination of types. A partial listing of paperboard grades is given in Table 19-1. The basic structure of multiply board with descriptive terminology is illustrated in Figure 19-1.

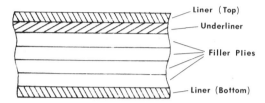

FIGURE 19-1 . Multiply board cross-section.

TABLE 19-1. Paperboard grades.

Linerboard - board having at least two plies, the top layer being of relatively better quality; usually made on a fourdrinier with 100% virgin pulp furnish (see Section 20.4).

Foodboard - board used for food packaging having a single-ply or multiply construction, usually made from 100% bleached virgin pulp furnish.

Folding Boxboard (Carton Board) - multiply board used to make folding boxes; top ply (liner) is made from virgin pulp, and the other plies are made from secondary fiber.

Chip Board - multiply board made from 100% low-grade secondary fiber.

Base Board - board that will ultimately be coated or covered.

Gypsum Board - multiply board made from 100% low-grade secondary fiber, used for the outer surfaces of plaster board.

Multiply boards are produced by the consolidation of two or more web plies into a single sheet of paperboard for use in the manufacture of rigid boxes, folding cartons and similar products. Traditionally, cylinder-type formers have been used to produce the individual plies; and such terms as cylinder forming, vat forming, and multi-vat cylinder machine have become almost synonymous with the manufacture of multiply paperboard products. In previous years, it would have been appropriate to give this chapter the title, "Cylinder Forming". However, the new generation of formers can no longer be grouped under this classification, and it is more suitable to consider cylinder forming within a broader context.

The main advantage of multiply forming is the ability to utilize inexpensive and bulky low-grade waste materials, mostly old newspapers and other post-consumer wastepapers, in the inner (filler) plies of the board, where low fiber strength and the presence of extraneous materials (inks, coatings, etc.) have little effect on board properties. Multiply board mills are typically located near large cities where they are in close proximity to both fiber supplies and markets. Unlike other paper products, multiply boards are often manufactured to a specified caliper rather than to a specified grammage.

Handling and pulping methods for secondary fibers are described in Chapter 14. The techniques of stock treatment (refining, screening, cleaning, etc.) are essentially the same as those utilized in papermaking. However, since a number of furnishes may be used concurrently on a single multiply machine, the stock preparation system is relatively more complex. For example, the top liner, underliner, filler plies and bottom liner may all require individual systems. Typically, a wide range of products are manufactured and the stock preparation system must have considerable flexibility.

19.1 MULTIPLY FORMERS

The cylinder former was originally patented in 1807 (for papermaking). In 1870 a number of these units were used in series to produce a multiply web, and from that time, cylinder forming machines have produced the majority of paperboard grades.

It was somewhat later that a conventional fourdrinier was modified by the addition of a secondary headbox, enabling a second ply to be applied on top of the first ply while still wet on the forming table. This technique has been widely applied to the manufacture of linerboard grades (see Section 20.4).

Although evolutionary development occurred in cylinder board mills in the first half of the 20th century, it became evident that significant advances in the papermaking area were not being matched in cylinder-mold equipment. As a result, some of the markets for lower weight boards were lost to fourdrinier-type machines during the 1950's. Since 1960 the situation has been reversed, and developments in multiply manufacturing technology have been impressive. During this period, a number of forming devices have been introduced which provide both better quality and higher productivity. Headbox arrangements are now available to provide a layered jet, thus producing a multi-structured sheet in one forming section.

Cylinder Former

The principle of web formation on a cylinder former is virtually identical to that of a gravity decker (see Section 9.6). A horizontal cylinder with a wire or plastic cloth surface rotates in a vat of dilute paper stock. Water associated with the stock drains through the mold and a layer of fibers is deposited on the cloth. The drainage rate is determined by the stock properties and by the differential head between the vat level and the white water level inside the mold. The fiber layer is continuously transferred to a moving felt by means of a soft rubber couch roll.

It is estimated that about 80% of existing vats are of the counterflow type, as illustrated in Figure 19-2. In this type of vat, the slurry flows opposite to the direction of cylinder rotation, causing sheet formation to begin at the point of highest consistency. Although relatively thick plies can be formed (i.e., up to 0.010 inch or 10 mils caliper, referred to as 10 "points" in the trade), a number of problems are associated with this method. Because the mold rotates opposite to the direction of stock flow, a portion of deposited stock is continuously washed off the mold, resulting in flocced fiber bundles that contribute to "blotchy" formation. The sheet also has extreme directionality (fiber orientation) caused by the method of fiber deposition. (N.B. Directionality is usually measured by the ratio of tensile values obtained on machine-direction and cross-direction samples. Values of 5.0 are typical for vat cylinder board, as compared to

values below 2.0 for fourdrinier papers.) This extreme directionality produces a high cross-direction stiffness, but a correspondingly low machine-direction stiffness.

Some of the problems associated with the speed differential between stock flow and cylinder mold are eliminated by use of the uniflow vat (Figure 19-3), where stock flow is in the same direction as cylinder rotation. However, improved formation and less directionality are achieved at the cost of lower ply weight. Formation starts at the point of lowest consistency, which results in lower initial deposit on the cylinder mold. Uniflow vats are best suited for plies up to only two points in caliper. On a multi-vat cylinder machine, a typical strategy is to use uniflow vats for top and underliner plies to provide appearance, and counterflow vats for the filler and bottom liner plies to provide the required caliper.

In spite of modifications, all conventional cylinder formers are severely speed-limited (up to about 110 m/min), and exhibit poor cross-machine grammage (basis weight) uniformity. Machine width also appears limited to about 4 m because of difficulty in draining the white water out of the ends of the cylinder.

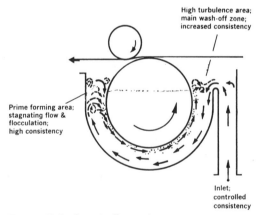

FIGURE 19-2. Counterflow vat.

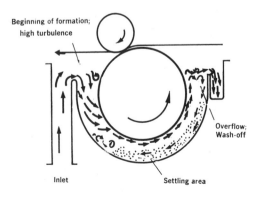

FIGURE 19-3. Uniflow vat.

Dry Vat

The concept of the dry vat developed from the realization that most of the fiber deposition on a vat former occurs during the initial period of cylinder immersion. By restricting the stock to a small part of the vat compartment, less wash-off of stock occurs and better formation is achieved without sacrificing much forming capacity. Some counterflow vats have been modified into dry vats by utilizing a specially-designed seal to isolate a selected portion of the vat. In cases where the vat is used exclusively as a dry vat, a half-vat can be used, as illustrated in Figure 19-4. This design facilitates maintenance and washups, and reduces the required spacing between forming units. Additional improvement in basis weight uniformity has been achieved with a modified inlet.

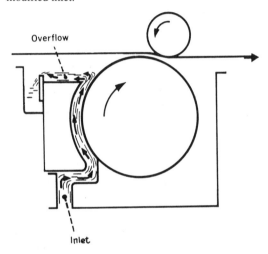

FIGURE 19-4. Adjustable dry vat.

Although vat formers might now appear outdated by more recent former developments, a machine built in 1986 for a Holland mill utilized vat formers (of unspecified design) at all nine forming positions (1). At about the same time, a similar machine was installed in a UK mill.

Suction Formers

A significant development in former design was achieved by further restricting the forming area and utilizing vacuum dewatering. The Rotoformer, introduced by the Sandy Hill Corporation in the early 1960's, employs an adjustable forming area controlled by a pond regulator (Figure 19-5). Initial formation takes place under a hydraulic head; adjustable suction boxes with increased vacuum are then used to obtain further fiber deposition and dewatering. Stock velocity can be controlled by the pond regulator to the same speed as the cylinder, thus obtaining greatly reduced directionality during sheet formation.

A large number of Rotoformers are in service, typically producing plies above 10 points. Speeds up to 300 m/min are reportedly possible when the rubber couch is replaced by a suction pickup.

Shortly after the introduction of the Rotoformer, a number of competitive designs became available including the Tampella suction former (Figure 19-6). The various designs differ primarily with respect to the stock inlet arrangements (which utilize modern headbox design concepts) and the number of vacuum compartments.

Pressure Formers

Beloit modified their design of suction former to produce the Hydraulic Former as shown in Figure 19-7. It had been found that the pressure differential was easier to control when the vacuum system was removed and hydraulic pressure increased within the forming lid. Subsequently, the hydraulic channel was simplified utilizing an adjustable throat as the flow control. The only other operating variables are the area of the roof (which helps to control forming pressure) and the location of the couch roll.

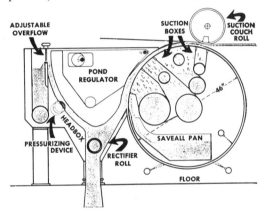

FIGURE 19-5. General configuration of Rotoformer (Sandy Hill Corp.).

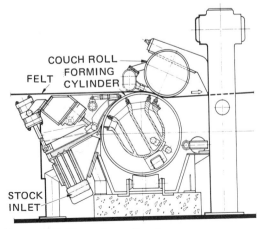

FIGURE 19-6. Tampella suction former.

A number of similar formers were developed by other manufacturers. The major advantage of a hydraulic-type forming unit is that it can be installed on an existing vat cylinder, thus providing a means of updating an older machine at moderate cost. While pressure formers are generally simpler in design, the operating speed and production rate are lower than that of vacuum formers. Some recent, more-sophisticated designs utilize both hydraulic pressure and internal vacuum to achieve increased dewatering and greater operating speeds.

On-Top Formers

A feature common to all the formers discussed thus far is that the sheet plies are picked up and carried on the underside of the felt. This transfer method restricts operating speeds and sheet dryness levels. Another problem is that the sheet must be

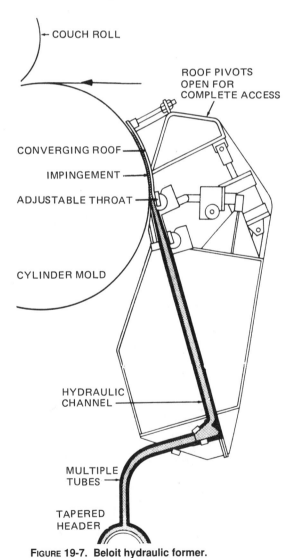

FIGURE 19-7. Beloit hydraulic former.

COUCH ROLL

ROOF PIVOTS OPEN FOR COMPLETE ACCESS

CONVERGING ROOF

IMPINGEMENT

ADJUSTABLE THROAT

CYLINDER MOLD

HYDRAULIC CHANNEL

MULTIPLE TUBES

TAPERED HEADER

righted before it enters the press section; the usual procedure is to run the forming section in the opposite direction to the rest of the machine, and turn the felt over with a long carryback section. Less often, the delicate sheet is transferred from one felt to another entering the press section.

The remaining formers to be discussed place the sheet on top of the felt or fabric. This approach eliminates the speed and dewatering limitations. But, housekeeping and maintenance aspects must be stressed since the sheet is more vulnerable to drips and spills.

Inverform

An English invention, the Inverform was developed in the late 1950's to overcome the quality and speed limitations of the cylinder former in manufacturing multiply paperboards. Basically, this system consists of a fourdrinier wire upon which are located a number of secondary headboxes with top-wire formers (see Figure 19-8). The top liner is formed onto the fourdrinier wire in the usual manner, and subsequent plies are laid down between the main wire and the top wire formers. Dewatering takes place mainly upwards, thus avoiding the problems of water removal through an increasingly thick base web.

Although the basic concept of the Inverform has not changed, a number of modifications have been made over the years to the top former. The Inverform demonstrated the ability to produce a sheet virtually equal to the fourdrinier sheet. However, only one North American mill elected to install this forming system because of high initial cost, and because the operation requires more skill and attention than other machines.

Inverform Sequels

The inverform development stimulated a proliferation of top former designs by various manufacturers, all aimed toward the production of high-grade multiply boards. The Tampella on-top former (Figure 19-9) is an adaptation of their conventional suction former. The Arcu-Former (Figure 19-10) is a further design modification utilizing two wires in the top former to achieve improved dewatering and higher speeds.

A large selection of single-wire and twin-wire top formers are now offered by all the major paper machinery manufacturers in North America, Europe and Japan.

Ultraformer (Kobayashi Engineering)

The development of the Ultraformer embodies a different approach to on-top forming. The basic Ultraformer, which retains the appearance of a cylinder former, is illustrated in Figure 19-11 (along with subsequent design modifications designated

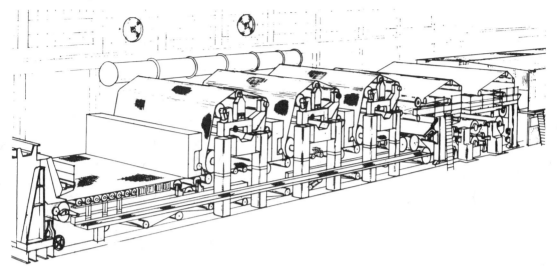

FIGURE 19-8. Arrangement of Inverform units.

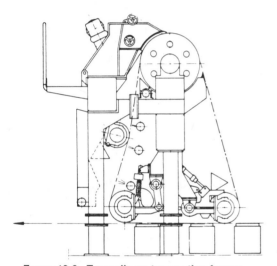

FIGURE 19-9. Tampella on-top suction former.

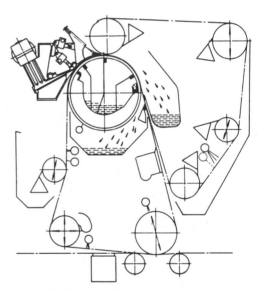

FIGURE 19-10. Tampella Arcu-Former.

Ultra C, Super Ultra, and Ultra Twin). Stock is introduced by a modern headbox onto the forming wire at the top of the cylinder. The web is then sandwiched between the felt and wire, where increasing pressure causes dewatering. A small, turbulent pond is maintained at the point of maximum dewatering to aid plybonding and formation. The felt and attached web leave the cylinder after about 200° of rotation, pass over a suction box, and proceed on to the next unit.

Although the basic Ultraformer produces a high-quality sheet, it is speed-limited because of centrifugal forces at the surface of the rotating cylinder. The modified versions are designed for operation over successively higher ranges of speed, but at considerable increase in cost and complexity.

Multiple Fourdriniers

In the 1930's, some mills in Europe produced board by arranging separate fourdrinier forming sections so that the individual webs could be combined while still wet. Although comparatively expensive in terms of capital and operating costs, this method produced high-quality multiply board at greatly increased production rates.

The concept of multiple fourdriniers re-emerged in the 1970's, and a large number of multiply machines with fourdrinier forming sections have been installed in the last 15 years, mostly in Europe (2). In order to support this approach toward boardmaking, machine designers have developed some flat or horizontal forming machines, similar to fourdriniers, with high-performance headboxes and excellent dewatering capabilities. A board machine with fourdrinier formers is illustrated in Figures 19-12.

Multiple-fourdrinier board machines are typically used for sheets of low to medium grammage. Practical factors normally limit the number of plies to four, so the type of board structure is somewhat limited. Usually virgin pulps or other clean furnishes are used. Multiple-level operation may be seen as an undesirable feature, but one advantage is that only conventional papermaking is involved.

Multi-Layered Headbox

The development of multi-channel headboxes (Figure 19-13 and 19-14) provides the opportunity to

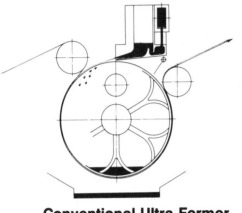

Conventional Ultra Former

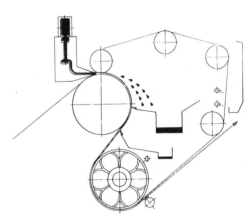

Ultra C Former

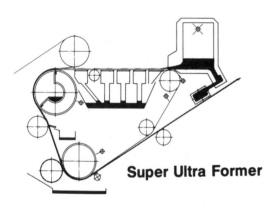

Super Ultra Former

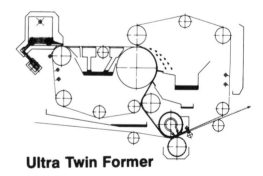

Ultra Twin Former

FIGURE 19-11. Ultraformer family tree (Kobayashi Engineering).

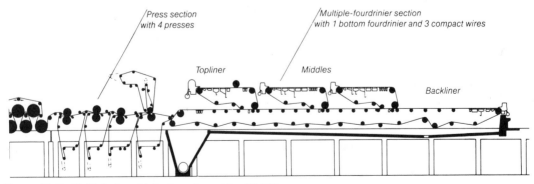

/Press section
with 4 presses

/Multiple-fourdrinier section
with 1 bottom fourdrinier and 3 compact wires

Topliner   Middles   Backliner

FIGURE 19-12. Multiple-fourdrinier board machine (Voith).

produce a layered structure within the discharge jet. In principle, this means that multiply sheets can be produced utilizing only one or two forming units. Especially in conjunction with high-volume two-wire formers, the prospect exists for a considerably more compact wet end to produce multiply boards in the low to medium grammage range.

As compared to conventional multiply methodology, more mixing occurs between layers of the headbox jet, and better plybonding is achieved. However, because of the intermixing, surface coverage of the top and bottom plies is relatively poor and a higher percentage of better quality fiber is required for the liners. This deficiency can be overcome by utilizing a separate former for either or both of the critical outer plies, as shown by the arrangement in Figure 9-15.

19.2 WATER REMOVAL

Pressing

With multiply board, particularly for heavy-weight grades, pressing should be gradual and gentle to avoid crushing and blowing between plies, and to preserve bulk and plybond. The general requirement is that the tensile stresses acting on the web be minimized and that each individual press nip be specifically designed for the amount of water to be removed at that press. The generalities of pressing were covered in Chapter 16; the following will touch on aspects which are peculiar to traditional-design multiply board machines.

As previously noted, the successive plies from conventional cylinder formers, suction formers or pressure formers are accumulated on the underside of a felt which usually runs in the direction opposite to the rest of the machine; the carrying felt is then turned over so that the sheet rests on top and is run back to the press section. An arrangement typical of many machines is illustrated in Figure 19-16. The

web leaving each forming mold cannot be much above 15% consistency without sacrificing plybonding ability. Therefore, a large quantity of water must be removed from the sheet while it is still supported by the carrying felt (i.e., bottom felt). Some water is removed by suction boxes positioned above the felt after each former; these also help to keep the sheet adhering to the felt.

Following pickup of all the plies, it is necessary to turn the felt over. A common method is to pass the felt around a suction turning roll (extractor roll) and then around a double-felted suction drum press. The extractor roll acts like a suction couch and can be quite effective in removing water. A vacuum box is also used in the suction drum roll prior to the press nip. Because of this preliminary dewatering, a pressure of 100 pli can usually be applied at the soft, double-felted nip without danger of crushing. While a common top felt up to the primary presses is the traditional arrangement, many machines now utilize a separate felt for the suction drum press, as illustrated in Figure 19-17.

The sheet is then carried into the primary press section where a series of low-pressure plain and

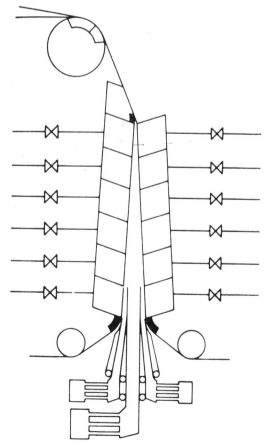

FIGURE 19-14. Tampella Contro-Flow former with multi-channel headbox.

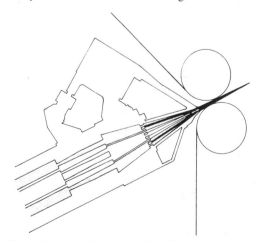

FIGURE 19-13. Beloit Converflow Strata-Flow headbox for producing a three-layered sheet.

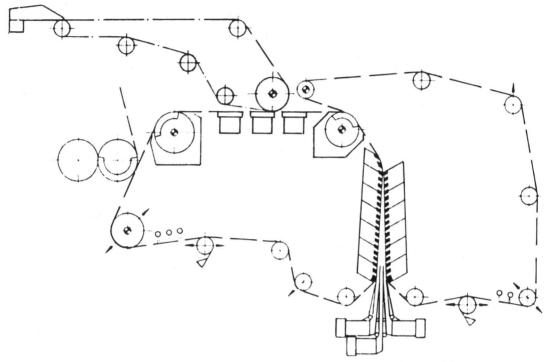

FIGURE 19-15. Arrangement for making a four-ply sheet; a three-ply Contro-flow former with a conventional top-ply forming unit.

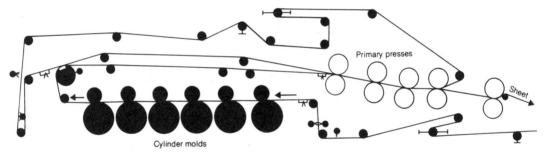

FIGURE 19-16. Typical arrangement of cylinder formers with attendant felt run.

suction nips dewater and consolidate the web. Typically, the top felt is used with all the primary presses, but the last one or two presses may be run single-felted to "harden" the nip a bit. Finally, the sheet is run through two or three suction or vented nips (i.e., the "main presses") at higher loading. The sheet is transferred following the first main press. The second and third presses are independently felted.

On cylinder-type machines, the bottom (or carrying) felt performs arduous service. The felt must be strong enough to drive the entire wet end, and at the same time, must be sufficiently permeable for good water removal. In addition, the side in contact with the sheet must have a reasonably fine surface to impart a satisfactory finish to the sheet. Considerable attention must be given to the care of this felt, which usually involves cleaning and conditioning with chemicals, showering including the use of "whippers", and dewatering with suction boxes.

Drying

Paper drying was described in Section 17.1. However, drying of multiply board requires special techniques because the board's thickness and composite furnish have a dramatic effect on the heat transfer rate from the surface into the center of the sheet, and on the steam diffusion rate from the center to the surface.

Due to the large number of dryer cylinders that were typically used (sometimes over 100), it was common practice on older board machines to utilize stacked arrangements up to 12 high to conserve space (Figure 19-18). This configuration did not allow easy removal of moisture-laden air, and the cylinders could not be felted. Therefore, dryer performance was poor.

The multiple-section double-tier arrangement common on paper machines is now standard for board machines. Most sections are felted, which increases drying rate and also minimizes cockling. Modern systems for condensate removal and pocket ventilation have made possible increased evaporation rates and a reduction in the number of drying cylinders required. A further reduction in the number of dryers has been achieved by using larger-diameter steam cylinders (i.e., 6 ft diameter instead of 5 ft).

With multiply board it is important to progressively increase the dryer surface

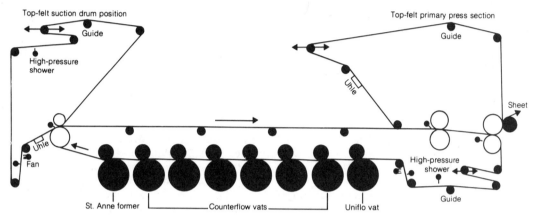

FIGURE 19-17. Arrangement of cylinder formers with double top felting.

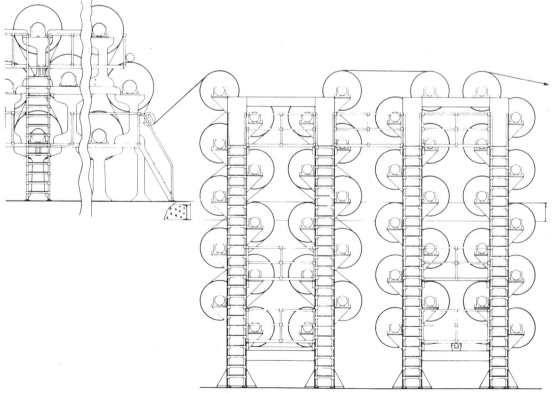

FIGURE 19-18. 40-dryer stack with 3-deck lead-in.

temperatures through the initial sections to avoid sheet picking and blistering. If steam generation is too rapid, pockets of steam can develop between plies, causing blowing and ply separation. Another critical stage is at the dry end where the moisture content is less than 20%; if the surface evaporation is more rapid than the rate of diffusion from the center, case hardening can result. The effect is to distort the sheet and reduce the drying rate.

Particularly where one side of the web has a different character than the other, each dryer section tier should have independent steam pressure control. Differences in evaporation rate between top and bottom liners will adversely affect the flatness and surface finish of the finished board.

19.3 SHEET FINISHING

The two basic methods of producing a surface finish on multiply board are the calender water doctor and machine-glazing cylinder.

Water Doctor Finish

Flattening and consolidating the sheet surface has traditionally been achieved utilizing two or three multiple-roll calender stacks equipped with a total of three or four water boxes for applying starch or other suitable solutions to seal the sheet and increase pick resistance. The sheet typically enters the section at about 3% moisture and leaves at about 6%. If the sheet enters at a higher moisture level, the loss of bulk can be excessive. The moisture pick-up depends on machine speed and the sizing of the sheet, but is usually limited to about 1% per doctor.

Machine Glazing Cylinder

Partial drying of the sheet while pressed against a highly-polished Yankee cylinder (large-diameter drying cylinder) is gaining ground in the board industry as a means of obtaining a smooth and glossy surface on one side of the board. This method generally retains more sheet bulk, and the surface obtained is especially suitable for subsequent coating. For best results, the dryer surface must be maintained in good condition and the sheet moisture content at the Yankee must be carefully controlled. This method is limited to relatively porous sheets not exceeding 450 g/m^2 in grammage because evaporated moisture must diffuse through the sheet structure to escape.

Due to the differential drying between the two sides, strains can become built into the board which may be released during subsequent converting operations, causing curl problems. Multi-furnish sheets are especially susceptible, and the drying rates in the other sections must be adjusted carefully to control the problem.

19.4 PLYBONDING

The bonding strength between the plies (plybond) of multiply board is important for all grades, and is absolutely critical for boxboard. Considerable attention has been devoted to maintaining and improving plybonding in recent years because of the changes in forming methods. The use of pressure formers has materially changed the characteristics of the individual plies, providing a better-formed web with more uniform cross-section profile and less machine-direction fiber orientation, but more two-sided with respect to fines distribution.

At the same time, many converting operations are becoming more demanding. For example, the gravure printing process typically subjects the sheet to high shear forces. A good understanding of the factors affecting plybonding and plybond failure is required to optimize manufacturing and handling techniques, particularly in the heavier-weight boards where the problems are magnified. Two major factors need to be considered:
1) the initial establishment of plybond;
2) the subsequent deterioration of plybond through the machine and converting operations.

Plybond is defined as the resistance to ply separation when a tensile force is applied perpendicular to the plane of the multiple plies. The bond is dependent on the mechanical interlocking of fines and fibrils at the interface, and the subsequent development of hydrogen bonds between the intimately-contacting fibers. A number of conditions have been identified for achieving maximum bonding:

1) Fines at the interface fill the voids and promote more intimate surface contact between plies. It has been shown that when the fines fraction is removed from one or both surfaces, the plybond value is reduced by one-half. This finding suggests that a distinct two-sidedness with respect to fines distribution will have an adverse effect on the potential bonding between plies.

2) Free water must be present at the pickup nip to give fluidity at the interface for migration of fines and fibrils. While a low consistency for both plies is considered ideal, it has been shown that, if one ply is below 9% consistency, the other may be as high as 18 - 20% without significantly reducing plybond strength.

3) With pressure formers, it has been found that additional refining and fiber development is required to achieve the bonding values experienced with counterflow vats.

4) It is well known that a large number of thin plies provides optimum plybond strength for a given total caliper. It is reasoned that in thin layers, the fiber orientation at the interface is the same as the bulk orientation and as layers become thinner, the plybond strength approaches the strength within a

single ply. In practice, a maximum ply caliper of 4 or 5 points is used as the basis of design for greatest plybond strength.

5) The higher the nip pressure applied by the couch roll and presses, the higher will be the plybond value within normal attainable limits. Unfortunately, nip pressure on couch rolls is severely limited by the amount of water present. It is only at the last two or three former positions that significant loading can be sustained because of wider distribution of the nip load.

Establishing plybond may be easier than preserving it. Deterioration of plybond through the machine and converting operations is due to shear forces acting on the board. Two mechanisms predominate:

1) Nip compression tends to "bead" the surface plies at the ingoing side of the nip. If one of the roll covers is relatively soft, the distortion due to compression is aggravated, particularly if the board wraps the soft-covered roll.

2) Bending of the board becomes progressively more severe as the sheet becomes drier, thicker or stiffer, or when the bending radius becomes smaller.

A major contributor to plybond degradation on the boardmaking machine is the calendering operation. To minimize the problems associated with bending, roll sizes must be increased to hold the stresses within tolerable limits. For example, field experience has indicated that on a machine with seven pressure formers, a 26-point board is the maximum that can be run safely through a calender with 12-inch diameter intermediate rolls.

Because inter-laminar shear rather than normal plybond appears to be the ultimate cause of ply failures, it is questionable if the existing standard testing method (perpendicular rupture force) is a relevant indication of sheet strength. Several alternative testing methods are now available where the sheet is passed through a loaded nip of opposing steel and rubber-covered rolls; the number of passes before failure is called the rolling shear strength index.

19.5 BOARD PROPERTIES

As noted previously, multiply board is often manufactured to a specified thickness in mils, commonly referred to as "points". It is advantageous for the producer to utilize a bulky (low-density) furnish to minimize sheet weight and reduce raw material cost. Therefore, a secondary furnish composed mainly of old newspapers is ideal for the inner plies.

In most papermaking applications, it is desirable to form fibers in such a way that their axes are parallel with the plane of the sheet and in random orientation within the plane. If this could be done perfectly, the sheet would be as dense as possible and would have equal strength in all directions. However, when the sheet needs to be bulky, it is helpful to have a large proportion of the fibers oriented perpendicular to the plane of the sheet. Cylinder molds, especially counterflow types, have an inherent advantage over fourdriniers in producing more perpendicular orientation of the fibers; unfortunately, more fibers are also oriented in the direction of cylinder rotation, which is usually considered a liability.

Perhaps the most important property for most multiply board grades is stiffness (i.e., resistance to bending). Each ply within the board structure makes a contribution to stiffness in proportion to the square of its distance from the center of the sheet and its own tensile stiffness. This means that the overall board stiffness is essentially a function only of the outer ply stiffness values. The center portion of the board gives bulk and acts as a spacer, thus increasing the contribution of the top and back liners. A typical distribution showing the relative contribution of each ply to the overall stiffness of a 7-ply board is given in Table 19-2.

TABLE 19-2. Relative contribution of each individual ply to the stiffness of seven-ply board.

Layer	% Contribution
top liner	38.5
filler	4.3
filler	0.3
filler	0.5
filler	3.1
filler	8.3
back liner	45.0

REFERENCES

(1) SUTTON, P. **Dollard Makes More Board for Blooms** *P&P International* (February 1987)

(2) GRANT, R. **Machine Technology: A Turning Point** *P&P International* (August 1986 Annual Review)

Chapter 20

Manufacturing Techniques for Specific Paper and Board Grades

Common technology is utilized in all papermaking and boardmaking operations, but due to the different physical and chemical characteristics of the fibrous and non-fibrous raw materials, and to the flexibility of the manufacturing process, the properties and qualities of the product can be varied over wide limits. This chapter will attempt to outline the basic production strategies for the most common paper and paperboard grades, and discuss some of the specific techniques utilized to carry out these strategies.

The number of paper and board grades produced in North America is so large it would virtually require an encyclopedia to detail the specifications and manufacturing methods for each product. In the 1992 issue of Lockwood-Posts Directory, over 200 specific types of paper products are listed. While the total grade mix is awesome, a relatively small number of commodity-type products account for a sizable percentage of North American tonnage. For example, newsprint accounts for almost one-third of paper production, while linerboard and corrugating together account for over half of the paperboard. A review of manufacturing techniques for several key products will provide insights into the production methods for a host of other products.

Even a general description of available paper grades is beyond the scope of this book. A few of the more common products are listed and briefly defined in Table 20-1. At the risk of oversimplification, these and most other paper products fall into three broad functional groups:
1) communicational, informational and literary purposes;
2) commercial (packaging), industrial and constructional use;
3) personal or sanitary purposes.
Products from each of these categories are represented in the following discussions.

TABLE 20-1. Definitions of common paper grades.

PRINTING GRADES

Newsprint - Machine-finished paper composed mainly of mechanical pulp, commonly used for printing newspapers.

TABLE 20-1. (cont'd)

Catalog - Basically lightweight newsprint, but usually contains fillers.

Rotogravure - Usually refers to uncoated newsprint-type sheet that is highly finished. May contain filler.

Publication - Supercalendered and coated magazine papers. Raw stock is composed mainly of mechanical pulp, but the best grades use chemical pulp.

Banknote, Document - High-grade permanent paper, usually made from rag furnish.

Bible - Lightweight, heavily loaded paper made from rag or chemical pulp furnish.

Bond, Ledger - High quality paper used for letterheads and records. Furnish is either rag or chemical pulp.

Stationery - Relatively soft and bulky paper of good appearance. Furnish is usually chemical pulp, but highest quality uses rag.

INDUSTRIAL GRADES

Bag - High-strength paper, usually made from highly refined unbleached softwood kraft.

Linerboard - Lightweight board commonly used as the liners for corrugated board. Also used for wrapping paper. Formed from high-yield unbleached kraft with a better quality top layer for printing.

Corrugating Medium - Used for the fluted inner plies of corrugated board. Usually prepared from high-yield semi-chemical hardwood pulp at nine "point" thickness.

Construction Paper - Newsprint-type sheet produced at higher grammage and bulk, typically used for kindergarten cutouts and artwork.

Greaseproof - Dense, nonporous paper made from highly refined sulfite pulp.

Glassine - Produced from greaseproof paper stock by dampening and heavy pressure during subsequent supercalendering. This glossy, transparent sheet is used for special protective wrappings, and is converted to wax paper.

TABLE 20-1. (cont'd)

TISSUES

Sanitary Tissues - This classificaction includes facial and toilet tissues, sanitary products, and table napkins. The primary feature is softness and absorbency. Contains a high percentage of lightly refined chemical pulp.

Condenser Tissue - Lightweight well-formed tissue (5 g/m²) made from highly refined kraft, used as a capacitor dielectric. Basically the same product is used as raw stock for carbonizing grades and (with wet-strength treatment) for tea bags.

Towelling - Creped absorbent paper usually made from lightly refined kraft with the addition of mechanical pulp. Fast absorbency and water holding capacity are prime requisites. Sometimes treated with wet-strength resins to prevent wet disintegration.

Wrapping Tissue - This designation covers a variety of tissues made for wrapping and packaging merchandise. The general requirements are strength, good formation and cleanliness. Grammage is in the 16 to 28 g/m² range.

20.1 NEWSPRINT

In terms of volume, newsprint is the most important member in the family of non-coated "wood-containing" printing papers, which also includes catalog and directory papers. The manufacturing methods are similar for all products in this group.

In the broadest sense, newsprint can be defined as any paper capable of being run through a modern high-speed printing press and producing an acceptable newspaper sheet at reasonable cost. The functional requirements of newsprint are pressroom runability, printability, good general appearance, and low price (see Table 20-2).

TABLE 20-2. Newsprint characteristics sought by newspaper publishers.

Runability - Ability to run the web through the presses without breaks.

Printability - Ability to accept and preserve the imprinted ink pattern with minimum rub-off, set-off, and show-through.

Appearance (Optical Properties) - Brightness, "whiteness", cleanliness, opacity.

Newsprint furnish is made up mainly of mechanical pulp and/or recycled deinked newspaper stock, commonly in admixture with a small percentage of lightly-refined chemical pulp. The only general additive is violet-blue dye to offset the natural yellow hue of most mechanical pulps. In some cases a small amount of mineral filler may be added.

The mechanical pulp contributes such valuable printability-related properties to the newsprint as absorbency, bulk, compressibility, opacity, formation (i.e., grammage micro-uniformity), etc. Unfortunately, the strength properties of the mechanical pulp may be insufficient to produce a sheet that runs well on the presses. For example, when stone groundwood is the only mechanical pulp used, the paper is usually reinforced with up to 30% chemical pulp, either semi-bleached kraft or unbleached sulfite. While giving the newsprint greater strength, the chemical pulp adversely affects printing characteristics.

Of all the printing grades, newsprint comes the closest to being a true commodity item. Virtually all the "standard newsprint" sold in North America is the same with respect to grammage (48.8 g/m²), brightness (58-60), general properties and delivered price. This is not to infer that differences in runability, printability and appearance do not exist between various suppliers. However, a newsprint manufacturer will usually find a market niche if the product meets minimum standards, especially in a situation of tight supply.

However, newsprint is no longer the monolithic product that it used to be. In Europe, most publishers print their newspapers on 45 g/m² newsprint (1), and some have gone down to 40 g/m² to help conserve forest resources. At the same time, certain newspapers incorporating 4-color printing (notably, USA Today) are utilizing higher-brightness, higher-opacity newsprint (sometimes referred to as hi-fi newsprint) that sells for a premium price.

Impact of Recycling

Newsprint producers are concerned about the trend among state governments to provide incentives for using recycled paper (2). Laws that tax virgin-fiber newsprint and set mandatory recycling levels for purchased newsprint are seen by states as a way to reduce the growing inventory of old newspaper stocks accumulated through curbside collection (see also Section 14.1). But in some cases, the legislation has not been realistic in allowing paper manufacturers enough time to adapt. Compliance is especially difficult (if not impossible) for Canadian producers who do not have access to large secondary fiber supplies.

However, industry observers are virtually unanimous in predicting that secondary fiber usage in newsprint will increase significantly over the long term. The garbage buildup is reaching crisis proportions in some areas and old newspapers are comparatively easy to segregate and recover. The technology to produce high-quality newsprint from 100% recycled fiber is already available, and secondary fiber usage is economically attractive (especially so while the inventory of old newspapers remains high).

Economic Factors

Assuming efficient paper machine operation at the highest practicable speed, a prime factor affecting newsprint machine profitability is the level of chemical pulp usage in the furnish. Semi-bleached kraft pulp is almost twice as costly to produce as a typical mechanical pulp component. On a modern high-speed newsprint machine, a reduction of 1% kraft (e.g., from 11% to 10%) represents increased annual profitability of approximately a quarter million dollars. Therefore, a prime strategy is to use only enough chemical pulp to meet minimum strength requirements, thus reducing furnish costs and improving printability (3).

Because newsprint is sold with respect to its average "as is" grammage, another prime consideration is to produce newsprint at the highest possible moisture content. If newsprint is produced near its optimum moisture level of 9 - 10% instead of, say 6 - 7%, a typical annual savings in furnish cost of a half million dollars is realized on a large high-speed machine. The customer is quite satisfied to receive newsprint at the higher moisture level because of improved runability. In particular, newsprint exhibits improved stretch at the higher moisture levels, which enables the sheet to better absorb shock energy without breaking. However, in order to build a uniform reel, operators of older newsprint machines often must over-dry the sheet to compensate for non-uniform pressing and drying (refer to Section 17.1). The effect of moisture content on runability is illustrated in Figure 20-1.

Runability

Pressroom runability can be considered a combination of all those sheet and roll properties which might cause the press to run at lower than necessary production. Poor runability is characterized by web breaks or slack tension which leads to wrinkles and poor register. The relationship between runability and paper strength properties has never been well defined, but pressroom operators feel that cross-direction tearing resistance provides the best indication of performance. Tearing resistance depends primarily on the relative amounts and qualities of the pulps used.

Breaks in the paper are usually attributed to defects in the sheet (often at the edge) which precipitate failure. Generally, if the "break end" is located, a flaw such as a shive, sliver, slime hole, etc. will be found. Many pressrooms attempt to

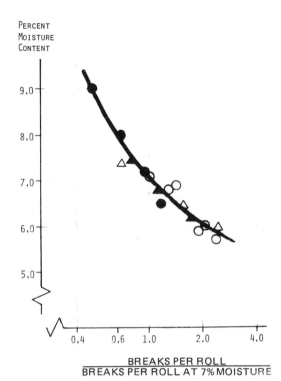

FIGURE 20-1. Effect of newsprint moisture content on "relative pressroom runability". (Note: Data from four ANPA pressroom systems have been plotted together by "regularizing" the data in the form of "relative pressroom runability".)

catalog the specific cause of each failure and maintain cumulative records, such as exemplified by Table 20-3. The comparative performance of various newsprint suppliers is also tabulated with respect to relative runability, measured in rolls per break. An actual long-term tabulation is shown in Table 20-4, with a coded designation for each supplier.

TABLE 20-3. Web break survey from a large pressroom.

	Paper Mills							
	A	B	C	D	E	F	Total	
Paper Defects								
Fiber cut	27	62	41	113	97	53	933	
Calender cut	5	7	5	11	9	4	41	49%
Moisture wrinkles	-	8	-	1	2	4	15	
Slime hole	6	14	15	30	9	7	81	
Roll Defects								
Bad splice	7	7	12	22	13	15	76	
Loose paper in roll	12	16	18	27	14	9	96	
Out-of-round roll	5	10	3	25	29	14	86	51%
Glue on roll end	20	10	11	117	39	15	212	
Faulty core	2	14	21	31	21	2	91	
Totals	84	148	126	377	233	123	1091	

TABLE 20-4. Three-year record of newsprint runability in a large pressroom.

Supplier	Rolls Run	Breaks	Rolls/Break
A	8,001	147	54
B	55,763	1224	46
C	61,693	1634	37
D	17,987	531	34
E	19,310	668	29
F	33,247	1605	21
G	21,406	1200	18
H	10,522	656	16

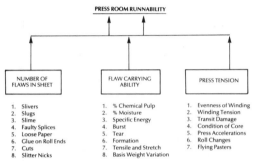

FIGURE 20-2. Factors affecting pressroom runability.

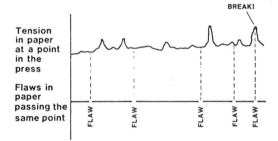

FIGURE 20-3. Newsprint break occurs when a flaw coincides with a tension surge.

In reality, breaks are most often due to a statistical aggregation of factors, as illustrated in Figure 20-2. While a flaw is always present, it must occur in combination with lack of "flaw-carrying ability" and/or high sheet tension in order for the sheet to break. If the press operator is aware in advance of a "weak spot" in the roll, he can easily carry it through the press by reducing speed and tension. However, newsprint breaks are extremely rare and anomalous events, occurring on average about once every 300 miles of web. In one long-term pressroom study in Sweden, it was found that the measured paper strength was always at least five times higher than the recorded web tension at break.

A more simplistic view of pressroom breaks is embodied in Figure 20-3. When a roll of newsprint is being run on the press, flaws in the sheet are passing a reference point while coincidentally the tension in the sheet fluctuates. The paper is normally capable of withstanding the highest tension peak as long as no flaw is present, while the average tension will not break a flawed sheet. It is only when a flaw coincides with a tension surge that a break occurs.

Another view of a newsprint break is shown as a diagnostic tree in Figure 20-4 (4). According to this diagram, a flaw is just one of several causes of low sheet strength or stress concentration which could precipitate a break when accompanied by high sheet tension.

Printability

Printability is a measure of the print quality obtained under normal conditions; it is dependent on the printing method used and on such sheet characteristics as smoothness, absorbency, moisture content, formation, opacity, and brightness. However, the concept of printability is rather complex and must be considered under different aspects.

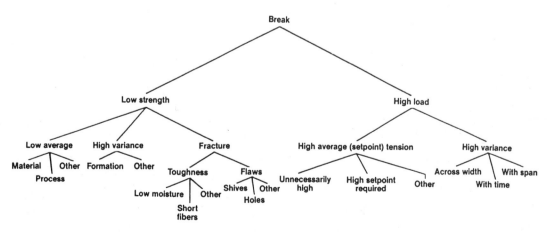

FIGURE 20-4. Diagnostic tree of factors contributing to web breaks.

Print density (one measure of printability) is the optical contrast between the printed and unprinted surface. For newsprint, this contrast is usually measured from a solid print and is dependent on the brightness of the paper. The density of a print can be raised by increasing the amount of ink on the printing plate.

The undesirable effect of the print being visible through the reverse side of the paper is called show-through or print-through. This factor depends on the opacity and porous structure of the paper. Unfortunately, a higher ink application to improve print density will also have the adverse effect of increasing show-through. (In extreme cases, the ink can migrate to the reverse surface, causing "strike-through".)

A higher ink application also has the effect of increasing set-off and rub-off. Set-off occurs when the still-moist ink smears off on the backing cylinder of the next printing unit, which then sets off the image onto the next page. Rub-off is when the ink rubs off on the reader's hands. These phenomena are related to the printing process used, and are caused by too-slow penetration of the ink into the paper or by inadequate drying. An optimum pore size distribution will usually allow a moderate ink application without excessive set-off or rub-off.

20.2 PUBLICATION GRADES

The category of publication grades comprises higher-quality wood-containing printing papers including uncoated supercalendered (SC) papers,

machine-finished pigmentized (MFP) papers, and lightweight/medium-weight coated (LWC & MWC) papers.

MFP papers are a new development in Europe, characterized by the application of a light pigmented coating at the size press (5). A decade ago, the choice for many applications was strictly between SC and LWC grades. But, the high price differential between SC and LWC caused many publishers to use SC papers when they really needed a slightly higher quality. MFP papers were developed to fill the quality niche between SC and LWC papers .

Coated publication grades are often differentiated by the numbers 1 through 5, as defined in Table 20-5. Number 1 represents the highest quality and number 5 the lowest. The normal grammage range is 47 to 103 g/m^2. Coat weights vary, but for the lightest-weight grades, the two-sided coating is 20 to 30% of the total sheet weight. Thus, the coating raw stock may be as light as 33 g/m^2. Magazine publications consume about 40% of the coated paper output. Because of rising postal rates, many publications have switched to a lighter-weight paper, especially for the number 4 and 5 grades (which constitute 94% of coated magazine papers).

In responding to the demand for a lighter-weight product, the papermaker is faced with the task of maintaining the same level of runability and surface characteristics while reducing the overall weight. At grammages below 58 g/m^2, it is usually necessary to shave weight from both the base sheet and coating. A higher percentage of chemical fiber is required in

TABLE 20-5. Characteristics of coated publication grades.

General characteristics	Typical use	Base stock	Coating	Brightness
No. 1. Enameled, double-coated, hand-sorted, expensive base and coating; high gloss; basis weight above 70 lb	Annual reports	High-brightness chemical pulp Heavily filled	High in TiO$_2$ High gloss Enameled Synthetics	82-88
No. 2. Double-coated, expensive base and coating	Expensive advertising	High-brightness chemical pulp Filled (clay)	High in TiO$_2$ Some clay Synthetics	78-82
No. 3. Single- or double-coated lower quality basesheet	Advertising	Chemical pulp Minor amounts of groundwood	Mostly clay, some TiO$_2$ Less expensive	76-82
No. 4. Lower cost, lower brightness	Magazine	Groundwood and chemical pulp, some clay filler	Less expensive Coating — clay and TiO$_2$	72-78
No. 5. Lower basis weight, high groundwood content	Directories, catalogs, magazines	Mostly groundwood or TMP, chemical pulp for runnability	Variable, some contain synthetics	68-72

the base sheet to maintain runability. Another approach toward higher sheet strength is to use wet-end additives such as starches, gums and wet-strength resins. TMP is also replacing groundwood in the lighter-weight number 4 and 5 grades.

Current forecasts continue to indicate that the market for LWC grades will expand faster than other sectors of the paper industry because of increased magazine advertising, higher living standards, greater leisure time, and a higher literacy level. While the future looks bright, the challenge for this sector is to achieve greater penetration of coated paper into the uncoated paper market.

20.3 SACK GRADES

In order to determine the requirements of paper used to make bags and sacks, a number of investigations were carried out in North America and Europe to find out how paper sacks fail. The results showed that tensile energy absorption (TEA) is the critical parameter when a full sack is dropped on end (butt drop). For flat drop resistance, both TEA and tear strength should be at a high level. When sacks are made of several plies of paper (multi-wall), a high stretch value is essential to distribute forces evenly among the layers.

TEA is the work required to break a test sample under tension, and is equal to the area under the stress/strain (load/elongation) curve. It is roughly proportional to the product of the tensile and stretch values at rupture.

To meet the strength requirements, sack grades are usually manufactured from well-refined unbleached softwood kraft pulp. Rosin sizing and/or starch is commonly added to the furnish to increase internal strength. Wet strength resins may be used for some grades. Good sheet formation is essential for uniform strength, and low headbox consistencies on the order of 0.2 -0.3% are typically employed; therefore, the wire table must be designed to handle the high drainage volume.

The most advantageous method of increasing TEA of sack grades is to increase stretch at rupture. Since a portion of the machine-direction sheet extensibility can be lost in the open draws at the wet end of the paper machine, it is common practice to utilize a suction pickup off the wire and to delay the first open draw to between the second and third presses.

A more significant increase in stretch can be obtained if the paper is allowed to dry without restraint. Since paper shrinks as it dries, a sheet that shrinks freely or under light tension will retain more stretch than a sheet prevented from shrinking by high tension. While some effort has been made to reduce tension on conventional steam-cylinder drying systems (e.g., by eliminating dryer fabrics), the scope for obtaining free shrinkage is relatively limited. On the other hand, practically free shrinkage

has been achieved by using air-impingement drying (6). Although different designs are available, Figure 20-5 illustrates an application with the same type of equipment commonly used for pulp drying (see Section 9.9). Air drying is typically utilized within the 50 to 85% sheet dryness range where the majority of shrinkage occurs. Figure 20-6 compares the load/stretch diagrams for steam-cylinder and air-impingement dried papers, demonstrating the increase in TEA obtained with the air drying system.

It is also possible to produce additional stretch by the process of creping (see Section 20.7). However, this treatment produces a visible surface modification which is considered a liability. The same effect on stretch, but without surface modification, is obtained by using a compactor device such as that shown in Figure 20-7. Here, the sheet is passed into a nip formed by independently-driven rolls of hard steel and soft rubber. The hard roll rotates with a peripheral speed matching the speed of the incoming paper web, while the soft roll is driven at slower speed. The rubber is stretched in front of the nip and the paper adheres to the rubber

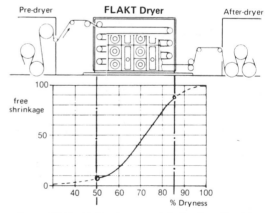

FIGURE 20-5. Air impingement drying is utilized in the interval of main shrinkage (Flakt).

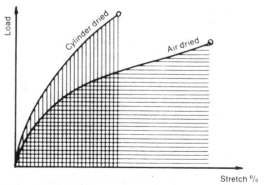

FIGURE 20-6. Paper stretch and TEA are dependent on the method of drying (Flakt).

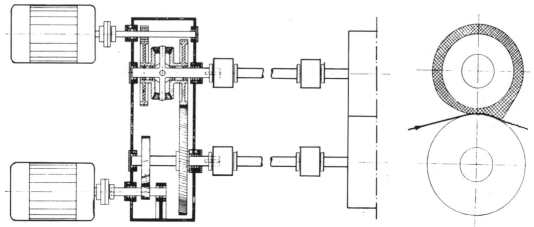

FIGURE 20-7. Schematic of compactor unit showing gearbox and rolls (Clupak).

and is forced to contract with it in the nip. The compactor is located between dryer sections, and the speed of subsequent sections is matched to that of the soft roll. The amount of forced shrinkage can be varied widely by adjusting the speed differential of the two rolls.

Although bag and sack paper shipments have been steadily declining since the late 1970's because of competition from plastic bags, there are indications that the trend has leveled out. Demand for kraft paper grocery bags in the 1990's is expected to benefit from environmental concerns about plastic bag disposal.

20.4 LINERBOARD

Linerboard is a relatively lightweight board commonly used as the outer plies of corrugated box stock and as wrapping paper. Standard liner is $205 g/m^2$ (42 lb/1000 ft^2), but other grades cover the range from 112 to 439 g/m^2. The market for linerboard is very large, and the total tonnage produced in North America exceeds that of newsprint by a large margin.

The most common product, duplex linerboard, has traditionally been manufactured from virgin high-yield kraft pulp on a fourdrinier machine utilizing two headboxes as illustrated in Figure 20-8. A modern hydraulic-type secondary headbox is shown in Figure 20-9. Recently, the industry began retrofitting existing fourdriniers with on-top formers or secondary fourdrinier formers to replace the secondary headboxes, and it is estimated that about 20% of North American machines had been converted by 1992. As well, almost half of North American linerboard machines have recently been equipped with high-intensity presses which are effective in providing higher sheet strength and reducing the dryer evaporative load. Many mills now utilize up to 25% of OCC (old corrugated containers) recycled stock in the bottom liner furnish.

A relatively dark and coarse high-yield kraft base sheet (bottom liner) is initially formed on the fourdrinier. A lighter, cleaner, better-quality layer (top liner) is formed on top of the base sheet to provide a good printing surface and better appearance. Enough top liner stock to fully hide the

FIGURE 20-8. Schematic of a duplex linerboard machine.

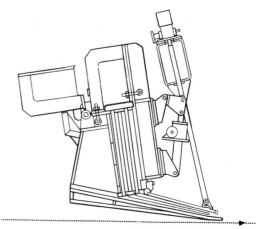

FIGURE 20-9. Hydraulic secondary headbox (Beloit Corp.).

base sheet (typically 20 - 30% of the sheet weight) is applied either from a secondary headbox (located above the suction box area of the forming table) or from an on-top former.

The primary requirements of linerboard are stiffness and burst resistance, along with good appearance and printability on one surface. These needs are well satisfied by using high-yield kraft pulp in the bottom liner and a lower-yield, well-refined unbleached kraft on the top liner. When specified, some mills are able to utilize bleached kraft for the top liner to give a cleaner and brighter sheet appearance. Internal starch is often added at the wet end to help meet stiffness targets; the top liner is usually lightly sized at the calender stack.

Pulping System for Linerboard

The most common pulping system for linerboard utilizes two complete kraft production lines for bottom and top liner stocks, respectively, as illustrated in the flow sheet of Figure 20-10. The base sheet pulp, cooked to 55 - 60% yield, is fiberized and refined out of the blow tank ("hot-stock refined"), washed, and given an additional refining step (called "de-shive refining") before

delivery to the paper machine storage chest. The top liner stock, cooked to about 48 - 50% yield, is fiberized and screened in the presence of hot residual liquor, and washed: the screen rejects are usually transferred to the base stock line to ensure exceptional cleanliness for the top liner. The operating conditions for the optimization of both hot stock refining and de-shive refining have been extensively investigated (7).

An alternative strategy is to employ a single pulping line and utilize a special fractionating screen system on the refined pulp (8). A clean, short-fibered portion is extracted to use as the top liner. A representative system is shown in Figure 20-11. This approach simplifies the pulping system with significant capital cost savings. However, it appears that overall yield must be sacrificed slightly to ensure that a suitable pulp fraction can be isolated for the top liner.

Both top and bottom liner stocks are given additional refining prior to use on the paper machine. Adequate refining is especially important for the top liner stock to ensure a well-formed, smooth surface for printing, and provide good coverage of the relatively dirty base sheet.

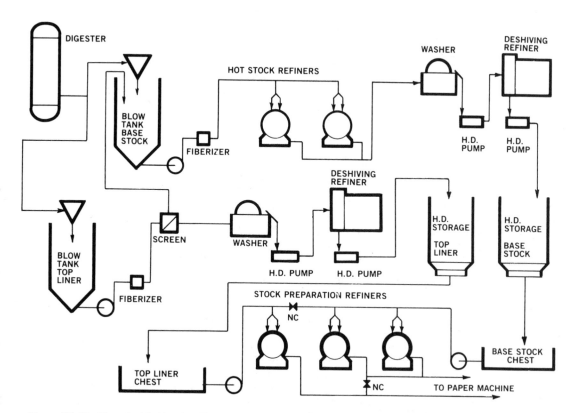

FIGURE 20-10. Flowsheet of kraft mill for dual cooking of bottom and top liner stocks. Fiberizers and de-shive refiners are installed in both systems. Hot stock refining is employed for base stock only.

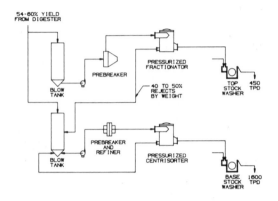

FIGURE 20-11. Flowsheet of kraft mill incorporating a fractionation step for separation of top and bottom linerboard furnish.

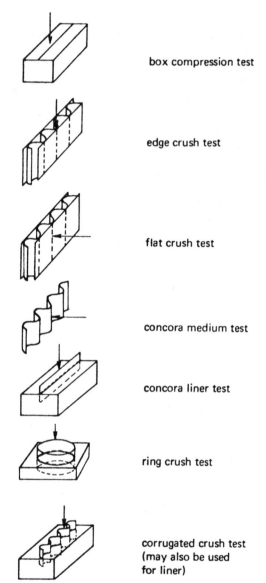

FIGURE 20-12. Various crush tests commonly applied to corrugating medium and linerboard.

20.5 CORRUGATING MEDIUM

Corrugating medium is a lightweight board used for the fluted inner ply of corrugated box stock. Standard 9-point medium (26 lb/1000 ft^2) accounts for over 80% of production, while other grades are manufactured within the weight range of 18 - 36 lb/1000 ft^2. A typical medium consists of 65% mixed-hardwood semichemical pulp cooked by neutral sulfite, carbonate or green liquor and 35% OCC (9). Some corrugating medium is made from 100% OCC and is known in the trade as "bogus medium".

The fluted layer provides much of the stiffness in corrugated box construction. Consequently, the major requirements of the medium are for stiffness and resistance to crushing (Figure 20-12). Mechanical strength properties such as tensile, burst, and tear are unimportant except in certain limited situations. Sheet finish and appearance can also be ignored in most instances. Generally, semichemical hardwood pulps are the ideal furnish for corrugated medium because of high stiffness and crush-resistance, although these pulps are notably weak with respect to the more common strength criteria.

One special consideration with respect to corrugating medium is "runability". Generally, this term refers to a paper's ability to run through papermaking and converting operations or printing presses without breaking; but for corrugating medium, runability specifically refers to the ability of the sheet to withstand the stresses and strains of the corrugating operation without fracture of the flutes. Two factors that contribute to good corrugating runability are high stretch and freedom from large shives or fiber bundles. Again, minimum restraint during drying is important to optimize the stretch properties of the sheet, while thorough stock refining is necessary to minimize shive content.

A mill manufacturing corrugating medium is relatively simple by industry standards. A representative process utilizing green liquor pulping is depicted in Figure 20-13. Both pulping and papermaking utilize standardized, no-frill equipment and operating procedures to produce the commodity-type product.

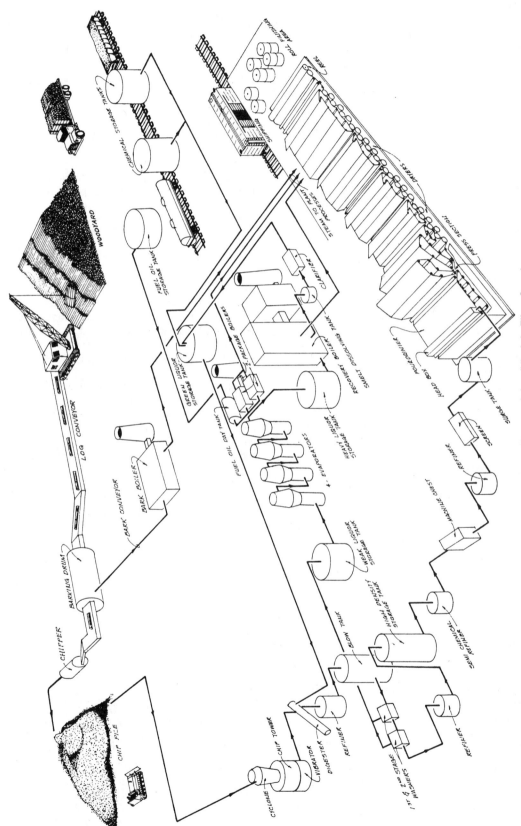

FIGURE 20-13. Layout of mill to produce corrugating medium utilizing green liquor pulping (Virginia Fibre Corp.).

20.6 FINE PAPER

The classification of fine paper refers to uncoated printing and writing grades which contain no more than 25% mechanical pulp in the furnish. Many furnishes are wood-free or limited to 10% mechanical pulp. Specific grades are offsets, tablet, envelope, bond, ledger, mimeo, duplicating, and various book stocks.

Since most fine paper machines produce a wide range of grades and grammages, flexibility of operation is an important factor in machine design. The headbox, forming section, and presses, in particular, must be capable of operating efficiently at speeds that may vary from 300 to 1000 m/min with different furnishes and frequent grade changes. Proper scheduling of grade runs is important to minimize the impact on operating continuity. For example, grades with the same furnish are run in sequence. The cleanest, brightest grades are run following a machine wash-up, while the lower-quality grades are scheduled just prior to a shutdown. Production scheduling is typically done with the help of a computer to ensure that machine time is used to maximum advantage. A process control computer is commonly employed to make the actual production change and minimize the transition tonnage.

Strength is usually not a limiting factor for fine paper grades, whereas good sheet formation and a well-filled surface are necessary attributes. As a consequence, sulfite, hardwood, and chemical sawdust pulps are preferred over long-fibered softwood kraft as papermaking furnish. In cases where a mill is dependent on softwood kraft supplies, it is usually necessary to refine the pulp in a manner to promote cutting of fibers in order to improve sheet formation. Greater inclusion of high-quality mechanical pulps in certain fine paper grades is being promoted by both pulp producers and papermakers, but the industry is still constrained by rather strict definitions as to what constitutes fine paper. A number of mills have introduced recycled printing and writing grades, mostly based on preconsumer waste, although some mills use 10 to 20% deinked post-consumer waste.

When using hardwood and sulfite pulps, the furnish is usually given only light to moderate refining. About 10 to 15% filler is added, and internal starch may be used in addition to surface sizing. Perhaps 50 to 60% of fine paper mills have now converted to an alkaline wet-end system. A dandy roll is frequently employed to improve sheet formation, apply patterning effects or produce water marks.

20.7 TISSUE GRADES

The designation "tissue" covers a wide range of low-weight sheets. Sanitary products comprise facial and bathroom tissues, towelling, and paper napkins. Industrial tissues include condenser, carbonizing, and wrapping grades. The lowest weights are about 5 g/m².

Because of the very low grammages for some products, and the loose structure for others, tissues cannot be produced on a conventional paper machine. Various tissue machine designs are used, but the traditional arrangement utilizes a fourdrinier forming section and a Yankee dryer, as illustrated in Figure 20-14. An important feature of all tissue machines is that the wet sheet is supported throughout the forming, pressing and drying processes. Tension is not applied to the sheet until it is dried.

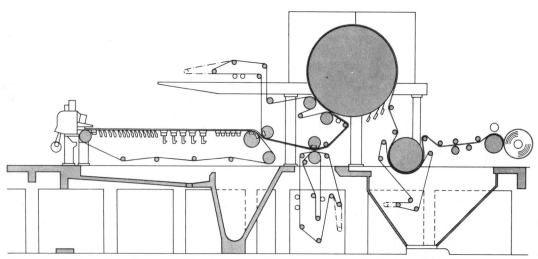

FIGURE 20-14. Fourdrinier tissue machine equipped with afterdryer and single-nip calender (Voith).

For the lighter tissues, a "lick-up felt" is sometimes used to transfer the sheet from the forming section to the press (instead of a suction pickup). The lick-up felt is relatively non-permeable and is run moist. The felt "kisses" the wire and the sheet preferentially adheres to the felt.

The Yankee cylinder is a large-diameter steam cylinder which serves as the major drying unit. The wet sheet is pressed tightly against the highly-polished surface of the cylinder and is transferred to it. The Yankee is enclosed by an air hood, and may employ high-velocity air impingement to increase the drying capacity. Some cylinders are used to impart a glazed finish to one surface of the sheet; if so, the dryer may be called an M.G. (machine-glazed) cylinder. (M.G. cylinders are also used for board finishing and cast coating applications.)

The sheet may or may not be calendered prior to reeling, depending on the grade. Some machines are not even equipped with calender rolls. For certain grades, calendering is done off-machine on a supercalender.

Creping

Many tissues are "creped" as they leave the Yankee to enhance bulk and stretch properties. The action of creping occurs due to the adherence of the sheet to the cylinder as it comes up against the square edge of the creping doctor, as illustrated in Figure 20-15. In general, the creping effect increases with a higher angle of contact (10).

The quality of the creped sheet is, in part, a function of the adhesion/release properties of the sheet, which are determined largely by the dryer surface coating. Normally, this coating is made up of the soluble organic and inorganic residues that are deposited onto the dryer surface as water flashes into steam. However, it is now recognized that a controlled coating will improve sheet quality and reduce the frequency of blade changes (11). Coating can be built on the Yankee either through wet end additives or direct spray onto the dryer surface.

Modern Tissue Formers

Modern twin-wire formers for high-speed tissue machines are gap roll formers where the drainage zone has a "C" or "S" shaped configuration. As a consequence, they are often grouped as "C" formers or "S" formers.

Figure 20-16 shows an example of a "C" former. Depending on the design, the forming roll can be solid for one-sided dewatering, or it can contain suction boxes to permit two-sided draining. A complete tissue machine equipped with an "S" former is shown in Figure 20-17.

Potential fouling of the fabrics with "stickies" is a consideration when running with furnishes containing secondary fiber. The contaminants may

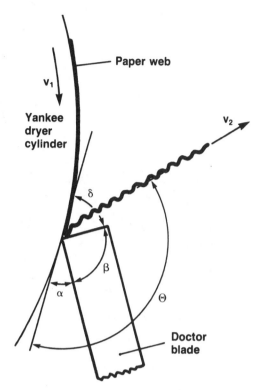

FIGURE 20-15. Creping doctor blade geometry.

preferentially build up on either the inner (conveying) wire or the outer (backing) wire depending on system geometry. Stickies on the conveying wire do not cause as many pin holes, because drainage is primarily through the backing wire. When a suction forming roll is used, some control can be exerted on the deposition of contaminants. Many tissue machines are now equipped with stratified headboxes which place the secondary fiber in the middle stratum where the contaminants are better contained within the sheet.

The first commercial gap former for tissue was the Crescent former, developed in the 1960's by Kimberly-Clark. This design (illustrated in Figure 20-18) is not a conventional "C" configuration because the sheet is formed between felt and wire rather than between two wires. Modern headbox and clothing designs have added to the original capability, and this former is of continuing interest.

Low-Density Forming

One of the most profitable and competitive sectors of the paper industry is the "soft" tissue market. Here, the customer pays for bulk and softness, and the producer finds himself in the happy position of being able to charge a higher price for a product which actually contains a lower weight of fiber.

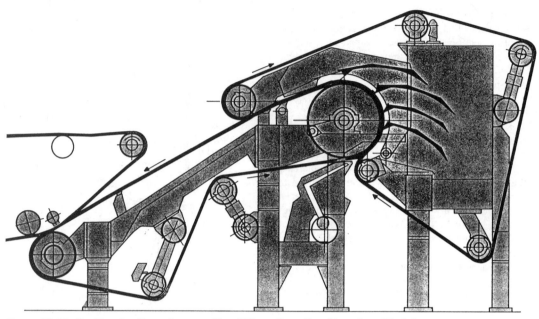

FIGURE 20-16. Double-wire tissue former with a "C" configuration (Sano Iron Works).

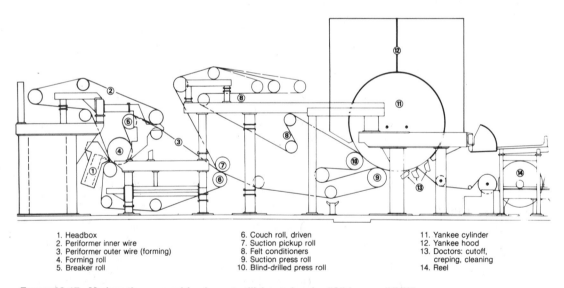

1. Headbox
2. Periformer inner wire
3. Periformer outer wire (forming)
4. Forming roll
5. Breaker roll

6. Couch roll, driven
7. Suction pickup roll
8. Felt conditioners
9. Suction press roll
10. Blind-drilled press roll

11. Yankee cylinder
12. Yankee hood
13. Doctors: cutoff, creping, cleaning
14. Reel

FIGURE 20-17. Modern tissue machine layout utilizing twin-wire "S" former (KMW).

Naturally, there has been considerable incentive to develop proprietary low-density manufacturing techniques which will give a competitive edge; and those companies engaged in soft tissue production are reluctant to divulge their methodology. Nonetheless, considerable information is available from technical and patent literature.

Perceived softness is a function of stiffness, bulk, fiber quality, and surface texture. The basic strategy behind modern low-density forming is to avoid compaction during the papermaking process. Pneumatic methods (suction and blowing) are used to remove water, and the porous sheet enters the dryer section at about 25% consistency, as compared to 40% when using mechanical pressing. Since pressing and sheet consolidation are eliminated,

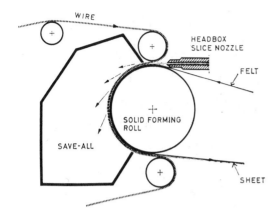

FIGURE 20-18. Crescent former.

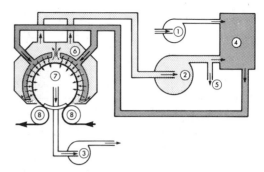

FIGURE 20-19. Schematic of through-dryer system (Papridryer): (1) makeup and combustion air; (2) circulation fan; (3) vacuum fan; (4) combustion chamber; (5) exhaust gases; (6) high-velocity hood; (7) vacuum cylinder; (8) paper guide rolls. Intensity of shading is indicative of air temperatures.

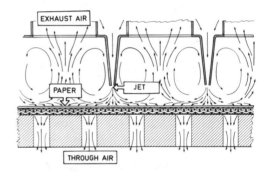

FIGURE 20-20. Diagram of small segment of through-dryer (Papridryer).

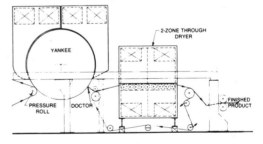

FIGURE 20-21. Placement of through-dryer after Yankee wet creping (AER Corp.).

hydrogen bonding is greatly reduced. To compensate, a typical low-density tissue furnish is made up of high-quality, lightly-refined, long-fibered pulps along with internal strength additives.

It will be noted from the above that in comparison to conventional tissue, the evaporative load is approximately doubled for each pound of fiber deposited (3.0 lb vs. 1.5 lb). But, since low-density techniques reduce fiber weight per unit of product by as much as 30%, the drying costs do not actually double for a unit of production. In fact, the extra drying cost is more than offset by raw material savings.

In conventional tissue manufacture, the wet sheet is pressed onto the Yankee steam cylinder at about 300 pli, thus compacting the sheet, providing intimate contact with the dryer surface, and promoting good heat transfer. However, Yankee cylinder drying is not ideal for soft tissues in view of the higher thermal load and because intimate contact with the sheet is no longer possible. More reliance is now placed on through-dryers to preserve bulk and achieve higher drying rates.

The process of through-drying is illustrated in Figures 20-19 and 20-20. Essentially, a hot, unsaturated gas is passed through a wet, porous material (e,g., the tissue web) by imposing a pressure differential across the material. As the hot gas passes through, heat is transferred to evaporate the water, and the gas leaves at a reduced temperature and with increased water vapor content. The intimate contact between the hot gas and the web, coupled with the large internal surface area of the porous material, produces exceptionally high drying rates.

When creping is done on the Yankee cylinder while the sheet is still wet (Figure 20-21), a more effective creping is possible (less reduction in strength). Some technologies do away with the Yankee altogether, and creping is done on smaller "creping rolls". Another arrangement where the through-dryer precedes the Yankee is shown in Figure 20-22.

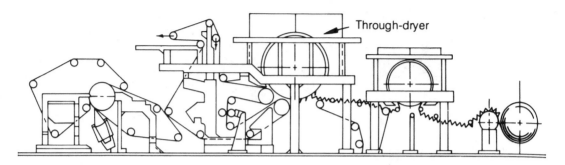

FIGURE 20-22. Wet crepe tissue machine schematic (by Honeycomb Systems Inc.).

REFERENCES

(1) GREINER, T.S. **The Trend Toward Lightweight Newsprint** *Tappi Journal* (August 1989)

(2) GARCIA, D.A. **Newsprint Producers Concerned As States Eye Taxes, Recycling Goals** *Pulp & Paper* (July 1989)

(3) SMOOK, G.A. **The Role of Chemical Pulp in Newsprint Manufacture** *P&P Canada 80:4:T123* (April 1979)

(4) ROISUM, D.R. **Runnability of Paper (Part 1 and 2)** *Tappi Journal* (January & February 1990)

(5) SUTTON, P. **New Method to Make a New Grade** *P&P International* (October 1986)

(6) GUSTAFSSON, R. **Airborne Dryers Can Help Mills Stretch Both Paper Quality and Fiber Resources** *Paper Trade Journal* (October 1, 1975)

(7) KURDIN, J.A. **Operating Conditions Critical in Refining High-Yield Chemical Pulp** *Pulp & Paper* (March 1981)

(8) DOBBINS, R.J. **High-Kappa Grocery Bag by Means of Stock Fractionation** TAPPI Alkaline Conference (September 1977)

(9) SMITH, W.E. **Using OCC in Container Board Grades** *Tappi Journal* (March 1986)

(10) OLIVER, J.F. **Dry-Creping of Tissue Paper - A Review of Basic Factors** *TAPPI 63:12:91* (December 1980)

(11) SLOAN, J.H. **Yankee Dryer Coatings** *Tappi Journal* (August 1991)

Chapter 21

Economics of Paper Machine Operation

Assuming a well-designed paper machine with a good market for its products, the major factors that affect the machine's productivity and profitability are:
• operating efficiency
• operating speed
• trim efficiency

Three terms are in common use to characterize paper machine operation. Time efficiency is the percentage of time that the machine is available for operation. Operational efficiency is the on-grade paper tonnage that was actually produced as a percentage of what could have been produced at the same speed in the available time. Absolute efficiency is the product of time efficiency and operational efficiency, and is also equal to the on-grade production as a percentage of the theoretical production in all the hours the mill is open, including maintenance time. The only parameter that really matters is absolute efficiency, and all figures cited in this discussion will be absolute efficiencies. Generally, the reader should be wary of machine efficiency figures that are not defined (1).

The effect of machine speed is obvious; if all other factors are unchanged, speed is proportional to the production of the machine. Speed is under the direct control of the paper machine operators. Operating efficiency is affected by many factors, but is often seen as a reflection of operating and maintenance discipline and methodology.

Trim efficiency is the amount of the sheet width that is actually cut into rolls or sheets as compared to the total salable width. This index is affected by how orders are scheduled, and is mainly under the control of the production planning department (2). Sometimes, for reasons outside the control of mill personnel, the machine width may not be compatible with the bulk of orders. For example, several newsprint machines installed in the late 1960's were designed to produce a salable 300-inch wide sheet, equivalent to five "standard" 60-inch rolls, or to various combinations of full, half, or quarter rolls. Shortly thereafter, a number of newspaper publishers decided to "save paper" by specifying slightly narrower rolls (say, 57 inches). Such a change in roll width specification makes it difficult for the production planning department to achieve a high level of trim efficiency.

Occasionally, a paper machine will have a high operating efficiency but overall productivity is low because operators are loathe to increase machine speed. In general, the machine should operate in the speed range that maximizes production, as illustrated in Figure 21-1. Although operating problems, maintenance, and clothing wear all increase with speed, these factors are subordinate to the objective of maximum productivity and profitability.

A paper machine operates at a characteristic production rate, which can be high or low compared to other machines making similar grades. To assess the relative performance of a particular machine, the operating rate in tons per day per inch of width can be calculated and compared with other machines, regardless of width. Table 21-1 provides a listing of 1987 maximum operating rates by grade.

Paper and paperboard machines generally operate near to the theoretical maximum level, which is imposed either by a limitation in the drive system or by lack of dryer capacity. In instances where the machine is speed limited, if operation is efficient at the present speed ceiling and is not otherwise limited, it is usually justifiable to upgrade the drives and accessories for operation at higher speed.

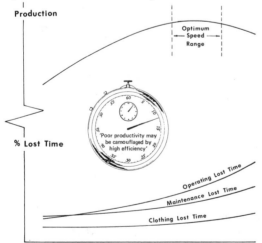

FIGURE 21-1. Effect of machine speed on paper machine efficiency and productivity.

TABLE 21-1. Maximum known operating rates, by grade (Don Ely, 1987).

Grade	Basis weight		Speed	t/d/in.
	lb/3000 ft	g/m²	fpm	
Newsprint	30	48.8	4503	2.70
Coating rawstock	27	40	4000	1.96
	30.7	50	3279	2.02
Printing/Writing	54.5	88	3650	2.58
	37	60.2	3354	2.48
	46.4	75.5	3050	2.92
Kraft paper, unbleached	50	81.4	2850	2.85
Kraft, linerboard	26	42.3	2550	3.98
	42	68.3	2290	5.77
	69	112.3	1700	7.29
	90	146.4	1260	6.80
Bleached kraft board	195	317.3	950	3.70
	310	504	640	3.97
Coated folding boxboard	148	240	1353	3.85
	172	280	1146	3.93
Liquid packing	130	212	1041	2.70
	470	465	330	3.10
Corrugated medium	78	127	2505	3.90
Wallboard	109	148	1716	3.08
Cylinder machine	177	288	760	2.70
Tissue, dry-creped	7.38	12	6890	1.02
	10	19	5300	1.06
Towel, dry-creped	8	13	6560	1.05
Towel, wet-creped	23.6	38.4	4600	2.17
	25	41	3000	1.49
	27	44	2800	1.52

21.1 FACTORS AFFECTING MACHINE EFFICIENCY

Daily machine efficiency is calculated as follows:

$$\% = \frac{\text{accepted production}}{\text{theoretical production}} (100)$$

where:
(1) accepted production is the cumulative scaled product at the end of a 24-hour period;
(2) theoretical production is the maximum tonnage that could be produced at the same speed, trim and grammage for a 24-hour period, calculated (in compatible units) as follows:

$$(\text{theor prod}) = (\text{speed})(\text{machine trim})(\text{grammage})(\text{elapsed time})$$

In metric units, the equation would utilize the following units and conversions:

$$\frac{\text{tonne}}{\text{day}} = \frac{m}{\text{min}} \times m \times \frac{g}{m^2} \times \frac{1440 \text{ min}}{\text{day}} \times \frac{\text{tonne}}{10^6 \text{ g}}$$

A number of factors are responsible for losses of production or operating time on a paper machine. A simplified grouping of items (representative of a typical newsprint machine) is given in Table 21-2. Obviously, more categories are required for operations of greater sophistication (e.g., involving size press, coating, and coat drying). For example, over 100 specific sources of lost time and lost production can be listed for a coated publication-grade machine.

TABLE 21-2. Sources of paper machine lost production.

Operating downtime
- breaks
- reeling and calender (stacking) problems
- start-ups
- grade changes

Maintenance downtime
- mechanical
- electrical
- instrument

Clothing repair or changeover
- forming fabric
- press felts
- dryer fabrics
- ropes (for carrying tail through dryers)

Loss of services
- power
- steam
- air

Slab-off losses
- paper cut off reel
- broke from winder

Off-grade production

Lost time and lost production are calculated as percentages of operating time and theoretical production, respectively. These percentile categories are additive; and if all losses are scrupulously accounted for, the total losses plus the operating efficiency should equal 100%. In practice, the measured losses do not fully account for the total loss of efficiency (i.e., 100 - efficiency). The difference, termed the "unaccounted loss", usually runs between 1 and 2%. A value as high as 4 or 5% on some machines is indicative of poor lost time accounting procedures, and could be masking some severe problems. The primary percentile categories of lost time and lost production are summarized in Table 21-3.

Operating lost time includes downtime or lost production from breaks, reeling problems, washups, grade changes, and startups. Clothing lost time includes all downtime for repair or changeover of forming fabrics, press felts, dryer fabrics, and ropes. Downtime due to loss of utility services or to lack of pulp (from trouble in the pulp mill) is totally outside

TABLE 21-3. Percentile categories of lost time and lost production.

% Operating Lost Time	=	Operating Downtime (100) / Available Time
% Maintenance Lost Time	=	Maintenance Downtime (100) / Available Time
% Clothing Lost Time	=	Clothing Downtime (100) / Available Time
% Services Lost Time	=	Services Downtime (100) / Available Time
% Off-Grade	=	(Gross Tonnage - Accepted Tonnage) (100) / Theoretical Tonnage
% Slab-Off Losses	=	Tonnage Cut Off Reel (100) / Theoretical Tonnage
% Unaccounted Losses	=	Calculated by Difference Between Total Losses and Known Losses

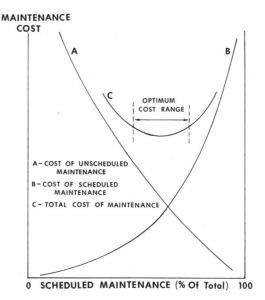

MAINTENANCE COST

A – COST OF UNSCHEDULED MAINTENANCE
B – COST OF SCHEDULED MAINTENANCE
C – TOTAL COST OF MAINTENANCE

0 SCHEDULED MAINTENANCE (% Of Total) 100

FIGURE 21-2. Effect of percentage scheduled maintenance on overall maintenance cost.

the control of the paper machine operators; but in established mills this category usually accounts for less than 1% reduction in machine efficiency.

Maintenance lost time includes all downtime for mechanical, electrical, or process control maintenance. Obviously, some maintenance should be carried out during scheduled shutdowns so that a number of concurrent jobs can be undertaken in a coordinated and efficient manner. However, if equipment rarely breaks down, the planned or preventive maintenance is probably being overdone. On the other hand, if a large percentage of maintenance is done on a breakdown basis, considerable time is wasted while assembling the work crew and from lack of planning and coordination. It is usually considered that an optimum balance between scheduled and unscheduled maintenance will result in the lowest overall maintenance cost and lowest downtime on the machine, as indicated qualitatively in Figure 21-2. According to one assessment (1), 75 to 90% of planned maintenance is the preferred range in which to operate.

Off-grade production is usually scaled, and is therefore quite well accounted for. Unfortunately, slab-off losses must be estimated, and a low figure here can easily increase the level of unaccounted losses.

Throughout the world, except in Germany and Japan, machine efficiencies have shown a downward trend over the past 20 years (1). It can be argued that the decline is due to a combination of lower grammage, higher machine speed, and lower-quality furnishes (e.g., more recycled fiber). More likely, the decline is due to long-term cost-cutting measures by management that have resulted in a general reduction in technical and engineering staff, inadequate training of operators and supervisors, and lack of planned maintenance. According to one Japanese expert, effective papermaking requires

close attention to every detail of lost production by a team of capable, imaginative people working closely together; to correct deficiencies, an action plan must be developed and implemented, and its effectiveness monitored.

21.2 LOST TIME ANALYSIS

A systematic method of documenting and analyzing paper machine lost time is required to locate specific areas of low efficiency and identify problems as they arise (3). While the benefits of lost time analysis are well recognized (as summarized in Table 21-4), the emphasis given to this activity by paper mill management varies within wide extremes. It must be stressed that the ultimate purpose of lost time analysis is the reduction of lost time and/or lost production. However, the enforced operating

TABLE 21-4. Benefits of lost time analysis.

1. Specifically defines the causes of lost production so that appropriate action can be planned.

2. Provides a management control. Brings to management's attention the cost of deficiencies (that might otherwise be tolerated).

3. Improves operating discipline by enabling the machine crew to realize how certain production deficiencies cause the machine to be non-productive.

4. Indicates to technical service and research groups the areas in which their activities can be usefully employed.

discipline and greater knowledge of what is going on also contribute significantly to improved machine operation.

All data for lost time analysis originates with automatic monitoring equipment and with the observations and assessments of the paper machine operating crews. Although an automatic paper break monitoring system will calculate the time loss due to web breaks at each section of the machine, the crew must provide certain explanatory information. The training of the machine personnel for these duties should be considered an important task of operating supervision. The actual recording of data during an operating shift or at the end of a shift is facilitated by providing the operators with suitable forms or data log formats.

The principal steps in lost time analysis are listed in Table 21-5. Although data must be tabulated on a daily basis, comparisons and trends based on operating periods of less than 3 months (i.e., quarterly) are probably not very useful. As an example of a trend plot, Figure 21-3 shows annual averages in two lost time categories (maintenance and operating) for three newsprint machines during 13 years of operation. Over this time span, increases in machine speed appear to have contributed to a higher incidence of operating problems, while no clear trend was evident with respect to maintenance.

TABLE 21-5. Steps in lost time analysis.

1. Document all types of machine lost time and lost production.

2. Tabulate data in a meaningful way.

3. Compare data with that of comparable machines to gain immediate insight into problems or deficiencies.

4. Plot data to show trends.

5. Focus on the specific problems that are identified.

Figure 21-4 shows a quarterly plot of maintenance lost time for a newsprint machine. In this example, the distribution between planned and breakdown maintenance, and progress toward a targeted level, can be readily seen. The apparent effect of actions undertaken in the fourth quarter of 1989 are obvious.

Breaks can be a significant source of operating lost time on high-speed lightweight machines. Where possible, every effort should be made to obtain the "break ends" for machine breaks, and these should be labeled and put into a special box for each machine. The break ends can be an important key to what is happening on the machine. A systematic examination of break ends, together with a

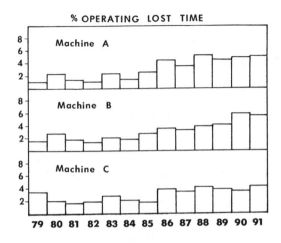

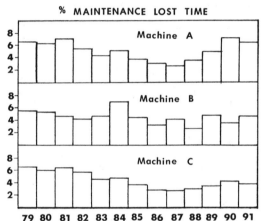

FIGURE 21-3. Annual averages of operating and maintenance lost time percentiles for three newsprint machines over a 13-year span.

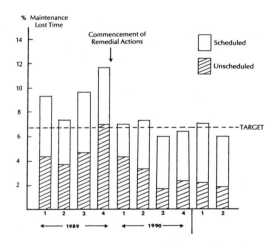

FIGURE 21-4. Quarterly plot of maintenance lost time.

knowledge of break location, will often pinpoint a problem before it becomes serious. Table 21-6 provides a summary of wet-end break characteristics for a twin-wire newsprint machine, while Figure 21-5 shows the distribution of the breaks by position across the machine. Figure 21-6 shows lost time for breaks by location for a machine with two coaters.

TABLE 21-6. Summary of wet-end break characteristics for a twin-wire newsprint machine.

	Number
Total breaks examined	87
Types of deposits:	
- accumulation of fines	44
- lumps	28
- miscellaneous	10
Location of deposit:	
- backing wire side	16
- carrying wire side	51
- indeterminate	15
Pickouts and delaminations	18
Edge breaks	27

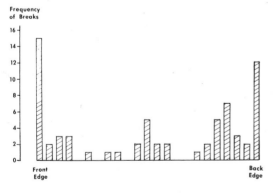

FIGURE 21-5. Distribution of break frequency across the machine.

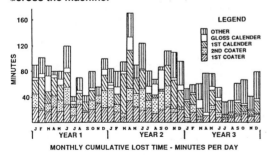

FIGURE 21-6. Lost time for breaks by location for a paper machine with two coaters.

In a large paper manufacturing company with several mills, a comparison of machine performance between mills can be helpful. Obviously, it is necessary that lost time be reported in a similar manner for all mills. Likewise, a comparison between companies is useful; but here, the terminology on which the lost time is stated must be clearly defined so that the necessary adjustments can be made to bring divergent systems to a common basis. Unfortunately, no standardized system for lost time reporting has been universally accepted at the present time.

One of the major benefits of lost time analysis is improved control. Management at all levels should be provided with the data revealed by lost time analysis in the form of continuous graphs displayed for easy reference. In gauging machine performance, it is useful to incorporate a target efficiency level for each machine based on a balanced judgement of what the machine is capable of achieving. This target level should not be based upon past performance, but on a realistic technical assessment of the situation. It must be recognized that the optimum efficiency may take several years to attain.

21.3 GRADE MIX VS. PROFITABILITY

Machines that manufacture a broad range of paper grades and grammages pose a special challenge with regard to optimizing profitability. The need for rapid and efficient transition between grades and the role played by the production planning department in the scheduling of the machine has already been noted. A general consideration with respect to multi-grade machines is to identify which grades are most

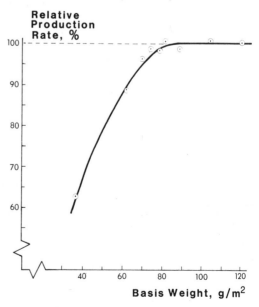

FIGURE 21-7. Effect of basis weight on fine paper machine productivity.

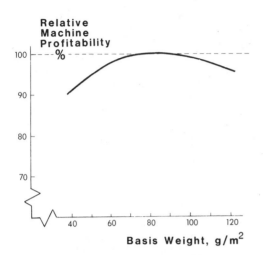

FIGURE 21-8. Effect of basis weight on fine paper machine profitability.

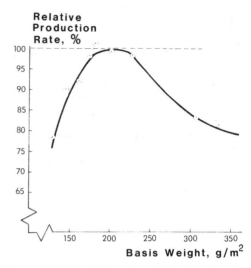

FIGURE 21-9. Effect of basis weight on linerboard machine productivity.

profitable and to emphasize sales of these "bread and butter" items. Sales of the remaining grades are used to "fill out the order book" and maintain machine productivity.

A fairly typical productivity curve for a fine paper machine is shown in Figure 21-7. Production rates for the lightweight items may be restricted by speed limitation of the drive system, wet-end runability, or after-size press dryer capacity. Increased machine speed will, of course, to a certain extent offset the effect of reduced sheet weight on production tonnage; however, the higher speed puts additional stress on the lightweight sheet which may cause runability problems, and moisture pickup at the size press is almost proportional to machine speed. Production in the higher weight range is typically limited by the capacity of the before-size press dryer section.

Since lightweight fine paper products command a higher per ton selling price, the profitability curve for this machine will have the shape shown in Figure 21-8. Optimum profitability is usually found in the mid-range of grammages. If orders are predominantly in the lower or upper weight ranges, it is usually worthwhile to implement changes on the machine to overcome whatever "bottleneck" is impeding productivity. Mills that operate two or three multi-grade machines are more ideally positioned to optimize profitability, since one machine can then be suitably equipped to maximize lightweight sheet production while another is tailored for heavier weights.

A less typical productivity curve for a linerboard machine is shown in Figure 21-9. On this machine, sizing solution is applied at the calender stack and production at higher speeds (i.e., lower weights) is limited by inter-calender drying capacity. Production

at higher weights is limited by fourdrinier wire drainage; the wire table utilizes mainly foils whose dewatering action is speed-dependent. For linerboard machine products, there is little price differential over the grammage range, so the profitability curve has the same general shape as the productivity curve.

Although maximum tonnage from a multi-grade paper machine is usually related to maximum profitability, there are exceptions to this rule, principally for machines producing lightweight specialty sheets. In such cases, it makes sense to operate the machine toward a profit target rather than toward a tonnage target.

REFERENCES

(1) MARDON, J., VYSE, R.N. AND ELY, D. **Paper Machine Efficiency...** *P&P Canada 92:12:87* (December 1991)

(2) URONEN, P. **Production Planning Systems for Integrated Paper Mills: Tasks and Methodology** *P&P Canada 82:3:T86-89* (March 1981)

(3) MARDON, J., VYSE, R.N. AND TKACZ, A.F. **Lost Time Analysis: the Key to Efficiency Improvement** *Tappi Journal 67:3:94* (March 1984)

Chapter 22

PROPERTIES AND TESTING OF PULP AND PAPER

Pulp and paper products are usually characterized and bartered on the basis of well-defined nomenclature and testing procedures. The objective is to ensure that all participants speak the same "language" when discussing pulp and paper products and test values.

Test procedures, along with some terminology, are described in a series of "standard methods" issued by the major pulp and paper organizations around the world. The principal promulgators of English-language standard methods are listed in Table 22-1. Standard methods from different organizations describing the same test procedure often vary with respect to particulars. However, as long as the source of the standard method is specified, pulp and paper transactions are carried out and scientific investigations are documented with a minimum of ambiguity and misunderstanding. In recent years, the ISO (International Organization for Standardization) has made progress in issuing a number of pulp and paper standards that are endorsed by a majority of countries.

TABLE 22-1. Producers of English-language standard methods.

TAPPI	- Technical Association of the Pulp and Paper Industry (USA)
CPPA-TS	- Canadian Pulp and Paper Association, Technical Section
SCAN	- Scandinavian Pulp, Paper and Board Testing Committee
ASTM	- American Society for Testing and Materials
BPBIF	- British Paper and Board Industry Federation
APPITA	- Technical Association of the Australian and New Zealand Pulp & Paper Industry

22.1 OBJECTIVES OF TESTING

Routine mill testing and monitoring programs can be identified with one or more of the objectives as

TABLE 22-2. Objectives of routine mill testing and monitoring programs.

Raw Materials Control	-	monitor incoming raw material quality
Process Control	-	control process within specified limits
Quality Control	-	monitor product quality
	-	minimize off-grade production
Pollution Control	-	control waste discharges
	-	control fiber losses
Process Monitoring	-	assess performance
	-	follow progress toward process objectives
	-	compare performance with other mills
	-	pinpoint problem areas
Monetary Control	-	establish cost breakdown
	-	provide monetary basis for consideration of alternative proposals
	-	pinpoint high cost areas

listed and defined in Table 22-2. Non-routine testing is usually carried out as part of an investigation into mill problems or as part of an effort to improve process efficiency.

For the technical control of mill operations and products, it is important that the tests carried out are sufficiently accurate and rapid to provide timely information. An effective testing program is vital to process control, on which high production and consistent quality are dependent. Unless small differences can be detected, the "momentum" of the process cannot be assessed, and the operator cannot take corrective action until gross changes have occurred.

The requirements of a mill control test (performed by operators who are not technically trained) are simplicity, speed, and inherent reproducibility. More complicated tests that must be performed in the laboratory by trained technical staff are not as useful for control purposes because of obvious time delays.

Unfortunately, some properties of interest in pulp and paper process control are difficult to measure. In many instances, one property must be estimated from measurement of a related property.

Because it may give spurious results from which incorrect decisions will be made, a poor test or an improperly-made test is sometimes worse than no test at all. An insensitive test method will allow detrimental changes in the process to occur without detection, while an erroneous test result will sometimes cause the operator to take the wrong course of action.

22.2 CHARACTERIZATION OF TESTING PROCEDURES

All testing procedures can be characterized by their sensitivity (or instrument readability), precision, and accuracy. These terms are defined and explained in Table 22-3.

TABLE 22-3. Definition of test characterization terms.

Sensitivity - the smallest change in the property being measured that can be detected, usually expressed as the smallest incremental difference.

Readability - the smallest incremental difference that can be read from an instrument. (Note: Readability can sometimes be improved by the use of a magnifying lens or vernier scale; but it is useless to have a readability that is better than the sensitivity.)

Precision - a measure of the variation that can be expected when repeated tests (replicates) are made on the same specimen or near-duplicate specimen.

Accuracy - the difference between the test value and the true value. (Note: It is not practical to measure accuracy if the "true value" must be determined by some idealized method.)

The distinction between precision and accuracy must be emphasized. Precision is an assessment of test reproducibility, but says nothing about the relationship of the test value to the true value. For

---

TABLE 22-4. Factors affecting the precision and/or accuracy of a test result.

- instrument readability
- instrument or test sensitivity
- sampling error
- procedural differences
- instrument calibration
- variations in correlation between measured property and desired property
- external factors

---

example, if an instrument is out of calibration or if a non-representative sample is being tested, the test result in either case may be precise, but will probably not be accurate. Factors affecting test precision and accuracy are listed in Table 22-4.

The precision of any test result can be defined statistically either in terms of a standard deviation or by statistical confidence limits. The standard deviation (denoted by the Greek letter sigma) is the root-mean-square deviation of the observed values from the average. A low value (relative to the mean) indicates that data points are tightly clustered around the mean, while a higher value indicates that the data are more scattered. The formula for the calculation of sigma is as follows:

$$\sigma = \sqrt{\frac{\Sigma (x - \bar{x})}{n-1}}$$

where: x = value for individual readings
$\bar{x}$ = arithmetic mean of all readings
n = number of test readings

The confidence level indicates statistically how often the true value will actually fall within a specified range. For example, the expression of an experimental value as 178 plus or minus 4 at the 95% confidence level means that, in 95% of all cases, the actual value falls within the range 174 to 182. The standard deviation and confidence level are directly related as shown in Table 22-5. A range of plus or minus 2 sigma is a good approximation of the 95% confidence level.

TABLE 22-5. Confidence level for various ranges of data in terms of sigma multiples.

+ or - 0.675σ	=	50%	confidence
+ or - 1σ	=	68%	"
+ or - 2σ	=	95.5%	"
+ or - 3σ	=	99.7%	"
+ or - 4σ	=	99.994%	"

It is often useful to convert the standard deviation into a percentage of the mean test value, which may be denoted either as % variation or % error. The standard deviation and % variation for some common groundwood tests are given in Table 22-6. The % variation usually provides a more convenient basis for comparing the relative precision of the various tests.

One way to improve test precision is to increase the number of test replicates. However, greater replication means a higher cost for technician testing. Therefore, the number of replicates specified for each standard testing method is a compromise between desired precision and testing cost. The

Table 22-6. Precision of common groundwood tests.

	$\bar{x}$	σ	% Variation
Canadian Standard Freeness, ml	102	4.0	3.9
Burst Index, kPa·m²/g	1.22	0.06	4.9
Tear Index, mN·m²/g	4.91	0.29	5.9
Breaking Length, km	3.50	0.11	3.1
Brightness, %	54.2	0.5	-
Bulk, cc/g	2.56	0.06	-

Table 22-7. Number of replicates specified by standard test methods.

	TAPPI	SCAN	APPITA
Caliper, tests (plies)	10 (5)	5 (4)	10 (8)
Tear (4-ply)	4	2	4
Burst	10	8	8
Tensile	10	8	8
Stretch	10	8	8
Folding Endurance	5*	5	16**

* or 10 or 20, depending on precision desired
** normally not followed in practice

number of replicates specified for some common strength tests is listed in Table 22-7. The effect of replication on the precision of these tests for a variety of pulp types is shown in Table 22-8.

It is not customary to include sigma limits or confidence limits with most test values. However, a qualitative inference of precision is shown by the significant figures in the reported value. Pulp and paper test values beginning with numbers 4 through 9 are typically rounded off to two significant figures, while values beginning with 1 through 3 are reported to 3 significant figures. An exception is made for

folding endurance (a notoriously imprecise test) which is reported to either one or two significant figures. In any overlapping series of test values, it is good practice to be consistent with the last significant figure.

Problems in Defining Precision

A statement of test precision should take into account the exact conditions for a test. Different circumstances may easily cause a wide variation in

Table 22-8. Precision of standard tests (F.H. Phillips).

	n	% Standard Error of Mean Value			
		Unbleached Hardwood		Unbleached Softwood	
		Unbeaten	Beaten	Unbeaten	Beaten
Caliper	5	0.30	0.27	0.59	0.29
	10	0.21	0.20	0.42	0.20
Tear	1	13.5	8.6	7.8	5.3
	2	9.6	6.1	5.5	3.7
	3	7.8	5.0	4.5	3.0
	4	6.8	4.3	3.9	2.1
Burst	2	11.7	4.8	4.3	3.2
	4	8.3	3.4	3.0	2.2
	6	6.7	2.8	2.5	1.8
	8	5.8	2.4	2.1	1.6
	10	5.2	2.2	1.9	1.4
Tensile	2	4.7	3.4	3.7	2.9
	4	3.3	2.4	2.6	2.1
	6	2.7	2.0	2.1	1.7
	8	2.4	1.7	1.9	1.5
	10	2.1	1.2	1.7	1.3
Stretch	2	10.3	6.1	4.9	4.6
	4	7.3	4.3	3.5	3.3
	6	6.0	3.5	2.8	2.7
	8	5.2	3.0	2.5	2.3
	10	4.6	2.7	2.2	2.1
Fold	2	25	16	18	13
	4	18	11	13	9
	6	15	9	10	8
	8	13	8	9	7

precision in the same test. For example, a set of paper caliper data may represent variation for:
• localized area of sheet,
• total sheet,
• different sheets from the same roll,
• different rolls across the reel,
• different sheets from a lot.
It should be noted that many tests on paper are destructive tests, which precludes the possibility of making duplicate tests on the same spot.

It is very important in defining the precision of a test method to specify whether it is between laboratories or within a single laboratory. If the test method is carried out within a single laboratory, it is necessary to specify the number of instruments involved and the number of operators. The smallest variation between tests will be found if the tests are carried out:
• by the same operator,
• on the same instrument,
• on the same sample,
• within a short time period.

Sampling

It is important that proper sampling procedures are followed if test results on a fragmentary sample are being used to characterize a large volume of material. For example, at the dry end of the paper machine only a small portion of the paper sheet is used to characterize the entire reel, and it is important that the sample or samples be properly weighted with respect to machine-direction and cross-direction variations. In the pulp mill it is always a challenging problem to obtain a representative sample of the chips being produced, delivered or conveyed to the digesters in order to determine their moisture content or characterize their quality.

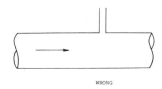

WRONG

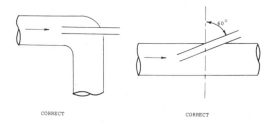

CORRECT   CORRECT

FIGURE 22-1. **Stock-sampling taps.**

A special problem in pulp and paper mills is to obtain representative stock samples for consistency measurement. It is helpful in this regard to install proper sample taps as illustrated in Figure 22-1, and to confirm their validity. The taps should be evaluated (1) by taking a rapid series of samples (e.g., perhaps 20 samples at 5-second intervals), measuring consistency on each sample, and plotting the results in chronological order. The variations seen will be composed of systematic variation and random variation (error). If systematic variation is obvious (i.e., exhibiting a pattern), a representative sample can be obtained by compositing several smaller samples. However, if random variation is dominant, an improved sample location may be required. Generally, a higher level of random variation can be expected for longer-fibered pulps at consistencies above 3.0%.

Assessing Inter-Mill Variability

To ensure that a pulp and paper mill's testing program is reliable and that its test results are meaningful, some basis of comparison within the company and/or between companies is necessary. Two methods are used to monitor testing variability between mills:

- In round-robin testing, nearly identical samples are sent to each mill or laboratory within a company for testing. The differences in test results are monitored on a regular basis to establish any existing or emerging bias in testing values.

- Interlaboratory reference systems are a more advanced form of the round-robin procedure. Here, samples along with provisional test values are supplied to each mill or laboratory by a recognized institution such as the National Bureau of Standards (in cooperation with TAPPI) or the Pulp and Paper Research Institute of Canada. Each mill or laboratory then reports their test values, and a compendium of values is issued by the coordinating institution. The true value is usually assumed to be near the arithmetic mean, and values far from the mean are indicative of poorly calibrated equipment or faulty methodology.

Since these programs are designed to check all aspects of testing variability (i.e., instrument calibration, procedure, operator technique, etc.), it is important that the assignment be given periodically to each of the testers. Also, where provisional values are provided, these must be withheld from the tester to ensure unbiased test results. Significant differences between the provisional values and the test results can alert the supervisor to the need for immediate corrective action.

22.3 PULP TESTING

A large number of testing methods are in common use to characterize pulps with respect to quality, processability, and suitability for various end uses. The more "fundamental" measurements provide the means to predict behavior, but are often too complicated to be applied in the mill laboratory. Functional tests are designed to measure specific properties. A summary of test methods is given in Table 22-9.

TABLE 22.9. Pulp Characterization Methods

Fundamental Measurements
- weighted average fiber length
- intrinsic fiber strength
- fiber coarseness
- specific surface and specific volume
- wet compactability
- chemical analysis
- filtration resistance
- fiber saturation point

Functional Tests
- kappa number
- cellulose solution viscosity
- drainability
- color and brightness
- cleanliness
- fiber classification
- wet web strength
- wet rupture energy
- water retention value
- wet compressability
- beater evaluation
        (strength of handsheets)

In general, considerable emphasis is placed on pulp strength tests. However, for many applications, strength is of secondary importance. In fine papers, for example, conformability is more critical. The usual strength tests on pulp handsheets, such as burst, tear, tensile, and folding endurance, should be supplemented by a knowledge of the fundamental properties to obtain a proper understanding of the pulp qualities. If a pulp is weak, this knowledge explains why and in what particular respect it is weak.

Chemical Testing of Pulps

Chemical tests are commonly used to determine the lignin (non-cellulosic) content of chemical pulps and to quantify the cellulosic portion with respect to average degree of polymerization (dp) and alpha cellulose content.

Kappa Number

Lignin and other non-cellulosic constituents of pulp fibers react readily with acidic permanganate. This reaction provides the basis for either the kappa number ("K number") or permanganate number ("p number") test. In both cases, a carefully weighed pulp sample reacts with a known volume of permanganate solution under controlled conditions, and the amount consumed is determined by back-titration. The kappa number test is gradually displacing the original permanganate number test because of its applicability over the full cooking range. Also, kappa number has a linear relationship with lignin content for pulps below 70% yield; the percentage lignin approximately equals K x 0.15.

The kappa number test is used in mill control work for two specific purposes:
- to indicate the degree of delignification occurring during cooking (i.e., as a control test for cooking),
- to indicate the chemical requirement for bleaching.

The two objectives are virtually never combined on the same test because of the large processing gap between cooking and bleaching. In mills with some pulp blending capability on the unbleached side, the reduction in kappa number % variation between cooking and pre-bleaching can be used as a measure of blending effectiveness.

Cellulose Solution Viscosity

A good indication of cellulose dp can be obtained by measuring the viscosity of a cellulose solution of known concentration. A number of reagents are recognized as being suitable solvents for this purpose, but cupriethylene diamine hydroxide (usually abbreviated CED or cuene) is now used almost universally because it dissolves cellulose rapidly and has good chemical stability.

It is well known that pulp strength factors deteriorate markedly when the dp falls below a critical level. The critical level depends on furnish and process conditions and must be defined for each specific situation. In most bleached chemical pulp mills, the cuene viscosity test is used to monitor the process for upsets and to flag off-quality pulp.

Alpha Cellulose

The long-chain molecular fraction of holocellulose (alpha cellulose), is defined chemically as that portion which resists solubilization in strong caustic solution. The chemical determination is carried out by extracting the pulp with caustic solution under specified conditions to dissolve shorter-chain cellulose and hemicelluloses. The soluble portion is then measured by oxidation with potassium dichromate, and the alpha cellulose fraction is calculated by difference.

Physical Properties of Pulps

Generally, all measurable pulp characteristics, excluding chemical properties, are grouped under the heading of physical properties. This heading includes properties that are defined both by fundamental measurements and functional tests.

Fiber Length

Traditionally, the fiber length distribution or weighted average fiber length of a pulp sample has been determined either by microscopic examination of a representative number of fibers, by means of a special drainage grid, or by screen classification of a sample into different length fractions. In the microscopic method, the magnified image of a known weight of fibers is projected onto a calibrated grid pattern; all the fibers are measured to produce a length distribution and the average fiber length is calculated mathematically.

In the grid method, a bronze ring equipped with a number of equally spaced blades is used in place of the screen in a standard sheet-making machine. A specified stock suspension is allowed to flow past the grid, and the weight of fibers retained on the grid is measured and reported as the "fiber weight length", a relative indication of average fiber length.

In the classification method, a dilute suspension of fibers is made to flow at high velocity parallel to screen slots, while a much slower velocity passes through the slots. In this way, the fibers are presented lengthwise to a series of selected screens with successively smaller mesh openings, and only the fibers short enough to bridge the openings pass into the next chamber. The pulp sample is typically divided into five length fractions.

The traditional methods of fiber length measurement are now rapidly being displaced by modern optical counting devices. For example, the Kajaani analyzer relies on the ability of cellulose to depolarize a polarized light source to measure the number of fibers in each of 144 separate length categories. A very dilute suspension of fibers (0.001% consistency) is poured into a small test chamber, then drained through a glass capillary tube under suction. The fibers are constrained to flow through the tube lengthwise and are measured by a series of photodiodes. The resultant signal is analyzed by a microprocessor to give statistics such as the average fiber length and frequency distribution.

Drainability

The resistance of a fiber mat to the flow of water is an important property with respect to pulp processing and papermaking. The classical method of determining this property in North America is by means of the Canadian Standard Freeness (CSF) tester as illustrated in Figure 22-2. The CSF is defined as the number of milliliters of water collected from the side orifice of the standard tester when a pulp suspension drains through the screen plate at 0.30% consistency and 20° C.

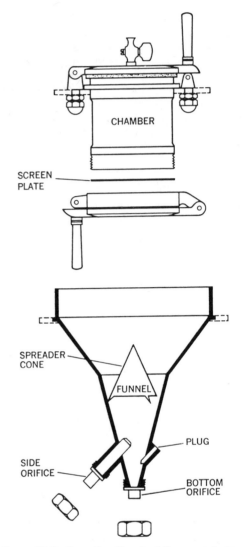

FIGURE 22-2. **Canadian Standard Freeness tester.**

Measurements of pulp drainage are known as freeness, slowness, wetness, or drain time according to the instrument or method used. If a pulp drains rapidly, it is said to be "free". If it drains slowly, it is said to be "slow". Freeness and slowness scales have an inverse relationship. The Schopper-Riegler slowness tester is the principal laboratory drainage testing device used in Europe.

Freeness is widely used as an indication of a pulp's suitability or quality. In most mechanical pulping processes, the energy input is adjusted to maintain the product freeness within a narrow range. Chemical pulps are often refined to a target freeness level before being added to the papermaking furnish. Studies have shown that the accumulating fines fraction (-200 mesh) is primarily responsible for reduced drainage. Removal of the fines fraction from a beaten pulp can restore the original drainability, while the pulp retains its beaten strength properties. This finding is sometimes used as an argument against the use of drainage measurements as an index of pulp quality.

Because the freeness determination is often used for process control, it is desirable to measure this property on-line. A number of automatic drainage measuring devices have been introduced and used with mixed success over the past two decades. A representative modern device is illustrated in Figure 22-3. This particular detector consists of a vertically oriented pipe incorporating a pressure sensor which regulates the intake and exhaust differential pressures independent of the stock line pressure. At the start of each measuring cycle, a small negative pressure causes the stock to move upward, eventually encountering the screen, whereupon fibers accumulate in a pad, and filtrate flows into the upper chamber. As the filtrate rises, it contacts a level detection probe which initiates a timing circuit; when filtrate nears the top of the detector tube, it contacts another probe which terminates the timing sequence. Elapsed time (which is relative to freeness) is displayed and transmitted as a signal. Upon completion of the rate determination, the pressure in the detector is increased to slightly above the stock line pressure, whereupon filtrate and fibers are returned to the stock line in preparation for the next cycle. Fresh water is also introduced during the exhaust phase to ensure detector tube cleanliness.

Dynamic Drainage

Although freeness measurements provide a basis for comparing similar pulps, the conventional tests do not simulate the type of drainage (i.e., with microturbulence and oriented shear) that occurs on the paper machine. For example, groundwood pulp gives a lower freeness than highly-beaten chemical pulp but shows faster drainage on the paper machine. The Britt dynamic drainage jar (illustrated in Figure 22-4) was developed for the purpose of studying stock drainage phenomena under conditions more closely approaching those of the paper machine. The barrel sits on a base that is designed to hold a perforated plate or sample of paper machine forming fabric. The propeller for stirring the stock sample is inserted into the barrel and is driven by a variable-speed motor capable of operating in the range from 0 to 3000 rpm.

Since its introduction in the early 1970's, a number of uses have been developed for the Britt jar. In addition to characterizing stock drainage on the forming wire, it is now a valuable tool for measuring first-pass retention, evaluating additives, and carrying out retention aid studies. A good introduction to the detailed procedures associated with various applications can be found in reference (2). In addition to the original version, a number of variations of the device have been developed.

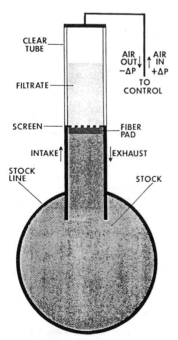

FIGURE 22-3. Continuous freeness sensor (Drainac III from Bolton-Emerson).

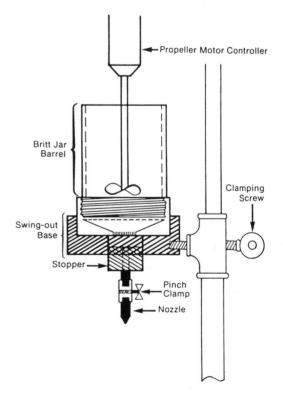

FIGURE 22-4. Diagram of Britt dynamic drainage jar.

Beater Evaluation

In the final analysis, the most important property of a pulp is its papermaking potential. This aspect can best be evaluated by beating or refining the pulp under controlled and reproducible conditions, then forming the pulp into standardized handsheets, and finally performing suitable physical tests on the conditioned handsheets.

The purpose of beating is to mechanically condition the fibers for papermaking. A more general term for mechanical working of pulp is "refining". The term "beating" actually denotes a specific type of refining action (see Section 13.2), but is now more commonly used to describe refining in the laboratory. Most laboratory beating methods utilize low-intensity energy application; they have a more selective action than mill refiners and produce results that normally cannot be duplicated in the mill.

There are two distinct techniques or objectives in laboratory beating. The idealized beating technique produces the greatest change in pulp properties for a given change in freeness. This method is used for research studies and is almost always used by market pulp mills as the basis of quality claims. The mill simulation beating technique more closely reflects what happens in the mill process. This technique is sometimes used by operators to evaluate new pulps and to provide a more meaningful basis for process control when using variable pulp sources.

A number of laboratory beating devices are in use around the world for routine testing and experimental work. A partial listing is given in Table 22-10. The two devices most commonly used in North America are the Valley beater and the PFI mill. The Lampen mill is more popular in Scandinavia, the United Kingdom and Australia. The Jokro mill is favored in Germany.

TABLE 22-10. Laboratory pulp beating devices.

Valley beater
PFI mill (Norwegian Paper Industry Institute)
Ball or Pebble mill
Kollergang
Lampen mill
Banning and Seybold beater
Noble and Wood cycle beater
Jokro mill
Aylesford beater
Bauer-Mead refiner (miniature disc refiner)
Escher-Wyss refiner (miniature Jordan refiner)

The Valley beater (shown in Figure 22-5) is essentially a miniature version of the Hollander beater. Although this device has a long tradition of use, it has some definite limitations and is gradually being displaced by the PFI mill. The principal

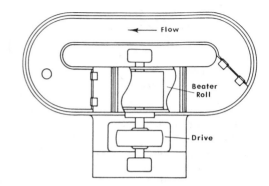

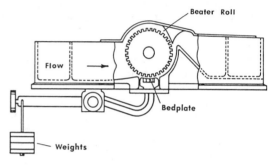

FIGURE 22-5. Valley beater.

disadvantage is the difficulty in obtaining reproducibility with respect to other Valley beaters and with respect to the same beater over long periods of time. The problem relates to variable wear patterns on the metal cutting bars.

The PFI mill (Figure 22-6) utilizes a grooved roll eccentric to a smooth trough on which the pulp sample at 10% consistency is distributed. The pressure between roll and trough can be varied by means of applied weights. Both the roll and "bedplate" rotate at high speed, but at different peripheral velocities; this action induces friction, rubbing and crushing of the fibers to produce the beating effect. With normal operation, there is no metal-to-metal contact and little wear on the edges. Consequently, the device rarely needs calibration. A relatively small amount of pulp is required (as compared to the Valley beater) to carry out a complete evaluation.

The Lampen mill consists of a hollow circular housing rotating around a horizontal axis at 250 rpm, in which a smooth spherical 10-kg ball is allowed to roll freely (Figure 22-7). The pulp sample is milled by the rotating ball at relatively low consistency (3.0% for softwood, 3.6% for hardwood), providing a slow, gentle refining action. The main variable affecting the uniformity of action is the micro-roughness of the ball and housing.

It should be emphasized that pulp beating is a

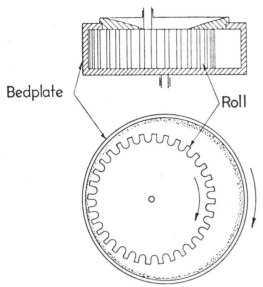

Bedplate

Roll

FIGURE 22-6. PFI mill, roll and housing.

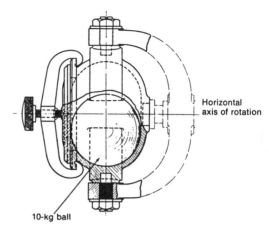

Horizontal axis of rotation

10-kg ball

FIGURE 22-7. Lampen mill.

mechanical process, and in itself does not represent a method for assessing pulp quality. It is the physical testing carried out in conjunction with beating that specifically provides the quality information. These tests show the relative rate of response of the pulp to the beating process, the maximum strength that is developed, and the relationships that prevail between the various physical parameters. The methodology for the forming, couching, pressing, drying and conditioning of pulp handsheets is covered in TAPPI Method T205. Methods for measuring various strength properties are covered in TAPPI Methods T200, T205, T220, T221, T224, T225, T227 and T231.

Pulp Strength Comparisons

The traditional technique for assessing the relative strengths of different chemical pulps is to compare strength test results at specified reference freeness levels, the typical levels being 500 and 300 CSF. Since the pulps are rarely tested at exactly these freeness values, the comparative values must be interpolated from the respective beater curves. This approach may show, for example, that one pulp has a higher tear value and a lower tensile value than another pulp; but it doesn't adequately define which is "stronger".

In recent years, tear-tensile plots have become the principal means of comparing pulps from different furnishes and processes (3). These plots are based strictly on strength parameters, not drainage behavior, and for that reason alone are intuitively more satisfactory for pulp strength comparisons. The plots also extend over several levels of bonding and therefore provide a more complete picture of

relevant pulp behavior. (In Chapter 4, Figure 4-5 shows strength differences with a series of tear-tensile plots, while Table 4-6 provides the traditional tabular comparison of strength data at a single reference freeness level.)

In order to produce a meaningful tear-tensile plot, it is necessary to carry out a 3 or 4-point beater run. For routine strength monitoring, such a labor-intensive exercise may be hard to justify. In fact, for comparing the strengths of chemical pulps that are produced using similar furnish and process conditions, a single-point method may be quite satisfactory. A number of mills use a "combined strength parameter" (a weighted value of tear plus tensile or tear plus burst) for routine strength monitoring and for strength comparisons between similar pulps (4).

22.4 PAPER TESTING

The wide diversity of paper grades with different functional properties necessitates a multiplicity of test methods. Some basic properties are important for all grades (e.g., grammage, caliper) and the relevant tests are used extensively. Other measurement methods have been specifically developed to assess the performance attributes of specialty products and their application is more limited.

Most of the properties that determine the usefulness of paper are not physical absolutes. Therefore, measurements are often dependent on the instrument or equipment used and on the details of the testing procedure. The situation is further complicated by the fact that paper is both visco-elastic and hygroscopic. Since paper has both plastic and elastic characteristics, any test causing deformation or destruction of the sample (e.g., tear, tensile) will give results that are dependent on the rate of application of force. Although most paper samples are conditioned to standard atmospheric

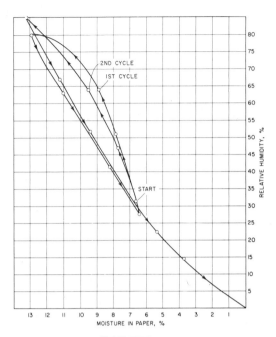

FIGURE 22-8. Moisture absorption and desorption curves for newsprint (old TAPPI TIS 017.06).

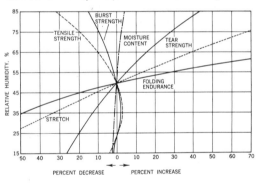

FIGURE 22-9. Effect of relative humidity on paper properties.

conditions prior to testing, the moisture content and physical properties will depend on the previous history.

Conditioning of Samples

The properties of a particular paper are greatly dependent on its moisture content. Paper, being a hygroscopic material, will seek an equilibrium moisture with the surrounding air. However, the final moisture content will depend on whether the sample approached the equilibrium from a more dry state or a more moist state. This behavior is shown by the moisture absorption and desorption curves for newsprint (Figure 22-8).

When reproducible results are required, paper samples must be conditioned in a standardized

environment. Standard conditions in North America are specified at 23° C and 50% relative humidity. At equilibrium, most paper products acclimated within this environment have a moisture content between 7% and 9%.

It is not possible to generalize on the effects of relative humidity on paper properties. The changes depend on the fibrous components, their degree of refining, additives, and sheet surface treatments. However, changes as shown in Figure 22-9 are reasonably typical.

Not all papers are tested at standardized conditions. When paper is produced and used in roll form, little change can occur in moisture content from the time the paper is manufactured until it is used. If strength is required for press runability, it must be inherent in the paper from the winder. Therefore, such paper products as newsprint are tested in the "as made" condition, and samples must be kept isolated from the ambient atmosphere until testing is carried out.

Two-Sidedness of Paper

Paper manufactured on a fourdrinier has a relatively coarse structure on the wire side of the sheet, and is more "closed" or finer-textured on the top side. This "two-sidedness" is caused by differences in fiber composition and by the obvious wire marking on the bottom surface. Paper formed on a twin-wire machine usually has less two-sidedness.

The wire mark usually consists of a pattern of diamond-shaped marks having their long diagonal in the machine direction. The extent of the wire mark penetration into the sheet varies greatly. In the case of newsprint with a total thickness of 90 microns, the penetration usually varies between 10 and 30 microns. The difference in texture between top and wire side is sometimes visible to the naked eye if the paper is folded over and the two surfaces are compared directly. The difference can be more readily seen if the paper is first submerged briefly in water or dilute caustic and then blotted. This procedure loosens the fiber structure and tends to overcome the smoothing effect of calendering.

Certain physical and optical tests may be required on both the top and wire sides, but often only one side is tested. Surface tests on newsprint are usually carried out only on the side that gives the poorest result. For example, brightness is measured on the top side while roughness is determined on the wire side.

The wire and top sides must be taken into account for certain end uses. In the case of forms to be printed on one side only, the best results are obtained by printing on the top side. When laminating high-quality papers (e.g., bristols), it is customary to put the wire sides together in order to keep the smoother surfaces on the outside. Postage stamps are usually printed on the wire side; they are gummed on the top side where the greater smoothness is helpful in attaining an even application of gum.

Directionality of Paper

Paper has a definite "grain" caused by the greater orientation of fibers in the machine direction and by the stress/strain imposed during pressing and drying. The directionality of paper must be taken into account in measuring physical properties. For such strength measurements as tensile, tear, and folding endurance, strips of paper are cut in both directions for testing.

To avoid confusion in the interpretation of directional strength test results, it should be kept in mind that the test directionality corresponds to the applied force, not to the direction of rupture. In the Elmendorf tear test, the applied force and the rupture are in the same direction; but in the tensile and folding endurance tests, the rupture is at right angles to the applied force. Tensile and fold are greater in the machine direction (MD), while tear is greater in the cross direction (CD).

Papers vary in their ratios of MD to CD strengths. Fourdrinier papers generally have from 1.5 to 2.0 times the tensile in the MD as in the CD. Cylinder machine papers can have much higher ratios, up to 5.0 or above. The ratio of MD to CD strength, particularly with respect to the zero-span tensile test, is interpreted as a relative measure of fiber alignment

The MD and CD strengths of machine-made paper are, of course, closely related to the strength of a randomly oriented sheet made from the same stock furnish. The random sheet will yield test values that are somewhere between the MD and CD values. It is generally found that the relationship between values is best defined by the geometric mean, i.e.,

$$X_{(Random)} = \sqrt{X_{(MD)} \cdot X_{(CD)}}$$

Due to headbox pressure fluctuations and irregularities in headbox slice delivery, there are corresponding variations in the makeup of fibers and orientation of fibers across the machine. These variations may be further affected by uneven drainage and non-uniform stressing during pressing and drying. Consequently, the physical properties of paper vary in both the MD and CD. To better define the MD to CD ratio for a given machine, it is necessary to sample and test paper at several locations across the machine.

It is relatively easy to identify the MD and CD orientation of a paper sample. When one surface of a small sample is wetted, the specimen will always curl, with the axis of curl in the MD (due to more expansion in the CD). When narrow strips are cut in both directions, it will be found that one strip (the MD) is stiffer than the other. The distinction can also be made by hand-tearing in both directions; the resistance to tearing is far less in the MD and it is also much easier to tear in a relatively straight line.

Physical Testing

Physical tests on paper can be conveniently divided into four groups:
1. Mechanical and strength properties.
2. Surface properties.
3. Optical properties.
4. Permeability to fluids (e.g., water, oil, air).

Some of the common tests in these groupings are listed in Table 22-11.

The grammage or basis weight of paper is determined by weighing a known area of paper. Caliper is measured using a micrometer with specified foot area and squeeze pressure on the sheet. From the basis weight and caliper measurements, either apparent density or bulk can be calculated.

Tensile strength is determined by measuring the force required to break a narrow strip of paper where both the length of the strip and the rate of loading are closely specified. The amount of stretch at rupture may be determined at the same time. Some modern testers provide a plot of the stress/strain curve and compute the area under the curve which is referred to as tensile energy absorption, a measure of paper "toughness". These testers also provide for measurement of creep under various tensile loadings.

Bursting strength is determined by clamping a paper sample over a rubber diaphragm through which pressure is applied at a gradually increasing rate, and noting the pressure at rupture. Folding endurance is measured by the number of folds sustained before rupture occurs when a sample is flexed through a specified angle under controlled tension. Stiffness is measured by the force required to bend a strip through a specified angle.

Tearing strength (or "internal tearing resistance") is normally determined with the Elmendorf apparatus which uses a falling pendulum to continue a tear in the paper sample when the force is applied perpendicular to the plane of the sheet; the loss of energy (as measured by the height of swing of the pendulum) is related to the force required to continue the tear. The Elmendorf tear test is recognized as a good measure of fiber strength within the sheet, but the applicability of this test as a gauge of pressroom runability, where failures occur in the plane of the sheet, is often questioned. Apparatus for carrying out in-plane tear testing is available, but the procedure is not widely utilized.

The two mechanisms of tear testing are illustrated in Figure 22-10.

Surface roughness (or smoothness) is usually measured by the air flow which occurs across a metal annulus in contact with the paper. Both the contact pressure of the annulus and the air pressure

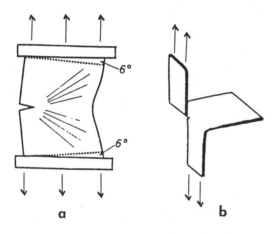

FIGURE 22-10. Illustration of in-plane tear (a) and Elmendorf tear (b).

within the annulus are carefully controlled. Air permeability (or air porosity) may be determined with the same instrument (Bendtsen or Sheffield) by measuring the air flow through a known area of paper when using a specified pressure differential.

The degree of surface sizing and internal sizing is reflected in its water absorbency. In the Cobb test, a tared paper sample is clamped under a short hollow cylinder, and water is added to the cylinder. After a selected time interval (usually 5 minutes), the water is poured off and the sample is blotted and reweighed. The water absorbed is determined in g/m^2.

The brightness, opacity, and color of paper are determined by taking reflectance readings at the appropriate wave lengths of light. Brightness is measured as the reflectance value (relative to a magnesium oxide standard) in the blue region of the visible spectrum (specifically, at a wave-length of 457 nm). Opacity is generally calculated as the "contrast ratio" between the reflectance value of a single sheet backed by a non-reflecting black surface and that of a pile of sheets of the same material. One method to characterize color is to take a series of reflectance readings over the visible wave-length range and produce a spectral reflectivity curve.

TABLE 22-11. Physical tests for paper.

	TAPPI	CPPA	SI Units
Mechanical and Strength			
grammage (basis weight, substance)	T410	D.3	g/m^2
caliper (thickness)	T411	D.4	mm
density	T411	D.4	g/cm^3
bulk (reciprocal of density)	T411	D.4	cm^3/g
folding endurance (MIT)	T423	D.17	-
tear factor	T470	D.9	$mN.m^2/g$
burst factor	T403	D.8	$kPa.m^2/g$
tensile strength (breaking length)	T404	D.6	Km
stiffness (Gurley)	T451	-	mN
stretch (elongation)	T457	D.7	%
tensile energy absorption (TEA)	T494	-	J/m^2
softness (Gurley-Hill)	T498	-	s/100 mL
Surface Properties			
roughness (Sheffield)	T479	-	mL/min
pick strength (Dennison wax)	T459	D.11	
erasing quality	T478	-	
abrasion resistance	T476	-	
Optical Properties			
brightness (Elrepho)	-	E.2	%
opacity	-	-	%
gloss	T424	-	%
color	-	-	
Permeability to Fluids			
sizing degree	T466	-	
oil resistance (Vanceometer)	RC278	-	
water absorbency (Cobb test)	T441	F.2	g/m^2
grease resistance	T454	F.6	
air resistance	T460	D.14	
water vapor permeability	T448	D.15	

Color can also be defined in terms of the relative amount of each of the three primary colors reflected from its surface.

Chemical Properties of Paper

The chemical properties of papers are determined by the process history of the fibrous raw materials and the types and amounts of non-fibrous additives. Chemical properties are important for certain grades of paper such as photographic papers, reproduction papers, anti-tarnish papers (for wrapping silver and steel), safety papers (treated to prevent counterfeiting), electrical papers, food-wrapping papers, and any paper requiring a high degree of performance. In general, these grades must be at the proper pH and/or have the requisite amount of acidity or alkalinity, and they must be free of harmful chemicals. Some grades require special chemical treatments.

A number of test methods have been devised to check the performance of chemical treatments or to detect chemical contaminants. For example, parchment wrapper for butter is commonly analyzed for copper and iron content; a copper content of more than 3 ppm or an iron content of over 6 ppm is believed to contribute to off-flavor. Another example is the test for flammability to ensure that "flame-proof paper" will not burst into flame when exposed to the ignition temperature; this property is an important attribute of such decorative papers as crepe streamers.

Effect of Aging

The physical and chemical properties of all paper products are adversely affected to some degree by aging. In order to compare and assess different papers with respect to aging characteristics, procedures have been developed to accelerate the changes that normally occur during aging. The results do not necessarily agree with a more natural aging process over a long period of time, but are probably valid as a relative indication. The usual method to simulate aging is to place a sample in an oven at 105° C. A period of 4 hours is recommended for brightness reversion (TAPPI SM-200), but other periods may also be used. The brightness loss between the initial and "aged" samples is taken as the measurement of reversion.

Folding endurance is known to be more affected by aging than other strength parameters. Therefore, it is customary to assess the "relative stability of paper" by the effect of heat on folding endurance. The standard procedure (TAPPI T-453) specifies a period of 72 hours at 105° C. The initial and heat-treated samples are then both conditioned and tested for folding endurance. The values are reported as "retention of folding endurance after heating" (i.e., the percentage of the original value).

22.5 AUTOMATED PAPER TESTING

In the 1980's, paper testing entered a new era with the advent of digitized bench-top instruments, automated off-line test management systems, and on-line process control systems. These developments by instrument and control system suppliers have provided the papermaker with the ability to monitor and control certain paper properties during manufacture and to generally utilize more accurate and timely information for quality control.

Paving the way for the automated testing systems was the development of digitized bench-top instruments which incorporate computer systems that automatically check the calibration of the instrument, record test values, perform calculations, and compute statistics. The most recent advances involve the use of robots in conjunction with the burst and tensile tests.

On-Line Testing

On-line (or on-machine) monitoring and control systems for grammage, moisture content, and caliper have gained universal acceptance (refer to Section 17.3). Reliable sensors are also available to measure brightness, color, gloss, opacity, smoothness, ash content, coat weight, and formation. On-line sensors are now available, as well, to nondestructively measure properties (elastic stiffnesses based on sound velocities) from which a number of sheet strength parameters can be inferred.

Although some industry people have predicted that on-line systems will eventually replace off-line

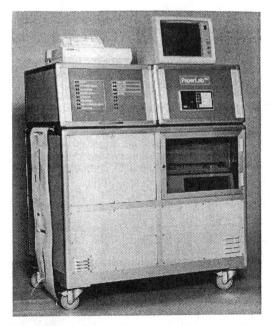

FIGURE 22-11. PaperLab automated testing system from Kajaani.

testing in paper mills, this appears unlikely to happen. The need for off-line testing will certainly decrease as on-line systems gain in credibility, but some off-line testing will always be necessary for investigative work and to validate the on-line sensors. Also, paper will continue to be sold on the basis of property specifications that must be confirmed by standardized test measurements.

Off-Line Testing

A number of fully automatic testing systems are now available to the paper industry utilizing advanced electronics and robotics, and many paper mills have made a commitment to convert to this technology. Perhaps the major incentive is the labor savings from the elimination of one or more round-the-clock testers. But also important is the potential of increased frequency of testing, a wider scope of testing, and the more rapid feedback of data to the mill floor in the form of control charts.

Although capability varies amongst the available systems and models, the system shown in Figure 22-11 is representative. Sheets or strips of paper are fed in one end, keys are punched to specify the number and combination of tests to be performed, and a few minutes later the test information is available either on the screen, as a printout, or as a graph.

REFERENCES

(1) CUTSHALL, K.A. **The Art of Obtaining a Representative Paper Pulp Slurry Sample** *Appita 21:4:21* (January 1968)
(2) **TAPPI Standard Method** T-261
(3) MacLEOD, J.M. **Comparing Pulp Strengths** *P&P Canada 81:12:T363* (December 1980)
(4) SMOOK, G.A. **Combined Pulp Strength Parameter Has Many Uses, Some Limitations** *Pulp & Paper* (September 1985).

Chapter 23

Introduction to Paper End Uses

After leaving the paper machine, many paper and paperboard products are subjected to finishing and converting operations before reaching the printer or consumer. Finishing involves a wide range of processing steps to prepare the product for shipment including winding, slitting, cutting, sorting, counting, cartoning, palletizing, and wrapping. Some of these steps could also be described as packaging operations.

In the broadest sense, converting includes all those remanufacturing steps which change the dimensions, shape, surface characteristics, or properties of the paper product. Obviously, finishing and converting operations overlap. In North America, finishing often denotes all the post-machine operations carried out within the confines of the paper mill; while converting refers to all the remanufacturing operations (except printing) carried out by the customer. Plants that utilize paper or paperboard to make such products as bags, envelopes, writing tablets, boxes, and paper towels are referred to collectively as converters. The packaging industry is the largest volume consumer and converter of paper and board products.

23.1 SHEET FINISHING

Where rolls are cut into finished sheets, a number of sequential operations are carried out in adjacent areas. Proper layout of equipment is important to minimize congestion, eliminate excessive handling of the product, and reduce loss through damage. The following processing steps are commonly utilized:
• decurling
• slitting the web into correct widths,
• rotary cutting the web strips into sheets,
• guillotine trimming the sheets,
• inspecting and sorting the sheets,
• ream wrapping the sheets,
• carton wrapping the sheets,
• skid loading the sheets.
Other operations less commonly applied are:
• conditioning the web to uniform moisture content,
• embossing.

Rotary Sheeting

Today, there are three predominant size ranges for sheeters:
• cut size: principally $8^1/_2$ inch by 11 or 14

• folio size: 11 x 17 inch up to 38 x 50 inch
• senior size: anything above folio size
In modern finishing plants, each of these size ranges is produced on specialized sheeters to maintain maximum efficiency and meet quality standards. Somewhat different approaches to production are characteristic of each category.

On today's sheeters, slitting and cutting are carried out on one machine. The rolls to be cut into sheets are placed on unwind stands which hold from one to a dozen rolls. The actual number of rolls depends in part on quality requirements and the relative difficulty of web control. From the unwind stand, the paper passes through slitters and onto the cutter, which commonly utilizes a revolving knife against a fixed bed knife. The length of the sheet is fixed by the speed of the revolving knife.

Sheeters are available in simplex and duplex configurations as illustrated schematically in Figures 23-1 and 23-2. In the duplex unit, the slit web is separated and each strip runs to a set of independently operated knives. After being cut, the sheets are conveyed on a moving belt to layboys, where the stacks are jogged into uniform piles. All layboys are equipped with counters so that a marker can be inserted after each accumulation of a specified number of sheets. On some modern layboys, the marker is inserted automatically. A more detailed sketch of a complete simplex sheeter with manual inspection station is shown in Figure 23-3.

Trimming

Where size dimensions or edge appearance are critical, paper is precisely trimmed following cutting. This technique produces a neat, squared stack with clean edges. The operation is carried out on a trimmer press, more commonly called a guillotine cutter. The stack is positioned for trimming by squaring up the stack with a movable side-stop and adjustable back-stop. A pressure bar is then lowered to prevent slippage and one edge is trimmed by the guillotine blade.

Each side of the stack is trimmed in turn. Most modern trimmer presses are equipped with automatic resetting of the back-stop position, thus providing exact dimensions for each trimmed stack and eliminating the time required for manual resetting. An

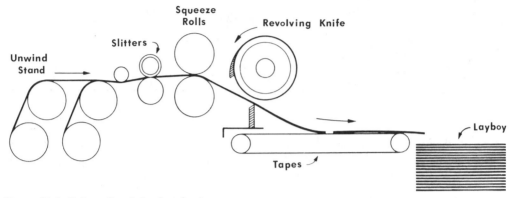

FIGURE 23-1. Schematic of simplex sheeter.

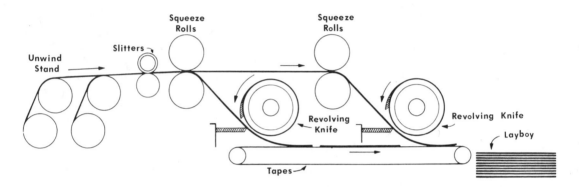

FIGURE 23-2. Schematic of duplex sheeter. One side of the split web drops into the first knife, while the other side is carried to the second knife. The two revolving knives have independent speed settings, making possible the production of two different-length sheets.

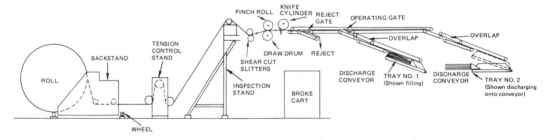

FIGURE 23-3. Cut-size sheeter line includes inspection station and provision for rejecting sheets containing defects.

air cushion enables the paper stack to be easily manipulated on the table; a multitude of high-pressure air jets are activated through miniature ball valves on the table surface by the weight of the paper.

Precision Sheeting

The trend within the industry is toward precision sheeting operations, where the paper is shipped directly from the sheeter, and the trimming step is eliminated. Very sophisticated equipment is required for precision sheeting, not only to meet the stringent dimension tolerance limits, but also to pile the sheets and deliver reams and/or skid loads. Large-size cut sheets especially have to be controlled and "mothered" every inch along the way to lay up in perfect piles (1). Precision sheeting equipment is expensive and can usually be justified only for high-tonnage operations.

Decurling

Printers require a flat sheet for proper running in the presses. Paper near the core of a roll is often a problem during sheeting operations because the paper has taken on a set curl. Uncorrected, this condition can account for up to 2% waste due to excessively curled sheets. The simplest design of decurler utilizes a flat "breaker bar" over which the sheet runs at a sharp angle (opposite to the direction of curl). In some operations, a round bar is substituted to prevent damage to coated surfaces. A small-diameter freely-rotating breaker roll is preferable when handling heavier sheets or materials prone to scratching or pressure marks.

Inspection and Sorting

Traditionally, high-grade printing sheets have been manually inspected and sorted into categories of satisfactory, second quality, and reject. Now, automatic control on the paper machine of such properties as grammage, moisture, caliper, color, and formation has obviated much of the need for subsequent inspection and sorting. Some mills resort to inspection of sheets only when the parent reels are shown to be outside specification tolerance limits with respect to the variables being automatically monitored or measured manually on reel samples. Optical fault detector systems of the type shown in Figure 23-4 can be used on some sheeters for automatic rejection of sheets containing defects.

Packaging of Sheets

The wrapping of paper sheets is a specialized process because of the flimsiness of individual sheets and the wide variety of stack sizes. The smaller sizes are usually conveyed in one-ream stacks from the trimmer (or from the precision sheeter) directly to an automatic wrapping machine which applies a vapor-barrier wrap and label. The packages are then boxed in standard cartons. A basic cut-size sheet production line is shown in Figure 23-5.

The larger sheet sizes require more manual attention. In some cases, they are packaged in cartons holding approximately 150 pounds and shipped singly. It is more common today to stack either sheets or packages of sheets onto pallets or skids. Skids are more rugged in construction, and have the advantage that they can be placed by some customers directly at the feed end of the printing press.

Conditioning

Conditioning is a treatment applied to high-grade printing papers to ensure flatness and good color register. This technique is commonly used in Europe, but has not caught on in North America, presumably because of the added processing cost. The principle is to add moisture to the roll web so that any expansion takes place before the paper is cut into sheets. Then, so long as the sheets are stored in a normal atmosphere, they will not expand or contract, and built-in stresses causing curl or cockling will be nullified.

A representative conditioning system is illustrated in Figure 23-6. The paper is subjected to high-velocity impingement of humidified air as it passes over successive drums. The driest areas in the sheet preferentially absorb moisture from the impinging air, and the resultant web has a higher and more

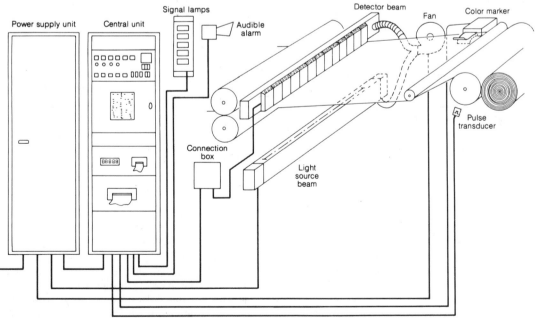

FIGURE 23-4. Representative fault detector system (Sentrol).

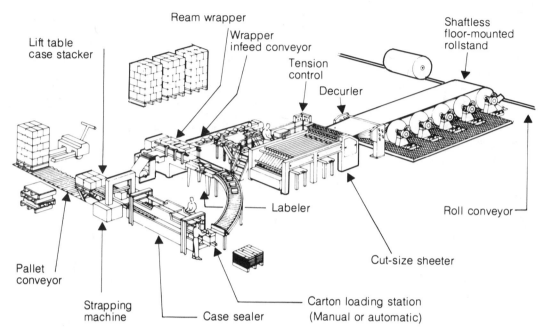

FIGURE 23-5. Basic cut-size sheet production line.

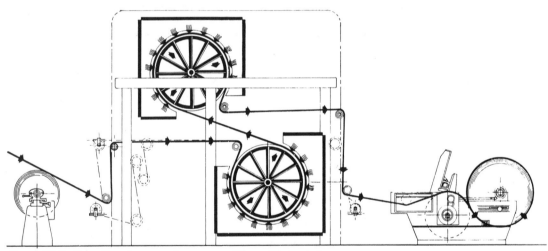

FIGURE 23-6. Operating diagram of paper conditioner (Swenson).

uniform moisture content and virtually no static charge. Another design of conditioner relies on a fine-mist water spray to add moisture.

Embossing

Embossing is a technique for imprinting a raised or depressed pattern on paper. It is used during paper finishing to provide (as an alternative to water marking) special finishes or identifying marks on business and printing grades. The operation is carried out by passing the paper between an engraved steel roll or plate and another roll or plate of soft or compressible material such as paper or cotton.

Embossing is more extensively used in converting plants for decorative effects, notable in the production of napkins, fancy wrapping papers and wall papers.

23.2 CONVERTING

It is roughly estimated that about 75% of North American paper and paperboard output is subjected to some type of converting operation before reaching the final customer. The exact percentage would depend on the definition of converting; but presently, no precise guidelines for inclusion are universally accepted by the industry. Printing of newspapers, books, and magazines is generally considered apart from converting; however many converting plants utilize printing as one stage of the converting process.

Some common converting operations are listed by groupings in Table 23-1. Most of these operations are carried out at independent plant sites. However, in line with increased plant integration, converting operations have been established in some paper and board mills, especially for the large-volume or high-profit items. Corrugated box plants utilize the highest tonnage of paper and paperboard, while other types of box and container plants collectively utilize the second highest tonnage. All of the groupings shown in Table 23-1 account for significant tonnage. Other converting operations are individually less significant, but collectively account for millions of tons each year.

TABLE 23-1. Major converting operations.

Boxes	- corrugated shipping containers - folding boxes - rigid boxes
Bags and Sacks	- grocery and variety - multiwall
Construction Boards	- laminated paperboards - impregnated or saturated boards
Containers	- sanitary food containers - fiber cans - fiber drums
Tissues	- toilet and facial tissues - towelling - table napkins
Wrapping Papers	- laminated wrappers - decorative wrappers
Business	- envelopes - business forms - data process forms - labels - file folders

It is not feasible in this introductory text to provide more than a brief overview of the converting industry. However, certain specific processing steps are common to a number of converting operations, and some of these will be discussed here. These techniques are usually grouped according to whether the paper is handled in roll form (called "wet converting") or made into the actual products (called "dry converting"). The major processing operations are listed in Table 23-2.

TABLE 23-2. Major processing operations during converting.

Wet Converting	- coating - laminating - corrugating - impregnating - embossing - wet creping
Dry Converting	- bag making - box making - envelope making - cutting and folding - hot pressing - die cutting - roll winding

Functional Coatings

Coating operations during converting are similar in principle to the on-machine and off-machine pigment coating operations carried out in the paper mill (see Section 18.2). A common objective with functional non-aqueous coatings is to provide a barrier to water, water vapor, air, or grease. Typical coating materials include waxes, asphalt, lacquer, varnish, resins and adhesives.

A wide variety of equipment is used to apply functional coatings, some of which is similar in design to equipment for applying sizing solutions and coating color. The extrusion coater (illustrated in Figure 23-7) is used primarily for polyethylene plastic coatings and does not have a counterpart in the paper mill. Here, the heated resin is extruded through a slot as a hot viscous film which is combined with the paper between a pair of rolls. The combining operation takes place so rapidly that a permanent bond is created between the plastic film and the paper.

A group of functional coatings known as hot melt coatings have gained wide acceptance in recent years. These coatings consist of mixtures of polyolefins and wax-like materials that have excellent barrier properties at low thickness levels. Hot melts are usually too viscous for conventional

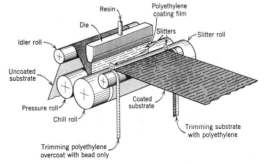

FIGURE 23-7. Extrusion coater for polyethylene.

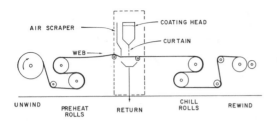

FIGURE 23-8. Schematic of curtain coater.

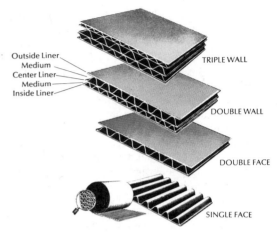

FIGURE 23-9. Corrugated board constructions.

coater equipment, but not viscous enough for the extrusion coater. Consequently, special equipment has been developed, such as the curtain coater shown in Figure 23-8. Here, the hot coating is pumped into the die slot from which it is discharged vertically downward in the form of a falling film or curtain. The sheet is passed through the curtain at high speed and the coating is deposited directly on the surface, where it solidifies by cooling.

Laminating

Laminating refers to any process in which two or more webs are fused together into a combined product. This operation is most often carried out to build up the thickness, strength and rigidity of the combined product. It is also used to produce a sheet with different properties on each surface or to produce a special type of barrier paper. In the latter case, the laminating agent is usually asphalt which serves both as adhesive and barrier material.

In a typical laminating operation, the adhesive is applied to one web by a conventional roll coater. The webs are then combined in a lightly-loaded press nip and, if necessary, passed around conventional drying cylinders. Often the combined product is too stiff to be collected in rolls and must be directly cut into sheets or blanks. The adhesive used is usually a water solution of starch, vegetable protein, or synthetic resin.

Corrugated Board

The manufacture and utilization of corrugated boxboard is a major industry in its own right. Because of its relatively low cost and high strength characteristics, corrugated board has become the prime structural material for the manufacture of boxes and shipping containers.

Corrugated board is manufactured in a number of constructions, as illustrated in Figure 23-9. The simplest form, known as single-face, is used for wrapping fragile items, not for making shipping containers. Double-face is the most common construction and utilizes a fluted medium sandwiched between two layers of liner. The terms inside liner and outside liner refer to the orientation with respect to the shipping container. Sometimes, a better-quality liner is used on the outside surface.

The double-wall and triple-wall configurations provide much stronger and stiffer corrugated boards. In order to optimize certain board properties, a number of variables can be altered within these basic constructions, including the grammages of the liner and medium, and the flute configurations. Four standard flute contours are produced (A-flute, B-flute, C-flute, and E-flute), each characterized by a different flute thickness and number of flutes per lineal inch.

Double-face corrugated board is fully constructed during one inclusive operation as illustrated schematically in Figure 23-10. The preheated and moistened medium passes through a flute former; then immediately picks up adhesive on the flute tips, and is brought into contact with a liner sheet to form a single-face construction. This combined web is accumulated on a "bridge" to permit re-threading or changing rolls without shutting down the entire operation. The exposed flute tips on the single-face web are then brought into contact with a second glue roll and immediately combined with the double-face liner; the double-face construction proceeds through a heating section to set the bond and is finally cooled and cut into box blanks. Details of a single-facer operation can be seen in Figure 23-11.

Although corrugating is a mature industry by conventional standards, a surprising amount of research is ongoing to improve existing processes and develop new products. A literature review indicates that the industry is currently concentrating on improved bonding between the fluted medium and the liners at the corrugator (2). A promising route to higher-strength and/or lower-weight corrugated board is by impregnating the combined product with polymeric materials (3), but this approach may be at odds with existing recycling strategies.

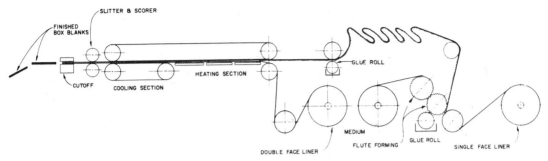

FIGURE 23-10. Schematic of double-face corrugated board manufacture.

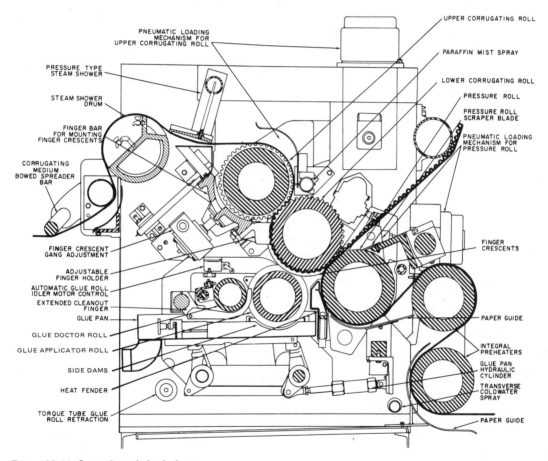

FIGURE 23-11. Operation of single facer.

Impregnating

Impregnating (or saturating) is carried out for a variety of products, including decorative laminates, gasket papers, oiled papers, and roofing felts. Tub sizing is sometimes considered to be an impregnating process (see Section 18.1).

For a typical impregnating process, the paper is unwound and passed through one or more pans of solution or melt; it is then dried or cooled by whatever method is appropriate. The impregnated sheet is either rewound into rolls or cut into sheets if the product is stiff or brittle.

Dry Converting

Such operations as bag making, box making and envelope making are carried out by highly-mechanized equipment that performs the necessary cutting, slitting, forming, folding and gluing steps.

A large number of products such as toilet tissue, waxed paper and adding machine paper are ultimately used in the form of rolls. Because of the large tonnage involved, complex automatic winding machinery has been developed for high-speed, high-volume operation. Mechanical handling is also employed for wrapping and boxing rolls.

A large group of sanitary products are used in the form of folded sheets, e.g., napkins and facial tissues. Two methods of handling are employed, depending on the product. In one case, the individual or multiply web is individually cut by a rotating knife and then folded. In the second process, a large number of webs are first folded, then gathered together and the entire wad is cut to package length.

23.3 PRINTING

In modern usage, the term printing can be applied to any process producing multiple reproductions. This discussion will be limited to the major direct or contact printing methods by which reproductions are made from a plate or image carrier on a press; these methods still account for over 90% of commercial printing. This is not to discount the growing importance of such non-contact methods as electrostatic and electrosensitive copying, thermal printing, ink jet printing, and laser printing.

A high proportion of paper and paperboard is printed in the final end product, and the printing industry is the largest collective consumer of paper products. The printing industry itself is a gigantic business with shipments in the United States estimated in excess of 100 billion dollars in 1990.

The papermaker needs to understand and appreciate the requirements of the printer in order to produce papers suitable for printing. Basically, two major factors must be considered - "runability" and "printability". Runability is the capability to get the sheet through the press without breaks. A poor running sheet will cause problems or necessitate slow operation of the press which adversely affects the economics of printing. The paper properties important to runability are strength, uniformity, and freedom from defects. It is also important from a runability standpoint that the paper rolls or pallets of paper sheets arrive at the pressroom in good mechanical condition.

Printability is the effect of the paper on the accurate reproduction of the printed image. Paper properties such as brightness, opacity, color, gloss, smoothness, porosity, and sizing are generally important, but the requirements vary with the printing process used. Typically, coated papers have better appearance than uncoated papers due to improved smoothness, brightness and gloss.

Printability characteristics are especially critical in the reproduction of half-tone prints. In printing, the reproduction of tone originals is made possible by converting the original continuous tone into a discontinuous pattern of dots of different sizes: large, well-joined dots in the shadows and small, separated dots in the highlights of the picture. In general, the finer the pattern of dots, the more the eye is fooled into perceiving the representation as continuous, and the greater is the detail that can be shown.

Printing Processes

Three basic methods are used for direct or contact printing as illustrated in Figure 23-12:
• from a raised surface (relief method)
• from a flat surface (planographic method)
• from recessed areas (intaglio method)

Four major printing processes based on these configurations are the basis for most commercial work:
• relief: letterpress and flexography
• planographic: lithography (offset)
• intaglio: gravure

The earliest form of printing (e.g., the Gutenberg press) embodied the letterpress principle utilizing raised hard type. Letterpress remained the dominant printing method up to the early 1960's when other methods began to displace it, mainly because of the poorer print quality associated with letterpress. Now, over 90% of newspapers are printed by lithographic and flexographic methods, while gravure is used increasingly for high-quality magazines, newspaper supplements, and advertising inserts or flyers.

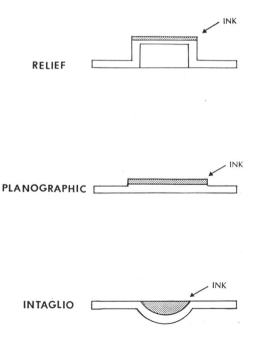

FIGURE 23-12. Print methods.

Letterpress

In the letterpress process, a hard raised surface is inked, and the inked image is transferred to the paper under heavy pressure. The ink, which consists mainly of 8-12% carbon black dispersed in oil, is literally pushed into the paper. For newspapers and most commercial printing, letterpress plates were traditionally made from hot metal. Photopolymer or shallow-relief plates began to replace the lead plates in the late 1960's, and by 1980 lead plates had virtually disappeared from use.

Although sheet-fed presses are still used by smaller printing firms, most newspapers, directories, and advertising flyers are printed with rotary web-fed presses. Rotary press equipment varies, but in all cases the image transfer occurs in the printing nip between the plate cylinder (containing the wrap-around plates) and the rubber-covered impression cylinder (Figure 23-13). Newsprint presses typically utilize two couples in series (i.e., a "perfecting" configuration) to print both sides of the paper in rapid succession, usually 8 or 16 pages on each side of the web. The typical perfecting configuration is shown in Figure 23-14.

The serious quality drawbacks of letterpress relate to the "squeeze-out" and "embossing" effects from the pressure of the raised surfaces. The ink on the raised surface cannot be instantaneously absorbed by the paper, and some ink oozes out around the high-pressure area. At the same time, the paper is compressed and it retains some of the deformation. Another limitation relates to the fineness of the dot pattern that can be used for half-tone reproductions. A fine pattern requires shallow relief, which tends to clog with lint and ink.

The traditional defects were most severe with metal plates. With the switch to shallow-relief photopolymer plates, overall print quality generally suffered and rub-off problems became more severe because printing pressure had to be reduced to avoid excessive deformation in the photopolymer layer.

Lithography

The lithographic process is based on the principle that oil and water do not mix. Materials and treatments are used in the preparation of the photo-developed printing plates that render the printing image areas receptive to oil-base ink and repellent to water, while the non-image areas are water-receptive and ink-repellent. On the web press, the wrap-around printing plate is attached to the plate cylinder; during rotation it comes in contact successively with rollers applying "fountain solution" and ink. The water solution wets the non-printing areas, preventing ink pickup except in the image-carrying areas.

Most lithographic presses utilize an intermediate roll covered by a rubber "blanket", which takes the

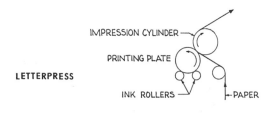

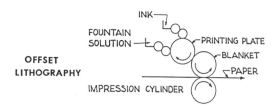

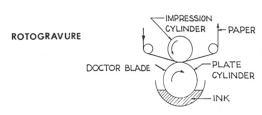

FIGURE 23-13. Configurations of printing couples.

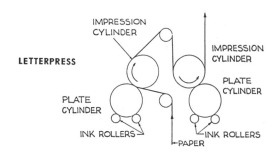

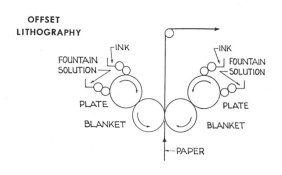

FIGURE 23-14. Perfecting press arrangements for letterpress and offset.

image from the printing plate and transfers it to the sheet as illustrated in Figure 23-13. This is known as the "offset" principle, and the term offset is sometimes thought to be synonymous with lithography. It must be stressed, however, that both letterpress and gravure are sometimes printed by offset, while not all lithographic printing is done offset. Most web-fed offset printing is carried out on presses in which two printing units oppose each other so that both sides are printed at the same time, as illustrated in Figure 23-14.

A number of advantages are attributed to the offset method, including longer plate life and better print quality. The rubber blanket, being elastic, can conform to irregularities in the paper surface, making it possible to produce good prints on fairly rough surfaces. The offset principle also makes possible the preparation of "positive image" printing plates. Non-offset methods require that the printing plate carries a "negative" or reversed image.

The viscous inks used for lithographic printing are "tacky", and they exert a significant pulling force on the surface of the paper as it is peeled from the printing blanket. Therefore, the papers used for offset printing should be free of lint and have good surface strength characteristics. Fiber accumulating on the blanket can be a major problem; the deposition will reduce print quality and may necessitate periodic shutdowns of the press to clean the blankets.

Gravure

In gravure printing, the image carrier is a copper plate or copper-plated cylinder which has been etched or engraved with microscopic cells. The cylinder or plate is flooded with a very fluid ink and the excess ink is doctored off. The ink in the cells is then transferred to the paper in the printing nip formed with a rubber-covered impression roll as illustrated in Figure 23-13.

Most gravure printing is done on rotary web-fed presses, and the process is commonly called rotogravure printing. Gravure provides excellent quality printing, but because of higher plate-making cost and longer make-ready time, the process is limited to high-volume work not subject to narrow deadlines. Once the press is set up, the process is easy to control. The gravure cylinder, if chrome-plated, can run for millions of impressions.

Gravure printing normally requires paper of exceptional smoothness to enable the surface to be uniformly pressed into contact with the liquid ink residing in the recessed cells. Operation with somewhat rougher paper is possible utilizing an electrostatic technique wherein an applied voltage to the web produces a distorted meniscus in the ink cells which bulges the ink into contact with the paper web.

Flexography

Flexography is a specialized form of relief printing characterized by highly fluid water-based inks, specially-designed polymer printing plates, and light pressure in the printing nip. This process was originally developed for use on packaging materials, but has been adapted for other applications, notably as a retrofit for existing letterpress newspaper presses. The key to the success of flexography is the flexo ink application system using an anilox roll, a specially-designed etched or engraved roll containing recessed cells. The anilox roll picks up ink from a trough or applicator roll, a doctor blade removes excess ink, and a uniform film is then transferred to the printing surface of the polymer plate as illustrated in Figure 23-15.

Flexography produces print quality intermediate between letterpress and offset, and it offers other advantages over letterpress, such as lower rub-off and see-through, and better compatibility with deinking processes. The flexography process is evolving and is expected to remain competitive with offset for newspaper printing retrofits and upgrades (4).

Other Processes

The di-litho process (illustrated in Figure 23-16) was developed in the 1970's specifically for newspaper letterpress conversions. Di-litho is a modified lithographic process that prints directly

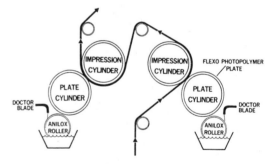

FIGURE 23-15. Flexographic press configuration for perfecting work.

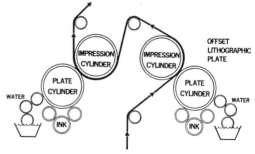

FIGURE 23-16. Configuration of di-litho perfecting printing units.

from the plate. Although, print quality is distinctly inferior to offset, a number of newspapers did convert until it became apparent that flexography or offset were better options.

In screen printing, the image is produced by applying ink through a fine porous screen onto paper under the screen. High ink applications are used compared to other processes. Screen printing is often done by hand for artistic effects. Commercial screen printing is comparatively slow and is reserved for such work as art prints, posters, greeting cards, menus and wallpaper.

In 1970, a lithographic printing plate which required no water during printing was introduced. The system, known as "driography", was removed from the market several years later because of persistent problems with inconsistent toning of the printing areas. Now, a Japanese firm has revived the process in somewhat modified form. Driography has great potential for simplifying the production of quality printing with minimum waste.

Platemaking

All modern printing plates are prepared by photomechanical means. The final layout (or makeup) of material is first photographed to produce a negative or positive image as required. This image is then exposed onto a plate covered with a light-sensitive coating. The plate is subsequently processed with the combination of chemicals required for a particular process.

Halftone images used in the layout have already been photographed through a "screen" which separates the continuous tones of a photograph or drawing into a discontinuous pattern of variable dots. The gradation of tones that can be simulated by this method is illustrated in Figure 23-17.

Colored halftone images are obtained by photographing through three separate filters, each corresponding to one of the additive primary colors. Color work is reproduced on the printing press by printing successively in three overlapping colors (usually with the "subtractive primaries" rather than the "additive primaries"). The steps in plate preparation for three-color printing are shown in Figure 23-18. For best quality color work, the three colors are supplemented with black to add density, detail and contrast to the reproduction. In the preparation of color tone plates, the halftone screen is rotated to a different angle for each color to prevent lines of dots from running together in the prints (see Figure 23-19).

Press Operation

Depending on the products being produced, a modern high-speed rotary press operation often involves many objectives and concerns in addition to obtaining good print quality, e.g.,

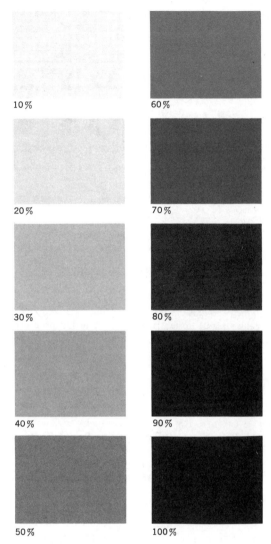

10%   60%
20%   70%
30%   80%
40%   90%
50%   100%

FIGURE 23-17. Gradation of tones as simulated by the same dot pattern, but using different sizes of dots.

- tension control from the unwind stand,
- automatically splicing (i.e., with a "flying paster") the new roll onto the tail of the old roll,
- tension control between printing couples,
- control of register,
- drying of ink,
- slitting, cutting, collecting, folding, stacking, etc.

Normally, the longer press leads require higher web tension for adequate control. If the press operator suspects that one source of paper provides a better running sheet, he/she will usually prefer to use that paper for the most difficult press runs. This factor must be considered if the pressroom maintains runability statistics for the various suppliers of paper rolls. Breaks usually occur as a result of defects in the sheet in combination with low paper strength or

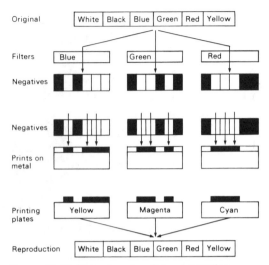

FIGURE 23-18. Steps in color plate preparation.

higher web tension. If the break "end" can be found, a defect will normally be evident at the point of rupture, usually a sliver or "shive". (Also refer to Section 20.1.)

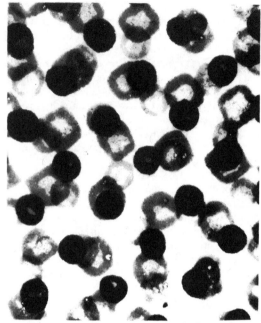

FIGURE 23-19. Detail of dot formation in four color printed image, showing different dot alignments .

REFERENCES

(1) GREINER, T.S. **Sheet Finishing Overview: Balancing Costs and Quality** *Pulp & Paper* (October 1986)

(2) INOUE, M. AND LEPOUTRE, P. **Bond Setting in the Corrugating Operation** *Tappi Journal* (December 1989)

(3) MILTZ, J., SEGAL, Y. AND ATAD, S. **The Effect of Polymeric Impregnation on the Properties of Paper and Corrugated Board** *Tappi Journal* (July 1989)

(4) SNIDER, E.H. AND THOMPSON, C.A. **Printing and Paper of the 21st Century** *P&P Canada 88:12:T447* (December, 1987)

Chapter 24

Process Control

All industrial processes manifest inherent variability which must be minimized if the plant operation is to yield uniform, high-quality product with efficient utilization of raw materials, manpower, and energy. The objective of process control is to maintain the process (and each element of the process) within well-defined limits of variation.

A time plot for a process variable (shown in Figure 24-1) illustrates acceptable control, except for one period of operation that is "out of control". For a process under manual control, the operator watches the trend of the data points, and makes adjustments to keep the process within limits. But sometimes, depending on the degree of process stability, accuracy of the data points, and narrowness of the limit lines, the operator may only be able to take corrective action after the variable has strayed outside the process limits.

Very few processes or process "elements" within the pulp and paper mill are still under manual control. Most processing steps are automatically controlled utilizing digital or analog equipment. Increasingly, computers are utilized to coordinate control and record-keeping aspects for the entire manufacturing sequence.

24.1 MEASUREMENT AND CONTROL

A typical automatic control loop consists of three basic components: sensor (with transmitter), controller, and control element. The interaction of these components is shown schematically in Figure 24-2.

An example of a control loop common to all pulp and paper operations is the consistency control loop shown in Figure 24-3. Here the sensor measures consistency in the stock line and transmits an appropriate signal to the controller; the controller compares the incoming signal to the set point and transmits an error signal to the dilution valve; the dilution valve opening is then attenuated in the direction of correcting the error.

Perhaps the most common control application with respect to all process industries is temperature regulation utilizing a steam valve as the control element. Other common applications are control of flow rate and tank level using a flow control valve and control of gas pressure using a pressure relief valve.

Sensors

The sensor is at the heart of any control loop. Automation cannot be properly implemented without

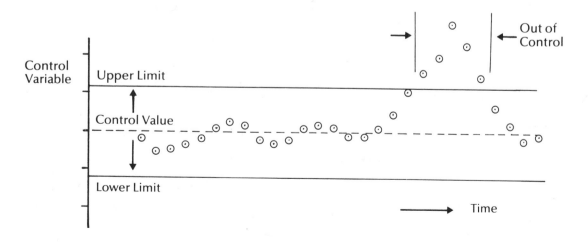

FIGURE 24-1. Time plot of control variable illustrating normal operation and out-of-control operation.

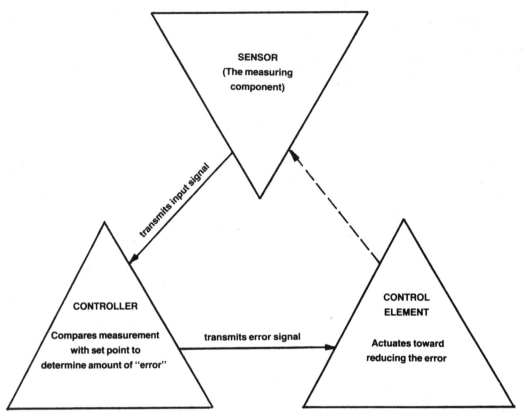

FIGURE 24-2. Schematic of control loop.

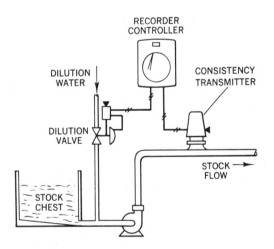

FIGURE 24-3. Consistency control loop.

a reliable indication of dependent variable behavior; and some operations within the pulp and paper industry have only recently become amenable to automatic control by virtue of new measurement techniques. Some control strategies must still rely on

manual testing results for input into the control system.

In general, few variables can be measured directly with process instruments and dependence must be placed on measurement of related properties. For example, the moisture content of paper can only be measured directly by laboratory methods; but it can be accurately inferred from measurement of such related properties as:
• resistance to electric current,
• dielectric resistance,
• absorption of microwave energy,
• absorption of infrared energy.
The principles of measurement for some of the variables commonly monitored or controlled in the pulp and paper industry are summarized in Table 24-1.

Controller

In almost every process, certain variables must be rigidly controlled at specified values in order to maintain efficient operation. The controllers are given these desired values in the form of set points. For a basic control loop, the set point may be keyed in at a control console, or entered by means of a set point knob on the controller itself. In either case, the

TABLE 24-1. Principles of measurement for various sensors.

Variable	Principle
Temperature	- fluid expansion - voltage inducement (thermocouple) - resistance measurement - thermister
Pressure	- mercury column (for standardization) - coil gauge (e.g., bourdon tube) - diaphragm - differential monometer
Differential Pressure	- twin diaphragm cell
Load	- filled load cell - strain gauge

Flow
(closed pipes)
	- differential pressure flow meter
	  (e.g., orifice plate, venturi
	  tube)
	- magnetic flow meter
	- turbine flow meter
	- rotometer

Flow (open channels)	- weir height - Parshall flume height
Liquid Level	- bubble tube - tank-mounted diaphragm - gamma radiation
Stock Consistency	- fluid viscosity - fluid shear
Fluid Density	- hydrometer - gamma radiation

setting corresponds to a dc electrical signal within the 4-20 ma range (or to a pneumatic signal within the 3-15 psig range).

The measured value is also converted to a 4-20 ma dc (or 3-15 psig) signal and is transmitted from the sensor to the controller. Comparison of the set point and input signals results in a difference, known as the error, which is the key actuating signal in the controller. The responding control action will depend on the direction and magnitude of the error, and in some cases on how fast the error is changing. The controller acts to reduce the error and bring the actual value toward the desired value by means of the output signal. The output signal is generally of the same type as the set point and input signals, namely 4-20 ma dc (or 3-15 psig).

Analog Model

If the controller is to carry out its function to automatically regulate a process variable at a desired value, the control loop must be adapted and tuned to the characteristics of the process. Three common modes of control are used in analog systems for the majority of process applications depending on response time and lag-lead requirements. The respective adjustable settings are known as proportional band or gain (P), integral or reset (I), and derivative or rate (D).

Proportional band is defined as the range of values of the process variable in which there is proportional control action, expressed as a percentage of the full scale range of the controller. For example, if the full scale range of a temperature controller is 0° to 200° C and the proportional control action is over a 50° C range, the proportional band is then 25%. If the proportional control action is over a 200° C range, the proportional band is then 100%. The width of the proportional band determines the amount of valve motion for any given change in the controlled variable.

A process may stabilize at a point other than the set point when using a simple proportional controller. The difference between the stabilization point and the set point is known as offset. Sometimes the offset is small and can be tolerated. When offset is undesirable, proportional plus reset (two-mode control) is required to modify the output signal and eliminate the offset. Reset action can be visualized as a shifting of the proportional band. The reset adjustment (tuning) is in the form of timing, and the reset rate is commonly expressed as repeats per minute. Reset is proportional to the magnitude of error and to the duration of error; its action accumulates as long as there is an error, but ceases when the error is reduced to zero.

Occasionally, a particular process will have a dead time and transfer lag in such a way as to cause unsatisfactory control with single or two-mode control. It is then necessary to add the third mode of control known as derivative or rate action. This third mode provides a continuous relationship between the rate of change of the process variable and the output signal. It provides an initially large over-correction to compensate for an unfavorable process lag. Derivative time is the interval by which derivative action advances the time of proportional action, and the derivative mode adjustment (tuning) is in a timing form.

Digital Controller

In the 1970's, instrument loops requiring the PID control function were best served by analog equipment. Computer control techniques such as time sharing and point sampling introduced deadtime and phase lag into the PID function which diminished the responsiveness and stability of the control system. These problems were overcome in the distributed control systems of the 1980's, through development of new algorithms and nonlinear compensation methods. The majority of controllers today are micro-processor based.

Modern digital controllers provide some significant advantages over analog systems. Firstly, the same control functions can be implemented with less hardware and with more reliable components. Perhaps the greatest advantage is that the control functions are not tied to hardware, but can be configured to implement a variety of functions. More sophisticated and better control strategies can be implemented at little cost once the basic controller hardware is installed. Also, the high accuracy and drift-free nature of digital control allows the user to operate his process nearer to its limits.

Smart Sensors

In the context of microprocessor-based controllers, the term "smart sensor" is often used. Generally a product with the designation "smart" will have certain capabilities of self calibration, internal diagnostics, setting of parameters from remote locations, compensation for variations in other variables, linearization, and simplified tuning of coefficients.

Programmable Logic Controller (PLC)

A programmable controller is a digitally operating electronic apparatus that uses a programmable memory for the internal storage of instructions that implement specific functions such as logic, sequence, timing, counting, and arithmetic to control machines and processes. This type of controller was developed in the 1970's and has a wide application in certain manufacturing operations. In the pulp and paper industry, these controllers are utilized mainly for discrete logic, safety interlocks and other sequencing operations.

While modern PLC's have the capability to control certain analog loops, they tend to interface poorly with computers because their design emphasis has not included algorithmic and configuration requirements. The inability to integrate with computer software has thus far severely limited their application to process control in the pulp and paper industry.

Feedback and Feed-Forward Control

A typical control loop utilizes feedback control. With this mode of control, the controller cannot act until an error has developed. In most control applications, the elapsed time between controller action and feedback signal is short, and satisfactory control is achieved.

For slow processes with long time delays, changes in load or composition may show up too late to enable the controller to hold the controlled variable within the desired limits. By adding feed-forward control, a correction can be made for anticipated changes of load or composition, thereby minimizing the transient error that would otherwise result. Feed-forward control by itself is functionally limited, and must be used in conjunction with a feedback loop to provide full control.

Control Valves

In the majority of control applications, an automatic valve is employed as the control element. The control valve is essentially a variable orifice used to regulate the flow of a process fluid in accordance with the requirements of the process. Proper selection of the valve with respect to design, size and materials of construction is critical to satisfactory performance and service.

Three features of control valves should be evaluated to ensure correct control response: capacity, characteristic and rangeability. The characteristic is the flow response to valve opening (percent of stem travel) over the full range. Different valve characteristics are illustrated in Figure 24-4. For most control applications, a linear-type response is desired. The rangeability is the ratio of flows (high to low) through which the control valve can give stable operation; a relatively low figure indicates that the valve characteristic is non-linear.

It is also important to select the proper type and structure of valve to avoid such service-related problems as corrosion, flashing, plugging, cavitation, noise, vibration and seat leakage. Some common types of valves are depicted in Figure 24-5.

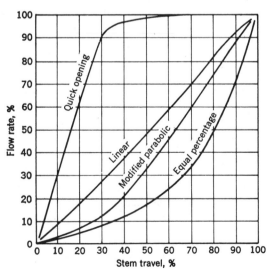

FIGURE 24-4. Flow characteristics of different valves.

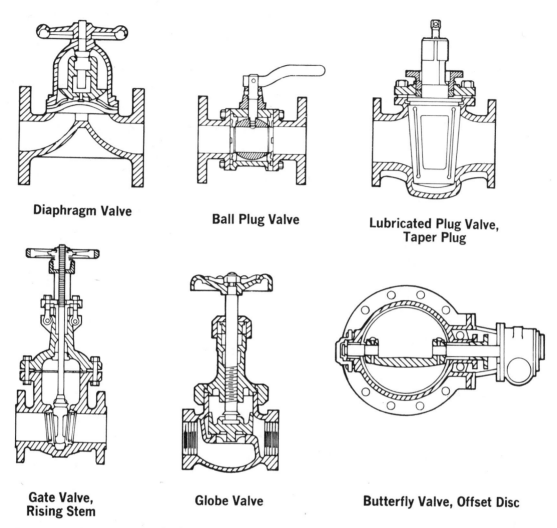

Diaphragm Valve

Ball Plug Valve

Lubricated Plug Valve, Taper Plug

Gate Valve, Rising Stem

Globe Valve

Butterfly Valve, Offset Disc

FIGURE 24-5. **Common types of valves.**

A typical automatic flow control valve with pneumatic actuator is shown in Figure 24-6.

The actuator on a control valve is normally an air-operated diaphragm or piston, which translates a signal from the controller into stem or vane movement. Springs oppose the force of air pressure to hold the plug or vane in position against the forces of fluid flow, and act to return the valve to a closed position when the air pressure is reduced.

Advanced Control Concepts

More complex control systems often use two or more measurements in a control loop. The basic form for multiple-variable control is cascade control. Other common systems are ratio control and cutback control.

With a cascade arrangement, the objective is to improve control of the primary variable by interlocking the primary controller with the controller for a secondary variable which influences the value of the primary variable. The output from the primary controller adjusts the set point of the secondary controller, thus helping to manipulate the primary variable. The secondary loop is typically introduced to improve control of processes marked by upsets and long time constants, and which cannot be satisfactorily controlled by a single three-mode controller. The secondary variable is not controlled in the usual sense, but is manipulated.

Ratio control is used when it is desired to control one variable in a fixed ratio to a second variable. In practice, a change in set point for a primary controller will actuate the ratio mechanism in the second controller, which adjusts the set point of the secondary variable to maintain the fixed ratio. This control principle is commonly used for paper mill stock proportioning systems.

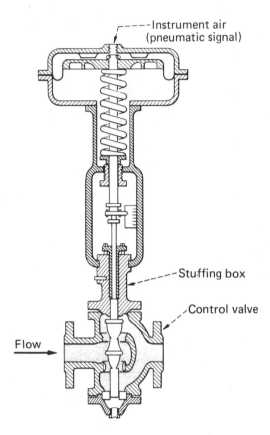

FIGURE 24-6. **Control valve with pneumatic actuator.**

Cutback (or auto selector) control is used when a single, final control element is to be manipulated to prevent a process variable from exceeding a preset limit. This control mode may be provided to ensure operator safety or protect equipment in case of process system malfunction.

Electronic vs. Pneumatic Instrumentation

Electronic and pneumatic instrumentation systems are both widely used, and each type provides good performance. Electronic systems cover virtually all industrial processes. Pneumatic systems can measure most variables, with a few exceptions such as pH and chemical composition. In many applications, there appears to be little difference in sensitivity and reliability between the two types of systems. However, electronic systems have the advantage of instantaneous response; and there is no limit to the distance that an electronic signal can be transmitted.

Generally, costs for electronic instrumentation are higher than for the pneumatic counterpart. For computer and microprocessor-based systems, electronic signals are required. While a pneumatic signal can be easily converted to an electrical signal using a transducer (and vice versa), the conversion imposes additional limits on control accuracy.

24.2 PROCESS CONTROL COMPUTERS

In the 1960's the digital computer was introduced into process control, but was slow to gain acceptance in pulp and paper mills because of high capital investment and because measuring techniques were not available to provide many of the inputs required for overall control strategy. The utilization of computers accelerated in the 1970's when their relative cost declined markedly at the same time that many new sensing techniques were introduced.

In the final analysis, the major reason for installing a computer is to be able to operate the process more efficiently and more profitably. Other benefits are the centralized interface which increases operator awareness of the process, and the provision of reporting documentation for management purposes.

Today's process computers and microprocessors utilize concepts of logic and handle complex control schemes including multiloop cascade systems, compensation for time delays, and non-linear and variable instrument parameters. They also provide for data acquisition and display, statistical analysis and establishment of mathematical models.

Evolution of Development

The first computer technology available for process control was the large main frame computer. To justify installation of this expensive unit, extensive (and often unrealistic) project objectives were developed. The purchasers were often unaware of the time and effort that was necessary to program their computer.

The earliest installations were used primarily to gather and record process data (i.e., monitor the process and act as data loggers), set off alarms (if necessary) and make calculations. Initially, these computers were not used to control process instruments directly, but they supplied the operator with the information he needed to properly adjust the set points of various control devices.

Supervisory Digital Control (SDC)

Supervisory control was a logical progressive step for a mill that was fully automated with analog control loops. In this concept, the computer receives the same sensor signals as the controllers and is able to provide a picture of the current status of the process variables. The computer periodically calculates (on the basis of input signals and programmed instructions) what the optimum values should be for selected variables (e.g., to produce maximum yield of product at the required quality level). It then changes the set points of the controllers to bring the process operation in line with the newly calculated conditions. Without the computer, these calculations were handled manually, usually with less frequency and with far less precision.

In a mill with existing controllers that are capable of accepting remote setpoints, SDC may still be a cost-effective approach to adding a computer control system. However, if the existing controllers must be replaced, a distributed control system will typically be the most viable option.

Direct Digital Control (DDC)

The next development was direct digital control where the computer replaced the functions of the analog controllers in the control loops, and added a level of supervisory control. With DDC, the loop concept is retained, but all loop variables feed to the computer. In each case, the computer calculates the necessary corrective action and transmits it directly to the control element. The major problem with DDC is total dependency on the computer, and it is necessary to provide backup options to ensure continuity of operation during servicing of the computer. The most satisfactory strategy has been to employ two digital computers, each with the full facility to control the process should a failure occur in the other.

The popularity of DDC was short-lived, and this concept is now seldom used. Dual computers are costly and quickly require upgrading or the addition of more dual computers as the control and programming tasks increase.

Minicomputer

With the development of smaller, lower-cost computers in the early 1970's, it was no longer necessary to put together an extensive control project to justify the purchase and installation of a system. The minicomputer was economically more suited than the mainframe to simple control strategies involving a few key loops. The special-purpose computer along with improved sensor technology was utilized most successfully around the paper machine. In spite of shortcomings, packaged control systems were successful because they concentrated on the key variables of any papermaking operation and controlled them.

However, by the late 1970's, the dedicated minicomputer approach was already at its limits. As the mini became more powerful, functions and features were added; and this brought to the minicomputers the problems of complexity and unreliability formerly associated with large computer projects. Expansions and retrofits were complicated, and inter-system communications were difficult and costly.

Distributed Control Systems (DCS)

Today's preferred mode of computer control is by deployment of numerous smaller computers in place of the large general-purpose computer. The key to DCS is to spread computing and controlling functions throughout the plant. This distribution puts the controlling "intelligence" close to the unit operations. Generally, smaller computers or microprocessors (designed and programmed for specific tasks) are easier to justify both technically and economically.

Many distributed systems utilize a hierarchy of controllers from single-loop microprocessor-based controllers up to multi-loop or interactive microprocessor-based controllers. This ability to select the appropriate sophistication of control technique for each specific process unit provides for the functional distribution of control throughout the entire process.

In a truly distributed system, the various process control units (which may be geographically separated) are linked together on a data bus or data highway, allowing speedy, efficient transfer of data back and forth between local computers (and with a central-operations computer, if used). This communication technique usually results in significant reduction in field wiring costs (compared to a central-control computer) since a single coaxial cable can link together as many as four or five hundred single control loops. In DCS, redundant communications are common, allowing for a back-up communications link to take over should the primary one fail.

While some weatherproof housing is required for the distributed controllers, the environmental conditions can be significantly more extreme than those required for the computer itself. If the computer is lost for any reason, the "intelligent" controllers continue to control the process. Thus, a redundant computer does not have to be purchased. Obviously, should a microprocessor-based controller fail, the degree of failure in terms of the number of loops affected is dependent on the design of the control system. Usually, those controllers on the most critical loops of the distributed system are backed-up by secondary units (redundant controllers).

Although a distributed system can be configured, if necessary, as a supervisory or direct digital control system, these control concepts are not applicable in the majority of cases because of the sophistication available within the microprocessor-based system at the single loop level. Essentially, each microprocessor-based controller has all the input, output and computing sub-systems of the computer itself.

Mill-Wide Control

While mini-computers and microprocessors have taken over many of the control functions, interest has re-focused on the central computer as an information and guidance system, especially for a paper mill or integrated operation. The objective would be to interconnect all the digital systems into a mill-wide network to provide effective management decision support tools. This integration of process control and decision making is sometimes referred to as the pulp and paper industry's realization of computer-aided manufacture.

In view of the complexity of pulp and paper mill systems and the constant efforts to improve overall mill performance, a sophisticated information and analytical system can be a valuable tool for both management and operating personnel. Typically, the central computer (as the final element of a "hierarchical" control structure) will utilize a high-speed printer to produce a number of reports on daily, weekly and monthly bases. While graphic display of an individual process (e.g., digester, lime kiln) is handled in the respective dedicated computer, the central computer usually provides overview displays. It also carries out material and energy balances, produces time-based curves (current and historical), and calculates production optimization strategies.

There are many concepts of millwide control. One conceptualization is illustrated in Figure 24-7 wherein the objectives, implementation responsibility, and typical response times are laid out (1). The basic building blocks at the bottom of the pyramid (level 1) are the instrumentation systems. Without a well-designed and maintained instrumentation/electrical system, any attempt at millwide control would be foolhardy. At the second level is unit process control and supervision of quality. The third level is production and energy management, which is usually referred to as the "millwide control" level. Finally, the highest level in this particular model is production scheduling. Here, the major objective is to schedule orders in a manner to achieve maximum utilization of the production facilities.

In spite of extensive lobbying by the proponents of mill-wide control, relatively few mill-wide systems had been implemented in North America by the early 1990's, perhaps because other concerns were perceived as being more urgent of attention.

Control Room

The move away from conventional analog instrumentation to computer control has revolutionized the control room. The most visible aspect is the video-based operator station. In the modern control room, the video screen (CRT) is the operator's window to the process, wherein the logical and functional presentations of process information permit a rapid and efficient two-way communication. Although the information on the screen is much the same as was formerly available on analog instruments, the method of depiction, the amount of information, and the accuracy are all enhanced.

Typically, the operator runs the process from an air-conditioned and humidity-regulated control center. Ideally the work station should be situated in an environment where the operator can concentrate on plant status, and not be distracted by high foot traffic, insufficient work space, or disorganized reference materials (2). All the displays and controls should be located at a central point within the reach and viewing range of the operator. A console should consist of four to eight CRT's installed in a cabinet configured for use by a seated operator (for example,

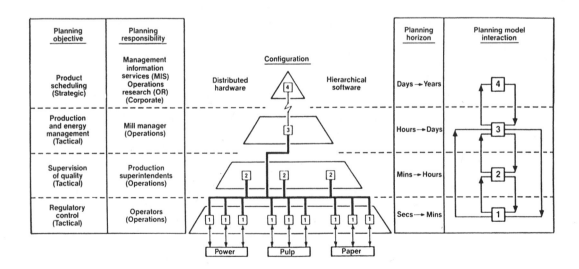

FIGURE 24-7. Example of a millwide system.

FIGURE 24-8. Video-based operator station (Foxboro Company).

see Figure 24-8). The console is usually semicircular in shape with one or two tiers of CRT's.

The keyboard is also a key part of the operator interface. By appropriate input, the operator changes the set points on the basic control loops, and pages through the different groups while sitting at the console, analogous to moving along a panel monitoring various analog instruments. Although most keyboards are standard units, customized keyboards may be designed for specific operator functions to simplify the operating system and provide greater security.

REFERENCES

(1) WAYE, D. **What is a Millwide System** *Tappi Journal* (April 1989)

(2) OTTINO, C. **Designing and Laying Out a Control Room** *Chemical Engineering* (September 1991)

Mill Services

When a pulp mill is being planned, a number of requirements must be met. Most important is an adequate wood supply. The other critical items are plentiful water and power. Labor can be brought in, and the infrastructure can be built, so in the final analysis water can become the key item.

25.1 WATER SUPPLY

As recently as 1965, a typical bleached kraft mill utilized 45,000 gallons of water for every ton of pulp produced; and a typical newsprint mill consumed 15,000 gallons for every ton of finished product. In the past 25 years these figures have been significantly reduced by virtue of more effective recycle of process water, more complete utilization of side streams, and recovery of contaminated condensates (1). Nonetheless, a dependable supply of good quality water remains an essential component for any conventional pulp and/or paper operation.

Water usage and quality should be monitored and the cost of the water service considered as with any other raw material. For many pulp and paper products, water quality is of vital importance, and the specification is tending to become more stringent as higher quality products are developed.

Water Supply Sources

Industrial water sources are broadly categorized as either surface water or groundwater. Surface water sources include lakes, rivers and reservoirs; these waters contain varying amounts of organic and inorganic contaminants depending on both seasonal factors and the characteristics of the terrain over which the water has flowed. The contaminants may be dissolved, suspended or colloidal. Bacteria are present. Some surface waters also contain municipal or industrial pollution.

Groundwater is taken from wells or springs. These waters usually contain relatively high concentrations of dissolved mineral matter, but low concentrations of suspended solids and organic matter. Typically, little color is present due to the cleaning and filtering action of the ground strata through which the water passes. A partial listing of common water impurities is given in Table 25-1.

Some mills are blessed with relatively pure sources of water that do not require treatment. However, in many situations it is advisable to treat the raw water

TABLE 25-1. Common impurities found in water.

Turbidity	- due to suspended solids.
Color	- usually associated with dissolved organic matter.
Hardness	- calcium and magnesium salts, a major cause of scaling in boilers and heat exchangers.
Silica	- SiO_2; causes scale in boilers.
Iron/Manganese	- can cause color problems on full bleach pulp.

to remove or control impurities that would otherwise adversely affect product quality or increase maintenance costs. Generally, the major consideration is to reduce the level of suspended material; and depending on the particular quality requirements, any of the following processes are commonly employed:
• settling or sedimentation
• filtering
• coagulation followed by sedimentation or filtering.

Other processes may be employed for all or part of the water depending on specific requirements. For example, boiler feedwater has rather stringent quality requirements (as discussed in Section 25.2). Supplemental processes include chlorination or ozonation, aeration, de-aeration, demineralization (ion exchange), and fine filtration.

Specific Treatment Methods

Sedimentation is any process that channels water slowly through a holding tank or basin, allowing time for the coarse particles to settle to the bottom before the water overflows. Typically, large circular clarifiers are used for this purpose.

Filtration is most commonly carried out using mixed media of sand and/or anthracite. Several filter units (perhaps of the type illustrated in Figure 25-1) are usually operated in parallel arrangements so that individual units can be periodically backwashed without reducing throughput. Typical media makeup is listed in Table 25-2.

TABLE 25-2. Typical media used in filters.

Sand and Gravel		Depth (inches)
top layer	20 to 35 mesh	26
next	6 to 10 mesh	4
next	1/4 to 1/8 inch	4
next	1/2 to 1/4 inch	4
bottom layer	1 to 1/2 inch	4
	Total	42
Anthracite		
top layer	3/32 to 3/64 inch	26
next	3/16 to 3/32 inch	4
next	9/16 to 3/16 inch	4
next	13/16 to 9/16 inch	4
bottom layer	1 5/8 to 13/16 inch	4
	Total	42

Coagulation (or flocculation) is used when it is necessary to remove finer suspended material and/or color from the raw water. Coagulating chemicals such as alum (aluminum sulfate) or lime (CaO) are added to form gelatinous precipitations ("flocs") which enmesh and absorb the suspended matter and color. The flocs usually have good settling characteristics and may be removed by sedimentation. A modern coagulation/sedimentation unit with internal sludge recirculation is illustrated in Figure 25-2.

Chlorination or ozonation of water supplies is carried out in many mills to reduce bacterial and slime growths. The chemical demand depends on the extent of oxidizable matter, but a small excess of chemical should be retained in the water for most effective treatment.

Several methods are available for removing hardness from water, but most commonly an ion-exchange resin is utilized. The water is passed through a pressure vessel which contains a filter bed of resin particles. The divalent calcium and magnesium ions are preferentially absorbed by the resin, typically in exchange for monovalent sodium (or hydrogen) ions. When the resin becomes saturated with divalent ions, the unit is taken out of service and is "regenerated", most commonly with concentrated brine solution (NaCl). The calcium and magnesium ions are removed from the system in the form of soluble chlorides, and the resin (after rinsing) is ready once again to effectively remove hardness from the water.

25.2 BOILER FEEDWATER

The previous section dealt with water treatment methods that are applied to the major incoming flows. These general methods are commonly referred to as "external treatments" to distinguish them from more-limited and specialized methods which are referred to as "internal treatments". A near universal requirement in pulp and paper mills is for internal treatment of boiler feedwater.

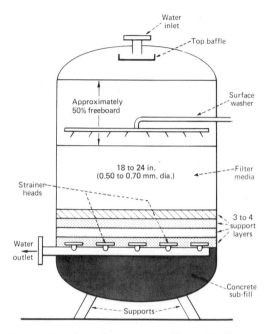

FIGURE 25-1. Downflow pressure filter, with filter media supported by layers of progressively coarser aggregate.

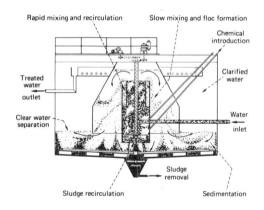

FIGURE 25-2. Solids-contact coagulant clarifier for raw water treatment.

The water entering the mill, regardless of the type and extent of external treatment, will usually contain contaminants that can cause problems in the boiler tubes or in the steam and condensate handling systems. The major objectives of feedwater treatment are to:
• prevent formation of deposits,
• reduce metal corrosion,
• prevent carryover of water (with entrained solids) into the steam section.

Prevention of Deposits

If not controlled, deposits in boiler tubes (Figure 25-3) will reduce heat transfer and restrict throughput. The conventional strategy for control is to precipitate dissolved mineral matter as a free-flowing sludge that can be removed with the "boiler blowdown". The specific steps employed are dependent on the contaminants present, but usually include several of the following:

- control pH at about 10.5 to precipitate magnesium hydroxide (and also to prevent acid corrosion),
- add sodium tripolyphosphate to precipitate calcium,
- add an organic sludge conditioner (usually a modified lignin, tannin, starch or synthetic polymer),
- add a chelating agent.

For boilers operating at pressures over 900 psig, complete deionization is a better method of hardness control (2). Here, the objective is to remove not only the calcium and magnesium, but all the ions (e.g., sulfates, fluorides, and carbonates). This is done in a two-stage ion exchange where the substituting ions are hydrogen and hydroxide, which combine to form water.

Control of Corrosion

The main sources of corrosion in boiler water are dissolved oxygen and acidity. (Corrosion in condensate return lines is mainly due to carbon dioxide). Acidic corrosion is easily controlled by neutralization with caustic soda (NaOH) or soda ash (Na_2CO_3). The optimum pH level is about 10.5. ("Caustic embrittlement" of tube surfaces can occur at higher pH levels.) Oxygen pitting can be controlled by mechanical deaeration (i.e., boiling off the dissolved air) or by use of chemical oxygen scavengers, such as sodium sulfite for low-pressure boilers and hydrazine for high-pressure boilers.

Return line corrosion from carbonic acid (due to dissolution of carryover carbon dioxide into the condensing stream) is usually controlled by adding neutralizing or filming amines to the boiler feedwater. The neutralizing amines function by chemically neutralizing the acidity, while filming amines produce a protective film on the metal surfaces.

Control of Boiler Water Carryover

If not controlled, solids associated with boiler water are carried with the steam and deposit throughout the steam system to cause various problems, including pitting and stress corrosion. Steam turbines and related components are especially sensitive to boiler water carryover.

The separation of steam from boiler water occurs in the steam drum (refer to next section on boilers). As steam bubbles explode from the surface, a fine mist is produced. The steam drum is typically equipped with baffles or more sophisticated internal separation equipment. However, no internal separator will remove all the mist, and the problems of mist entrainment can be quite severe if any foaming occurs at the water/steam interface. In many instances it is worthwhile to add defoaming chemicals to the boiler feedwater to reduce carryover.

25.3 BOILER OPERATION

A boiler system provides the means for converting fuel energy into steam. Steam is a convenient form of energy for performing work and transferring heat; and it can be readily converted into electrical energy by means of a turbine. The boiler itself consists of a series of drums and tubes which hold and transport water and steam under pressure. The overall boiler system includes a furnace (where fuel and air are mixed in correct proportion), along with the necessary fans and auxiliaries for metering of air and fuel, and for controlling the combustion reaction. Examples of conventional boiler systems utilizing hog fuel are shown in Figures 25-4 and 25-5.

Boilers are fabricated in a wide range of sizes and designs depending on the pressure, temperature and quantity of steam produced, and the characteristics of the available fuels. The furnace can be designed for solid, liquid or gaseous fuels including coal, bark and hogged woodroom refuse (hog fuel), municipal waste, oil, spent liquor, and natural gas. Most steam for the pulp and paper industry is produced in large field-erected boiler units at pressures of at least 600 psig to allow for efficient generation of electrical power. However, small shop-fabricated "package boilers" are sometimes used to produce supplemental steam for seasonal demand variations or to accommodate swing load situations.

Circulating Fluidized Combustion (CFC)

Circulating fluidized combustion is a new and commercially available technology that is providing flexibility of fuel usage and efficient combustion of

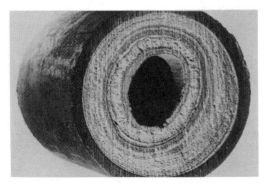

FIGURE 25-3. Scale deposits in boiler water line.

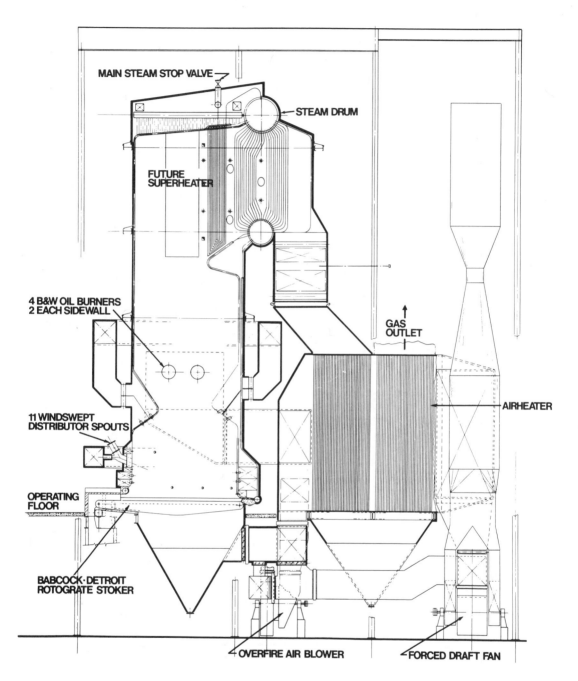

MAIN STEAM STOP VALVE

STEAM DRUM

FUTURE
SUPERHEATER

4 B&W OIL BURNERS
2 EACH SIDEWALL

GAS
OUTLET

AIRHEATER

11 WINDSWEPT
DISTRIBUTOR SPOUTS

OPERATING
FLOOR

BABCOCK·DETROIT
ROTOGRATE STOKER

OVERFIRE AIR BLOWER

FORCED DRAFT FAN

FIGURE 25-4. Wood waste-fired boiler (Babcock and Wilcox).

inexpensive low-grade and/or high-moisture fuels. CFC also has the ability to control gaseous emissions from sulfur-bearing fuels at low levels by burning these fuels in combination with limestone. The ash and residue from the process are dry and easily landfilled.

One design of CFC power plant is shown schematically in Figure 25-6. The basic technique is to

force the combustion air through the fuel and inert material (either sand or limestone) at a velocity sufficient to turbulently expand the mixture and promote near complete combustion, followed by recapture and recirculation of the larger particles. This recycle technique, along with ash removal, heat transfer, and refractory surfaces in appropriate locations are the key elements in the CFC system design.

Combustion Process

The combustion of a typical organic fuel causes the pyrolytic breakdown of complex organic molecules into atoms of carbon and hydrogen followed by their reaction with oxygen from the air. The principal chemical reactions and their heat release values are as follows:

| | BTU value per |
| Chemical Reaction | lb of combustible |

carbon to carbon monoxide

$$2C + O_2 \longrightarrow 2CO \qquad 4,400$$

carbon to carbon dioxide

$$C + O_2 \longrightarrow CO_2 \qquad 14,600$$

hydrogen to water

$$2H_2 + O_2 \longrightarrow 2H_2O \qquad 62,000$$

Carbon will combine with oxygen to form either or both CO and CO_2 depending on how much oxygen is available. Since formation of CO_2 generates greater heat, it is important to provide sufficient air for complete combustion.

Air is actually a mixture of gases, predominantly nitrogen and oxygen with small amounts of carbon dioxide and water vapor, and trace amounts of other gases. Oxygen and nitrogen are present in approximately the following proportions:

	Volume Percent	Weight Percent
Oxygen	21	23
Nitrogen	79	77

Nitrogen provides some benefit in controlling the rate of the combustion reaction, but also carries away heat in the exit gases.

The thermal efficiency of a boiler is measured by the ratio of heat pickup in the steam to the total heat value of the fuel burned. Depending on boiler design and loading, the efficiency can vary between 65 and 90%. Many strategies are available for automatic continuous utilization and control of boilers (4).

Description of Boilers

In the most simple package boiler designs, the combustion reaction takes place in a furnace under the boiler vessel and the hot gases traverse submerged tubes, thus heating the water to raise steam. These "fire tube" designs are suitable only where saturated steam up to a pressure of 150 psig is sufficient.

Most boilers are of the "water tube" type, where water circulates within the tubes and heat transfer occurs from the hot combustion gases on the outside of the tubes to water on the inside. Designs utilizing either natural or controlled circulation are available. The water is initially heated in the tubes nearest to the combustion flame; it then rises to the "steam drum" where vapor is separated from liquid. The liquid travels down connecting tubes to a lower drum, usually called the "mud drum". Here, the sludge formed from the concentration of mineral impurities is removed by manual or automatic "blowdown".

The steam drum is larger in diameter than the mud drum to provide space for the separation of steam and water. Low-pressure boilers require larger diameter steam drums than high-pressure boilers (for the same steam delivery per foot of drum length) because of the greater specific volume of the steam. The steam drum is equipped with internal baffles and other separation devices to prevent carryover of water with the steam (see Figure 25-7).

The purpose of the superheater is to raise the temperature of the steam to some specified level above the saturation temperature. Superheat is essential to provide good "cycle efficiency" during electrical generation and to minimize moisture in the last stages of the turbine (to avoid blade erosion). The bank of tubes serving as the superheater is positioned relatively near the fire to obtain good heat absorption. Tube spacing in these locations is dictated by the "slagging tendency" of the ash in the fuel.

The original purpose of "water walls" was to protect the refractories used in the furnace construction, and to cool any ash particles below their fusion point to prevent slagging in the subsequent tube banks. In modern boilers, the furnace "water cooling" is made part of the boiler circulation sequence.

Depending on the size and operating characteristics of the boiler, the economics will usually favor some type of "secondary heat-absorbing surface", i.e., an economizer and/or air heater. At higher boiler pressures, the temperature of the combustion gas must be kept high at the superheater to effectively transfer heat. Since most of the latent heat of vaporization is added to the water in the furnace tube walls, the loss of heat with the exit gases would be excessive without further recovery. Heat can be returned to the unit either in the boiler feedwater by means of an economizer tube section or in the combustion air by using an air heater.

While the reduction of exit gas temperature to a low level is attractive from an efficiency standpoint, this reduction is limited to the point where condensation of vapors from the combustion process occurs. These condensed vapors usually contain sulfuric and sulfurous acids or other highly corrosive substances. It may be desirable to control the exit gas temperature at a level considerably higher than the condensation temperature in order to positively prevent both corrosion and deposit formation.

Most boilers operate at a slight draft (sub-atmospheric pressure) to minimize danger to

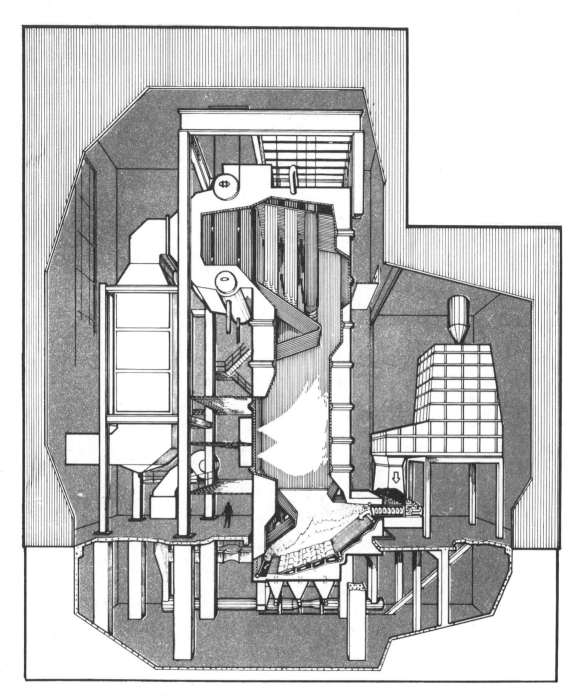

FIGURE 25-5. Bark-fired boiler (Gotaverkin).

operators. Therefore, some ambient air unavoidably infiltrates into the furnace; usually between 5 and 10% of the process air enters in this manner.

Bark/Hog Fuel

The escalating cost of fossil fuels in the 1970's, particularly of gas and oil, focused attention on more efficient utilization of bark and other wood waste materials (hog fuel) as an energy source for the pulp and paper industry. Up to that time, hog fuel was generally regarded more as a nuisance than as a resource. Many mills had preferred to discard wood waste materials as landfill rather than reduce boiler capacity by displacing cheap oil.

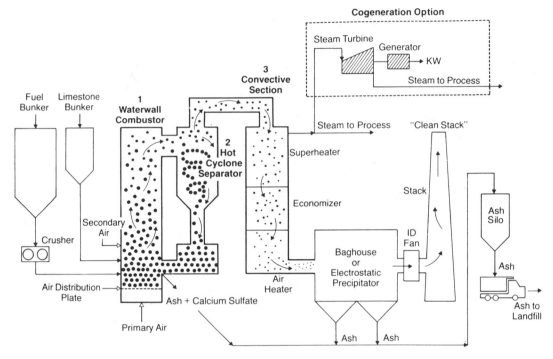

FIGURE 25-6. Circulating fluidized combustion power plant (N.W. Dunlap).

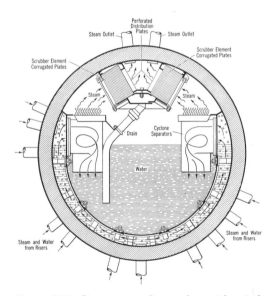

FIGURE 25-7. Steam separating equipment located in the steam drum.

A high moisture content has been the major problem associated with effective utilization of hog fuel, especially when wet debarking methods are used. The moisture reduces both the effective heat content of the fuel (refer to Figure 25-8)) and the output of the boiler. Boiler furnaces have since been modified to obtain better combustion and higher boiler efficiency with relatively wet wood waste materials. At the same time, much effort has been made to improve the action of presses and other mechanical dewatering devices where free water is present. One arrangement for handling and dewatering wet bark utilizing a double-nip roll press is shown in Figure 25-9. However, the most promising approach toward better utilization is through drying of the fuel.

Most schemes for drying hog fuel utilize heat from the boiler exhaust gases. In effect, the hog fuel acts as a "secondary heat-absorbing surface", replacing or augmenting the economizers or air heaters. A representative system is shown in Figure 25-10. Another approach toward hog fuel drying is to use a separate drying system fired by gas, oil, or dried wood waste, as depicted in Figure 25-11.

25.4 STEAM UTILIZATION

Steam utilization in pulp and paper mills can be categorized as follows:
- for power generation (primarily turbine) which is usually back-pressure,
- for heating loads from which condensate is recovered,
- for heating loads without condensate recovery.

Power is produced in a turbine by a series of expansion steps which converts the potential energy of high-pressure steam into mechanical energy to drive rotating equipment. This principle is illustrated in the schematic drawing of a multi-disc radial turbine shown in Figure 25-12. Here the steam flows from the center sequentially, outward-inward-outward. This design utilizes a large number of rings of small diameter to ensure that the leakage areas are small and the efficiency high.

The shaft of the turbine is typically coupled to an alternating-current generator to produce electrical power. Turbines are also commonly used as direct drives for equipment with high unit loading such as the recovery furnace induced-draft fan or the paper machine. The quantity of energy extracted in the turbine is proportional to the difference in the energy of the steam at the inlet and outlet.

Pulp and paper mills utilize substantial amounts of both electrical (or mechanical) energy and low-pressure process steam. Therefore, it makes sense to extract electrical energy from the high-pressure boiler steam as the pressure is reduced to the level required for process demands. Typically, some steam is passed out of the turbine at an "intermediate pressure" (e.g., 160 psig) and the balance at a "low pressure" (e.g., 60 psig) as shown schematically for a kraft mill energy system in Figure 24-13. The relative amounts of steam are dependent on mill requirements; obviously, the lower pressure steam should be used whenever possible to optimize power generation. This "cogeneration" system would be ideal if the steam requirements exactly matched the electricity requirements. In many situations, however, the mill requires additional electrical energy, which is usually purchased from the local utility company.

Additional power could also be produced by generating more steam in the power boiler than required for process use and condensing the excess steam. Although "condensing power generation" produces more electrical energy for a given quantity of steam than "back-pressure power generation", the former method inefficiently utilizes fuel energy because the latent heat in the steam is transferred to cooling water. The relative thermal efficiencies for the two methods are compared in Figure 25-14.

In a typical pulp and/or paper mill, about one-half of the process steam is used "indirectly" in tube banks, drying cylinders, and heating jackets. In these applications, the condensate is available for return to the boiler house. A condensate return system involves capital costs for pumps, tanks, traps and piping, but is easily justified because of heat credits and savings in raw water treatment. Where contamination of the condensate is possible (as, for example, with evaporators or black liquor heaters), some method of monitoring the condensate must be employed to prevent return of a fouled stream.

Direct use of steam is relatively more wasteful

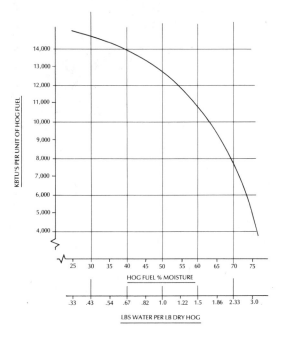

FIGURE 25-8. Expected steam generation from mill wood waste at varying moisture content.

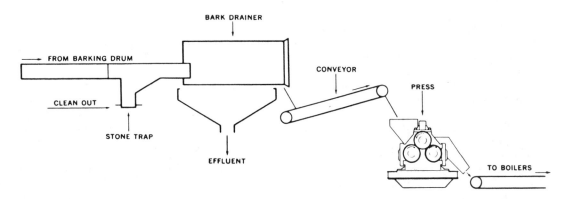

FIGURE 25-9. Typical arrangement for handling wet bark (Fulton Iron Works).

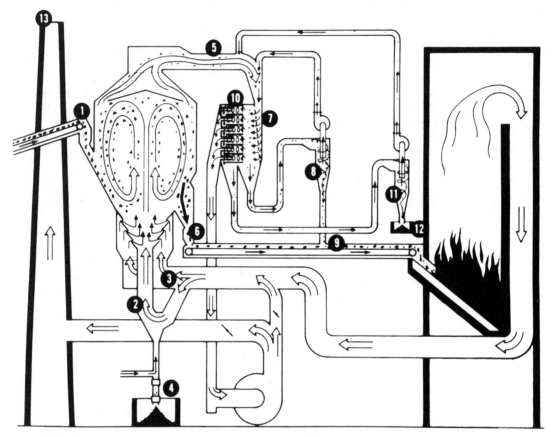

1. Wet bark conveyed to dryer
2. High-velocity gas stream
3. Main gas stream
4. Stone and scrap discharge
5. Gas and combustible fines leave dryer
6. Dried bark conveyed to boiler
7. Precollector separates combustible fines
8. Secondary cyclone for combustible fines
   from precollector
9. Combustible fines join dried bark enroute
   to boiler
10. Multicyclone for final separation
11. Secondary cyclone for flyash from
    multicyclone
12. Flyash removed from system
13. Stack emission cleaned to code

FIGURE 25-10. In-line bark drying system (Bahco Industri AB).

because of the loss of condensate. However, for such services as heating high-consistency stock or cooking starch, direct steaming is the most reasonable choice from an operational standpoint. Also, a direct system is invariably cheaper to install because heat exchangers, pumps and condensate-return piping are not required.

25.5 ELECTRICAL DISTRIBUTION

The electrical power utilized in pulp and paper mill operations is typically a combination of purchased power and self-generated power from hydro and thermal facilities (see previous section). A few fortunate mills are self-sufficient in power, whereas a number of small operations are totally dependent on power supplied by the local utility grid.

The charges for purchased electrical energy usually depend on three key parameters:
• peak power.
• load factor.
• power factor.
The peak power is the highest rate used for a specified minimum period (typically ten minutes). The load factor is the ratio of average power to peak power, usually expressed as a percentage. Power factor is a measure of the degree to which the alternating current and voltage are out of phase; if in phase, the power factor is 100%, but typical mill operation is between 92 and 95%. Actual power (watts or kilowatts) is equal to the product of the

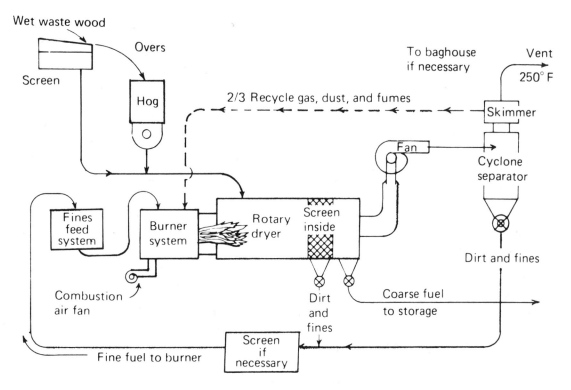

FIGURE 25-11. System for drying hog fuel utilizing a rotary dryer (R.C. Johnson).

FIGURE 25-12. Multi-disc radial turbine (Stal-Laval Turbin AB).

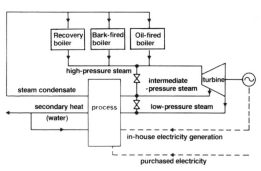

FIGURE 25-13. Typical energy system for a kraft mill.

voltage, amperage (i.e., current), and the load factor.

Power is the rate at which electrical energy is consumed. Energy is therefore the product of (power) x (time) and is usually expressed as kilowatt-hours (kwh). Average power can be calculated by determining the number of kwh for a specific period and dividing by the number of hours.

Since peak power has a significant impact on purchased energy costs, it is normally desirable to control the peak to a realistic maximum. In mills with grinders, it is good practice to maintain load control using one or more of the grinders as "load-shedding" devices when the peak is exceeded.

A representative electrical power distribution system is illustrated in Figure 25-15. The incoming voltage from the local power grid may be as high as 115 kilovolts (kv). This power is received at a main substation, and is usually "stepped down" by means of a transformer to 13.2 kv for distribution to substations around the mill. Further reductions to the working voltages take place near the point of usage. The largest motors (for service above 1500 kw) operate directly off 13.2 kv; 4,140-volt motors are used for intermediate service, and 550-volt motors are used below 150 kw.

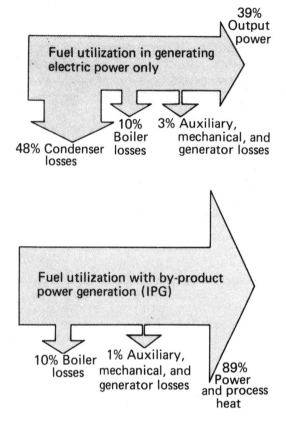

FIGURE 25-14. Energy balances for condensing and back-pressure power generation.

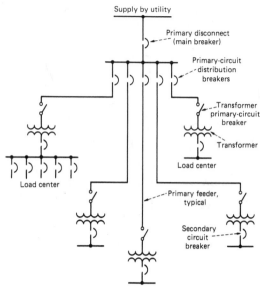

FIGURE 25-15. Representative electrical distribution system.

25.6 ENERGY MANAGEMENT

Energy costs are a significant factor in the manufacture of pulp and paper, and optimal utilization and conservation of energy is a serious concern to the pulp and paper industry. A number of aspects must be considered for effective energy management (5):

• utilization of available energy sources, including alternate fuels,
• improvement in thermodynamic efficiency (i.e., obtaining more usable energy from fuels),
• reduction of losses and waste,
• recovery and reuse of energy whenever possible,
• modification of processes to consume less energy.

Planning for a new mill is based on cost/benefit analysis and is relatively straightforward. Such planning is made with due regard to the high cost of energy and to future cost escalations and demands for conservation. However, in an older mill, the operators may still be tied to a system based on different premises (e.g., on the premise of cheap, expendable energy).

Improvement of energy management in an existing plant must be based on an accurate characterization of the process and operation. A basic requirement is an energy balance (or "energy audit") detailing all the energy inputs and distributions throughout the operation. It then becomes a challenging and informative exercise to analyze these streams and look for alternative process options that provide savings in energy or reductions in the cost of energy.

Energy Sources

The high cost of oil and natural gas, as well as potential shortages of these fuels, has necessitated the utilization of alternative fuels. Over the last decade, the industry has moved sharply away from dependence on oil toward more effective utilization of spent liquors and hogged woodroom refuse, as shown by the American Paper Institute's data on energy sources for the U.S. industry (Table 25-3). Several mills are investigating the utilization of forest thinnings and forest "residuals". In some situations, consideration is being given to municipal waste and dried peat as potential energy sources.

Thermodynamic Efficiency

Many older mills are operating with steam cycles and equipment that are far from today's optimum. In general, higher steam pressures and temperatures will dramatically improve the thermal efficiency of a steam plant. However, within the confines of a given steam cycle, improvements can be made by limiting boiler losses, controlling back pressure at the minimum level dictated by process demands, and

TABLE 25-3. Estimated energy sources for the U.S. pulp and paper industry, 1972, 1979, 1990 by percentages.

	1972	1979	1990
Purchased steam/power	5.4	6.7	7.3
Coal	9.8	9.1	13.7
Fuel oil	22.3	19.1	6.4
Natural gas	21.5	17.8	16.4
Hogged wood and bark	6.6	9.2	15.4
Spent liquor solids	33.7	37.3	39.4
Self-generated power	0.6	0.8	1.2

maximizing condensate returns. Overall tight control of a boiler system is greatly facilitated by a modern process control system (4).

Reduction of Losses and Waste

A mill-wide energy balance will usually show that the bulk of energy leaves the mill in relatively few streams (e.g., paper machine dryer exhaust, various warm effluents). From this simplified analysis, the most advantageous approach toward reduction of energy losses will become apparent. Generally, the paper machine exhaust flows deserve careful study; excessive steam consumption for paper drying is not unusual. The emphasis with regard to effluent streams should be on reducing the use of water, since a lower volume of water will significantly reduce heating requirements. Closing up of the paper machine system and counter-current washing in the bleach plant are two practical ways to lower water usage.

Recovery and Reuse of Energy

The pulp and paper industry already employs a high degree of energy recovery and reuse, but further improvement is possible in most older mills. The economics for recovery become more favorable as the cost of fossil fuels escalates. Areas of the mill that offer especially good returns on waste heat recovery equipment are lime kiln product coolers, paper machine economizers, blow heat recovery, and bleach plant effluent heat exchangers. The utilization of waste heat in the boiler exhaust gases to dry wet bark is a special case which was discussed in Section 25.3.

Process Modifications

It does not appear that new process technology will provide any magical solutions to the industry's energy problems. In the near term, pulp and paper mills will use equipment and processes already available. Nonetheless, a number of operations can be designed or modified for lower energy consumption. Several examples have already been covered in this brief discussion of energy management. An excellent example of a design concept to minimize electrical energy can be found in the woodroom/chip handling area, where belt conveyors are displacing pneumatic systems for transporting chips. According to one source, the belt conveyor motor consumes about one-eighth the energy required by a blower to move the same quantity of chips.

25.7 CORROSION CONTROL

Corrosion is a factor in every pulp and paper mill which contributes to costly downtime and equipment replacement. Most metal components eventually deteriorate in use due to the chemical nature of the process applications. Since corrosion is impractical to eliminate, the best industrial strategy is to control it within tolerable limits. Major savings in operating and maintenance costs are possible by anticipating problems, and by providing proper process design and equipment before construction begins. Nevertheless, corrosion problems will invariably arise after the mill has been built and is operating.

Corrosion is the deterioration that occurs when a material reacts with its environment. While all materials can be affected under certain conditions, this brief discussion will concentrate on the major problem of metal corrosion.

Although many pulp and paper companies provide staff specialists to assist with specific problems, it is generally the responsibility of mill technical and engineering people to control corrosion. The mill technologist should be able to recognize corrosion, understand how it is caused, and be able to measure its severity. He or she should also know what inspection techniques are available and be able to interpret and apply corrosion information.

Forms of Corrosion

Corrosion occurs in numerous forms, but its classification is usually based on one of three factors:

1. Whether the corrosion is "wet" or "dry". Most corrosion involves the presence of liquid or moisture. Dry corrosion usually involves reaction with a high-temperature gas.

2. Mechanism of corrosion; either electrochemical or direct chemical reactions.

3. Appearance of corrosion; either uniform or localized.

Localized corrosion is further distinguished as either macroscopic or microscopic; in the latter case, the amount of metal dissolved is minute and considerable damage can occur before the problem becomes visible to the naked eye. Various types of corrosion are illustrated in Figure 25-16.

Uniform attack over large areas is the most common form of corrosion. Knowledgable selection of materials and protection methods (e.g., various coatings) can control it. Uniform corrosion is the

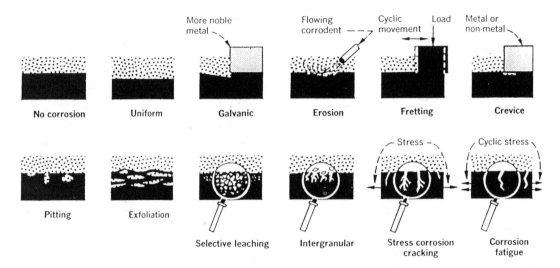

FIGURE 25-16. Different forms of metal corrosion.

easiest form to measure, and unexpected failures can usually be avoided by regular inspection. The various forms of localized attack are more difficult to predict and control.

All types of corrosion require attention, but particular importance must be placed today on the dangers of stress corrosion cracking and corrosion fatigue. Stress corrosion cracking involves the combined action of a tensile stress and a corrosive environment. The environment is often quite specific to the metal or alloy in question, and the effect usually cannot be predicted. The stresses arise from residual cold work, welding, or thermal treatment, or they may be externally applied during service.

Corrosion fatigue cracking is a special type of stress corrosion cracking which occurs under cyclic stress conditions. Failure may occur at cyclic levels significantly lower than those considered safe for operation under noncorrosive conditions. Stress corrosion and corrosion fatigue cracks can propagate rapidly through metals, yet escape visual detection. As with other forms of microscopic localized corrosion, the metal may appear to be perfectly sound until unexpected failure occurs. Many catastrophic failures of pressure vessels and rotating equipment are due to stress corrosion and corrosion fatigue cracking. Some notable problems in the pulp and paper industry have occurred with boilers, continuous digesters, paper machine suction rolls, and refiner discs.

Certain bacteria play a role in papermill corrosion processes, especially those aerobic and anaerobic types whose metabolisms require some form of sulfur. These bacteria are thought to promote electrolytic corrosion by acting as cathodic depolarizers. Different types of bacteria may cohabit in a microenvironment created by fungi and algae. If bacteria are involved in corrosion, it may be necessary to use biocides, surfactants and/or chlorinators, in addition to conventional corrosion control methods.

Corrosion Testing

It is often desirable to undertake corrosion measurements to monitor corrosion rates occurring in the mill, evaluate alternative materials, or study the mechanisms of corrosion. The most common measurements are weight-loss tests where a clean coupon is weighed, exposed to a corrodent for a specified time, cleaned to remove corrosive products, and reweighed. Corrosion rates in the U.S. are commonly reported in mils/yr.

There is no single standard size or shape for corrosion-test coupons. They usually weigh from 10 to 50 g and preferably have a large surface to mass ratio. A wide range of methods are used for supporting corrosion test samples, and special mountings are used to evaluate galvanic and crevice corrosion resistance and stress corrosion cracking.

Corrosion Control

There are basically four ways to control corrosion: by engineering design, materials selection, control of process variables and protection. Some of the more common process variables that affect corrosion are listed in Table 25-4.

Typically, corrosion increases with the concentration of the corrodent; but this is not always the case, and the effect often depends on the concentration range. pH is usually an important factor, and it is often helpful to raise pH to reduce corrosivity. Many metals and alloys are resistant to

TABLE 25-4. Process variables that affect corrosion.

- concentration of corrosive agent
- pH
- temperature
- impurities
- degree of aeration
- fluid velocity
- inhibitors

alkalis but readily attacked by acids. Temperature control is a major method of reducing corrosion. Higher temperature almost always increases the corrosion rate. According to one estimate, corrosion in the paper machine approach system doubles for every 20° F rise in temperature.

An inhibitor is a substance that reduces corrosion when relatively small amounts of it are present in the corrodent. Inhibitors are classified by their composition (organic or inorganic), mechanism of action (cathodic or anodic control), or form (solution or vapor). These agents can sometimes effectively control corrosion, especially in relatively closed systems.

By applying external electrical circuits, various types of electrochemical corrosion can be controlled. In some cases, an opposing current (cathodic protection) can be used to nullify corrosion; in other cases, the metal's potential can be adjusted (anodic protection) so that the metal still corrodes, but more slowly. Anodic protection has been successfully applied to white and green liquor storage tanks and continuous digesters.

REFERENCES

(1) MINER, R. AND UNWIN, J. **Progress in Reducing Water Use and Wastewater loads in the U.S. Paper Industry** *Tappi Journal* (August 1991)

(2) SENDELBACH, M.G. **Boiler-Water Treatment - Why, What and How** *Chemical Engineering* (August 15, 1988)

(3) DUNLAP, N.W. JR **Circulating Fluidized Combustion Power Plant Technology of the Future - Today!** Presented at 1987 TAPPI Annual Meeting

(4) LIPTAK, B.G. **Improving Boiler Efficiency** *Chemical Engineering* (May 25, 1987)

(5) MANNISTO, H. **Mill Energy Audit Can Point Out Opportunities for Conservation** *Pulp & Paper* (January 1981)

Chapter 26

Water Pollution Abatement

26.1 INTRODUCTION

An important consideration of modern pulp and paper mill design and operation is to minimize losses from the process and to treat mill effluents so that their impact on the environment is minimal and essentially non-polluting. The goal is to preserve environmental quality for the benefit of present inhabitants and future generations. Environmental concerns are translated by regulatory agencies into specific criteria for limiting or controlling mill discharges.

Abatement programs within the pulp and paper industry are costly to implement and entail significant operating expense; but these costs may be partly offset in some cases by better retention of fiber in the process or improved energy conservation. The incremental cost attributed to pollution abatement is generally accepted as a necessary cost of doing business, and is passed along to the customer in the form of higher-priced products. There can be little argument against sensible waste management, and no responsible spokesperson is likely to advocate a return to "laissez-faire" discharges. However, debate is ongoing between industry and governmental agencies as to what level of discharge is required for environmental protection.

Definition of Pollution

Water pollution can best be defined as any change in the condition of water which is detrimental to some beneficial use. In each situation, the beneficial uses of the water and the conditions affecting these uses must first be determined. Then, the significance of any changes (i.e., degree of pollution) can be assessed.

In human terms, the highest water use is for drinking purposes. Pollution from domestic sewage will impair water potability due to bacterial infection. Substances of industrial origin may adversely affect the taste and odor of drinking water. Other problems such as foaming, radio-activity, toxic substances, or heavy metal ions have been encountered in specific situations.

Fish habitats are most often affected by a reduction of dissolved oxygen or by toxic substances in the water. Deposition of solid particles can also affect the viability of fish populations.

Water utilization for industrial purposes depends on the specific requirements. Some industries need high quality water relatively free of suspended solids, organic substances, or inorganic salts. Although incoming raw water is frequently treated to remove suspended material (refer to Section 25.1), other types of pollution may be more troublesome. For example, dissolved organic substances (especially colored compounds) can adversely affect a wide range of industrial uses.

The legal definition of pollution has varied within North America and around the world. In some instances, reaction to public pressure has resulted in narrow definitions and legislated limits on industrial discharges which have been unnecessarily stringent and idealistic. After a period, this type of legislation is often found to be difficult to enforce, and a more practical definition is brought in that takes into account economic factors and the site-specific utilization of the receiving waters. Legislation based on a realistic assessment of how industrial discharges impact on the environment have usually been more effective in controlling pollution.

Characteristics of Receiving Waters

The relevant attributes of a typical receiving water (e.g., river, lake, estuary, etc.) are listed in Table 26-1. Most regulatory statutes specify that little or no change in these characteristics will be detectable as a result of effluent discharges.

TABLE 26.1. Characteristics of Receiving Waters.

- dissolved oxygen
- pH
- toxicity
- suspended solids
- temperature
- floatable solids
- foam
- taste
- odor
- nutrient concentration
- productivity (population of microorganisms)

Dissolved oxygen is essential to the survival of fish and all other useful organisms in the water, including those responsible for waste assimilation. When natural waters become seriously deficient in dissolved oxygen, their ability to support life is impaired and they can become foul-smelling due to the action of anaerobic organisms.

At a temperature of 20° C, natural waters can contain about 9 ppm of dissolved oxygen. The ability of water to dissolve oxygen is inversely related to temperature, and is also affected by salinity, with less oxygen uptake as salinity rises. The relationships are illustrated in Table 26-2. Oxygen enters solution in water principally through the air/water interface; the rate of transfer depends on such factors as surface agitation and initial oxygen concentration. The further the concentration is away from saturation, the higher is the driving force for absorption.

TABLE 26-2. Concentration of dissolved oxygen in fresh water and in moderately saline water (5 g/liter chloride) as a function of temperature.

	Oxygen Concentration, ppm	
Temperature, °C	Fresh Water	Saline Water
0	14.6	13.8
5	12.8	12.1
10	11.3	10.7
15	10.2	9.7
20	9.2	8.7
25	8.4	8.0
30	7.6	7.3

The level of dissolved oxygen in water at any time represents the equilibrium between a number of factors, such as absorption, respiration, photosynthesis, and decomposition of organic compounds. The oxygen demand of organic waste while being "assimilated" by bacteria and fungi is of primary concern with respect to the discharge of mill effluents into receiving waters. Virtually all naturally-occurring organic materials are "biodegradable" by organisms which occur in surface waters. Ultimately, these compounds will be reduced to carbon dioxide and water. The chemical equation for the biological breakdown of a "simple compound" like sugar shows that each molecule of sugar requires six molecules of oxygen for its complete assimilation:

$$C_6H_{12}O_6 \ + \ 6O_2 \longrightarrow 6CO_2 \ + \ 6H_2O$$

Biological assimilation takes place over a period of days and weeks, the rate of which is primarily dependent on the nature of the waste, the water temperature, and the concentration of oxygen. Relatively simple, soluble compounds such as sugars and short-chain alcohols are consumed quite rapidly, while more complex structures such as tannin and lignin derivatives are broken down very slowly.

Generally, organic suspended solids are relatively slow to degrade.

Most measurements of dissolved oxygen concentration are made by electrode instruments. These instruments should be calibrated periodically against a reliable chemical test such as the classic Winkler method.

Measurement of Pollution

The major categories of water pollution which are of concern to the pulp and paper industry are effluent solids, oxygen demand, toxicity and color. Abatement efforts generally focus on the reduction of solids and oxygen demand. Fortunately, typical mill wastes are weakly toxic by conventional measurement, and are usually essentially non-toxic following biological treatment. Color is of critical concern only when the dilution factor in the receiving water is low and light penetration is affected sufficiently to have an impact on plant growth in the water system.

Effluent solids are categorized and measured as detailed in Table 26-3. Settleable organic solids are especially objectionable in natural waters because they form "sludge blankets" which snuff out the bottom-dwelling flora and fauna, and interfere with the feeding habits of fish. Entrapment of fibers or other suspended solids in the gill tissue of fish can cause stress, secondary infection, and possibly suffocation. Bottom accumulations of organic material rapidly become depleted in oxygen, and the aerobic organisms die off. Anaerobic bacteria and other undesirable life forms take over and continue biological action; however, the end products are now methane and hydrogen sulfide which are released into the atmosphere as bubbles of gas.

The oxygen demand of an effluent stream may be measured or estimated by a number of methods, some of which are defined in Table 26-4. The most widely used test is the five-day biochemical oxygen demand (BOD_5), in which a sample of effluent is allowed to consume oxygen by the action of microorganisms. The classical method is to determine oxygen concentration on both an initial sample and on a duplicate sample following incubation at 20° C over a period of five days, the difference in oxygen concentration being taken as the BOD. Since the initial sample can retain only about 9 ppm of oxygen, the effluent must be appropriately diluted so that oxygen depletion during the test does not exceed much more than 7 ppm (i.e., about 70 - 75% depletion).

Biochemical oxidation is a slow process, and complete breakdown may take up to 100 days. However, for most effluents, it has been shown that 60 - 70% of the oxidation takes place in the first five days. If properly carried out, the five-day BOD test gives a good indication of the effect an effluent is

TABLE 26-3. Definition and measurement of effluent solids.

Total Dissolved Solids (TDS)	- obtained by evaporating a known sample of filtered effluent to dryness and determining the weight of residue.
Total Suspended Solids (TSS)	- obtained by measuring the increase in dry weight of a tared filter paper after passing a known quantity of effluent sample.
Total Solids	- total of suspended and dissolved solids (the sum of the previous two determinations); also obtained by evaporating a known effluent sample to dryness and determining the weight of dried residue.
Settleable Solids	- portion of the suspended solids that will settle out after a specified time (usually one hour).
Floatable Solids	- portion of suspended solids that float on the surface.
Nonsettleable Solids	- portion of suspended solids that will not settle or float, usually colloidal or near-colloidal size.
Volatile Solids	- solids in any of the above categories that are volatized at 600° C (taken as an estimate of organic material).
Fixed Solids	- solids in any of the above categories that are left as a residue after ignition at 600° C (taken as an estimate of inorganic material).

TABLE 26-4. Oxygen depletion tests.

Symbol	Test	Measurement
BOD_5	biochemical oxygen demand	oxygen consumption by microorganisms.
IOD	immediate oxygen demand	oxygen consumption by strong chemical reducing agents such as sulfur dioxide, sulfides, and hydrosulfites.
COD	chemical oxygen demand	oxygen demand as measured by direct chemical methods, usually employing dichromate or permanganate.
TOD	total oxygen demand	instrumental method for determining the total amount of oxidizable material.
TOC	total organic carbon	instrumental method for determining the amount of oxidizable carbon.

likely to have on the oxygen balance of any natural receiving water. However, it must be stressed that the application of BOD data is not simple, and knowledge of the behavior of the receiving water (e.g., temperature, degree of natural aeration, etc.) is necessary to predict the impact of the effluent.

Because the BOD test requires a five-day period, much interest has been focused on methodology to provide more rapid results. A number of related chemical and instrumental methods have been developed, some of which are listed and described in Table 26-4. However, there is no inherent correlation between BOD and any of these other tests, as typified by the BOD vs. TOC data shown in Figure 26-1. For some specific effluents, a reasonable correlation has been found (e.g., Figure 26-2), but the rapid characterization of one or two effluent streams does not provide the basis to control the operation of plant abatement facilities.

It now appears that a dependable biosensing instrument has been developed that correlates well with the BOD_5 test (1). The heart of the instrument is a temperature-controlled bioreactor in which biomass grows on thousands of small plastic rings. These rings are kept in constant turbulence by means of a circulation pump. A two-pump system feeds diluted wastewater at a constant rate into the bioreactor and also causes an equal outflow. The biomass in the reactor feeds on the nutrient in the sample stream, thereby consuming oxygen. Dissolved oxygen measurements taken at the inflow and outflow of the reactor system are reported to the

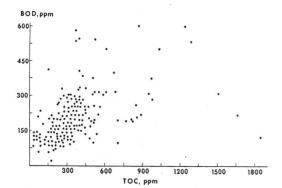

FIGURE 26-1. Plot of TOC vs. BOD₅ for combined outfall samples from seven bleached kraft mills (Howard & Walden).

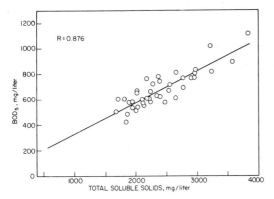

FIGURE 26-2. Plot of total dissolved solids vs. BOD₅ for a series of decker filtrate samples (W.D. South).

system computer. The computer, in turn, controls the dilution ratio by means of two measuring pumps in such a way that the difference between inflow and outflow dissolved oxygen is maintained at 3 ppm. The computer calculates the BOD value based on the required dilution ratio.

As already noted, severe toxicity is normally not a problem for pulp and paper mill effluents. However, a number of pulp effluent constituents have been identified as toxic (e.g., resin acids, unsaturated fatty acids, chlorinated phenolics), and periodic testing is required to ensure that the treated effluent is within specified limits. For this purpose, bioassays are commonly performed where fingerling fish (or other suitable aquatic animals) are exposed to a known concentration of effluent and the survival rate is measured after a specified time period (usually 96 hours).

The main limitation with conventional toxicity testing is that only acute lethal effects are found. Sub-lethal cumulative effects over long periods on growth and reproduction have traditionally not been considered, nor have potential mutagenic/carcinogenic properties of the effluents. Regulating bodies are now becoming more concerned with sub-lethal effects, and future emphasis will likely be on the elimination or removal of known contaminants, such as chlorinated dioxins and furans (see Section 11.3).

Types of Treatment - Overview

Improved fiber retention and better in-plant utilization of raw materials are effective means of reducing or controlling mill discharges. Fiber and soda are contaminants in the effluent, but these constituents also represent a costly loss to the mill. Therefore, all actions taken to "tighten up" the process will have the dual benefits of lower raw material cost and reduced pollution loading.

Perhaps the single most effective stratagem for reducing in-plant losses has been to recycle and reuse mill process waters. In recent years, the amount of water consumed by pulp and paper mills has been dramatically reduced. Although the trend will continue, buildups of temperature and impurities act to limit the ultimate degree of "closure". In addition to reducing the quantity of contaminants, a reduced volume of effluent also facilitates subsequent external treatment steps.

External treatment is usually by means of sedimentation to remove suspended solids ("primary treatment") and biological oxidation to remove BOD ("secondary treatment"). Prior to leaving the mill, the various effluent streams are segregated according to treatment requirements. Some effluents need only primary treatment; others require only secondary treatment; while a few streams must have both primary and secondary treatment before discharge.

Any treatment beyond primary and secondary treatment is usually termed "tertiary treatment". Sometimes a clarification stage following secondary treatment is called a tertiary stage, but this terminology is not universally applied. Perhaps the only valid tertiary treatment now being undertaken in a few mills is for the removal of effluent color.

26.2 SOURCES OF POLLUTANTS AND IN-PLANT ABATEMENT

The major sources of effluent pollution in a pulp and paper mill complex are the following:
- water used in wood handling/barking and chip washing,
- digester and evaporator condensates,
- white waters from screening, cleaning and thickening,
- bleach plant washer filtrates,
- paper machine white water,
- fiber and liquor spills from all sections.

In mills using wet-drum or hydraulic debarking equipment, the effluent from these operations can be a significant source of suspended solids, BOD and color. Recycle of this water following screening and clarification helps to reduce the volume requiring treatment. Some mills utilize hot caustic extraction effluent (from the bleach plant) to increase bark-removal efficiency (especially from frozen wood) and to reduce the overall volume of color-containing water. The effluent load from wet barking can be eliminated by converting to dry-barking methods.

All condensates from digester vapors and low-solids evaporator effects are concentrated sources of BOD; some of these condensates also contain reduced sulfur compounds. These low-volume flows should be steam-stripped to remove methanol, reduced sulfur gases, and other volatiles, and thus reduce their BOD content prior to discharge. Steam stripping is essentially a multistage distillation separation using direct steam as the heat source. (see Figure 26-3). The hot overhead gases are usually cooled sufficiently to condense water, and the remaining noncondensibles are burned in the lime kiln or power boiler.

Unbleached white water from screening and cleaning operations is typically a large-volume source of effluent. In kraft and sulfite mills, this effluent contains the unrecovered lignin from the washing operation and, depending on washing efficiency, can have a significant BOD. All screen room effluents contain suspended fiber. The screening department is a prime area for recycle of white water and/or reuse of selected evaporator condensates or bleach plant filtrates.

Most of the BOD and color from bleaching operations are contained in the chlorination and alkaline extraction effluents. The volume of these effluents can be reduced by optimal application of countercurrent washing techniques (see Section 11.10) and/or by recycle of chlorination filtrate (see Section 11.3). A marginal reduction in BOD and color is provided with an oxygen-augmented extraction stage. However, the most dramatic in-plant reduction in BOD discharge is achieved with oxygen delignification (see Section 11.4).

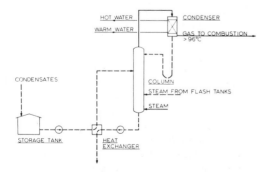

FIGURE 26-3. Typical system for steam-stripping contaminated condensates.

The paper machine produces large volumes of white water, which in many cases are of sufficient quality to be reused in other sections of the mill. As a minimum requirement, all white water not used as furnish dilution should be sent to a saveall (either filter or flotation type) for recovery of solids prior to discharge. In many cases, additional uses can be found for clarified water in cleaning showers and seals (see Section 16.7).

In a modern mill design, it makes economic sense to provide for recovery of fiber and liquor spills, and in so doing to eliminate shock loadings on the external effluent treatment facilities. In the causticizing area, a sump should be available for collecting spills from lime mud storage and liquor clarifiers. A dump tank is necessary for intermediate storage during tank cleanouts and to provide holding capacity for overflows. Collection sumps are also required in the power group (recovery and evaporators) for spills from black liquor storage tanks. In areas where stock spills are likely to occur, access to a clean fiber salvage sump should be provided.

26.3 STANDARDS AND REGULATIONS

Up to 1970, stream quality standards in the United States were largely the responsibility of individual states. The federal government became dominant in 1970 through passage of the National Environmental Policy Act which instituted the concept of environmental impact statements and established the Environmental Protection Agency (EPA). In 1972, the Federal Water Pollution Control Act Amendments were passed which provided effluent limitation guidelines and authority to levy fines for noncompliance. This act stipulated a step-wise schedule for meeting conventional discharge criteria, the first target level by 1977 being equivalent to "best practical technology" (BPT), and the second target level by 1983 being equivalent to "best available technology economically achievable" (BAT). Recognizing the greater effectiveness in contemporary plant layout and design, the effluent limitations stipulated for new plants were generally similar to BAT. A representative comparison of the original BPT, BAT, and new mill guidelines is given in Table 26-5. Although the act was subsequently amended to extend the deadline for compliance, most U.S. mills by 1992 were probably operating at or near the BAT level.

In the early 1980's, the scope of regulations expanded to include BAT guidelines for control of "priority pollutants", thus focusing attention on a range of substances with known toxic or sub-toxic effects. A number of specific compounds were identified that could be measured at relatively low concentration. Among these were a number of byproducts of the chlorine bleaching process.

TABLE 26-5. EPA guidelines for effluent limitation, lb/ton of product (maximum for 30-day averages).

	5-Day BOD			Suspended Solids		
	BPT	BAT	New Mill	BPT	BAT	New Mill
unbleached kraft	5.6	2.7	3.1	12.0	3.7	7.5
market bleached kraft	15.8	7.1	5.3	31.7	5.2	5.8
market bleached sulfite	41.7	20.1	9.3	53.3	6.9	5.8
paperboard from wastepaper	3.0	1.3	1.5	5.0	1.6	4.0
nonintegrated fine paper	8.5	2.7	2.7	11.8	1.4	2.8

Recently, the EPA has expanded the list of priority pollutants, and also defined a new target level referred to as "maximum achievable control technology" (MAT).

The U.S. federal regulations that deal with environmental protection seem to change every three or four years. The task of keeping up-to-date with current legislation and EPA policy directives will be a constant challenge to mill managers and technical personnel (2). Increasingly, specialists may be required to provide assistance in working out programs of compliance in many cases.

It is equally difficult to keep up-to-date with the environmental situation in Canada where both the federal and provincial governments are active on the regulatory front (3). Federally, pulp and paper mills are subject to the general pollution control provisions of the Fisheries Act and the Effluent Regulations promulgated under the act. In addition, each of the provinces has its own environmental legislation. One distinction is that the provinces have almost exclusive responsibility and authority for air pollution control. While overlapping jurisdiction has led on occasion to conflicting requirements, the two levels of government generally coordinate their activities. The requirements for effluent discharges in most cases are very similar to those in the United States.

Water quality guidelines in both the United States and Canada are typically administered by means of the permit system, which lays down the specific criteria and schedule of compliance for each industrial operation. Guidelines are usually applied uniformly, but with an awareness of the site specifics. Where sensitive environmental conditions exist, more stringent limitations may be imposed. By the same token, where it can be demonstrated that no deleterious effect is detectable in the receiving water, a less stringent limitation on mill discharges may be tolerated, or a more relaxed schedule for full compliance may be allowed. Standards set for an entire industry are equitable from the standpoint of relative pollution control costs, and they are easy to administer. But it must be recognized that the assimilative capacity of the receiving water is different in each case.

26.4 ENVIRONMENTAL MONITORING

Compliance with government standards is the most obvious and immediate priority of a mill environmental monitoring program. Other purposes include performance checks of the various treatment stages, determining the impact of effluent discharge on the receiving water, and gathering information for process control. Monitoring usually encompasses the entire data collection effort, including sampling, flow measurement, testing and analyses.

The most accurate part of any monitoring operation is usually the analytical laboratory, while the least accurate part (and the part that is frequently neglected) is the sampling. In any application, the capability of a proposed sampling device to collect a representative sample from a waste stream should be carefully assessed. In practice, up to 30% variability has been found to exist between samplers and sampling methods, depending on the fluctuations in concentration and flow. Attention to proper selection and installation of the sampler, and careful planning of the sampling program will significantly increase the reliability and accuracy of the data collected (4).

The minimum information required for mill effluents includes water usage in each department, individual sewer flows, and analysis of each major sewer stream for suspended solids, dissolved solids, BOD, pH, and toxic effects, if applicable. For the treatment facilities, it is necessary to know the inlet loading of suspended solids and BOD, and the corresponding values following treatment. And, of course, the combined effluent at the point of discharge must be carefully monitored with respect to all relevant properties.

The effect of mill effluent on the receiving water must be determined by field studies involving sampling near the point of discharge and at various locations distant from the mill site. In a lake or marine situation, a grid of sampling stations is established radiating from the effluent discharge points, with control stations located in unaffected areas. In a river situation, sampling stations are located at intervals downstream from the mill with control areas upstream from the mill site (refer to Figure 26-4).

Conventional physical and chemical tests carried out on the receiving water provide only transitory,

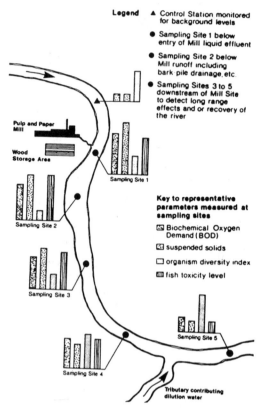

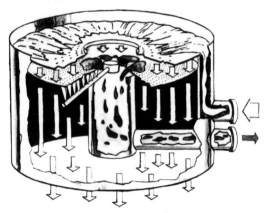

FIGURE 26-5. Operation of screen is shown schematically. Effluent flows radially toward the center and through screen perforations. Fiber retained on the screen is lifted off by a rotating shower and floated toward the center outlet (AES).

FIGURE 26-4. Example of a generalized monitoring program for a river system (K. Schiefer and D. Lush).

short-term data. To acquire knowledge of long-term effects, biological monitoring is necessary. Examination of a representative aquatic community provides a more complete picture of the quality of the environment and the effects due to mill effluents.

Biological monitoring can be accomplished in a variety of ways. In one method, artificial substrates are placed along the bottom of a river bed or water body in several locations selected to reflect changes in habitat or effluent concentration. In unpolluted waters, these substrates will become quickly populated by a diverse number of bottom-dwelling invertebrates. In the presence of toxic substances, species are progressively eliminated in relation to their sensitivity or tolerance to the pollutants. By evaluating the biological diversity on various substrates, a biologist or trained technician can monitor changes in the environment and infer effects on the entire water body.

26.5 PRIMARY TREATMENT

Primary treatment generally refers to the methodology for removing suspended solids from mill effluents. In the pulp and paper industry, solids

removal is always accompanied by some reduction in BOD and toxicity; but these latter categories of pollution usually must be further reduced by biological treatment (i.e., secondary treatment).

Screening is often used as a preliminary step to remove relatively large floating or suspended particles from an effluent stream. Sometimes the objective is to salvage fiber from selected effluents for return to the process, while reducing the loading to the main clarification stage. One design of gravity screen used for this purpose is depicted in Figure 26-5. More complete removal of suspended solids by fine screening is impractical because of the blinding or plugging action of the solids.

The two principal methods of primary treatment employed in the pulp and paper industry are gravity sedimentation and dissolved-air flotation. Sedimentation or gravity settling is by far the most common process used because it is relatively insensitive to variations in flow and solids concentration, and requires little attention and maintenance. Flotation processes are generally more efficient in removing solids, but are more expensive to operate.

Sedimentation

Sedimentation can be carried out in any holding pond or impounding basin that provides enough retention time for settling to take place. Generally the equipment of choice is a large circular, mechanically-cleaned clarifier of the type illustrated in Figure 26-6. Whatever the design, the sedimentation unit must provide a sufficient period of quiescent flow to enable the specified percentage of settleable solids to drop to the bottom. In the mechanically-cleaned units, the solids are usually

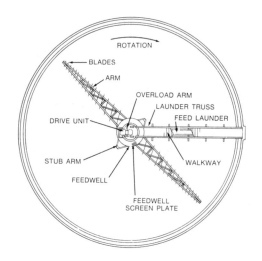

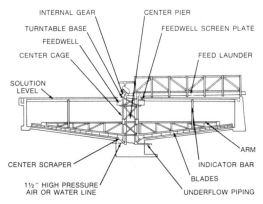

FIGURE 26-6. Circular, mechanically-cleaned clarifier (Dorr-Oliver).

raked toward a center sump utilizing a center-mounted sludge scraper. Depending on the characteristics of the solids, the underflow may be pumped anywhere between 1.5 and 6% solids.

Four functional zones can be identified within a clarifier with respect to design and operation:

1) The inlet zone provides a smooth transition for the influent as it moves from the feedwell into the steady radial flow required in the settling zone.

2) The settling zone provides the area and depth for settling and thickening without interference from other zones.

3) The outlet zone provides a smooth transition for the effluent as it moves out of the settling zone and over the discharge weir.

4) The sludge zone receives and removes settled material without interfering with the settling process.

The clarifier actually performs two functions: it clarifies the liquid passing through it, and it concentrates the solids. The clarifying action is primarily a function of clarifier diameter; surface

loadings of 800 to 1000 gal/day/ft^2 are typical. The thickening action is more a function of clarifier depth; a minimum depth of 11 feet is generally recommended for clarifiers ranging from 30 to 65 feet in diameter, while a minimum depth of 15 feet is more desirable for tanks over 200 feet in diameter. A floor slope toward the center of 1 to 12 is typically used to help move the sludge to the sump.

The design of a clarifier cannot be based on theoretical principles alone. Test data on settling rates for the particular wastewater to be treated are necessary for proper sizing of the unit. Nonetheless, some understanding of the factors that affect settling can help to optimize any clarifier operation (5). Sedimentation theory reveals that settling is affected not only by particle density, but also by particle shape and size. Therefore, where required, coagulants or flocculants can be added to the influent stream to increase particle size and thereby increase the rate of settling. Since non-settleable solids can also be made settleable through flocculation, much higher efficiencies of suspended solids removal are possible through use of additives.

Improved clarifier efficiency or increased capacity can also be achieved by means of inclined tubes or channels installed in the settling zone. These auxiliary devices serve to effectively increase the hydraulic path at the same time that convection currents are virtually eliminated.

Flotation Clarifiers

Dissolved air flotation is a solids-removal process where fine air bubbles become attached to the suspended particles, thus reducing the density of individual particles and causing them to float to the surface. The separated solids then form a floating layer that is a mixture of solids and air bubbles. The buoyant force exerted by the entrapped air also acts to compact the solids into a smaller volume before the floating layer is skimmed off.

Air is usually introduced to the inflowing wastewater in a pressurized mixing chamber. When the influent enters the nonpressurized flotation unit, the super-saturated solution releases air in the form of very fine bubbles which become attached to the suspended particles. To get maximum results from a flotation unit, a flocculant aid such as alum must be added along with the air.

A flotation clarifier of recent design is illustrated in Figure 26-7. This unit features a deep, baffled flow shaft for the wastewater. Air is injected on the downcomer side at the bottom of the shaft where a portion is dissolved by hydrostatic pressure. Undissolved air rises against the downward flow, contributing to water saturation. Rising water on the other side of the shaft is thereby saturated with air, but free of undissolved air bubbles. As hydrostatic pressure declines, microscopic bubbles are released

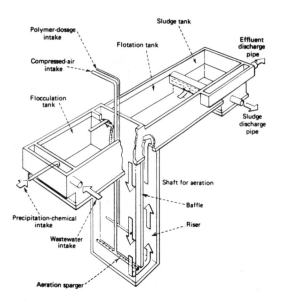

FIGURE 26-7. Flotation clarifier featuring a deep flow shaft for air dissolution and gradual release of bubbles from solution (Environmental Systems Div., AB Electrolux).

from solution and become attached to the suspended solids.

Flotation methods can achieve high levels of suspended solids removal, up to 98%. However, the significant costs for compressor power and flocculating chemicals have generally mitigated against this method except where high efficiency is required or where space is limited. The more common application of flotation separation technology within the pulp and paper industry is for white water clarification (see also Section 16.7).

26.6 SECONDARY TREATMENT

Secondary or biological treatment is nothing more than a duplication of nature's own purification process, except that it is carried out under contained and controlled conditions, and usually at accelerated rates.

Under aerobic conditions, microorganisms (mostly bacteria and fungi) consume oxygen to convert organic waste into the ultimate end products of carbon dioxide and water. An important aspect of most biological oxidation processes is to provide sufficient aeration and mixing to prevent anaerobic conditions from developing. Aerobic biological oxidation can be accomplished by various means, depending on the characteristics of the effluent, the area available for external treatment, and the required degree of BOD removal. Each method depends on sustaining a viable population of specially-adapted microorganisms. In some cases, it is necessary to add

nutrient chemicals (containing nitrogen and phosphorous) to stimulate metabolic activity.

Although anaerobic digestion is often used in the treatment of sanitary wastes, this type of process was employed infrequently in the pulp and paper industry up to 1980. However, during the 1980's, the trend toward reduced process water consumption and increased yields in mechanical and secondary pulping systems have resulted in more concentrated wastewater streams that are more suitable for anaerobic treatment. In environments where molecular oxygen is absent, anaerobic microorganisms become dominant and utilize chemically-bound oxygen. The end products of anaerobic processes are reduced chemical forms such as methane and hydrogen sulfide. While the high-sulfur content of many wastewaters mitigates against anaerobic treatment, it now appears that some sulfur-free pulping wastewaters can be treated cost-effectively by this method (6,7).

A mass-time curve of aerobic biological oxidation is shown in Figure 26-8. Initially, with an unlimited food supply, bacterial growth is restricted only by the ability of the microorganisms to reproduce. During this initial rapid growth phase, a large portion of the soluble BOD is converted to "biomass" (i.e., microorganisms and their metabolic byproducts), while a smaller portion is oxidized to carbon dioxide and water. This "bioconversion" is capable of converting 30 to 70% of the BOD into insoluble material, depending on the respiration of the particular microorganisms present. As the food supply becomes more limited, the rate of bioconversion slows down. When the food supply is almost depleted (the endogenous phase), the cell mass uses its stored food supply (auto-oxidation by endogenous respiration) to stay alive. Generally, the so-called "high rate" biological processes remove BOD utilizing bioconversion and subsequent removal of biomass. Extended treatments remove BOD by more complete oxidation.

The simplest form of aerobic treatment is the oxidation lagoon, which depends on natural means to diffuse air into the effluent. This system uses shallow basins that cover very large areas. If the depth exceeds 3 to 4 feet, anaerobic microorganisms will become active in the lowest levels. Substantial BOD reductions are achieved by this method with little operating skill or attention required, but long retention times are necessary and few mills can afford to utilize such large areas of land. Typically, a 30-day retention is required for 85 to 90% BOD removal. Approximately 20 acres of lagoon must be used for each million gallons per day of effluent to be treated.

Aeration Lagoon

The aeration lagoon (Figure 26-9) utilizes continuous, mechanical aerators or diffusors of

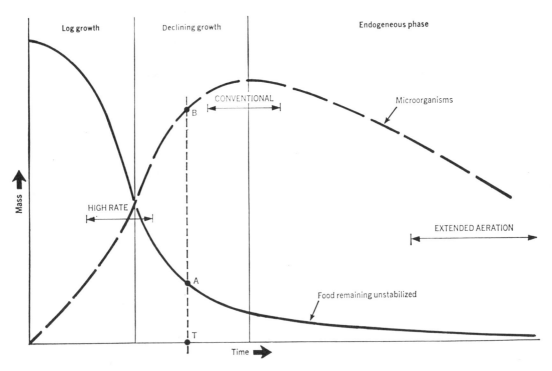

FIGURE 26-8. Mass-time relationship for typical biological treatment.

FIGURE 26-9. This oxidation lagoon at the Weyerhaeuser mill in Everett, WA utilizes 27 40-hp aeration units, and provides five days of retention. (Aqua Aerobic System)

various types (Figure 26-10) to solubilize large amounts of oxygen into the effluent. Since a significant concentration of oxygen is maintained in the effluent (at least 0.5 ppm), the biological activity is relatively high, and retention times of 3 to 5 days are typical. Since the aeration action provides continuous mixing, the depth of the lagoon can also be increased to 25 feet or more. Therefore, the area required for an aeration lagoon is far less than for a simple oxidation lagoon. The design and selection of a proper-size diffused-air or surface-agitator system is a matter of understanding oxygen transfer principles and then applying some common sense (8).

Where space is available, the aeration lagoon is generally the preferred method of biological oxidation, especially for large wastewater volumes of relatively low BOD concentration. The process is tolerant of load variations and usually produces a low level of residual biological floc. Capital and operating costs are generally about one-half that required for the activated sludge system.

Activated Sludge

The activated sludge system (illustrated schematically in Figure 26-11) is the most popular high-rate method of treatment, and is most commonly utilized in situations where land is not available for an aeration lagoon. The essential feature is the

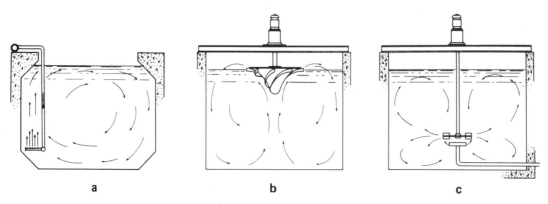

FIGURE 26-10. Illustrating flow patterns for (a) diffused air, (b) mechanical aeration, and (c) combination turbine and diffuser-ring system.

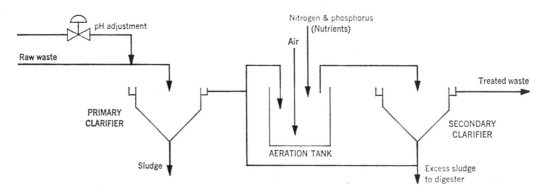

FIGURE 26-11. Schematic of activated sludge process.

development of a microbial floc held in suspension in an aeration tank or mixing chamber. Clarified wastewater is fed continuously to this tank (usually sized for 3 to 8 hours of retention), and the activated sludge solids multiply as dissolved organic waste is metabolized. At the same time, effluent is continuously drawn off to a clarification unit where solids are separated. A certain portion of these solids are recirculated back to the aeration tank to provide a high floc density, while the remainder are concentrated and disposed of by landfill or incineration.

Compared to an aeration lagoon, the activated sludge process has some disadvantages. The process is far more sensitive to changes in the character of the waste, and the pH must be carefully controlled. The nutrient requirements are higher, and settling aids are usually required for proper clarification of the outflow water before discharge. If the mill shuts down for any reason, provision must be made to feed the system during the outage to ensure a viable population of organisms upon resumption of operations. In general, the activated sludge system requires more attention and a higher level of skill to maintain the operation.

Biological Filter

High-rate treatment can also be obtained in a biological filter system. Biological filtration is a technique for promoting contact between a free-flowing wastewater and a stationary growth of microorganisms in the presence of atmospheric air. The system does not depend on filtration in the usual sense for its effect.

In the conventional trickling filter (Figure 26-12), microorganisms are cultivated on such porous media as crushed rock or plastic modules. The wastewater flows down through the bed and contacts the layer of microorganisms (zoogloeal film). The entire system is maintained in an aerobic condition by the free passage of air through the unit. The microorganisms consume the soluble organic constituents of the wastewater, primarily for the production of new cell tissue (biomass). Eventually, the cell tissue reaches a critical size and is sloughed off to be carried with the outgoing flow into a subsequent clarification step.

Biological filter systems (such as the trickling filter) have generally proven to be unsuitable for pulp and paper wastewaters, exhibiting even greater

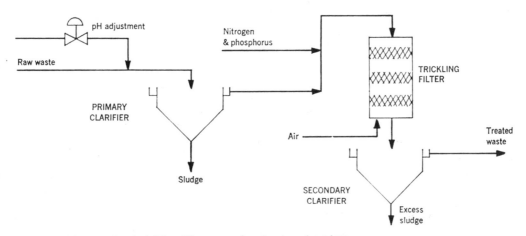

FIGURE 26-12. Schematic of trickling filter secondary treatment system.

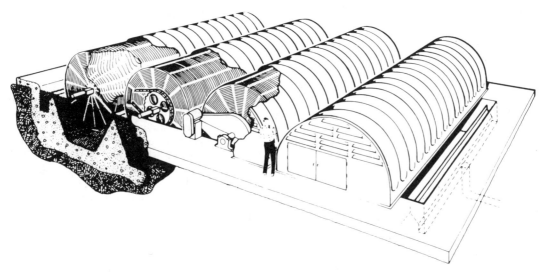

FIGURE 26-13. Rotary disc biological filter plant.

instability to waste variability than activated sludge systems. However, the rotary disc system (a more recent design of biological filter) has been successful on a number of wastewaters and appears to deserve consideration for future applications. In this system (Figure 26-13), a biological growth is cultivated on the disc surfaces. A series of closely-spaced corrugated plastic discs are anchored to a rotating shaft and supported above a trough through which the wastewater is channeled at a controlled rate. The lower 30 to 40% of each disc extends into the wastewater, while the upper portion is exposed to air. Thus, an area of biological growth alternates between submergence to absorb food and exposure to air for oxidation. Any number of rotating assemblies can be used in series to achieve the desired degree of BOD removal.

26.7 COLOR REMOVAL

The brown color of pulp mill effluents originates with the wood handling, chemical pulping and bleaching operations, and is due to the presence of tannin and lignin derivatives. These compounds are slow to biodegrade and are usually little affected by conventional secondary treatment methods. Changes in plant design (e.g., elimination of wet debarking, bleach process modifications) have been successful in lowering effluent color, but the problem remains serious in situations where the dilution factor in the receiving water is low.

A number of approaches toward external treatment are being utilized, including lime, alum or polymer coagulation, membrane processes, activated carbon adsorption, and filtration through soil substrata ("rapid infiltration"). These methods are capable of

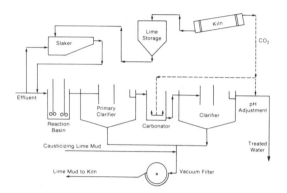

FIGURE 26-14. Lime coagulation system for color removal.

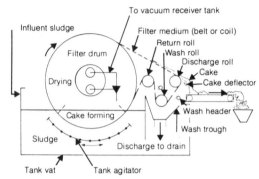

FIGURE 26-15. Vacuum belt filter for thickening sludge.

providing varying degrees of color removal, but the costs of treatment are high in each case. Research toward a more cost-effective method for color removal is continuing.

The most common approach to color removal has been by means of lime coagulation, and a number of process variations are used. A representative process is illustrated in Figure 26-14. Here, lime is added to the wastewater prior to the primary clarifier where it greatly improves settling rates and removes substantial color. The overflow from the clarifier is carbonated (using CO_2 from the kiln lime stack gas) in order to precipitate residual soluble lime and color, and is then clarified again. The lime sludge from both clarifiers is thickened and reburned in the lime kiln.

Coagulation processes are now preferentially applied to wastewaters following secondary treatment because of more complete color removal. It has also been found that the biological process itself causes the formation of additional color.

The rapid infiltration process appears to be a good choice for color removal at mill sites with suitable topography and subsoil characteristics (9). In this method, the wastewater is percolated through porous subsoil strata before reaching the receiving water. The color is removed by adsorption and precipitation onto soil particles. Because the soil strata becomes rather rapidly saturated, it is necessary to operate a number of basins intermittently, allowing time between dosages for regeneration of the adsorbing surfaces.

26.8 SOLIDS HANDLING

The sludge from primary treatment and/or high-rate biological treatment clarification must be concentrated before final disposition, whether by landfill, soil application, or incineration. In the 1960's and 1970's the dewatering operation was carried out as a two-stage process, utilizing filtration

or centrifugation as the first stage (up to 15 - 25% solids), followed by pressing with a V-press or screw press to achieve a final solids content of 35 to 40%. The new generation of equipment is capable of taking sludge and thickening it directly to 40 - 45% solids in one operation.

In the past, a sludge solids content of 20% might have seemed adequate for some purposes. However, greater concern about proper solid waste management has dictated toward higher solids levels. There are a number of potential benefits:
• reduced volume and transport costs,
• easier handling,
• reduced environmental impact when landfilling,
• improved heat value when incinerating.

Proper selection of dewatering equipment depends on both the initial solids content and dewatering characteristics of the sludge. Generally, primary sludges are relatively easy to dewater, while biological sludges are extremely difficult. A common method of handling difficult sludges is to blend them with freer-draining sludges; occasionally bark or sawdust is added as a filter aid. More recently, the chemical industry has developed polymer additives which aid in flocculating and dewatering.

Conventional vacuum drum filters have limited application in sludge dewatering because of progressive blinding of the filter media. On the other hand, the continuous belt drum filter (shown in Figure 26-15) has been extensively utilized because the endless belt travels off the drum where high-pressure water sprays clean both sides of the filter cloth to control the blinding problem.

The horizontal-belt filter press has also been a popular device for sludge dewatering. A number of designs are currently being offered by equipment suppliers, one of which is illustrated in Figure 26-16. All designs utilize two merging fabric or coil belts that enclose the cake and pass it through a series of roller arrangements where increasingly greater

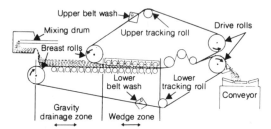

FIGURE 26-16. Horizontal belt filter system for thickening sludge.

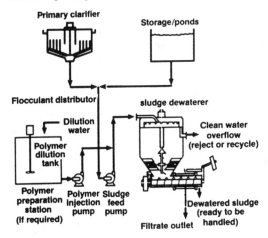

FIGURE 26-17. Sludge dewatering system (Alsthom, Inc.).

pressure is applied. Before the belts merge, a portion of the water is drained through the lower belt.

A more recent approach to sludge dewatering is illustrated in Figure 26-17. The dewatering unit uses a two-step process to compact the sludge. First the sludge enters a vertical/conical tank that serves as a gravity thickener. The thickened sludge is then fed to a dewatering-compaction stage in a horizontal screw extruder. Water is squeezed out at two points through a screen, which converges to a valved outlet. The dryness of the product is controlled to a certain extent by adjusting the outlet pressure and by changing the polymer addition rate.

Landfilling

Landfilling is usually the ultimate disposal method for waste sludges. Although incineration is often worthwhile, it should be considered primarily as a volume-reducing step. Generally, little heat value is realized from burning the sludge, and the admixture of sludge with hog fuel or bark has a negative impact on steam-generating capacity.

The primary environmental problem associated with landfilling is the potential for leachate contamination of ground and surface waters. This concern has prompted governmental regulations requiring leachate control and ground water

monitoring. Although the U.S. federal government has developed criteria for solid waste disposal facilities, state regulatory agencies are primarily responsible for developing and enforcing regulations (10). Generally, disposal in such areas as wetlands, flood plains, perma-frost zones, water sheds, and critical habitats is prohibited, and stricter controls are being applied in all areas. As a minimum, the landfill designs will be required to conform to site-specific hydrologic conditions. Some states will require the use of double-lined landfills with leachate collection and treatment. The higher cost of future landfilling will encourage mills to reduce the volume of solid waste discharges.

REFERENCES

(1) FIRTH, B.K. AND CIESLEK, P.R. **Biosensing for the Protection of Water Quality** *Tappi Journal* (April 1990)

(2) HANLEY, R.W. **Setting the Scene for Environmental Compliance in the Last Decade of the Twentieth Century** *Tappi Journal* (October 1990)

(3) CROAL, P., ET AL **Impact of the Canadian Environmental Protection Act on the Recycled Paperboard Industry** *P&P Canada 90:12:T496* (December 1989)

(4) RUSSEL, D.L. **Monitoring and Sampling Liquid Effluents** *Chemical Engineering* (October 20, 1980)

(5) SEIFERT, J.A. **Selecting Thickeners and Clarifiers** *Chemical Engineering* (October 12, 1987)

(6) GARNER, J.W. **Tighter Mill Effluent Regulations Make Anaerobic Option More Viable** *Pulp & Paper* (February 1991)

(7) PEARSON, J. **Major Anaerobic Plants Start Up** *P&P International* (March 1989)

(8) BUSCH, A.W. **A Practical Approach to Designing Aeration Systems** *Chemical Engineering* (January 10, 1983)

(9) SWANEY, J.M. **Rapid Infiltration Effluent Color Removal System Becomes Operational at Skookumchuck** *P&P Canada 86:2:T47* (February 1985)

(10) STEIN, R.M., ET AL **Mills Search For New Waste Disposal Methods as Landfill Policy Tightens** *Pulp & Paper* (September 1989)

Chapter 27

Air Pollution Abatement

Air pollution may be defined as the presence in the ambient atmosphere of substances in concentrations sufficient to interfere directly or indirectly with personal comfort, safety, or health, or with the full use and enjoyment of property. For an air pollution problem to exist, there must a receptor who responds unfavorably to an airborne agent. The receptor may be animal, vegetable, material of any kind, or a human being; and the pollution effects may be categorized as health, economic, nuisance, or aesthetic.

Although manufacturing operations account for only about 20% of measurable air pollution in North America, this segment receives the greatest public attention due to its high visibility and because characteristic odors are associated with certain industrial processes. Limitations on industrial air emissions (as with water discharges and land disposals) are set by government regulation. In the United States, the basic statutory framework is set forth in the 1970 Clean Air Act and in numerous subsequent amendments. The amendments build on and, in most instances, make the previous regulations more stringent. The provisions of the act are administered by the EPA through a permit system. In Canada, each province retains primary jurisdiction for setting and enforcing limits on atmospheric pollutants.

Air pollution abatement efforts in the pulp and paper industry are concerned with the control of gaseous and particulate emissions. All types of airborne emissions may be partly controlled by operating strategies or by modifications to the production process. However, some type of external treatment of discharging gas volumes (i.e., stack effluents) is also required in most pollution control applications. In particular, odor control continues to be the most daunting task facing the industry. Although the odorous materials (principally reduced sulfur gases) are not dangerous in airborne concentrations, community resentment tends to focus on the obvious and prevailing odor.

Elimination of odors is not as straightforward as the elimination of particulates and/or odorless gases. Odors are difficult to measure and a tolerable or acceptable level is not easily defined. The task is to reduce the amount of odorant at the point of emission by whatever methods are available, and then disperse the emitted material to the maximum extent so that it is far less concentrated by the time it reaches any point where people can smell it. Hopefully, the concentration will be below the "detection threshold", which for most reduced sulfur gases is around one part per billion (ppb).

Up to the 1970's, tall chimneys were used as pollution control devices to disperse contaminants into the atmosphere and thereby keep the concentrations at ground level within permissable levels. While limits are now applied primarily to stack discharges, it is still important to disperse these effluents to minimize their impact on the surrounding area. A high stack still plays an important role with respect to odor control.

Ambient air quality in communities close to a mill site are greatly affected by meteorological factors. Although little can be done for existing mills, the meteorological characteristics must be carefully considered in the selection of a new mill site. Account should be taken of prevailing wind conditions and topographic features that affect local inversions and plume diffusion. The combined effects of source factors, meteorological factors, and receptor factors are illustrated in Figure 27-1.

27.1 SOURCES OF AIR POLLUTANTS

The major types and sources of air pollutants of prime interest to the pulp and paper industry are summarized in Table 27-1. Although water vapor is the most prominent emission from a typical pulp and paper operation, it is considered a pollutant only when the plume restricts visibility or modifies climate.

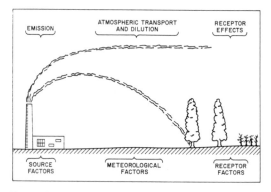

FIGURE 27-1. Interaction of factors in the air pollution process.

TABLE 27-1. Major air pollutants.

Type	Source
fine particulates	principally soda fume from the kraft recovery furnace
course particulates	mainly "fly ash" from hog fuel and coal-fired boilers
sulfur oxides	especially from sulfite mill operations
nitrogen oxides	from all combustion processes
reduced sulfur gases	from kraft pulping and recovery operations
volatile organic compounds	noncondensible gases from digester relief and spent liquor evaporation

The most severe air pollution problems are usually associated with kraft pulp mills. As a consequence, this section will concentrate on the major sources of pollution from a typical kraft mill process, with only occasional reference to other pulping methods or to papermaking operations. The potential emission sources from the kraft cycle are illustrated in Figure 27-2.

Fine particulates are emitted primarily from the recovery furnace. Micron-size particles (fume) are formed by sublimation of various soda compounds (e.g., Na_2SO_4, Na_2CO_3, NaCl) as the furnace gases are cooled. In the 1980's the EPA has placed greater limitations on fine particulate discharges because of their ability to bypass the body's respiratory filters and penetrate deeply into the lungs where the bloodstream can rapidly extract the toxic elements (1).

Coarse particulates are mainly derived from hog fuel or coal-fired boilers due to carryover of ash and

char. The lime kiln and dissolving tank vents are also significant sources of particulates. Emissions from the lime kiln consist both of relatively coarse particulates from lime mud entrainment and fume due to sublimation of sodium compounds.

Sulfur oxides are a minor problem in kraft mills. The recovery and power boilers are both potential sources of sulfur dioxide (SO_2), but the emission levels are generally below the nuisance level. SO_2 is the principal emission from sulfite mills, but because the odor threshold is about one thousand times higher for SO_2 than for reduced sulfur gases, odor problems are not of the same magnitude.

Within the pulp and paper industry, relatively little attention has been directed toward control of nitrogen oxides, which are byproducts of high-temperature combustion processes. However, both NO and NO_2 are known to have photochemical impact in the atmosphere and are classified by the EPA as priority pollutants. As such, the EPA has set rather stringent emission limits. Fortunately, with respect to recovery boilers, power boilers and lime kilns, it appears that formation of nitrogen oxides can be adequately controlled without the requirement of external treatment by operating these combustion units with minimum flame temperature and limited excess air.

Kraft mill odor is principally due to four reduced sulfur gases: hydrogen sulfide, methyl mercaptan, dimethyl sulfide, and dimethyl disulfide. These gases are collectively referred to as "total reduced sulfur", and the discharges are usually called "TRS emissions". Some characteristics of the individual gases are given in Table 27-2. The major sources of TRS emissions are digester blow and relief gases, multiple-effect evaporator noncondensibles, and recovery boiler exhaust gases. Measurable concentrations are also found in the brown stock washer and seal tank vents, liquor storage vents, smelt dissolving tank vent, lime kiln exhaust, black liquor oxidation exhaust, and slaker exhaust. Just recently, oxidation and aeration lagoons have also been identified as possible contributors of TRS emissions (2). The noncondensibles from digesters and evaporators are small-volume, high-concentration sources, and are therefore relatively easy to treat, usually by incineration. The high-volume recovery furnace gases are more difficult to treat, and best results are achieved by reducing emissions at the source by such means as black liquor oxidation (see Section 10.1), low-odor recovery boiler design (see Section 10.3), and proper operation of the recovery boiler (see Section 27.3).

Volatile organic compounds other than those containing sulfur can also be emitted from several kraft process sources. The most concentrated sources are the noncondensible gases from the digester and evaporator operations. Typical constituents are

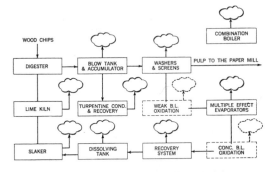

FIGURE 27-2. Potential air emission sources in a kraft pulp mill.

TABLE 27-2. Characteristics of kraft mill reduced sulfur gases.

Compound	Chemical Formula	Type of Odor	Aprox. Odor Threshold
hydrogen sulfide	H_2S	rotten eggs	1 ppb
methyl mercaptan	CH_3SH	rotten cabbage	1 ppb
dimethyl sulfide	CH_3SCH_3	vegetable sulfide	10 ppb
dimethyl disulfide	CH_3SSCH_3	vegetable sulfide	10 ppb

alcohols, terpenes and phenols. These gases are themselves mild odorants, but their main action is to enhance the effect of the sulfur gases. Some of these organic gases also undergo photochemical reactions in the atmosphere. Control by means of incineration is relatively straightforward.

27.2 MONITORING AND TESTING

A monitoring and testing program for air quality control generally consists of two specific phases - source sampling and ambient sampling. As a minimum requirement, source sampling of individual stack effluent streams is carried out to measure specific emission rates and determine if regulatory limitations are being met. In many situations, testing of the ambient atmosphere is also necessary to verify community air quality and confirm that ground level concentrations of specific contaminants are within allowable limits.

Source Sampling

Source sampling and testing may be carried out for a number of specific reasons (Table 27-3), but the principal objective is to provide feedback information for control of emissions. Not so long ago, virtually all stack monitoring was done manually, but today a number of on-line, automatic, continuous monitoring devices are being successfully used, mainly for gaseous emissions. Particulate sampling is not as readily adaptable to automatic methods.

A detailed description of manual testing methodology is beyond the scope of this text. Most determinations involve careful measurements of both gas flow rate and pollutant concentration as the basis of calculating emission rates. When testing for particulates, a special requirement is that the gas sample be drawn into the sampling probe at the same velocity as the bulk gas travelling in the duct or flue; this procedure, known as isokinetic sampling, ensures that a representative sample is obtained. An example of a modern particulate sampling train is shown in Figure 27-3.

It has long been recognized that grab-sample type source measurements do not provide sufficient information for process control. These measurements are usually taken during periods of relatively stable operation, and therefore give no indication of the higher rates of emission that occur during upset conditions. Continuous emission data is definitely needed, and consequently a number of measuring devices have been introduced by instrument suppliers in the past two decades in an effort to meet the requirements of the industry.

Continuous emission monitors (CEMs) underwent rapid development in the 1970's and 1980's. Reliable instrumentation is currently available for many gaseous contaminants including TRS, sulfur dioxide, nitrogen oxides, hydrocarbons, carbon dioxide, and carbon monoxide. Some representative techniques are illustrated in Figure 27-4. All CEMs require a probe or inlet device, a sample conditioning unit, a detector-analyzer, and a data recording unit. To minimize physical or chemical interference during analysis, particulates should be removed from the sample gas flow, and condensation of moisture must be prevented.

TABLE 27-3. Reasons for source sampling.

1. Obtain reliable emission data to provide basis for control strategy.
2. Assess whether an emission is within compliance of regulatory standard.
3. Assess economic impact of chemical or product loss.
4. Determine inlet loading for a proposed collection device.
5. Determine efficiency of existing collecting device in terms of its inlet and outlet loading.
6. Monitor one or more constituents in an exit stream as a means of process control (e.g., O_2 and/or CO for a combustion process).
7. Obtain specimen material for determination of physical or chemical properties.

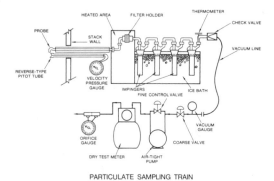

PARTICULATE SAMPLING TRAIN

FIGURE 27-3. Particulate sampling train (Joy Manufacturing Co.).

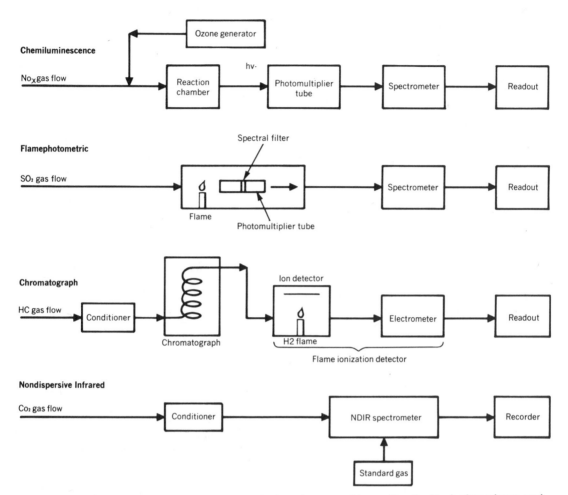

FIGURE 27-4. Representative techniques for monitoring nitrogen oxides, sulfur dioxide, hydrocarbons, and carbon dioxide.

CEMs for particulates have been under development for two decades, and a number of installations have been made within the industry. One analytical technique utilizes filtration and beta-radiation attenuation, which has the advantage that it depends only on the weight of collected material, not on its chemical or physical properties. Another technique for soluble particulates involves scrubbing a gas sample with deionized water and analyzing the solution with a sodium specific ion electrode or conductivity measurement. However, a totally satisfactory particulate CEM in terms of cost, reliability, durability and acceptance by authorities is not yet available.

Variable light transmission within a flue is effective for continuously measuring visible emissions, but this method does not indicate a weight rate of emission. Most CEMs of this type use photoelectric techniques as illustrated in Figure 27-5. The major problem with this approach is the

possibility of coating the lens or detector with particulates.

While CEMs provide useful information, their intimate contact with flue gases laden with particulates, moisture and corrosive condensates cause predictable problems. Sophisticated optics and electronics that are subjected to vibration, heat and a harsh environment require regular attention from service personnel.

Ambient Sampling

Measurement of the ambient pollution in the local pulp and paper mill community and outlying impact areas can prove invaluable in dispelling misconceptions about the magnitude and extent of a mill's specific contribution to environmental pollution. For this purpose, a variety of sampling devices are available, representing several levels of sophistication. A minimum of eight sampling stations is normally used, with the prevailing wind being a primary consideration for sample site selection. Ideally, the sampling locations should be convenient and accessible to electric power, but not extremely noticeable (to reduce the likelihood of vandalism).

Information obtained from wind measurements is best summarized in the form of a "wind rose". In the

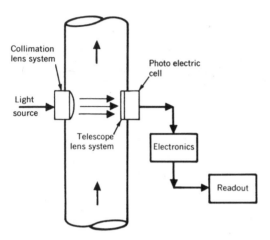

FIGURE 27-5. One type of visible-emission detector.

most basic type of wind rose (Figure 27-6), the lengths of the lines for each of 16 wind directions represent the percentage of time that the wind is coming from that direction. Another type of wind rose (Figure 27-7) shows the distribution of mean wind speeds and directions (from which the wind comes) in terms of four wind speed ranges and eight points of the compass.

The ambient measurements of primary interest are usually settleable particulates (i.e., "dustfall"), suspended particulates, reduced and oxidized sulfur gases, and corrosion effects. Dustfall is measured by means of an open-top bucket of known surface-collecting diameter which is allowed to remain at the sampling station for a month. The amount of solids is then measured and the dustfall is calculated, typically in units of tons per square mile per month.

Suspended particulates are measured by means of a high-volume sampler, which uses a vacuum-cleaner type blower to draw large metered volumes of air through a filter on which the particulates are collected. Filters are usually weighed after a 24-hour sampling period to determine the amount of accumulated material. Concentrations are reported as micrograms per cubic meter.

The traditional means of measuring ambient sulfur gas concentrations was to expose a surface coated with lead oxide paste for a period of up to four weeks. Sulfur dioxide and other sulfur gases reacted with the lead paste to form lead sulfate, and the amount formed was subsequently determined by chemical means. This rather crude method has largely been replaced by a host of instrumental and chemical methods. Both manual and continuous devices are available for measuring sulfur dioxide and TRS gases in the ambient atmosphere.

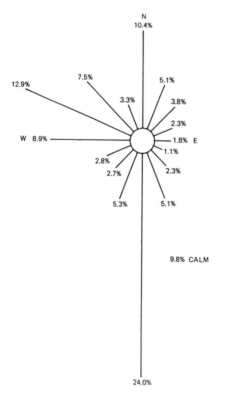

FIGURE 27-6. Basic wind rose showing percentage of time the wind comes from each of 16 directions.

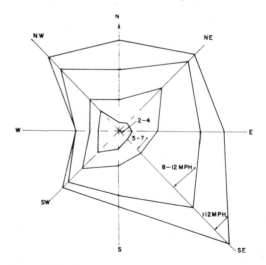

FIGURE 27-7. Wind rose showing distribution of wind direction and speed ranges for each of 8 directions.

Corrosion rates can be measured by atmospheric exposure of coupons of mild steel, zinc, copper, aluminum, or other selected material. After exposure for a known period of time, the coupons are cleaned and weighed as an indication of corrosion, and metal loss is usually reported in $mg/cm^2/month$.

27.3 IN-PROCESS ABATEMENT

In-process air pollution abatement steps are usually aimed at odor reduction. The major strategies under this heading are related to the design and operation of the recovery boiler system, and the collection/incineration of noncondensible gases and vapors from cooking and evaporation operations.

Recovery Boiler Operation

The objectives and mechanics of black liquor oxidation are covered in Section 10.1. This odor abatement step is necessary only in mills that still utilize direct-contact evaporators. The oxidation converts sulfides into thiosulfates, and thereby prevents the hot flue gases from stripping TRS out of the black liquor during the contacting process. The newer-generation "low-odor recovery boilers" omit the direct-contact evaporator, and thereby eliminate a major source of TRS emissions. However, black liquor oxidation or low-odor design has no effect on TRS emissions from the combustion zone of the furnace.

A number of studies were undertaken in the early 1970's (e.g., Reference 3) to determine the effect of recovery furnace operating variables on the emission of odorous sulfur compounds from the combustion zone. It was determined that TRS emissions from this source could be controlled at a low level by strict adherence to operating discipline. Most investigators agree on the following guidelines:
1. Furnace loading should not exceed a well-defined "critical level". On some older recovery boilers, the critical level was found to be about 115% of rated solids loading.
2. Primary air should not exceed 65% of the total air supplied to the furnace (i.e., at least 35% should be secondary and tertiary air).
3. Excess oxygen should be controlled within the 2.0 to 2.5% range. TRS emissions increase below 2% while above 2.5% there can be problems with formation of SO_3 and sticky particulates.
4. Liquor sulfidity should be maintained at the lowest practicable level, definitely below 30%.
5. Liquor solids concentration for firing should be at least 62%.
6. The concentration of inert inorganic compounds (e.g., Na_2CO_3, $NaCl$) in the black liquor should be minimized.
7. The black liquor nozzle spray should be finely dispersed.

8. High turbulence in the secondary air is helpful. Basically, the presence of TRS in the flue gas is caused by incomplete combustion in the upper (oxidation) zone of the furnace. Besides sufficient air for oxidation, an adequate "penetration" or mixing action is required.

It is well established that the sulfur concentration in the black liquor has a profound effect on TRS formation. As sulfur losses are controlled by in-process measures, the liquor sulfidity level tends to rise, thus increasing the driving force for TRS emissions. To keep a ceiling on sulfidity, the makeup rate of sulfur chemicals must be reduced. Many mills have virtually eliminated saltcake (Na_2SO_4) as a soda makeup chemical because of its high sulfur content, and are using such sulfur-free chemicals as caustic soda ($NaOH$) and soda ash (Na_2CO_3).

Over the last two decades, evaporator and recovery boiler manufacturers have improved their designs to achieve higher solids liquor for firing and better mixing within the combustion zone of the furnace. These improvements serve to reduce TRS emissions at the same time that they increase thermal efficiency.

Collection and Incineration of Noncondensibles

Noncondensible gases and vapors from digester relief, digester blow, and liquor evaporation (containing both TRS and volatile organic compounds) are usually combined into one gas header for subsequent treatment. In some systems, the gases from the digester/evaporator condensate steam stripper (see Section 26.2) are also added. A single induced-draft fan is often sufficient for drawing gases from all sources and injecting them into the lime kiln or power boiler.

Noncondensible gas flows are fairly constant in mills with continuous digesters. Where batch digesters are used, some method of equalizing the flow is necessary. Two basic types of gas holders are used to provide surge capacity so that gases can be metered at a constant rate. The vaporsphere system (Figure 27-8) utilizes a plastic diaphragm that rises and falls to maintain a constant pressure in the surge compartment. The floating head gas holder (Figure 27-9) utilizes a "floating head" that rises and falls as gases enter or leave the tank.

In some mills, the gases are scrubbed with caustic to absorb sulfide prior to incineration; but in most systems the gases are burned without further treatment, usually in the lime kiln. The incineration destroys the organic compounds and converts the TRS to SO_2, which may then be partly absorbed if the lime kiln is equipped with an exhaust scrubber. Every precaution against potential explosion hazards must be taken in the design and operation of a noncondensible incineration system. Even though the

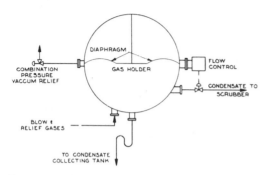

FIGURE 27-8. Spherical type gas holder.

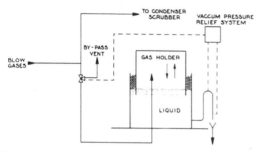

FIGURE 27-9. Floating-head type gas holder

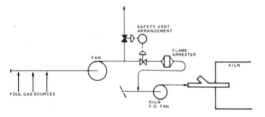

FIGURE 27-10. Standard burning arrangement for incineration of noncondensibles.

pipeline flow is usually too "rich" to be flammable, such safety devices as flame arresters and rupture discs are commonly used throughout the system. A minimum burning arrangement for a kiln is illustrated in Figure 27-10.

27.4 CONTROL EQUIPMENT

Strictly speaking, there are three types of air pollutants: particulates, gases/vapors (odors), and liquid droplets (e.g., mists and aerosols). However, liquid droplets are so easy to remove that they are not normally considered a pollution problem. In fact, pollution control equipment often produces liquid entrainment, which is then easily removed from the gas stream in a subsequent process step. In practical terms, only two types of pollutants need to be considered, namely particulates and gases/vapors.

The main categories of equipment for removing particulates and gas/vapor pollutants from effluent

gas streams are listed in Table 27-4. The proper choice of equipment in each case is determined by careful assessment of the performance requirements, process and safety considerations, and economic factors. Scrubbers are designed for removing either particulates or gases/vapors; and are usually not well suited for removing both types of pollutants simultaneously. The principal information required for system design is summarized in Tables 27-5 (for gases/vapors) and 27-6 (for particulates).

TABLE 27-4. Air Pollution Control Equipment.

Gas/Vapor Removal	Particulate Removal
catalytic combustion	mechanical collection
thermal incineration	fabric filtration
adsorption	gravel bed filtration
wet scrubbing	("dry scrubbing")
	electrostatic precipitation
	wet scrubbing
	hybrid designs

TABLE 27-5. Information required before selecting equipment for particulate removal.

Dust or Fume:	chemical identification
	concentration in gas stream
	particle size characterization
	required removal efficiency
	conductivity
Gas Stream:	volumetric flow rate
	temperature
	dew point
	flammability?
	presence of gaseous contaminants?

TABLE 27-6. Information required before selecting equipment for gas/vapor removal.

Gas/Vapor Contaminant:	chemical identification
	concentration in gas stream
	required removal efficiency
	sufficiently valuable to recover?
	combustible or noncombustible?
Gas Stream:	volumetric flow rate
	temperature
	moisture content
	flammability?
	presence of particulate contaminants?

Combustion Process for Destroying Gaseous Pollutants

Many gaseous air pollutants can be destroyed by combustion, including organic materials and TRS emissions. In order to completely destroy the pollutants, it is necessary to add sufficient air and either maintain a high temperature of 650 - 800° C (i.e., thermal incineration) or use a catalyst at more moderate temperatures of 300 - 500° C (i.e., catalytic combustion). The basic design requirements are the same, namely to provide thorough mixing with air and raise the temperature to the required level for complete oxidation.

Although capital investment is somewhat higher, catalytic systems designed specifically for destroying pollutants are much cheaper to operate than comparable thermal systems. One factor to be considered is whether fouling agents are present in the gas stream that would reduce the functionality of the catalyst. In spite of their effectiveness and operating economy in many applications, catalytic systems are little used in the pulp and paper industry because thermal incineration sources such as the lime kiln and power boiler are on site and can be adapted for the destruction of pollutants (e.g., refer back to Section 27.3).

Adsorption of Gases and Odors

Adsorption is the collection of gas molecules on the surface of a solid adsorbent material, such as activated carbon or silica gel. Adsorption is a physical process, making use of attractive forces between the adsorbent surfaces and selected gas molecules. When the adsorbent sites are saturated, the adsorbent bed is "regenerated" by passing steam or hot air through it. The concentrated regenerative stream must then be taken to a recovery or disposal facility for final treatment.

Adsorption has rarely been applied for removing pollutants in the pulp and paper industry, but this technique could be useful under certain conditions. Generally, adsorption has found its major applications when the pollutant is present in very dilute concentrations or when the gas or vapor is sufficiently valuable to be recovered.

Wet Scrubbers

A wet scrubber removes dust or gaseous contaminants from a gas stream through intimate contact with a suitable absorbing or wetting liquor, most commonly water. The collection efficiency is usually a function of the pressure drop across the device and the relative difficulty of contaminant removal. To achieve a specified collection efficiency, the required pressure drop (a measure of the work performed by the fan) varies inversely with the size of particulates and/or the solubility of gaseous contaminants (see Figure 27-11 and 27-12).

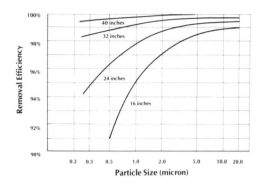

FIGURE 27-11. Particulate removal efficiency in a scrubber as a function of particle size and pressure drop.

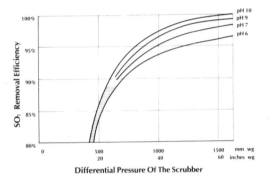

FIGURE 27-12. SO₂ removal efficiency in a scrubber as a function of water pH and pressure drop.

Therefore, when removing micron-size particulates or low-solubility gases, a large pressure drop is required. The pressure drop across commercial scrubbers varies from 3 to over 100 inches water gauge, depending on requirements. Obviously, scrubbers with high pressure drops are expensive to operate.

Wet scrubbers are most appropriately used for cleaning hot, moist gases. During operation, wet scrubbers may produce significant corrosion as a result of mixing air, water and contaminants. Mild steel scrubbers can be coated or lined with various materials to resist corrosion. In many cases, it may be worthwhile to use stainless steel or fiberglass-reinforced plastic as materials of construction.

Manufacturers of scrubbers offer a bewildering array of designs. (Typically, the manufacturer claims that a given efficiency can be achieved at a lower pressure drop.) To overcome some of the confusion, it is possible to group the devices into a number of categories, as has been done in Table 27-7. It is beyond the scope of this section to discuss the relative merits of the various designs; however, the operating characteristics are described for a few of the more common types.

TABLE 27-7. Categories of scrubbers.

Plate Column	- Liquid travels down vertical column over a series of plates. Gas travels up through contacting devices on each plate.
Stationary Packing	- Liquid travels down vertical column over packing material, which provides large surface area. Gas travels up through packing.
Moving Bed	- Similar to stationary packing design, except that the packing is light and the gas velocity serves to keep the packing bed in a fluidized state.
Fibrous Packing	- Beds of inert fiber (e.g., plastic, spun glass) provide contact surface for liquid and gas. Most designs utilize a horizontal configuration.
Preformed Spray	- Particles or gases are collected on liquid droplets that have been atomized by spray nozzles. Either horizontal or vertical gas flowpaths are used.
Gas-Atomized Spray	- The moving gas stream serves to atomize the liquid into droplets and accelerate the droplets for greater contact (e.g., Venturi scrubbers).
Centrifugal	- A spinning motion is imparted to the gas passing through a cylinder. The walls are wetted down or sprays are directed through the rotating gas stream.
Baffle & Secondary Flow	- Gas flow direction and velocity are changed by means of solid surfaces. Particle collection onto wet surfaces is caused by centrifugal action.
Impingement & Entrainment	- A self-induced spray is produced when gas impinges on or over the liquid surface. The atomized droplets then act as collecting surfaces.
Mechanically Aided	- A motor-driven device between the inlet and outlet of the scrubber body is used to provide contact surface. Typically, particles are collected by impaction onto wet fan blades as the gas moves through the device.
Combination	- Any combination of the principles already described.

The plate column scrubber (Figure 27-13) and stationary packing scrubber (Figure 27-14) are traditional designs successfully used for multiple-stage gas absorption. These designs are less suitable for removing particulates due to problems with buildups and plugging. All the other types are more specifically designed for particulate removal.

A representative scrubber of the impingement-entrainment type is illustrated in Figure 27-15. In this design: 1) the dust-laden gases sweep over the surface of the scrubbing liquid and underneath the base of the collecting tubes, 2) each inlet bonnet imparts a cyclonic, spinning action to the gas mixture, which shears the water into a dense atomized spray as the gases continue up the tubes, 3) at the discharge of the tubes, the gas-liquid mixture impacts against a curved deflector, which acts 4) as an entrainment eliminator.

Perhaps the most common and versatile scrubber designs for particulate removal are those employing a Venturi throat entry (see Figure 27-16 for typical example). As the gas enters the upper portion of the Venturi throat, it meets with scrubbing liquor introduced tangentially or through spray headers. The walls of the throat converge to a minimum diameter determined by the specific application. In general, the smaller the diameter of convergence, the greater will be the pressure drop and the higher the efficiency for removing sub-micron size particles. As the gas converges in the throat, its velocity increases, and the particulates in the gas stream impact the slower-moving water droplets. The dust-laden droplets are then carried into the cyclonic separator where the water is separated from the gas. To ensure that no liquid entrainment remains in the effluent gas, some units also utilize a wire-mesh mist eliminator.

Figure 27-17 shows a representative combination scrubber design employing a Venturi throat for primary impaction of particulates, followed by a moving bed section for removal of agglomerates.

Mechanical Collection of Particulates

The simplest type of particulate removal device is the mechanical or inertial collector, where the gas flow is made to follow a path that particles, because of their inertia, cannot easily follow. As a class, mechanical collectors are inexpensive to install, with low to moderate operating costs, but are limited in application to the removal of relatively large

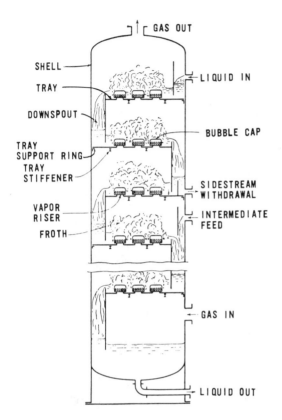

FIGURE 27-13. Tray tower scrubber with liquid-gas contact elements.

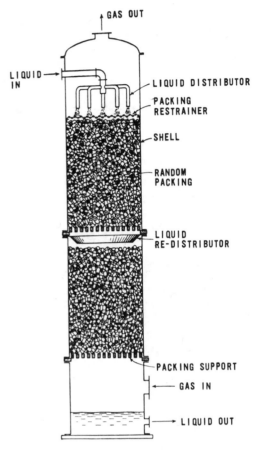

FIGURE 27-14. Vertical column scrubber with stationary packing.

particles. The simplest form of mechanical collector is the settling chamber (Figure 27-18), in effect, a spacer of large cross-section where the gas velocity is reduced allowing heavy particles to settle out. Little pressure drop is involved, but separation capability is limited to particles above 50 microns.

In other types of mechanical collection equipment, the gas undergoes continuous direction change. The most familiar and widely-used mechanical collector is the cyclone, one example of which is illustrated in Figure 27-19. Dust-laden gas enters tangentially at the top of a cylindrical, tapered shell and the gas is forced to follow a spiral downward path. Particles are thrust outward by centrifugal force and work their way downward, where they usually discharge into a hopper. The gas is forced into a center vortex and is carried out the top. The tighter the flow spiral, the greater will be the centrifugal force acting on the particles.

The highest particulate-removal efficiency for cyclonic collectors is achieved with small-diameter units. For many applications, the collector consists of a large number of small-diameter cyclone tubes (Figure 27-20) within a single housing (Figure 27-21), where the gas flow is split equally to each tube. With this type of arrangement, particles as small as 5 microns can be removed at efficiencies approaching 95%.

Two methods are used to increase the efficiency of cyclonic collectors, but each involves the further processing of a secondary gas flow. For existing installations, one technique is to allow a significant gas flow (perhaps 10%) to discharge with the dust into a separator compartment within the hopper. For new installations, an effective technique is to "shave off" the dust-rich layer on the outer boundary of the outlet tube, allowing the main gas stream to continue on to the fan and stack (see Figure 27-22). The secondary gas flow from either hopper evacuation or secondary shave-off, is typically put through a small medium-energy scrubber where it is cleaned of small particles. Mixing of the wet scrubber exhaust with the main gas stream then provides a nonsaturated stack effluent.

Fabric Filters

Fabric filters collect solid particulates by passing gas through cloth that most particles cannot

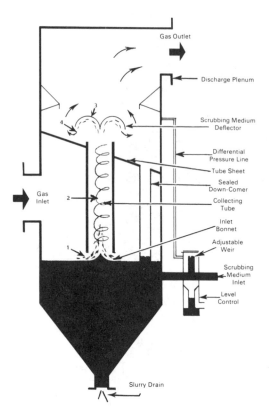

FIGURE 27-15. Scrubber utilizing impingement-entrainment principle (Zurn Air Systems).

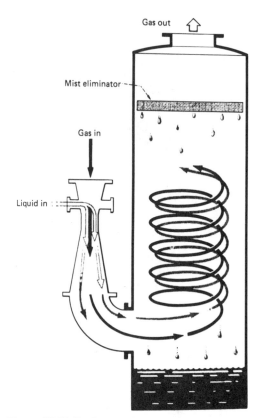

FIGURE 27-16. Typical Venturi scrubber equipped with cyclone separator and mesh mist eliminator.

penetrate. Many designs of fabric filter are offered by equipment suppliers, one of which is illustrated in Figure 27-23. Typically, these devices will remove particles down to 0.1 microns with high efficiency using moderate pressure drop. However, application is limited to the removal of dry, nonagglomerating particulates at flue gas temperatures of 300° C or less.

As a layer of collected material builds on the filter media, the pressure differential required for continued flow increases. Therefore, the accumulation must be periodically removed. Most large "baghouses" are sectionalized into a number of filter compartments that are cleaned on a rotating cycle without affecting the continuity of operation. The cleaning mechanism may be by means of mechanical shaker, reverse flow of air, flushing jet of air, or any combination of these methods. The dust that is discharged from the filter media falls into a hopper section and is removed by an air lock, screw conveyor or other device.

Gravel Bed Filter

The "dry scrubber" concept of a particulate collector utilizes a moving bed of granular filter media. The representative design shown in Figure 27-24 consists of a vertical vessel containing two concentric, louvered cylindrical tubes. The annular space between the tubes is filled with pea-size gravel that acts as the filter medium. The dust-laden exhaust gases pass through the filter medium and the particulates are removed by impaction. To prevent excessive dust buildup on the face of the filter, the medium column is slowly moved downward where the particulate-laden layer is continuously removed. The gravel is cleaned (usually by a shaker screen) to remove the particulate matter, and then transported by bucket elevator to the top of the unit for use in another cycle.

The gravel bed filter is a versatile collection device that is tolerant to a wide range of operating conditions. The performance of the unit can be adjusted to a certain extent by changing the particle size of the filtering medium and/or the recirculation rate. However, the required pressure drop is higher than for comparable fabric filters.

Electrostatic Precipitator

The electrostatic precipitator (ESP) can be an extremely efficient particulate collector. Modern units

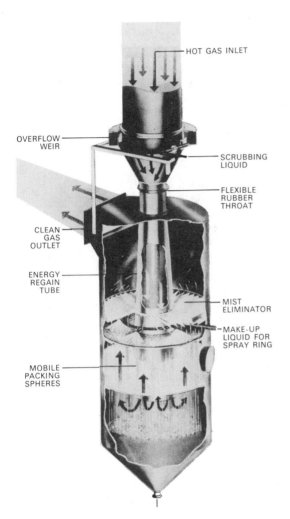

HOT GAS INLET

OVERFLOW WEIR

SCRUBBING LIQUID

FLEXIBLE RUBBER THROAT

CLEAN GAS OUTLET

ENERGY REGAIN TUBE

MIST ELIMINATOR

MAKE-UP LIQUID FOR SPRAY RING

MOBILE PACKING SPHERES

FIGURE 27-17. Representative combination scrubber (UOP Air Correction Div.).

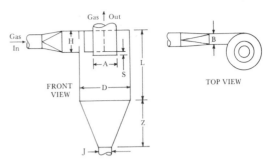

FIGURE 27-19. Cyclone separator showing key design dimensions.

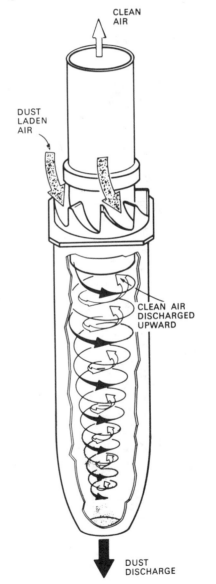

CLEAN AIR

DUST LADEN AIR

CLEAN AIR DISCHARGED UPWARD

DUST DISCHARGE

FIGURE 27-20. Cut-away of dust collector tube (UOP Air Correction Div.).

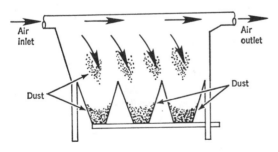

Air inlet

Air outlet

Dust

Dust

FIGURE 28-18. Settling chamber has low pressure drop, but also low collection efficiency.

are typically removing greater than 99.9% of dust particles down to 0.1 microns. Although equipment is massive and expensive to install, operating costs are modest because, unlike other pollution-control

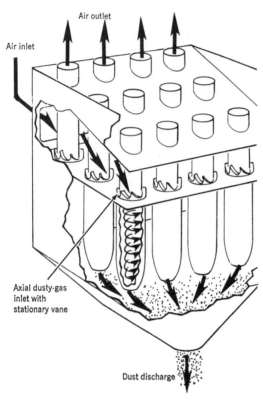

FIGURE 27-21. Typical arrangement of multi-tube collector.

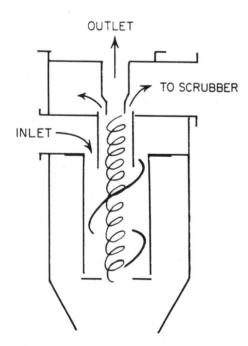

FIGURE 27-22. Secondary shave-off from a cyclonic collector.

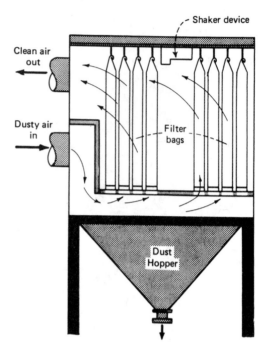

FIGURE 27-23. Baghouse filter with shaker device.

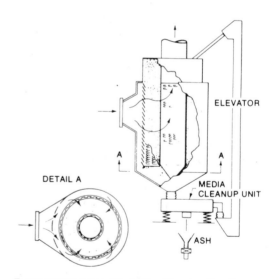

FIGURE 27-24. Gravel bed filter (Combustion Power Co.).

devices, the precipitator works only on the particles to be collected, not on the entire gas stream. The ESP has long been the equipment of choice for removing soda fume from recovery boiler exhaust. Now, ESPs are increasingly being used, as well, to clean up lime kiln and bark boiler exhaust gases.

A 1985 evaluation of lime kiln pollution control devices (4) yielded the comparison shown in Figure

27-25 between capital and annual operating cost for an ESP and wet scrubber. Generally, the gas flow to be treated must be large in order to justify the initial expense of an ESP. At the higher gas flow levels, the savings from lower operating cost rapidly offset the higher capital cost.

The ESP utilizes high voltage to create a negatively-charged field through which the dust-laden gas passes. The particles pick up a negative charge and are drawn out of the gas flow by positively-charged collecting plates. These plates are periodically rapped by mechanical action to discharge accumulations into a hopper (dry bottom) or into a pool of liquor (wet bottom). The usual arrangement of wires and plates is illustrated in Figure 27-26. A typical operating unit is shown in Figure 27-27. While the applied voltage is on the order of 100 kv, the electric current is only about 50 milliamps (ma) per 1000 square feet of collecting surface. The voltage gradient across the gap between wires and plates is typically about 4 or 5 kv per cm.

The collection efficiency of an ESP is related to the time the particles are exposed to the electrostatic field and to the resistivity of the dust particles. The exposure time is a design variable determined by the cross-sectional area of the precipitator and its length in the direction of gas flow. Resistivity is a measure of how readily the particulate will pick up a charge; a high resistivity indicates a high impedance to charge transfer across a gas layer. Surprisingly, particulate matter having a moderate resistivity is typically easier to collect in an ESP than material having either extremely high or low values.

Hybrid Collectors

A number of recently-introduced collection devices utilize electrostatic principles to augment performance. Of particular interest to the pulp and paper industry are the electrified filter bed and the electrostatically augmented scrubber.

The operation of the electrified filter bed (an adaptation of the gravel bed filter) is illustrated in Figure 27-28. In this design, the incoming particulates pick up an electrical charge from the high-voltage ionizing discs, and the dust particles are then captured by the electrically polarized gravel bed. Collection efficiency is said to be greatly enhanced due to incorporation of the electrostatic principle (5).

In one type of electrostatically augmented scrubber (or "wet electrostatic precipitator"), water is sprayed into the particulate-laden air as it passes through a high-voltage section; both particulates and droplets are charged with the same polarity, and they

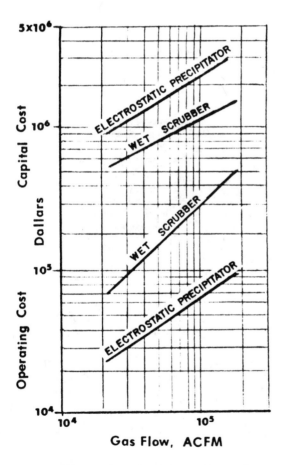

FIGURE 27-25. Installed cost and annual operating cost for wet scrubbers and electrostatic precipitators (1985, E. Pollock & R.T. Walker).

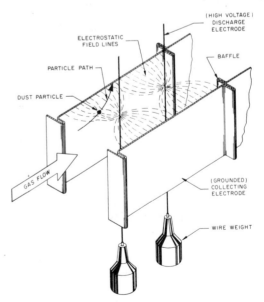

FIGURE 27-26. Electrostatic charging of dust particles.

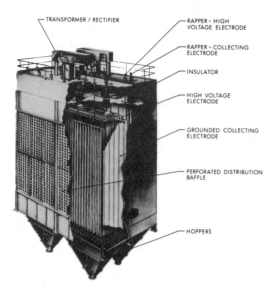

FIGURE 27-27. Cut-away of electrostatic precipitator (UOP Air Correction Div.).

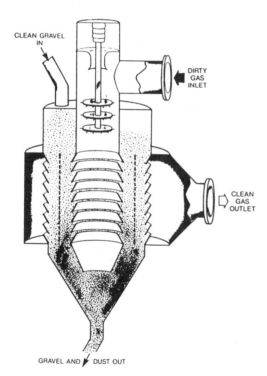

FIGURE 27-28. Schematic of electrified filter bed.

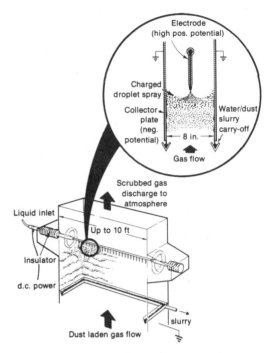

FIGURE 27-29. Wet electrostatic precipitator (TRW Inc.)

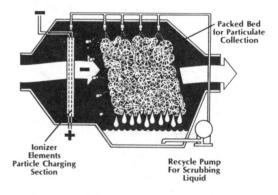

FIGURE 27-30. Hybrid of wet electrostatic precipitator and packed-bed scrubber (Ceilcote Co., Air Pollution Control Div.).

then migrate to collecting electrodes where they are removed. In the design illustrated in Figure 27-29, charged particulates enter a chamber where they are collected by water droplets carrying an opposite charge. The water droplets then migrate to the collection surface where they coalesce and drain,

leaving the collecting surface free of buildup. In variations of these two designs, the collection surfaces are kept free of buildup by a continuous stream of water. In a third hybrid design (illustrated in Figure 27-30), the charged particulates are captured either by liquid droplets or a packing material.

REFERENCES

(1) PINKERTON, J.E. **Significance of PM10 Regulations for the Forest Products Industry** *Tappi Journal* (August 1988)

(2) ESPLIN, G.J. **Total Reduced Sulfur Emissions From Effluent Lagoons** *P&P Canada 90:10:T398* (October 1989)

(3) CREIGHTON, D.M. **TRS Measurement: Guide to More Efficient Boiler Operation** *Pulp & Paper* (April 1971)

(4) POLLOCK, E. and WALKER, R.T. **Energy Usage in Air Pollution Control Equipment** *P&P Canada 87:1:T11* (January 1986)

(5) ALEXANDER, J.C. **Control of Particulate Emissions from Bark and Wood Fired Boilers With the Electrified Filter Bed** *P&P Canada 87:3:T83* (March 1986)

INDEX